Open source software in life science research

Published by Woodhead Publishing Limited, 2012

Woodhead Publishing Series in Biomedicine

Published by Woodhead Publishing Limited, 2012

Published by Woodhead Publishing Limited, 2012

Published by Woodhead Publishing Limited, 2012

Woodhead Publishing Series in Biomedicine: Number 16

Open source software in life science research

Practical solutions in the pharmaceutical industry and beyond

EDITED BY
LEE HARLAND AND MARK FORSTER

Oxford   Cambridge   Philadelphia   New Delhi

Published by Woodhead Publishing Limited, 2012

Woodhead Publishing Limited, 80 High Street, Sawston, Cambridge, CB22 3HJ, UK
www.woodheadpublishing.com
www.woodheadpublishingonline.com

Woodhead Publishing, 1518 Walnut Street, Suite 1100, Philadelphia, PA 19102–3406, USA

Woodhead Publishing India Private Limited, G-2, Vardaan House, 7/28 Ansari Road,
Daryaganj, New Delhi – 110002, India
www.woodheadpublishingindia.com

First published in 2012 by Woodhead Publishing Limited
ISBN: 978-1-907568-97-8 (print); ISBN: 978-1-908818-24-9 (online)
Woodhead Publishing Series in Biomedicine ISSN: 2050-0289 (print); ISSN: 2050-2097 (online)

Typeset by RefineCatch Limited, Bungay, Suffolk
Printed in the UK and USA

For Anna, for making everything possible

Lee Harland

Thanks to my wife, children and other family members, for their
support and understanding during this project

Mark Forster

Published by Woodhead Publishing Limited, 2012

Contents

Published by Woodhead Publishing Limited, 2012

List of figures and tables

Figures

Published by Woodhead Publishing Limited, 2012

Tables

Foreword

Twelve years ago, I joined the pharmaceutical industry as a computational scientist working in early stage drug discovery. Back then, I felt stymied by the absence of a clear legal or IT framework for obtaining official support for using Free/Libre Open Source Software (FLOSS) within my company, much less for its distribution outside our walls. I came to realize that the underlying reason was because I do not work for a technology company wherein the establishment of such policies would be a core part of its business. Today, the situation is radically different: the corporate mindset towards these technologies has become far more accommodating, even to the point of actively recommending their adoption in many instances. Paradoxically, the reason why it is so straightforward today to secure IT and legal support for using and releasing FLOSS is precisely because I do not work for a technology company! Let me explain.

In recent years I have perceived a sea change within our company, if not the industry. I recall hearing one senior R&D leader stating something to the effect of 'ultimately we compete on the speed and success of our Phase III compounds' as he was making the case that all other efforts can be considered pre-competitive to some degree. This viewpoint has been reflected in a major revision of the corporate procedure associated with publishing our scientific results in external, peer-reviewed journals, especially for materials based on work that do not relate to an existing or potential product. Given that my employer is not in the software business, the process I experience today feels remarkably streamlined. Likewise, in previous years I would have been expected to file patents on computational algorithms and tools prior to external publication in order to secure IP and maintain our freedom to operate (FTO). The prevailing strategy today, at least for our informatics tools, is defensive publication.

The benefits of publication to a pharmaceutical company in terms of building scientific credibility and ensuring FTO are clear enough, but what about releasing internally developed source code for free? A decade ago my proposal to release as open source the Protein Family Alignment

Annotation Tool (PFAAT) [1] was met by reactions ranging from bemusement to deep reluctance. We debated the risk associated with our exposing proprietary technology that might enable our competitors, at a time when 'competitive' activities were much more broadly defined. Moreover, due to our lack of experience with managing FLOSS projects, it was difficult to assure management that individuals not in our direct employ would willingly and freely contribute bug fixes and functional enhancements to our code. Fortunately in the case of PFAAT, our faith was rewarded, and today the project is being managed by an academic lab. It continues to be developed and available to our researchers long after its internal funding has lapsed. In many key respects our involvement with PFAAT foreshadowed our wider participation in joint precompetitive activities in the informatics space [2], now with aspirations on a grander scale.

It has been fantastic to witness the gradual reformation of IT policies and practices leading to the corporate acceptance and support of systems built on FLOSS in a production environment. I imagine that the major factors include technology maturation, the emergence of providers in the marketplace for support and maintenance, and downward pressure on IT budgets in our sector. For a proper treatment of this subject I recommend Chapter 22 by Thornber. From an R&D standpoint, the business case seems very clear, particularly in the bioinformatics arena. The torrent of data streaming from large, government-funded genome sequencing centers has driven the development of excellent FLOSS platforms from these institutions, such as the Genome Analysis Toolkit [3] and Burrows-Wheeler Aligner [4]. Other examples of FLOSS being customized and used within my department today include Cytoscape [5], Integrative Genomics Viewer [6], Apache Lucene, and Bioconductor [7]. It makes sense for large R&D organizations like ours, having already invested in bioinformatics expertise, to leverage such high-quality, actively developed code bases and make contributions in some cases.

Looking back over the last dozen years, it is apparent that we have reaped tremendous benefit in having embraced FLOSS systems in R&D. Our global high performance computing system is based on Linux and is supported in a production environment. The acceptance of the so-called LAMP (Linux/Apache/MySQL/PHP) stack by the corporate IT group sustained our highly successful grassroots efforts to create a company-wide wiki platform. We have continued to produce, validate, and publish new algorithms and make our source code available for academic use, for example for causal reasoning on biological networks [8]. It has been a real privilege being involved in these efforts among others, and with great optimism I look forward to the next decade of collaborative innovation.

Enoch S. Huang
Newton, Massachusetts
April 2012

References

[1] Caffrey DR, Dana PH, Mathur V, et al. PFAAT version 2.0: a tool for editing, annotating, and analyzing multiple sequence alignments. *BMC Bioinformatics* 2007;8:381.

[2] Barnes MR, Harland L, Foord SM, et al. Lowering industry firewalls: pre-competitive informatics initiatives in drug discovery. Nature Reviews Drug Discovery 2009;8(9):701–8.

[3] McKenna A, Hanna M, Banks E, et al. The Genome Analysis Toolkit: a MapReduce framework for analyzing next-generation DNA sequencing data. *Genome Research* 2010;20(9):1297–303.

[4] Li H, Durbin R. Fast and accurate long-read alignment with Burrows-Wheeler transform. *Bioinformatics* 2010;26(5):589–95.

[5] Smoot ME, Ono K, Ruscheinski J, Wang PL, Ideker T. Cytoscape 2.8: new features for data integration and network visualization. *Bioinformatics* 2011;27(3):431–2.

[6] Robinson JT, Thorvaldsdóttir H, Winckler W, et al. Integrative genomics viewer. *Nature Biotechnology* 2011;29(1):24–6.

[7] Gentleman RC, Carey VJ, Bates DM, et al. Bioconductor: open software development for computational biology and bioinformatics. *Genome Biology* 2004;5(10):R80.

[8] Chindelevitch L, Ziemek D, Enayetallah A, et al. Causal reasoning on biological networks: interpreting transcriptional changes. *Bioinformatics* 2012;28(8):1114–21.

About the editors

Dr Lee Harland is the Founder and Chief Technical Officer of ConnectedDiscovery Ltd, a company established to promote and manage precompetitive collaboration within the life science industry. Lee received his BSc (Biochemistry) from the University of Manchester, UK and PhD (Epigenetics and Gene Therapy) from the University of London, UK. Lee has over 13 years of experience leading knowledge management and information integration activities within major pharma. He is also the founder of SciBite.com, an open drug discovery intelligence and alerting service and part of the open PHACTS (*http://openphacts.org*) initiative to create shared public–private semantic discovery technologies.

Dr Mark Forster is team leader for the Chemical Indexing Unit, within the Syngenta R&D Biological Sciences group. He received his BSc and PhD in Chemistry from the University of London. He has over 25 years of experience in both academic research and in the commercial scientific software domain. His publications have been in diverse fields ranging from NMR spectroscopy, structural biology, simulations, algorithm development and data standards. Mark has been active in personally contributing new open source scientific software, encouraging industrial uptake and donation of open source, and organising workshops and conferences with an open source focus. He currently serves on the scientific advisory board of the open PHACTS and other projects.

About the contributors

Roman Affentranger, having obtained his PhD on the development of a novel Hamiltonian Replica Exchange protocol for protein molecular dynamics simulations in 2006 from the Federal Institute of Technology (ETH) in Zurich, Switzerland, worked for three years as postdoctoral scientist for the Group of Computational Biology and Proteomics (Prof. Dr X Daura) at the Institute of Biotechnology and Biomedicine of the Autonomous University of Barcelona, Spain. In 2010, he joined Douglas Connect (Switzerland) as Research Activity Coordinator, where he worked on the EU FP7 projects OpenTox and SYNERGY. At Douglas Connect, Roman Affentranger is currently involved in the scientific coordination and project management of ToxBank, in particular in the setup of project communication resources, the organisation and facilitation of both ToxBank-internal and cross-project working group meetings, the planning of project meetings and workshops, and in dissemination and reporting activities.

Laurent Alquier is currently Project Lead in the Pharma R&D Informatics Center of Excellence at Johnson & Johnson Pharmaceuticals R&D, L.L.C. Laurent has a PhD in optimisation techniques for Pattern Recognition and also holds an engineering degree in Computer Science. Since he joined J&J in 1999, Laurent has been involved in projects across the spectrum of drug discovery applications, from developing chemo-informatics data visualisations to improving compounds logistics processes. His current research interests are focused on using semantic data integration, text mining and knowledge-sharing tools to improve translational informatics.

Teresa Attwood is a Professor of Bioinformatics, with interests in protein sequence analysis that have led to the development of various databases (e.g. PRINTS, InterPro, CADRE) and software tools (e.g. CINEMA, Utopia). Recently, her interests have extended to linking research data with scholarly publications, in order to bring static documents to 'life'.

Erik Bakke is a Senior Software Engineer at Entagen and works out of the Minneapolis, MN office. He began his career in 2008 working extensively with enterprise Java projects. Coupling that experience with a history of building web-based applications, Erik embraced the Groovy/Grails framework. His interests include rich, usable interfaces and emerging semantic technologies. He maintains a connection to the next generation of engineers by volunteering as mathematics tutor for K-12 students.

Colin Batchelor is a Senior Informatics Analyst at the Royal Society of Chemistry, Cambridge, UK. A member of the ChemSpider team, he is working on natural language processing for scientific publishing and is a contributor to the InChI and Sequence Ontology projects. His DPhil (physical and theoretical chemistry) is on molecular Rydberg dynamics.

Michael R. Berthold, after receiving his PhD from Karlsruhe University, Germany, spent over seven years in the US, among others at Carnegie Mellon University, Intel Corporation, the University of California at Berkeley and – most recently – as director of an industrial think-tank in South San Francisco. Since August 2003 he holds the Nycomed-Chair for Bioinformatics and Information Mining at Konstanz University, Germany, where his research focuses on using machine-learning methods for the interactive analysis of large information repositories in the life sciences. Most of the research results are made available to the public via the open source data mining platform KNIME. In 2008, he co-founded KNIME. com AG, located in Zurich, Switzerland. KNIME.com offers consulting and training for the KNIME platform in addition to an increasing range of enterprise products. He is a past President of the North American Fuzzy Information Processing Society, Associate Editor of several journals and the President of the IEEE System, Man, and Cybernetics Society. He has been involved in the organisation of various conferences, most notably the IDA-series of symposia on Intelligent Data Analysis and the conference series on Computational Life Science. Together with David Hand he co-edited the successful textbook *Intelligent Data Analysis: An Introduction*, which has recently appeared in a completely revised, second edition. He is also co-author of the brand-new *Guide to Intelligent Data Analysis* (Springer Verlag), which appeared in summer 2010.

Erhan Bilal is a Postdoctoral Researcher at the Computational Biology Center at IBM T.J. Watson Research Center. He received his PhD in

Computational Biology from Rutgers University, USA. His research interests include cancer genomics, machine-learning and data mining.

Ola Bildtsen is a Senior Software Engineer at Entagen and works out of the Minneapolis, MN office. He has a strong background in rich UI technologies, particularly Adobe's Flash/Flex frameworks and also has extensive experience with Java and Groovy/Grails building web-based applications. Ola has been working with Java since 1996, and has been in a technical leadership role for the past seven years – the last four of those focused in the Groovy/Grails space. He has a strong background in Java web security and is the author of a Grails security plug-in (Stark Security). Ola holds a BA in Computer Science from Amherst College, and a MS in Software Engineering from the University of Minnesota.

Christopher Bouton received his BA in Neuroscience (Magna Cum Laude) from Amherst College in 1996 and his PhD in Molecular Neurobiology from Johns Hopkins University in 2001. Between 2001 and 2004, Dr Bouton worked as a computational biologist at LION Bioscience Research Inc. and Aveo Pharmaceuticals, leading the microarray data analysis functions at both companies. In 2004 he accepted the position of Head of Integrative Data Mining for Pfizer and led a group of PhD-level scientists conducting research in the areas of computational biology, systems biology, knowledge engineering, software development, machine-learning and large-scale 'omics data analysis. While at Pfizer, Dr Bouton conceived of and implemented an organisation-wide wiki called Pfizerpedia for which he won the prestigious 2007 William E. Upjohn Award in Innovation. In 2008 Dr Bouton assumed the position of CEO at Entagen (*http://www.entagen.com*), a biotechnology company that provides computational research, analysis and custom software development services for biomedical organisations. Dr Bouton is an author on over a dozen scientific papers and book chapters and his work has been covered in a number of industry news articles.

Nick Brown is currently an Associate Director in the Innovative Medicines group in New Opportunities at AstraZeneca. New Opportunities is a fully virtualised R&D unit that brings new medicines to patients in disease areas where AstraZeneca is not currently conducting research. His main role is as an informatics leader, working collaboratively to build innovative information systems to seek out new collaborators and academics, access breaking science and identify potential new drug repositioning opportunities. He originally received his degree in Genetics from York University and subsequently went on to receive his masters in Bioinformatics. He joined AstraZeneca as a bioinformatician in 2001, developing scientific software and automating toxicogenomic analyses. In 2004 he moved to the Advanced Science & Technology Labs (ASTL) as a senior informatician, developing automated tools including 3D and time-series imaging algorithms as well as developing the necessary IT infrastructure for high-throughput image analysis. Recently he has been partnering with search vendors to drive forward a shift in how we attempt to access, aggregate and subsequently analyse our internal and external business and market information to influence strategic direction and business decisions.

Craig Bruce is a Scientific Computing Specialist at AstraZeneca. He studied Computer-Aided Chemistry at the University of Surrey before embarking on a PhD in Cheminformatics at the University of Nottingham under the supervision of Prof Jonathan Hirst. Following the completion of his PhD he moved to AstraZeneca where he works with the Computational Chemistry groups at Alderley Park. His work focuses on providing tools to aid computational and medicinal chemists across the company, such as Design Tracker, which reside on the Linux network he co-administers.

Michael Burrell is IT Manager at The Sainsbury Laboratory. He graduated with a BSc in Information Technology from the University of East Anglia and has worked extensively on creating and maintaining the computer resources at The Sainsbury Laboratory since then. Michael constructed and maintained a high-performance environment based on IBM hardware running Debian GNU Linux and utilising Platform LSF. He has extensive experience with hosting server based software in these high-performance environments.

Meiping Chang is a Senior Staff Scientist at Regeneron Pharmaceuticals. Meiping received her PhD in Biochemistry, Biophysics & Molecular Genetics from University of Colorado Health Sciences Center. She has worked in the field of Computational Biology within Pharmaceutical companies in the past decade.

Aileen Day (née Gray) originally studied Materials Science at the University of Cambridge (BA and MSci) from 1995 until 1999, and then obtained a PhD (computer modelling zeolites) at the Chemistry department, University College London. During her postdoctoral research she adapted molecular dynamics code to calculate the lattice vibrational phonon frequencies of organic crystals. As a Materials Information Consultant at Granta Design Ltd (Cambridge, UK), she developed materials data management databases and software to store, analyse, publish and use materials test and design data. Since 2009 she has worked in the Informatics R&D team at the Royal Society of Chemistry developing RSC publications, educational projects and ChemSpider, and linking these various resources together.

David P. Dean is a Manager in Research Business Technology with Pfizer Inc. David received his BA (Chemistry) from Amherst College and MS (Biophysical Chemistry) from Yale University and has been employed at Pfizer for 20 years supporting Computational Biology and Omics Technologies as a software developer and business analyst.

Mark Earll graduated from The University of Kent at Canterbury in 1983 with an honours degree in Environmental Physical Science. After a short period working on cement and concrete additives, he joined Wyeth Research UK where he developed expertise in chiral separations and physical chemistry measurements. In 1995 Mark moved to Celltech to continue working in physical chemistry and developed interests in QSAR and data modelling. In 2001 he joined Umetrics UK as a consultant, teaching and consulting in Chemometric methods throughout Europe. In 2009 Mark joined Syngenta at Jealott's Hill International Research Centre, where he is responsible for the metabolomics informatics platform supporting Syngenta's seeds business.

Kirk Elder is currently CTO of WellCentive, a Population Healthcare Intelligence company that enables new business models through collaborative communities that work together to improve the quality and cost of healthcare. He has held senior technology leadership positions at various companies at the forefront of revolutionary business models. This experience covered analytics and SaaS solutions involving quality measure, risk adjustment, medical records, dictation, speech recognition, natural language processing, business intelligence, BPM and B2B solutions. Kirk is an expert in technology life-cycle management, product-to-market initiatives, and agile and open source engineering techniques.

Brian Ellenberger is currently the Manager of Software Architecture for MedQuist Inc., the world's largest medical transcription company with a customer base of 1500 healthcare organisations and a transcription output of over 1.5 billion lines of text annually. He has over 15 years of Software Engineering experience, and eight years of experience in designing and engineering solutions for the healthcare domain. His solutions span a wide range of areas including asset management, dictation, business process management, medical records, transcription, and coding. Brian specialises in developing large-scale middleware and database architectures.

Dawn Field received her doctorate from the University of California, USA, San Diego's Ecology, Behavior and Evolution department and completed an NSF/Sloan postdoctoral research fellowship in Molecular Evolution at the University of Oxford, UK. She has led a Molecular Evolution and Bioinformatics Group at the Centre for Ecology and Hydrology since 2000. Her research interests are in molecular evolution, bioinformatics, standards development, data sharing and policy, comparative genomics and metagenomics. She is a founding member of the Genomic Standards Consortium, the Environment Ontology, the MIBBI and the BioSharing initiative and Director of the NERC Environmental Bioinformatics Centre.

Ben Gardner is an Information and Knowledge Management Consultant providing strategic thinking and business analysis across research and development within Pfizer. He led the introduction of Enterprise 2.0 tools into Pfizer and has delivered knowledge-management frameworks that enhance collaboration and communication within and across research and development communities. More recently he has been working with information engineering colleagues to develop search capabilities and knowledge discovery solutions that combine semantic/linked data approaches with social computing solutions.

Roland C. Grafström is a tenured Professor in Biochemical Toxicology, Institute of Environmental Medicine, Karolinska Institutet, Stockholm, Sweden, since 2000, and visiting Professor, VTT Technical Research Centre of Finland, since 2008. Degree: Dr Medical Science, Karolinska Institutet, 1980. His bibliography consists of 145 research articles and 200 conference abstracts and he has a CV that lists leadership of large scientific organisations, arrangement of multiple conferences and workshops, 200 invited international lectures, and roughly 1000 hours of graduate, undergraduate and specialist training lectures. Roland received international prizes related to studies of environmental and inherited host factors that determine individual susceptibility to cancer, as well as to the development of alternative methods to animal usage. His research interests include toxicity and cancer from environmental, man-made and life style factors; molecular mechanisms underlying normal and dysregulated epithelial cell turnover; systems biology, trancriptomics, proteomics and bioinformatics for identification of predictive biomarkers; application of human tissue-based *in vitro* models to societal needs and replacement of animal experiments.

Niina Haiminen is a Research Staff Member of the Computational Genomics Group at IBM T.J. Watson Research Center. Dr Haiminen received her PhD in Computer Science from the University of Helsinki, Finland. Her research interests include bioinformatics, pattern discovery and data mining.

James Hardwick is a Software Engineer at Entagen and works primarily out of the Minneapolis, MN office. He began his career in 2006 working on a variety of enterprise Java projects. In 2009 he received his master's degree in Software Engineering from the University of Minnesota. While involved in the program James fell in love with Groovy & Grails thanks in part to a class taught by Mr Michael Hugo himself. His core interests include rapidly building web-based applications utilising the Groovy/ Grails technology stack and more recently developing rich user interfaces with Javascript.

Barry Hardy leads the activities of Douglas Connect, Switzerland in healthcare research and knowledge management. He is currently serving as coordinator for the OpenTox (*www.opentox.org*) project in predictive toxicology and the ToxBank infrastructure development project (*www. toxbank.net*). He is leading research activities in antimalarial drug design and toxicology for the Scientists Against Malaria project (*www. scientistsagainstmalaria.net*), which was developed from a pilot within the SYNERGY FP7 ICT project on knowledge-oriented collaboration. He directs the program activities of the InnovationWell and eCheminfo communities of practice, which have goals and activities aimed at improving human health and safety and developing new solutions for neglected diseases. Dr Hardy obtained his PhD in 1990 from Syracuse University working in the area of computational science. He was a National Research Fellow at the FDA Center for Biologics and Evaluation, a Hitchings-Elion Fellow at Oxford University and CEO of Virtual Environments International. He was a pioneer in the early 1990s in the development of World Wide Web technology applied to virtual scientific communities and conferences. He has developed technology solutions for internet-based conferencing, tutor-supported e-learning, laboratory automation systems, computational science and informatics, drug design and predictive toxicology. In recent years he has been increasingly active in the field of knowledge management as applied to supporting innovation, communities of practice, and collaboration. With OpenTox he is leading the development of an open, interoperable, semantic web platform whose goal is to satisfy the needs of the predictive toxicology field through the creation of applications linking resources together for data, algorithms, models and ontologies.

Lee Harland is the Founder and Chief Technical Officer of ConnectedDiscovery Ltd, a company established to promote and manage pre-competitive collaboration within the life science industry. Lee received his BSc (Biochemistry) from the University of Manchester, UK and PhD (Epigenetics and Gene Therapy) from the University of London, UK. Lee has over 13 years of experience leading knowledge management and information integration activities within major pharma. He is also the founder of SciBite.com, an open drug discovery intelligence and alerting service.

Martin Harrison is an Associate Principal Scientist at AstraZeneca, where he supports oncology discovery projects at Alderley Park as a computational chemist. His work also focuses on providing tools to aid design teams (computational, physical, synthetic, medicinal chemists and DMPK colleagues) and he uses these tools in his own projects. He developed Design Tracker and is now product owner of this global tool. In this capacity he manages the development effort and prioritises enhancement requests from the users. Previous to AstraZeneca he worked for Tularik Ltd and Protherics/Proteus Molecular Design Ltd as a computational chemist. He completed his PhD in Computational Chemistry at Manchester University under the supervision of Prof Ian Hiller in 1997.

Richard Head is the Director of Research & Development for Genomics and Pathology Services, and Director of the Genome Technology Access Center (GTAC) at Washington University School of Medicine. Prior to Washington University, he spent 14 years in pharmaceutical research. Most recently he was the leader of Computational Biology in the Inflammation & Immunology and Indications Discovery research units at Pfizer's St. Louis Laboratories. His research has primarily focused on the application of high throughput genomic technologies, and the subsequent data interpretation, to multiple stages of drug discovery and development.

Ed Holbrook is an experienced IT professional, including five years with Astra Zeneca where he led a number of development and implementation projects, ranging barcode scanner systems to a global publication management system referencing over 50 million documents. Ed has worked as a project manager, team leader, architect, lead analyst and trainer. He has advised and assisted a number of organisations in the installation, upgrade and development of their systems, has presented at a number of national and international conferences, and has delivered a number of public training courses. Ed is currently leading the development and implementation of systems to help mobilise workers access their office-based systems via their mobile phones. When not at work, Ed enjoys taxiing his children to their various activities, running a church kids' club, and instructing skiing at the local dry ski slope.

Jolyon Holdstock studied for a BSc in Biochemistry and Physiology at Southampton University followed by a MSc in Molecular Endocrinology at Sussex University and a PhD elucidating mechanisms regulating intracellular signalling in pituitary gonadotropes at the Royal London Hospital Medical College. Subsequently he took a postdoctoral fellowship in the Department of Experimental Pathology at Guy's Hospital elucidating mechanisms controlling chromatin structure. Increased time spent in front of the computer forged an interest in bioinformatics and he made the switch joining the computational biology group at Oxford based startup Oxagen. Moving to another Oxfordshire located company, Oxford Gene Technology, he is currently leading the microarray probe and NGS custom bait design team.

Mike Hugo is a Senior Software Engineer at Entagen and works out of the Minneapolis, MN office. He has extensive experience with Groovy, Grails, Search and Semantic Web technologies. In addition to his expertise with these technologies he has more than seven years of enterprise Java web development, design, and technical leadership experience. He's the author of several Grails plug-ins (including Code Coverage, Liquibase Runner, Build Info and Hibernate Stats), has been a featured author in *GroovyMag* and makes regular presentations at developers' meetings.

Mike holds a master's degree in Software Engineering from the University of Minnesota and is an adjunct faculty member of the University of Minnesota Software Engineering Center.

Bernd Jagla, after receiving his PhD from the Free University Berlin, Germany in the field of bioinformatics, did a PostDoc at the Memorial Sloan-Kettering Cancer Center (MSKCC) in New York, USA in molecular biology. He then joined a bioinformatics startup company in Heidelberg, Germany developing clinically relevant workflows using an in-house workflow management system, before being recruited back to MSKCC where he pioneered some bioinformatics approaches for siRNA design and high-content high-throughput imaging. He continued his work with high-throughput confocal microscopes now integrating chemical compounds and images in the open microscopy environment (OME/OMERO) at Columbia University, New York, USA. He then became quality assurance manager for an integration platform (geWorkbench) developed at Columbia University. He is currently working as a bioinformatician at the Institut Pasteur in Paris developing KNIME nodes and workflows solving problems related to next-generation sequencing.

Nina Jeliazkova received a MSc in Computer Science from the Institute for Fine Mechanics and Optics, St. Petersburg, Russia in 1991, followed by a PhD in Computer Science (thesis 'Novel computer methods for molecular modeling') in 2001 in Sofia, Bulgaria, and a PostDoc at the Central Product Safety department, Procter & Gamble, Brussels, Belgium (2002–2003). Her professional career started as a software developer, first at the oil refinery Neftochim at Burgas, Bulgaria, then at the Laboratory for Mathematical Chemistry, Burgas, Bulgaria (1996–2001). She joined the Bulgarian Academy of Sciences in 1996 as a researcher and network engineer at the Network Operating Centre of the Bulgarian National Research and Education Network. She is founder and co-owner of Ideaconsult Ltd and is technical manager of the company since 2009. She participated in a number of R&D projects in Bulgaria, Belgium, and EC FP5-FP7 programs. She is an author and project leader of several open source applications for toxic hazard estimation and data and predictive model management (Toxtree (*http://toxtree.sf.net*), AMBIT (*http://ambit.sf.net*), (Q)SAR Model reporting format inventory (*http://qsardb.jrc.it/qmrf/*), and had a leading role in the design and implementation of the OpenTox framework.

Published by Woodhead Publishing Limited, 2012

Misha Kapushesky is the Founder and Chief Executive Officer of GeneStack Limited, a company providing a universal platform for genomics application development. Prior to this Misha was Team Leader in Functional Genomics at the European Bioinformatics Institute, where he researched, taught and developed algorithms, tools and databases for managing and analysing large-scale multiomics datasets. Misha received his BA (Mathematics & Comparative Literature) from Cornell University, USA, and PhD (Genetics) from the University of Cambridge, UK.

Richard Kidd, Informatics Manager, is an ex-chemical engineer and has worked at the RSC for many years on the underlying data and production processes behind its publications. Richard's team developed the Project Prospect semantic enhancement of its journals, based on text mining and manual quality assurance to link together similar compounds and subject terminology, and is currently integrating publications and tools with the RSC's public access ChemSpider database. He is currently Treasurer of the InChI Trust and is also involved in the IMI Open PHACTS project.

Frans Lawaetz is Entagen's Chief Systems Architect working out of the Boston, MA office. He has over ten years experience in Linux, software stacks, grid computing, networking, and systems automation. Frans has a Bachelor of Arts from La Salle University.

Rob Lind has worked in the field of biological sciences since completing his BSc in Applied Biology and PhD in insect neurosciences in 1998 at the University of Bath. During his PhD, Rob worked alongside Iain Couzin in the Biological Science Department who was using computer vision techniques to better understand the collective behaviour of ants for his PhD. After joining Jealott's Hill International Research Centre to work in insect control, Rob collaborated with Iain to build up an imaging system that could be implemented in the screening for new insecticides since 1999. This imaging system evolved and demonstrated its utility to understand the response of insects to xenobiotics and the advantages that a computer vision system offered over human observation became apparent. The application of image analysis now extends into every corner of the Syngenta business, and, in 2010, the Syngenta global network for image analysis within the company led by Rob was set up to capitalise on the cross learning and sharing of ideas between scientists. For this network to operate successfully, a common platform was needed and a freely available, flexible and open source solution to perform image analysis was chosen, namely ImageJ.

Dan MacLean is Head of Bioinformatics at The Sainsbury Laboratory, Norwich, UK. Dan received a BSc and MRes in Biological Sciences from the University of Manchester and a PhD in Plant Molecular Biology from Darwin College, University of Cambridge. Dan carried out postdoctoral research in Plant Bioinformatics at Stanford University, California, USA before returning to the UK. Dan's areas of expertise include genomics, in particular assembly and alignment with high-throughput sequencing data, genetic polymorphism detection, homology search and automatic sequence annotation, transcriptomics (microarray and RNA-seq), statistical analysis, mathematical and systems-biology approaches to modelling and visualising data and large-scale phosphoproteomics analysis. Dan has pioneered and implemented a novel biologist-first bottom-up support model that prioritises the needs of the scientist and makes support provision flexible and scalable in the Big Data era.

Eamonn Maguire completed a BSc in Computer Science at Newcastle University, UK, followed by an MRes in Bioinformatics at the same university. He is currently the Lead Software Engineer of the ISA software suite at the University of Oxford, UK, where he is also undertaking a DPhil in Computer Science focused on visualisation of biological data and meta-data; he also contributes to the development of the BioSharing catalogue. His research interests lie in bioinformatics, software engineering, visualisation and graphical user interface design and development.

James Marsh is Postdoctoral Research Fellow in the School of Chemistry at The University of Manchester, with interests in collaborative systems, scientific visualisation and the semantic web. His current research involves modelling and visualising relationships between academic publications and related data.

Catherine Marshall is a Senior Manager and Strategic Business Partner for the Translational and Bioinformatics group at Pfizer. Her primary focus is assisting Pfizer's Research organisation in defining and implementing technology solutions in target selection and validation and computational biology. Catherine received her BSc in Management and Computer Science from Boston College, US and MSc in Computer Science from Rensselaer Polytechnic Institute, US. Catherine has over 20 years of experience managing and leading software development teams through the entire Software Development Life Cycle, SDLC. She has experience working in the pharmaceutical, finance, energy services, and insurance industries.

Thorsten Meinl is currently working as a postdoc in the Computer Science Department at the University of Konstanz. He graduated from the University of Erlangen-Nuremberg in 2004 and received his PhD in 2010 on the topic of 'Maximum-Score Diversity Selection' from the University of Konstanz. His current research topics are optimisation with metaheuristics and parallel computing. He is also interested in chemoinformatics and has worked on a couple of projects with several pharma companies. He is also affiliated with KNIME.com and has been working on KNIME for more than five years.

Ted Naleid is a Senior Software engineer at Entagen's Minneapolis, MN office. He an active contributor to the Groovy and Grails open source communities and is the author of a number of popular Grails plug-ins including the redis, build-test-data, jasypt encryption, and markdown plug-ins. He frequently speaks at conferences and user groups. Before Entagen, Ted was the Technical Architect at Bloom Health, where he built and led a team of developers through a successful acquisition by three of the largest health insurers in the country. He also has extensive experience with the Semantic Web and was one of the first developers at the semweb startup Evri in Seattle, WA.

Nguyen Nguyen is a Senior Software Engineer at Entagen and works out of the Minneapolis, MN office. He began his career in 2000 working on various technologies from Linux driver development to web application development. Nguyen holds a BS in Computer Science, and a MS in Software Engineering from the University of Minnesota.

Ketan Patel is currently a Healthcare Solutions Consultant with Oracle Health Sciences working on translational medicine solutions. Prior to this post he has led multiple teams and projects both at Pfizer and at Lilly working on translational bioinformatics in the fields of oncology, diabetes and inflammation over a nine-year period. Dr Patel holds a PhD in Bioinformatics from the University of Oxford and an MSc in Artificial Intelligence from the University of Edinburgh.

Steve Pettifer is a Senior Lecturer in Computer Science with interests in distributed systems, computer graphics, human/computer interaction and visualisation. His current research involves the application of semantic technologies to the design of user interfaces, with a particular focus on improving access to scholarly publications and biochemical data.

Simon Revell has been in the IT industry for 15 years covering a variety of technical lead, project management and consultancy roles on projects for customers in Europe, Asia and the US. In recent years he has become internationally recognised for his groundbreaking work implementing Web 2.0 tools and approaches for collaboration and information sharing for customers within an industry-leading Fortune 500 firm.

Philippe Rocca-Serra, PhD, received his doctorate in Molecular Genetics from the University of Bordeaux, France, in collaboration with the University of Oslo, Norway. Involved in data management for the last ten years (eight spent at the EMBL-EBI, UK) he has working expertise in ontology development, experimental design, curation and collaborative standards' development. Currently Technical Project Leader at the University of Oxford, UK, he is the primary developer of the ISA-Tab format and the ISA software's user specifications, and lead contributor to the BioSharing initiative, promoting standards' awareness and re-use. He is also a core developer of ontology for biomedical investigation and board advisor for the identifiers.org project.

Susanna-Assunta Sansone, PhD, received her doctorate in Molecular Biology from the Imperial College, London, UK. She is a Team Leader at the University of Oxford, UK, with over ten years' experience in project management focusing on curation, ontologies and software for data management. She has developed a significant expertise in the area of standardisation for the purpose of enabling reporting, sharing and meta-analysis of biological, biomedical and environmental studies. She has designed the ISA project and led the development of the software suite. She is a central player in the grass-roots data standardisation movement. Co-founder of the MIBBI and BioSharing initiative, she sits on the Board of several standardisation initiatives, including the Genomic Standards Consortium and the Metabolomics Standards Initiative.

Annapaola Santarsiero received her BSc in Software Engineering from the Politecnico of Milan, Italy. She is currently finalising an MSc in Communications Engineering at the same university where projects have included work in developing analysis workflows for miRNA and DNA microarray data at the Mario Negri Pharmacology Institute, Milan. She has developed the initial BioSharing catalogue and currently is a web developer in the Zoology department at the University of Oxford. Her main interests lie in web design, usability and human–computer interaction.

Ola Spjuth is a researcher at Uppsala University with main research interests in developing e-Science methods and tools for computational pharmacology and especially drug metabolism and safety assessment. He is advocating the use of high-performance computing, web-based services, and data integration to address the challenges of new high-throughput technologies, and is the deputy Director of Uppsala Multidisciplinary Center for Advanced Computational Science (UPPMAX, *www.uppmax. uu.se*). Dr Spjuth obtained his PhD in 2010 from Uppsala University, Sweden on the topic of pharmaceutical bioinformatics. After this he did a postdoc at Karolinska Institutet, Sweden, in the field of data integration and cancer informatics. He is the author and project leader of the Bioclipse (*www.bioclipse.net*) project, which is a free and open source integration platform and graphical workbench for the life sciences. He is also currently serving as coordinator for the Swedish national initiative for Next Generation Sequencing storage and analysis (UPPNEX, *www. uppnex.uu.se*), as well as coordinating the Swedish e-Science Research Center (SeRC, *www.e-science.se*) in the Complex Diseases and Data Management communities.

Claus Stie Kallesøe has been working with data management and informatics in pharma research since 2000. Claus finalised his MSc (pharm) in 1997 with a thesis in synthetic medicinal chemistry and was then engaged at H. Lundbeck A/S as section head of the compound management group and in 2000 appointed Head of Department of Discovery Informatics. From 2003 to 2005 Claus also held a position as Executive Assistant to SVP of Research Dr Peter Høngaard Andersen. In 2008 Claus was appointed Chief Specialist/Director of Global Research Informatics. The informatics group at Lundbeck mainly use open source tools for their in house development and the group promotes the sharing of tools and ideas. Claus is also engaged in several pre-competitive activities. He was PM for the second round of IMI knowledge management calls leading to the current OpenPhacts project, he holds a seat in the PRISME forum and on the Board of Directors of the Pistoia Alliance. Claus also holds a Diploma in Software Development and an E-MBA from INSEAD in France.

David Stokes is Global Lead for Information Systems Consulting and Compliance at Business & Decision Life Sciences, an international

Consultancy and Systems Integrator. Originally qualifying as a process control engineer in the early 1980s, he has more than 25 years' experience in validation and compliance of information systems in the life sciences industry and is a leading participant in industry organisations such as ISPE®/GAMP. His pragmatic and cost-effective risk-based approach to applying established principles of computer systems validation to newer information systems technologies such as open source software, cloud computing, agile software development life cycles and middleware/service oriented architecture is widely recognised, and he is a regular speaker at international conferences and has written a wide range of published papers, articles and books on the subject of information systems compliance.

Chris Taylor, PhD, received his doctorate in Population Genetics and Speciation from the University of Manchester, UK. A member of the Natural Environment Research Council's Environmental Bioinformatics Centre based at the EMBL-EBI, UK, he is a coordinator of the MIBBI project. He has also taken various roles in a wide range of domain-specific standardisation projects, such as the Proteomics Standards Initiative.

Simon Thornber is a senior IT consultant at GlaxoSmithKline. Simon has worked in the life science industry for over 15 years, moving from Technology Development, to BioInformatics, to Cheminformatics, and then into R&D IT. He is also currently the work-group chair for the Pistoia Alliance 'Sequence Services' project (*www.pistoiaalliance.org*).

Dave Thorne is a postdoctoral research fellow in Computer Science, interested in the design of semantically rich software systems, specifically for the biosciences. He has worked for the past half-decade on modelling and visualisation (e.g. CINEMA, Ambrosia, Utopia) and now focuses his expertise on applying these techniques to the scholarly literature.

Valery Tkachenko has graduated from the Lomonosov Moscow State University, receiving his MSc in Chemistry and BSc in Computer Sciences. He is currently Chief Technology Officer of ChemSpider at the Royal Society of Chemistry. Over the course of the last 15 years he has participated in the development of a number of successful enterprise projects for large pharmaceutical companies and the public domain, including PubChem. He is the author of over 20 peer reviewed papers and book chapters.

Published by Woodhead Publishing Limited, 2012

Aristotelis Tsirigos is a research staff member of the Computational Genomics Group at IBM T.J. Watson Research Center. Dr Tsirigos received his PhD in Computer Science from the Courant Institute of New York University (with Prof Dennis Shasha). His research interests include cancer genomics, algorithms and data mining.

Filippo Utro is a postdoctoral researcher as a member of the Computational Genomics Group at IBM T.J. Watson Research Center. Dr Utro received his PhD in Mathematics and Computer Science from the University of Palermo, Italy. His research interests include algorithms and data structure, bioinformatics, pattern discovery and data mining.

Phil Verdemato is a software engineer at Thomson Reuters. He has a BSc in Biochemistry, a PhD in Biophysics and a Masters in Bioinformatics, and has spent the last ten years developing software for life sciences and financial institutions.

David Wild is Assistant Professor of Informatics and Computing at Indiana University School of Informatics. He directs the Cheminformatics Program, and leads a research group of approximately 15 students focused on large-scale data mining and aggregation of chemical and biological information. David has developed a teaching program in cheminformatics at the university, including an innovative distance education program, most recently resulting in an online repository of cheminformatics teaching materials. He has been PI or CoPI on over $1.4m of funding, and has over 30 scholarly publications. David is Editor-in-Chief (along with Chris Steinbeck at the EBI) of the *Journal of Cheminformatics*, and works as an editorial advisor or reviewer to many journals. He is involved in several cheminformatics organisations including being a trustee of the Chemical Structure Association Trust and a member of the American Chemical Society. David has helped organise many conferences and symposia in this field, and has recently acted as an expert witness in cheminformatics. He is also the director of Wild Ideas Consulting, a small scientific computing company specialising in informatics and cheminformatics.

Antony Williams is the VP of Strategic Development at the Royal Society of Chemistry and the original founder of ChemSpider, one of the world's primary internet resources for chemists. He holds a BSc Hons I from the University of Liverpool and obtained his PhD from the University of London focused on Nuclear Magnetic Resonance. He held a postdoctoral position at the National Research Council in Ottawa, Canada before joining the University of Ottawa as their NMR Facility Manager. He was the NMR Technology Leader at the Eastman Kodak Company in Rochester, NY for over five years before joining Advanced Chemistry Development (ACD/Labs) where he became Chief Science Officer focusing his cheminformatics skills on structure representation, nomenclature and analytical data handling. He is known as the ChemConnector in the chemistry social network and is a blogger and active participant in Wikipedia Chemistry.

Egon Willighagen studied Chemistry at the University of Nijmegen in the Netherlands (1993–2001) with a major in chemometrics at the Department of Analytical Chemistry (Prof LMC Buydens) studying the unsupervised classification of polymorphic organic crystal structures. He continued his studies in the same group on the relation of representation of molecular knowledge and machine learning during his PhD (ISBN:978-90-9022806-8). Taking advantage of his extra-curriculur research in (open source) cheminformatics, he worked on methods to optimise the amount of information gained from pattern recognition methods in the fields of Quantitative Structure–Activity Relationships (QSAR), supervised clustering and prediction of properties of organic crystal structures, the general reduction of error introduced in exchange of chemical data, and on improving the reproducibility of data analysis in this field in general. This research was partly performed at the European Bioinformatics Institute (Prof J. Thornton) and the University of Cambridge (Prof P. Murray-Rust). After his PhD research he continued his efforts on reducing the error introduced by data aggregation and cheminformatics toolkits with postdoctoral research in Cologne (Dr C. Steinbeck), and Wageningen University and the Netherlands Metabolomics Center (Dr R. Van Ham) in the area of metabolite identification, and after that in proteochemometrics at Uppsala University (Prof J. Wikberg). After a short project in text mining for chemistry back at the University of Cambridge (Prof. P Murray-Rust), Willighagen started in 2011 a post-doc at the Institute of Environmental Medicine of the Karolinska Institute (Prof B. Fadeel, and Prof R. Grafström) working on applications of cheminformatic in toxicology. In January 2012 he will start a postdoc position at Maastricht University to work on the EU FP7 project Open PHACT with Dr C. Evelo. Willighagen is editorial member for two journals in the field cheminformatics and scientific computing, and member of advisory boards of various international organisations, and has been awarded for his contributions to open science in cheminformatics.

Timo Wittenberger studied Biochemistry at the Universities of Tuebingen and Tucson and received his PhD from the Center for Molecular Neurobiology at the University of Hamburg, Germany. He then joined Altana Pharma as bioinformatician and later was responsible for the research IT at the same company. In 2007 he joined Genedata, a leading bioinformatics company delivering innovative solutions for data analytics and R&D workflows, where he is managing professional services and research collaborations.

Introduction

Sharing the source code of software used in life science research has a history spanning many decades. An early example of this being the Quantum Chemistry Program Exchange (QCPE), founded by Richard Counts at Indiana University in 1962, with the aim of distributing the codes then available in the domain of quantum chemistry. It also came to cover other areas of chemical science such as kinetics and NMR spectroscopy. Even in the 1980s, code distribution over networks was not common, so that programs were typically distributed on cartridge tapes or other media usable by the workstations of the era. Although some codes may have been available only in binary form, it was normal to make Fortran or C source code available for compilation on the receiving platform.

For code to be 'open source', the license that covers the code must allow users to study, modify and redistribute new versions of the code. This is often referred to as libre, or 'free as in speech'. It is often the case that the source code is available at zero cost, which may be referred to as gratis, or 'free as in beer'. In English, we often use the single word, free, to cover both of these distinct types of freedoms. For a more detailed explanation see the Wikipedia entry 'Gratis versus libre'. Software that is accompanied by these freedoms may be described as Free/Libre Open Source Software, with the acronyms FOSS or FLOSS. A fuller description of the philosophy and aims of the free software movement can be found at the website of the Free Software Foundation (*http://www.fsf.org*), an organisation founded in 1985 by Richard M. Stallman, which aims to protect and promote the rights of computer users with regard to software tools. The open source philosophy fits well with the sharing of data and methods implicit in scientific research, the floss4science website (*http://www.floss4science.com/*) supports an effort to apply and utilise the open source tools and principles in scientific computing.

The best known and widely used example of open source software is perhaps the Linux Operating System (OS). This became available in the early 1990s with Linus Torvald's creation of the Linux kernel for 386 based PCs, combined with the GNU tool chain (C compiler, linker, etc.) produced by the free software foundation. Today, this software stack, as well as relatives and derivatives of it, are very pervasive. They exist in devices ranging from set top boxes, smart-phones and embedded devices, through to the majority of the world's largest supercomputers. Open source is 'powering' not only life science research but also the digital lives of countless millions of people.

It is beyond the scope of this editorial to cover the landscape of open source licenses in great detail, the open source initiative website provides definitions and a listing of approved licenses (*http://opensource.org/*). It is perhaps important to understand the permissive and restrictive nature of some open source licenses, as well as the idea of copyleft as opposed to copyright. Copyright covers the rights of the author over his/her work and is a long-established and well-understood legal concept. Copyleft seeks to establish a license that maintains the freedoms given to a work by the original author in any derivative, such as new versions of computer source code. The licenses used for open source may be restrictive in that they require the source code of derivative works also be made available, an example of this being the GNU public license v2 (GPLv2). This seeks to ensure code made available as open source cannot later be closed and made proprietary. Other license types may be very permissive in not requiring the source code for derivative works, one example of this being the BSD license. For further details on these licenses and others, see the web page: *http://opensource.org/licenses/alphabetical*. Whatever license is used, the intent of creators of open source is to freely share their work with a wider audience, promoting innovation and collaboration.

A key aspect of scientific research is that experimental methods, analysis techniques and results should be both open to scrutiny and reproducible. The issue of reproducibility in computational science has recently been discussed (Peng, *Science* 2011;334:1226–7; Ince et al., *Nature* 2012;482:485–8). Open source tools and resources support this need and ethos in a very direct way, as the code is manifestly open for inspection and recompilation by those with the technical background and interest to do so. Consequently, this directly supports the reproducible nature of computational analysis of data. The free availability of code that can be shared among researchers without restrictive licenses allows analysis methods to be more widely distributed. It is also the case that many challenges facing life science researchers are common, be they in the academic, governmental or industrial domains, or indeed working in the pharmaceutical, biotechnology, plant science/agrochemical or other life science sectors. There is a growing awareness (Barnes et al.,

Nature Reviews Drug Discovery 2009;8:701–8) that pre-competitive collaborations are an effective and perhaps necessary response to the challenges facing the pharmaceutical and other sectors. As research and IT budgets are static or shrinking and data volumes expand greatly, these new ways of working are central to finding cost-effective solutions. It is clear that open source development performed in a collaborative manner, feeding code back to the open source community, is very consistent with this new paradigm. The benefits of industrial collaboration exist not only in the domain of software tools but also in the data vocabularies that describe much of our industrial research data, a recent publication discusses this further (Harland et al., *Drug Discovery Today* 2011;16:940–7). Conferences and meetings that help to link the open source and industrial communities will be of great benefit in promoting this environment of pre-competitive collaboration. In 2011 a Wellcome Trust funded meeting on 'molecular informatics open source software' was organised by Forster, Steinbeck and Field (*http://tinyurl.com/MIOSS2011WTCC*), and drew together many key developers, software experts and industrial and academic users seeking common solutions in the domain of chemical sciences, drug discovery, protein structure and related areas. This meeting created a possibly unique forum for linking the several threads of the open source community, building opportunities for future collaboration. Future meetings on this and other life science research areas would be valuable.

Importantly, not all free software is open source; the authors choosing a model to gain broadest use of the software (by making it free) but retaining control (and perhaps competitive advantage) of the source code. These models often morph into 'freemium' models where basic functionality is free, but more specialised or computationally intensive functionalities require payment. However, as eloquently described by Chris Anderson (*Free: How today's smartest businesses profit by giving something for nothing*. Random House Business Publishing), this and many other models for 'free' have proliferated thanks to the rapid growth of the internet. In addition to splitting free versus paid customers based on functionality needs, a division is often made based on non-profit status. This therefore, offers an additional mechanism for consumers to gain easy access to often quite complex software, which traditionally might have meant significant expense.

A phrase sometimes heard in connection with free and open source is 'What about support?'. Although many open source codes can be obtained at low or no cost, albeit with no obligation of support from the developers, there is often a thriving community of developers or other users that can answer questions of varying complexity through web based forums. Such support may sometimes be more responsive and insightful than the help mechanisms of purely commercial and closed source offerings. In addition, new business models have emerged, offering professional paid support for code that is completely or fully open source. Perhaps the best known example is that of Red Hat, offering support and update packages to customers for a fee. This business model has grown to such an extent that in 2012, Red Hat has become the first open source-focused company with one billion dollars of revenue. These 'professional open source' business models are increasingly used within the life science domain, offering researchers the 'best of both worlds'.

This book seeks to describe many examples where free and open source software tools and other resources can or have been applied to practical problems in applied life science research. In researching the book we wanted to identify real-life case studies that business and technology challenged faced by industrial research today. As we received the first drafts from the chapter authors, it became clear that the book was also providing a 'window' on the software and decision-making practices within this sector. As is often the case with large commercial organisations, much of the innovation and creativity within is never really observed outside of the company (fire)walls. We hope that the book helps to lift this veil somewhat, allowing those on the 'outside' to understand the strategies companies have taken, and the reasons why. Many examples demonstrate successful deployment of open source solutions in an 'enterprise', organisation-wide context, challenging perceptions to the contrary. These successes highlight the confidence in the capabilities of the software stack and internal/external support mechanisms for potential problem resolution. A further conscious decision was to mix contributions by tool consumers with chapters by those that produce these tools. We felt it was very important to show both sides of the story, and understand the rationale behind the decisions by developers to 'give away' the systems developed through their hard work.

Our coverage starts with use-cases in the areas of laboratory data management, including chemical informatics and beyond. The first chapter illustrates how free and open source software has become woven into the fabric of industrial research. Claus Stie Kallesøe provides a detailed study of the development rationale and benefits of an end to end enterprise research data system in use within a major pharmaceutical company and based almost entirely on open source software technology. His approach and rationale makes a compelling case around the value of free and open source in today's industrial research environment. The field of chemical toxicology is important to researchers in any company with a focus on small molecule or other bioactive molecule discovery. A powerful and flexible infrastructure for interactive predictive chemical toxicology has been developed by Egon Willighagen and other authors. Their chapter explains how the Bioclipse client and OpenTox infrastructure combine to meet the needs of scientists in this area. The creation of flexible and efficient tools for processing and dissemination of chemical information are covered by Aileen Day and co-authors, who have pioneered new methods in the field of academic publishing and freely share this knowledge through their chapter. Open source tools also find application in the fields of agrochemical discovery and plant breeding. Mark Earll gives us insight into the use of open source tools for mass spectroscopy data processing with specific application to metabolite identification. Key application examples and methodology are described by Rob Lind who covers the use of the powerful ImageJ free software for image processing. Some software tools have very wide applicability and can cover a range of application areas. Thorsten Meinl, Michael Berthold and Bernd Jagla explain the use of the KNIME workflow toolkit in chemical information processing as well as next generation sequence analysis. Finally in this area, Susanna Sansone, Philip Rocca-Serra and co-authors describe the ISA (Investigation-Study-Assay) tools, these allow flexible and facile capture of metadata for diverse experimental data sets, supporting the standardisation and sharing of information among researchers.

The book then covers case studies in the genomic data analysis and bioinformatics area, often part of the early stages of a life science discovery pipeline. The field of computational genomics has been well served by the release of the 'GenomicTools' toolkit covered in the chapter by Aristotelis Tsirigos and co-authors. The multifaceted nature of 'omics data in life science research demands a flexible and efficient data repository and portal type interface suitable for a range of end-users. Ketan Patel, Misha Kapushesky and David Dean share their experiences and knowledge by describing the creation of the 'Gene Expression Atlas'. Industrial uses are also discussed. Jolyon Holdstock shares the experience of creating an 'omics data repository specifically aimed at meeting the needs of a small biotechnology organisation. A similar small organisation experience is shared by Dan MacLean and Michael Burrell who show how a core bioinformatics platform can be created and supported through the use of FLOSS. Although the institution in question is within academia, it is still an 'applied' research facility and will be of great interest to those charged with building similar support functions in small biotechnology settings.

There are numerous mature and capable open source projects in the domain of information and knowledge management, including some interesting, industry-centric takes on the social networking phenomenon. Craig Bruce and Martin Harrison give insight into the internal development of 'Design Tracker', a highly capable support tool for research hypothesis tracking and project team working. Next, Ben Gardner and Simon Revell describe a particularly successful story regarding the use of wikis and social media tools in creating collaboration environments in a multinational industrial research context. In the following chapter by Nick Brown and Ed Holbrook, applications for searching and visualising an array of different pharmaceutical research data are described. The contribution of the open source text-indexing system, Lucene, is central to many of the system's capabilities. Finally, document management, publication and linking to underlying research data is an important area of life science research. As described in the chapter by Steve Pettifer, Terri Attwood, James Marsh and Dave Thorne, the Utopia Documents application provides a 'next-generation' of tool for accessing the scientific literature. Not only does this chapter present some exciting technology, it also describes an interesting business model around the support of 'free' software for commercial customers.

Our next section concerns the use of semantic technologies, which have grown in popularity in recent years and are seen as critical to success in the 'big data' era. To start, Laurent Alquier explains how the knowledge-sharing capabilities of a wiki can be combined with the structure of semantic data models. He then describes how the Semantic MediaWiki software has been used in the creation of an enterprise-wide research encyclopaedia. Lee Harland and co-authors follow on the Semantic MediaWiki theme, detailing the creation and capabilities of intelligent systems to manage drug-target and disease knowledge within a large pharmaceutical environment. Although the MediaWiki-based software demonstrate some very useful semantic capabilities, the more fundamental technologies of RDF and triple-stores have the potential to transform the use of biological and chemical data. To illustrate exactly this point, David Wild covers the development of the Chem2Bio2RDF semantic framework, a system which provides a unique environment for answering drug-discovery questions. This is followed by Ola Bildtsen and colleagues who describe the 'TripleMap' tool, that connects semantic web and collaboration technologies in a visual, user-friendly environment.

A critical and highly expensive stage in the pharmaceutical development pipeline is that of clinical trials. These areas are highly regulated and software may need complex validation in order to become accepted. Kirk Elder and Brian Ellenberger describe the use of open source tools for 'extreme scale' clinical analytics with a detailed discussion of tools for business intelligence, statistical analysis and efficient large-scale data storage. In a highly regulated environment such as the pharmaceutical domain, software tools must be validated along with other procedures for data generation and processing. David Stokes explains the needs and procedure for achieving validation and regulatory compliance for open source software. The final chapter in the book is written by Simon Thornber and considers the economic factors related to employing FLOSS in developing, deploying and supporting applications within an industrial environment. Simon's chapter asks the question 'is free really free?' and considers the future potential business model for FLOSS in industry. The central theme of 'hosted open source' is reminiscent of the successful Red Hat model and argues that this model can drive increasing industry adoption of certain tools while also promoting pre-competitive collaborations.

Our own experiences of using FLOSS in industry show that this paradigm is thriving. The range of applications covers the spectrum of scientific domains and industry sectors, something reflected in the diverse range of chapters and use-cases provided here. Use of open source is likely to grow, as pre-competitive initiatives, including those between industry and public sector are further developed. We consider that industry has much to offer the open source community, perhaps contributing directly to, or leading the development of some toolkits. In some companies the climate of internalisation and secrecy may still prevail, but as knowledge is shared and the benefits of open source participation are made clear one may expect the philosophy to spread and become more embedded. Even without contribution of code or project involvement, simple use of FLOSS may support the movement, by encouraging the use of open data standards and formats in preference to closed binary formats (or 'blobs').

What is perhaps most important is that the ready availability of FLOSS tools, and ability to inspect and extend, provide a clear opportunity for life science industry to 'try things'. This reduces the entry barrier to new ideas and methods and stimulates prototyping and innovation. Industry may not always contribute directly to a particular project (although in many cases it does). However, growing industry awareness of FLOSS, increasing usage and participation bode well for the future, a mutually beneficial interaction that can catalyse and invigorate the life science research sector. We hope that readers of this book get a true sense of this highly active ecosystem, and enjoy the ability to 'peek inside' the operations of industry and other groups. We are eternally grateful to all of the authors for the hard work putting each story together. We also thank Enoch Huang for providing the foreword to this work, providing an important perspective as both a consumer and producer of FLOSS tools. Finally, we are grateful to the publishers and all who have helped us bring this unique volume to a wide audience.

Mark Forster and Lee Harland
Summer 2012

## 1

Building research data handling systems with open source tools

Claus Stie Kallesøe

Abstract: Pharmaceutical discovery and development requires handling of complex and varied data across the process pipeline. This covers chemical structures information, biological assay and structure versus activity data, as well as logistics for compounds, plates and animals. An enterprise research data handling system must meet the needs of industrial scientists and the demands of a regulatory environment, and be available to external partners. Within Lundbeck, we have adopted a strategy focused on agile and rapid internal development using existing open source software toolkits. Our small development team developed and integrated these tools to achieve these objectives, producing a data management environment called the Life Science Project (LSP). In this chapter, I describe the challenges, rationale and methods used to develop LSP. A glimpse into the future is given as we prepare to release an updated version of LSP, LSP4All, to the research community as an open source project.

Key words: research data management; open source software; software development; pharmaceutical research; Lundbeck; LSP; LSP4All.

1.1 Introduction

All pharmaceutical company R&D groups have some kind of 'corporate database'. This may not originally be an in-house designed knowledge base, but is still distinct from the specific area tools/databases that companies acquire from different software vendors. The corporate database is the storage area for all the 'final' pre-clinical results that companies want to retain indefinitely. The corporate database holds data from chemistry, biology, pharmacology and other relevant drug discovery disciplines and is also often a classic data warehouse [1] in the sense that no transactions are performed there – data is fed from the other databases, stored and retrieved. The system in this chapter is partly an example of such an infrastructure, but with a somewhat unique perspective.

The system is not only a data warehouse, however. Final/analysed data from other (specialist) tools *are* uploaded and stored there. Additionally, it is the main access point for data retrieval and decision support, but the system does a lot more. It forms the control centre and heart of our data transactions and workflow support through the drug discovery process at Lundbeck [2]. Lab equipment is connected, enabling controlled file transfer to equipment, progress monitoring and loading of output data directly back into the database. All Lundbeck Research logistics are also handled there, covering reagents, compounds, plates and animals. The system is updated when assets enter the various sites and when they get registered, and it stores location information and handles required re-ordering by scientists.

The system also supports our discovery project managers with 'project grids' containing compounds and assay results. These project grids or Structure Activity Relationship (SAR) tables are linked to the research projects and are where the project groups setup their screening cascade, or tests in which they are interested. Subsequently, the project groups can register both compounds and assay results to generate a combined project results overview. The grids also enable simple data mining and ordering of new tests when the teams have decided what compounds should be moved forward. To read more about corporate pharmaceutical research systems see references [3, 4].

How is the system unique? Is this any different from those of other companies? We believe it is. It is one coherent system built on top of one database. It covers a very broad area with data concerning genes, animals and compounds in one end of the process all the way to the late-stage non-GLP/GMP [5] exploratory toxicology studies. With a few exceptions, which are defined later, it is built entirely with open source software. It is therefore relevant to talk about, and fits well with the theme of this book, as a case in which a pharmaceutical company has built its main corporate database and transaction system on open source tools.

Published by Woodhead Publishing Limited, 2012

Corporate sales colleagues would likely call our system something like enterprise research data management or 'SAP [6] for Research'. We simply call it LSP – which is short for the Life Science Project.

1.2 Legacy

It is difficult to make a clear distinction between before and after LSP, as the core part of the database was initiated more than 10 years ago. Internally, LSP is defined as the old corporate database combined with the new user interface (UI) (actually the full stack above the database) as well as new features, data types and processes/workflows. The following section describes what our environment looked like prior to LSP and what initiated the decision to build LSP.

Lundbeck has had a corporate database combining compounds and assay results since 1980. It has always been Lundbeck's strategy to keep the final research data together in our own in-house designed database to facilitate fast changes to the system if needed, independently of vendors.

Previously, research used several closed source 'speciality' software packages with which the scientists interacted. In chemistry these were mainly centred around the 'ISIS suite' of applications from what used to be called MDL [7]. They have since been merged into Symyx [8], which recently became Accelrys [9]. As an aside, this shows the instability of the chemistry software arena, making the decision to keep (at least a core piece of) the environment in-house developed and/or in another way independent of the vendors more relevant. If not in-house controlled/ developed, then at least using an open source package will enable a smoother switch of vendor if the initial vendor decides to change direction.

The main third-party software package in the (*in vitro*) pharmacology area was ActivityBase [10], a very popular system to support plate-based assays in pharma in the early 2000s. Whereas the ISIS applications were connected to the internal corporate database, ActivityBase came with its own Oracle database. Therefore, when the chemists registered compounds into our database the information about the compounds had to be copied (and hence duplicated) into the ActivityBase database to enable the correct link between compounds and results. After analysis in ActivityBase, the (main) results were copied back into our corporate database. Hardly efficient and lean data management!

Of course, the vendors wanted to change this – by selling more of their software and delivering the 'full enterprise coverage'. Sadly, their tools were not originally designed to cover all areas and therefore did not come across as a fully integrated system – rather they were a patchwork of individual tools knitted together. A decision to move to a full vendor system would have been against Lundbeck's strategy, and, as our group implemented more and more functionality in the internal systems, the opposite strategy of using only in-house tools became the natural direction.

Workflow support is evidently a need in drug discovery. Scientists need to be able to see the upstream data in order to do their work. Therefore, 'integration projects' between different tools almost always follow after acquisition of an 'of the shelf' software package. The times where one takes software from the shelf, installs and runs it are truly rare. Even between applications from the same vendor – where one would expect smooth interfaces – integration projects were needed.

As commercial tools are generally closed source, the amount of integration work Lundbeck is able to do, either in-house or through hired local programmers with relevant technology knowledge is very limited. This means that on top of paying fairly expensive software licences, the organisation has to hire the vendor's consultants to do all the integration work and they can cost £1000/day. If one part of the workflow is later upgraded, all integrations have to be upgraded/re-done resulting in even more expensive integration projects. Supporting such a system becomes a never-ending story of upgrading and integrating, leaving less time for other optimisations.

At that time the internal Research Informatics UI offering was a mixture of old Oracle Forms-based query and update forms and some Perl web applications that were created during a period of trying to move away from Oracle Forms. However, this move towards fully web-based interfaces was put on hold when Lundbeck decided to acquire Synaptic Inc. [11] in 2003, leading to prioritisation of several global alignment and integration projects into the portfolio.

Clearly, this created an even more heterogeneous environment, and an overall strategic and architectural decision, enabling us to build a coherent affordable and maintainable system for the future, was needed.

1.3 Ambition

With the current speed of technology developments, to think that one can make a technology choice today and be 'prepared for the future' is an illusion. Under these circumstances, how does one choose the right technology stack?

Very few have the luxury of basing a technology decision purely on objective facts. Firstly, it takes a long time to collect these facts about the possible options (and in some cases the objective truth can only be found after coding the application in several technologies and comparing the results, and this is very rarely done). Secondly, it is a fact that a technology decision is always taken in a certain context or from a certain point of view, making these kinds of decisions inherently biased. Finally, according to research described in the book *Sway* [12] we all, in general, overrate our rationality in our decision making.

Therefore, any decision is likely to be based on a combination of objective facts, current environment, current skill sets in the team, what the infrastructure strategy/policy allows and whether we like it or not; our gut feeling! Despite our best efforts and intentions, in the end the right choice becomes clear only after a couple of years, when it is apparent which technologies are still around, have been adopted by a significant number of groups and are still in active development.

With the above decision limitations in mind, the following are the criteria used when deciding on technology for the future global Lundbeck research system.

- Multi-tier architecture – an architecture where the client, running on local PCs, is separated from the application server, which is separated from the database server. In our case, the servers should be located only at Lundbeck HQ in Denmark, keeping the hardware setup as simple and maintainable as possible.

- Web-based UI – a web-based solution that would be easy to deploy on all the different kinds of PCs and languages used around the world. A lightweight web client should still be fast enough from any location with the servers placed in Denmark.

- Open source – open source tools give us the confidence that we are likely always to find a solution to the problems we encounter, either by adding a plug-in if one is available or by developing what is needed in-house.

- Oracle – as the existing data are stored in a relational database based on Oracle, we need a technology with support for this type of database.

In addition, there are also more subjective, but no less important, criteria to consider.

- Fast to develop – a web framework that is fast to learn/get started with and fast when developing, relative to other technology choices.

- Active development – tools having a large and active development community.

- Easy to deploy – with only limited requirements on the user platform in terms of Operating System, browser etc., hence reducing the probability and impact of desktop platform upgrades.

In conclusion our ambition was to build a web-based research enterprise data management system, covering the pre-clinical areas at Lundbeck, entirely on open source tools with the exceptions that the legacy commercial database and chemical engine would stay.

1.4 Path chosen

One could argue that an in-house system such as LSP is commodity software that does not offer anything more than some commercial packages, and therefore should be bought 'off the shelf'. It is argued here that there was (is) no single commercial system available which supports all the areas of functionality required across the pipeline described in the introduction. Consequently, without our own in-house system Lundbeck would need to acquire several different packages and subsequently combine these via continued expensive integration projects as mentioned earlier. Such an approach is very likely not the best and most cost-effective approach and certainly does not give the scientists the best possible support. Additionally, the maintenance of many single applications would be a burden for the department. Thus, the approach at Lundbeck is instead to build one coherent system. Everything that is built needs to fit into the system and hence all projects are extensions of the system rather than individual applications for chemistry, biology and pharmacology.

Having a one-system focus is not without conflicts in today's business environment with a heavy focus on projects, deliverables and timelines (not to mention bonuses). If developers are forced to deliver on time and on budget they will likely ask for agreed deliverables, which means a small(er) tangible piece of code (application) on which all stakeholders can agree. Firstly, if we can agree up front what needs to be built we are not talking about something new (hence no innovation), and, secondly, if the timeline and/or budget is approaching the developers will cut down on the functionality. The question is whether software developed under these circumstances will match the real requirements.

The one-system approach allows searches across data from different areas without any integration, data warehousing or data duplication. We

Published by Woodhead Publishing Limited, 2012

can maintain the environment with a small group of people as maintenance and development becomes 'one and the same'. There is only a need for one group – not a group to develop and another to maintain. Since 2004, the Research Informatics group has consisted of seven people (six developers and the head) placed at two sites supporting around 400 researchers, as well as data input from five to ten CRO partners around the world.

The aim is to get more coding and support for less money and people. The Informatics group is placed in the centre of a network from where we import and utilise the relevant tools from the network for incorporation into our LSP platform. This setup is a combination of 're-use and build' as very little is coded from scratch. Instead, we use libraries, tools and frameworks that others develop and maintain while we concentrate on adding domain expertise and the one-system vision to combine all tools in the right way for us.

1.5 The 'ilities

In system engineering (software development) two different types of requirements exist – the functional and the non-functional. The non-functional, as the name indicates, does not concern what the system needs to do for the user but rather *how* it does it. There are many more or less important non-functional requirements like scalability, availability and security often collectively called 'the 'ilities' [13].

Interestingly there is a tendency for management, IT, consultants etc., to put the 'ilities forward when talking about outsourcing, cloud computing, open source software or anything else new that might disrupt the 'as is' situation and shift people out of the comfort zone or potentially cause a vendor to lose business. This is a well-known sales strategy described as FUD (Fear, Uncertainty and Doubt) marketing.

The part of the non-functional requirements that are viewed as important by an organisation must apply equally to open as well as closed source software. But this does not always seem to be the case. Often the open source tools may receive more scrutiny than the closed source commercial ones. It is not a given that a commercial closed source tool is more scalable, reliable or secure than other pieces of software. Certainly poor open source tools are available, as are good ones – just as there are poor commercial tools and good commercial tools.

There is a tendency in the pharmaceutical industry to aim too high in terms of the non-functional requirements. Naturally, the tools deployed need to be scalable and fast. But one must be sure to add the relevant amount of power in the relevant places. As an example, what happens if the corporate database is unavailable for a couple of days? Scientists can continue their lab experiments, can read or write papers and prepare their next presentation. The research organisation does not come to a complete stop. Is 99.99% availability really needed and does it make sense to pay the premium to get the last very expensive parts covered? The experience of our group suggests otherwise, we assert that many of the relevant open source tools are sufficiently stable and scalable to meet the needs of pharmaceutical research.

After these considerations – and some initial reading about the different tools, to understand how they really worked, and try to get a feel for the level of activities in the different communities – it was decided to go with the stack outlined in Figure 1.1 and described in detail below.

1.5.1 Database

When deciding on the LSP stack, a database change was ruled out. The Oracle database is one of the exceptions from the fully open source stack mentioned in the introduction. It would be natural to build LSP on top of an open source database but at the time a deliberate decision was made to continue our use of Oracle. As the Lundbeck corporate IT standard

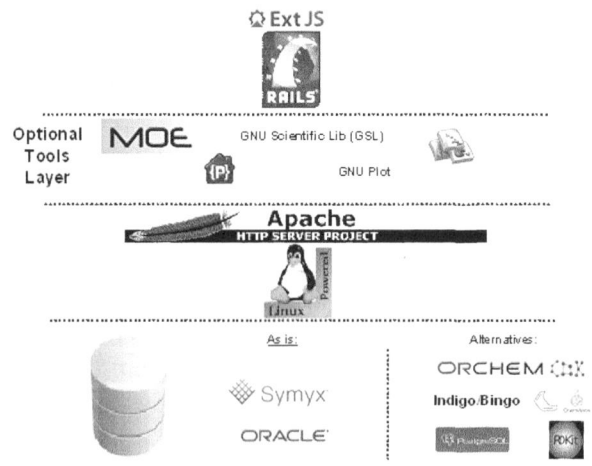

Figure 1.1   Technology stack of the current version of LSP running internally at Lundbeck

database, we are able to utilise full support and maintenance services from IT, not to mention the many years of experience and existing code we have developed. However, the main reason for not moving to an open source database was that this would trigger a switch of the other non-open source piece, the chemistry engine from Symyx/Accelrys, which did not support any of the open source databases. Changing a chemistry engine requires a lot of testing to make sure the functionality is correct and also a lot of change management related to the scientists who need to change their way of working. We did not have the resources to also handle a chemistry engine change on top of all the other technology changes and hence decided to focus on the remaining technologies.

1.5.2 Operating system (OS)

Based on our internal programming expertise, it was natural to choose Linux [14] as the OS for the web server component. It ensures the ability to add the functionalities/packages needed and Linux has proven its value in numerous powerful installations. Linux is used in 459 of the top 500 supercomputers [15] and as part of the well-known LAMP-stack [16].

In terms of Linux distributions the department's favourites are Debian [17] based. However, it was decided to use the one delivered by our Corporate IT group, RedHat Enterprise Linux [18], as part of their standard Linux server offerings. Service contracts for RedHat were already in place and we could still install the tools we needed. Corporate IT currently run our databases and servers, handle upgrades and back-ups while we concentrate on adding tools and perform the programming.

1.5.3 Web server

The choice of web server might have been even easier than choosing the OS. Firstly, Apache [19] has for several years been the most used web server in the world and is, according to the Netcraft web server survey [20] July 2011, hosting 235 million web sites. Secondly we already had experience with Apache as we used it to support our Perl environment.

1.5.4 Programming language

We did not choose Ruby [21] as such; we chose the web framework (see next section), which is based on the Ruby language. Before deciding to use the web framework and hence Ruby, we looked quite carefully into the language. What do people say about it? What is the syntax? Does it somehow limit our abilities especially compared to C [22], which we earlier used to perform asynchronous monitoring of lab equipment and similar low-level TCP/IP communication. Our initial investigations did not lead to any concerns, not even from the experienced colleagues trained in many of the traditional languages.

Ruby may have been on the hype curve [23] at this time but it is not a new language. It has been under careful development over more than 15 years by Matz [24], with the aim of combining the best from several existing languages.

1.5.5 Web framework

Several years of Perl programming had resulted in a lot of code duplication. The new web framework needed to be an improvement from our Perl days, providing clear ideas for separation and re-use rather than copy code.

The Ruby on Rails (RoR) (http://rubyonrails.org/) framework had been getting a lot of attention at the time, with a focus on the following concepts.

- DRY – Don't repeat yourself.
- Convention over configuration.
- REST – or Representational State Transfer, which is the basis for a RESTful architecture [25], seen by many as the best way to implement web applications today.

Ruby on Rails was initiated by David Heineman Hansson (DHH) [26] as part of the Basecamp [27] project. In general, the framework looked very productive and the vision/thinking from the people behind RoR/37Signals about software development fit very well with our internal thinking and way of working. As it appeared easy to use and there seemed to be high adoption and very active development, we decided that the future research informatics development at Lundbeck should be based on RoR.

Although one may download RoR freely, the choices of *how* to serve the graphical interface, organise modules and workflows, pass data between modules and many other considerations are open and complex. Thus, our web framework is more than just RoR, it is a Ruby-based infrastructure for defining, connecting and deploying distinct data service components.

Published by Woodhead Publishing Limited, 2012

1.5.6 Front-end

UI design is inherently difficult. It takes a lot of time and is rarely done well. One can appreciate good design but it is not as easy to create. The tools need to look good or 'professional', otherwise the users will judge the functionality harshly. A nice looking front-end is required in addition to correct functionality. RoR comes with a clean but also very simple UI, but an interface with more interactivity was required (implemented using AJAX/javascript) to enable the users to work with the data.

The javascript library Extjs [28] delivers a rich UI framework with pre-build web tools such as grids, forms and tables that can be used to build the special views needed for the different data types. Extjs also creates a platform on which many external javascript tools such as structure viewers can be incorporated via a 'plug-in' mechanism. Although Extjs is an open source library, it is not free for all due to licence restrictions on commercial use. However, on open source projects, developers can use Extjs for free under a GPL [29] type licence [30].

1.5.7 Plug-ins

LSP also utilises different plug-ins for some special tasks. As an example, the GNU Scientific Library (GSL) [31] is used as support for simple maths and statistics calculations. Many of the standard calculations performed on science data are fairly straightforward and well described, and can be implemented with an open source math library and some internal brain-power. With these tools, IC50-type sigmoidal curve fitting has been implemented using the flot [32] javascript library for graphics rendering in the browser allowing scientists to see and interact with the data (Figure 1.2).

1.5.8 Github

All the LSP code is hosted on Github [33], a low-cost (free for open source projects) and effective site hosting software projects using the Git source code version control system [34]. Git was, like Linux, initially developed by Linus Torvalds [35]. It is fully distributed and very fast, and many of the tools selected made use of Git, and were already hosted on Github. It would therefore be easier to merge future, new code and technologies into the code stream by also using Git. Despite the fact that

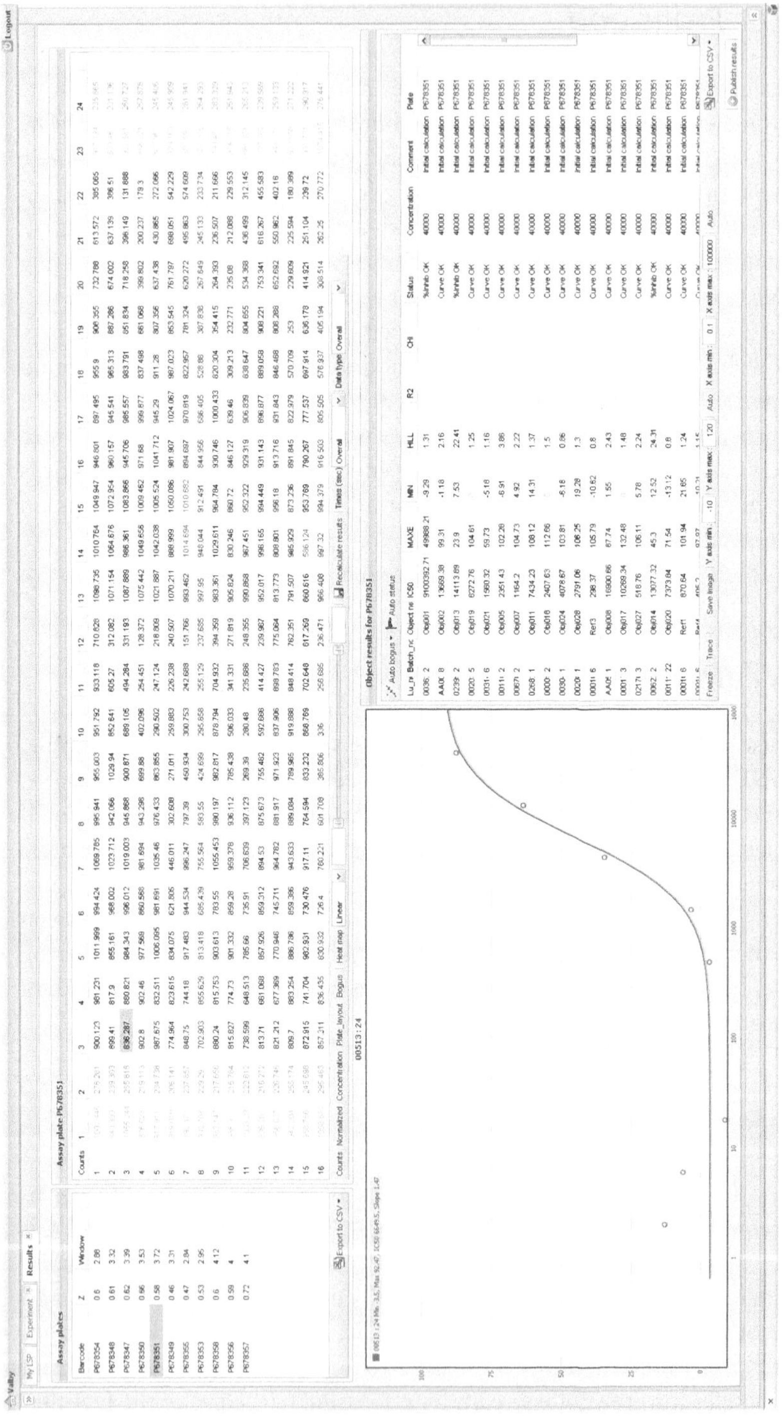

Figure 1.2 LSP curvefit, showing plate list, plate detail as well as object detail and curve

it added another complexity and another new thing to learn, it was decided to follow the RoR community and move to Git/Github.

As mentioned, most, but at the moment not all, parts of the stack are open source tools. An Oracle database is used, whereas structure depictions and searches use tools from Symyx/Accelrys. MOE [36] is also not a FLOSS tool. At Lundbeck it is used for some speciality compound calculations where the result is displayed in LSP. MOE is not essential for the running of LSP, but Oracle and the Symyx/Accelrys cartridge (for chemical structures) currently are.

1.6 Overall vision

LSP is designed to span the entire continuum, from data storage and organisation, through web/query services and into GUI 'widgets' with which users interact. The core of the system is a highly configurable application 'frame', which allows different interface components (driven by underlying database queries) to be connected together and presented to the user as business-relevant forms. Meta-data concerning these components (which we call 'application modules') are stored in the LSP database, enabling us to re-use and interlink different modules at will. For example, connecting a chemical structure sketcher, pharmacology data browser and graph widget together for a new view of the data. It is this underlying framework that provides the means to develop new functionality rapidly, with so much of the 'plumbing' already taken care of. With this in place, our developers are free to focus on the more important elements, such as configuring the data storage, writing the queries and deploying them in user-friendly interface components.

1.7 Lessons learned

The informatics department consists of experienced programmers, who tend to learn new tools, languages etc., proactively by reading an advanced-level book and starting to work with the new skills, i.e. learning by doing. RoR was no exception. The team acquired some of the books written by DHH and started the development. It is evident, as can also be seen on the RoR getting started page [37] that RoR is a really quick and easy way to build something relatively simple.

In order to achieve the intended simplicity, RoR expects simple database tables following a certain naming scheme – if the model is called Compound the table behind should be Compounds in plural. This is convention over configuration. If one already has a big database in place that does not follow this naming scheme, some of the ease of use and smartness of RoR disappears. This was true in our case. Tough architecture discussions therefore started when choosing between changing the database to fit RoR and adapting RoR to our existing setup. The database was the only remaining piece of the old setup, so it was desirable that this should not enforce the mindset of the old world while using technologies from the new world, which might have happened in this case.

Building LSP *is* more than just downloading and installing the key components. They have to be built together in a meaningful way in which the different parts handle the tasks they are designed to do, and do best, in the relevant context. But it is not always clear where to make a clean separation of the layers to achieve the best possible solution. In addition, programmers may often disagree. RoR follows the Model-View-Controller [38] paradigm. In the RoR world this, simply explained, means a model that gets the data from the database, a view that takes care of the design of the page displayed to the user and a controller that glues the pieces together. Although it would have been preferable to follow that layout, it was again realised that our mature database would force some tough architecture decisions. The database contains many views as well as Oracle PL/SQL procedures. Should the procedures be replaced with models and controllers? Would it be better to have a simple database and then move all sorts of logic into the RoR layer? After some consideration, the real world answer is that 'it depends'!

Oracle procedures are good for transactions inside the database. Taking data out of the database, manipulating them in the RoR layer and putting them back into the database did not seem a very efficient approach. On the other hand, it is preferable not to be single-database-vendor centric. Moving to a simpler table/procedure structure would likely make it easier to port to other databases if/when that becomes relevant.

Similarly, how to best combine RoR and Extjs was debated. As mentioned, RoR delivers a way to build interfaces but using Extjs also meant finding a way to elegantly knit the two layers together. A decision needed to be taken regarding incorporating the Extjs code into the view part of the RoR MVC setup or keeping it separate. Integrating Extjs into RoR would tie the two parts very closely together and although that might have advantages, it would make it more complicated to replace one of the layers if something better were to appear at a later stage. While deciding how to integrate Extjs with RoR, we also needed to learn how to use the Extjs library itself most effectively.

It took time to agree/understand how to best use and integrate the layers of the stack and we did not get to that level of understanding without a longer period of actual coding. The current LSP code base may not currently be structured in the best possible way for long-term support and development, similar code structure issues are known and understood by other developers in the community. The future direction of RoR and Extjs are consistent with the Lundbeck team's views of the 'right way' to structure and enhance the code in future. The team behind Extjs recognised that they needed to implement more structure in their library in order to support the various developers. A MVC structure is therefore implemented in version 4 of Extjs. Hence a clean installation of the new version of our stack is being considered, along with re-building LSP as that will be easier, better and/or faster than trying to upgrade and adapt our initial attempt to fit the new version of the stack.

1.8 Implementation

It is likely that there is never going to be a 'final' implementation. The development is going to continue – hence the name Life Science *Project* – as technologies and abilities change as do the needs in terms of more data types, analysis of the data, better ways of displaying the data, etc.

Overall, the current system looks very promising. Our group is still happy with the choices made regarding the tools to use and LSP does work very well. It is therefore entirely possible, and in fact very beneficial and cost-effective, to develop a research enterprise data management system purely on open source tools. All parts of the technology stack continue to develop, and Ruby, and especially RoR and Extjs, are moving forward at high speed and adding a lot of new valuable features.

Neither scale nor speed or any other of the 'ilities seem to be an issue with our implementation and number of users. Occasional slowness in the front-end is due to (1) very big/complex database queries that need optimisation or can be handled with more power on the database or (2) Javascript rendering on the clients. This is not an issue in general but certain (older) PCs with lack of RAM and/or certain browsers do not render quickly enough. On a standard laptop running Windows and using Firefox, LSP works very well.

Using Javascript (a Javascript library) as the UI has also turned out to be a very sound idea. Javascript can be challenging to work with, but, handled correctly, it can be used to create many relevant tools [39]. There are a lot of useful javascript-based visualisations and plug-ins on the web that are easy to incorporate into LSP, making the front-end attractive as well as interactive with several different ways to display the relevant data to the scientists.

1.9 Who uses LSP today?

Years of experience have shown that creating something new for users, in a new technology that is only slightly better than what they currently use is not the road to success. Users do not seem to think the new technology is as exciting as do the development team! For the LSP deployment it was decided to only involve the users when LSP would give them something new or greatly improve what they already had.

The first piece deployed to a larger group was our interactive (real time) MedChem Designer (Figure 1.3). The medicinal chemists responsible for designing pre-clinical compounds are able to draw structures that the system will submit to a property calculator and return the results in easy readable visualisations. The response time is instant, meaning as soon as the chemists change the structure in any way the properties will be updated. This was a tool Lundbeck did not previously have, it was liked by chemists, leading to almost immediate and full adoption in the chemistry groups.

The informatics group handles all logistics in research (compounds, plates, animals, etc.) but the chemistry reagent logistics were still being handled by an old ISIS-based application that was performing poorly. The existing in-house compound logistics database functionality was re-used and a new front-end with up-to-date functionality was created in LSP, also incorporating a new workflow that would support reagent ordering by local chemists but shipment directly to outsourcing partners. Because of the huge increase in usability and functionality, and by migrating systems over a weekend, an easy deployment with full adoption was achieved.

One benefit of investing time in re-building a fairly small reagent application was also that it would enable a move of compound/reagent registration into LSP. This was not an expressed need from the chemists but was required to enable direct registration of compounds from our outsourcing partners into the LSP database in Denmark. Shortly after the reagent store deployment, the LSP compound registration form was published as a Lundbeck extranet solution, where the chemistry partners

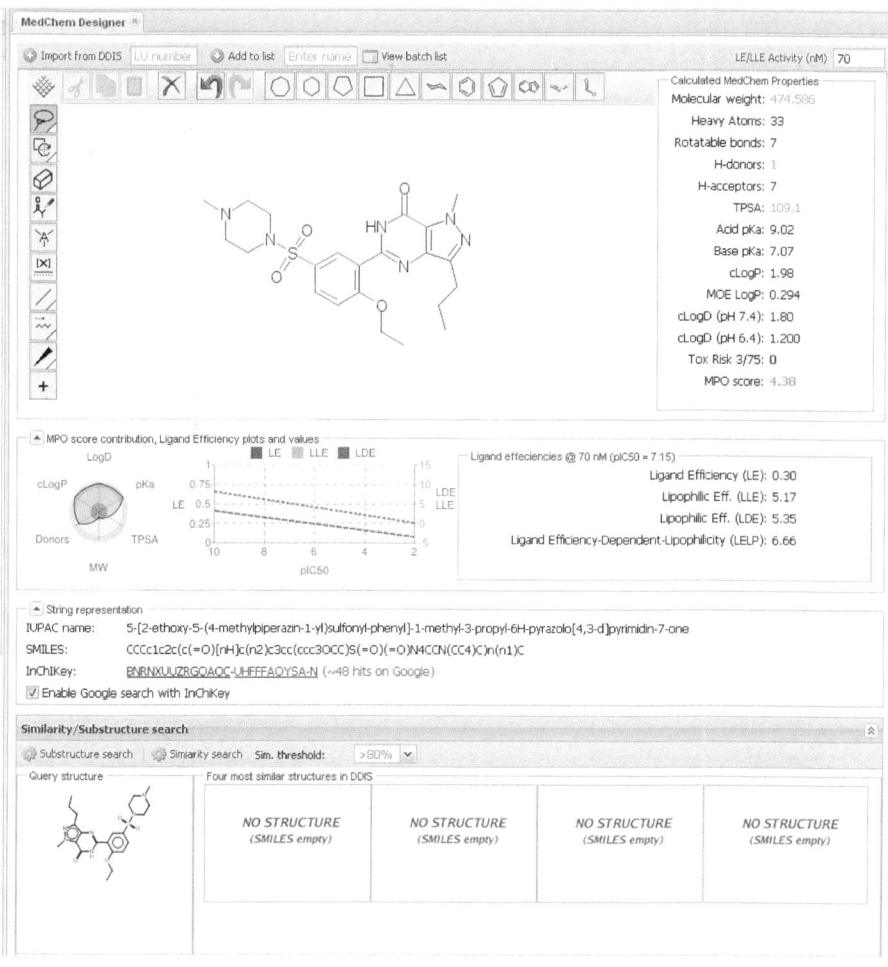

Figure 1.3 LSP MedChem Designer, showing on the fly calculated properties and Google visualisations

can access and use it directly (Figure 1.4). This is now deployed to five different partner companies placed in three different countries on several continents around the world and works very well. This platform therefore enables Lundbeck to very quickly and cost-effectively implement direct CRO data input, something basically all pharmaceutical companies need today. The LSP partner solution, LSP4Externals, is a scaled-down (simpler) version of the internal solution. Security considerations resulted in the deployment of a separate instance, although using the same database. A row level security implementation makes it possible to give

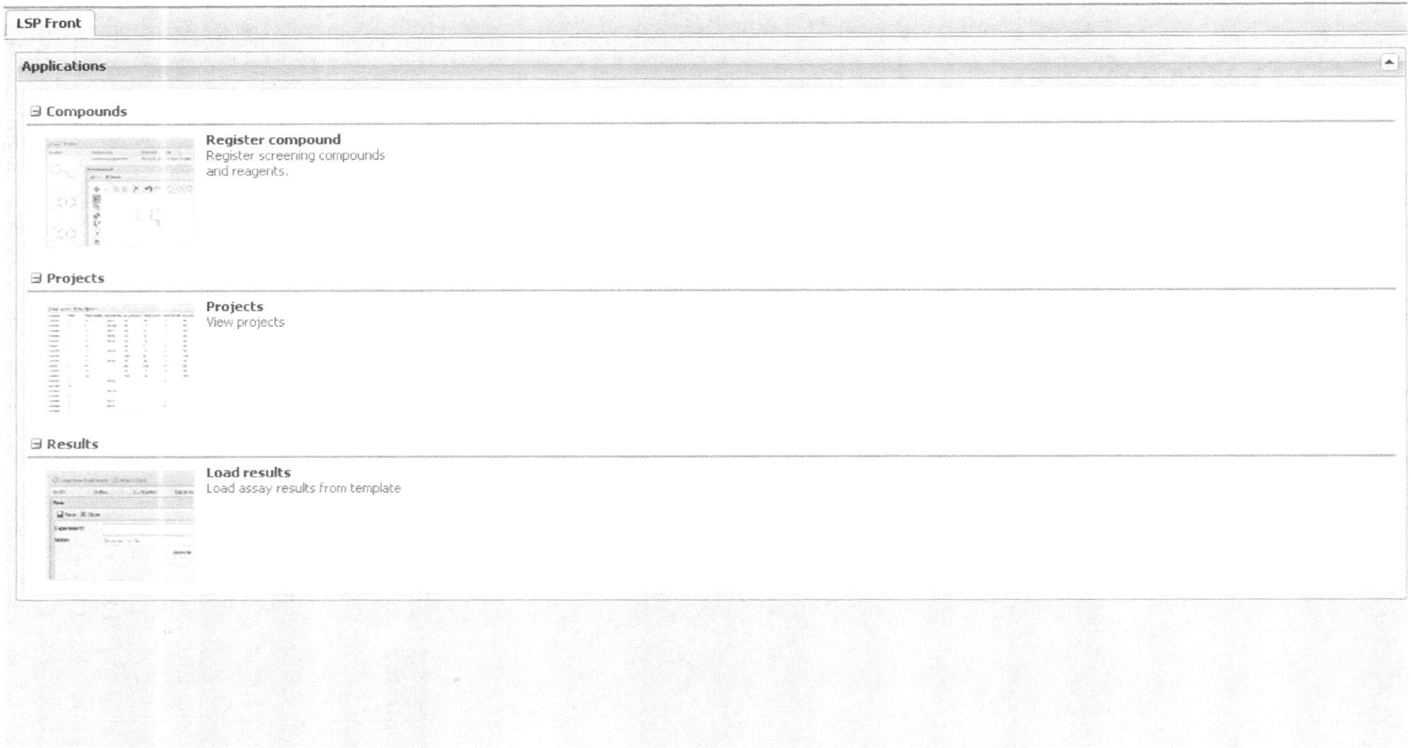

Figure 1.4 LSP4Externals front page with access to the different functionalities published to the external collaborators

the different partners access to subsets of compounds and/or assays in the projects as relevant.

Although assay registration was implemented fairly early in LSP, it was not deployed as a general tool. Assay registration functionality existed in the old systems, although it was not as effective and the new LSP version therefore did not initially give the users significant new functionality. The LSP version of assay registration would introduce a major, but necessary data cleanup. Unfortunately, an assay information cleanup introduces more work, not functionality, to the scientists, so 'an award' was needed that would show the user benefits of this exercise. The award was a new assay query tool that enabled searches across parameters from all global research assays and therefore made it easier for the scientists to find all assays working against a certain target, species, etc.

This started the move towards using LSP for assay registration and retrieval. Another important push came with external (partner) upload of data and collaboration on project grids/SAR tables. In order to make sure the external partners can understand and use the Lundbeck data, the scientists need to clean up the assay information.

The last part of existing legacy functionality to be moved into LSP and then improved, is our project overview/SAR table grids, displaying the combined view of compounds and results in our projects. This should be the centre/starting point for all data retrieval and analysis as well as transactions in the form of new assay or compound ordering. Currently, the project overview with the ability to see all history data on a single compound has been ported to LSP (Figure 1.5), while ordering is in the plans.

1.10 Organisation

IT/Informatics departments normally consist of managers, business analysts, project managers, applications specialists, etc. as well as developers (if still available in-house). The departments therefore tend to become large and costly while most of the individuals do not develop or maintain applications but rather manage resources. However, the biggest issue with these types of organisations are the translations, where valuable information gets lost. There are translations from scientists to business analysts, from analysts to project managers, from project managers to developers – or more likely from project managers to the outsourcing partner project managers and then to the developers. The handovers between all these individuals are normally performed via a User

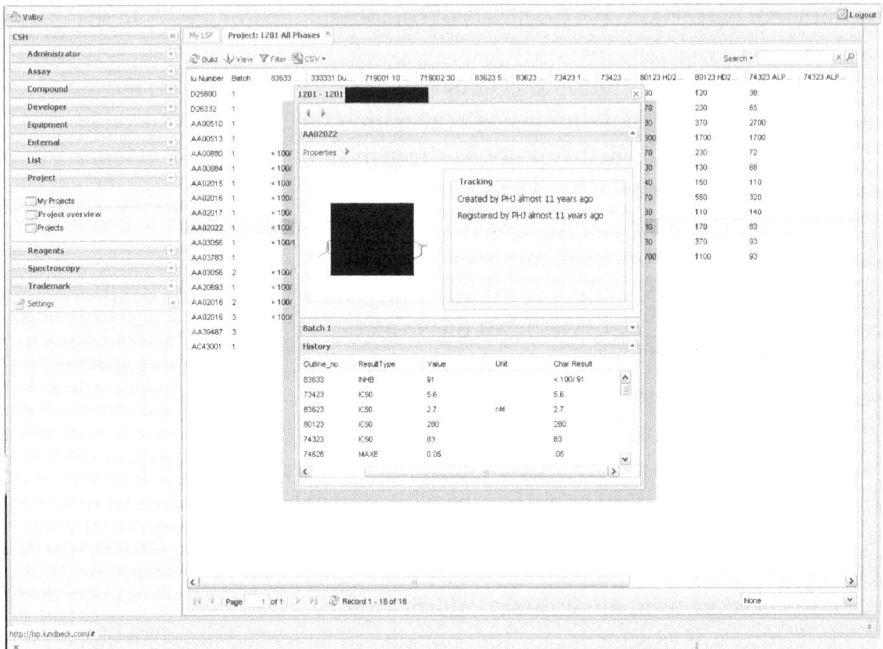

Figure 1.5 LSP SAR grid with single row details form

Requirement Specification (URS). The developer therefore has to build what is described in the document instead of talking directly to the users and seeing them in action in the environment in which the system is going to be used.

Lundbeck Research Informatics uses another approach, without any business analysts or any other type of translators and without any URS documents. All members of the team are code-producing and in order for them to understand what is needed they will visit the relevant scientists, see them in action and learn their workflow. This way the programmers can, with their own eyes, see what the scientists need to do, what data they require and what output they create. When asked to identify their requirements, users may reply 'what we have now, only bigger and faster'. Users naturally do not see a completely new solution they did not know existed, when they are deep into their current way of doing things. Do not build what they want, but what they need. It is evident that users are not always right!

This way of working naturally requires skilled people with domain knowledge as well as programming skills. They all constantly need to monitor the external communities for new relevant tools, incorporate the tools into the environment in a meaningful way and build systems that make sense to scientists. We therefore try to combine different types of 'T-shaped' [40] people. They are either experienced programmers with science understanding or scientists with programming understanding.

Our group operate as a small, close-knit team, ensuring that 'everybody knows everything' and therefore avoid any unnecessary alignment meetings. Meetings are normally held in front of a white board with the entire team present and concern workflow or front-end design, so all members know what is going on, what is being decided and everybody can contribute ideas.

The method is not unique and was not invented by us, although it may be unique in the pharma world, in Lundbeck Research Informatics it is well established and feels natural. Similar approaches are being used by several successful companies within innovation, design and software development like IDEO [41] and 37signals [42]. In many of the large open source projects that require some form of coordination, all participants are also still producing and/or at least reading and accepting the code entering the code base.

In addition, the mode of action is dependent on empowerment and freedom to operate. If the informatics group needed to adhere to corporate technology policies or await review and acceptance before embracing a new tool, it would not be as successful. At Lundbeck the group reports to the Head of Research and hence is independent of Corporate IT and Finance as well as the individual disciplines within Research.

The group has been fortunate to experience only a single merger and therefore has had the opportunity to focus on the one-system strategy for a longer period, with the environment and system kept simple enough for us to both develop and maintain at the same time.

1.11 Future aspirations

We do not expect, and do not hope, that the Life Science Project will ever end. As long as Lundbeck is actively engaged in research there will be a constant need for new developments/upgrades with support for new data types, new visualisations and implementation of new technologies that we do not yet know about.

Currently, we are looking into the use of semantic technology [43]. What is it and where does it make most sense to apply? Linking our internal databases via Resource Description Framework (RDF) might enable us to perform searches and find links in data that are not easily found with the current SQL-based technologies. But, as mentioned here, a lot of our research data are stored together in one database so for that piece RDF might not be appropriate at the moment. The large amounts of publicly available external data that scientists use in their daily work, on the other hand, are good candidates for linking via RDF [44]. In order to continue to follow the one-system strategy, we are looking to display these linked external data sources in LSP alongside the internal data so the scientists can freely link to the external world or our in-house data.

With the new cloud-based possibilities for large-scale data storage and number crunching [45], a very likely future improvement to LSP will be use of a cloud-based setup for property calculations. In such a setup the scientists would submit the job and LSP would load the work to the cloud in an asynchronous way and return when the cloud calculator is done. The first cloud-based pilot projects have been initiated to better understand what the best future setup would look like.

Another new area to move into could be high-level management overview data. As all information is in one place, it would not be difficult to add more management-type views on the data. If overall research/ discovery projects have meta-data like indication, risk, phase added, we could then deliver portfolio-like overviews, which management currently produce by hand in Powerpoint. Chapter 17 by Harland and colleagues is relevant here, using semantic MediaWiki in managing disease and target knowledge and providing an overview to management. We also know how many compounds have been produced for the individual projects in what time period, by whom working for which partner and located at which site. This could give the project teams and management quick access to data about turnaround time and other efficiency metrics at the same front-end at which they can dive into raw data.

Lundbeck Research Informatics is participating in the Innovative Medicines Initiative (IMI) OpenPhacts project [46]. Our main contribution to the project is likely to be the front-end, which basically will be a scaled-down version of LSP, which we can further develop to fit the needs of the IMI project. The data in the project will be based on RDF and a triple store (Figure 1.6), so we will hopefully be able to improve the way LSP handles this type of data by means of the project and use that experience for our internal RDF data handling.

As true believers in open source software development, we feel obliged to give something back to the community. We therefore plan to open source the LSP platform development, under the name LSP4All, so others can download it and use it for data management in their labs, but naturally also because we thereby hope that others will join the team and help develop LSP further to the benefit of all future LSP users.

Published by Woodhead Publishing Limited, 2012

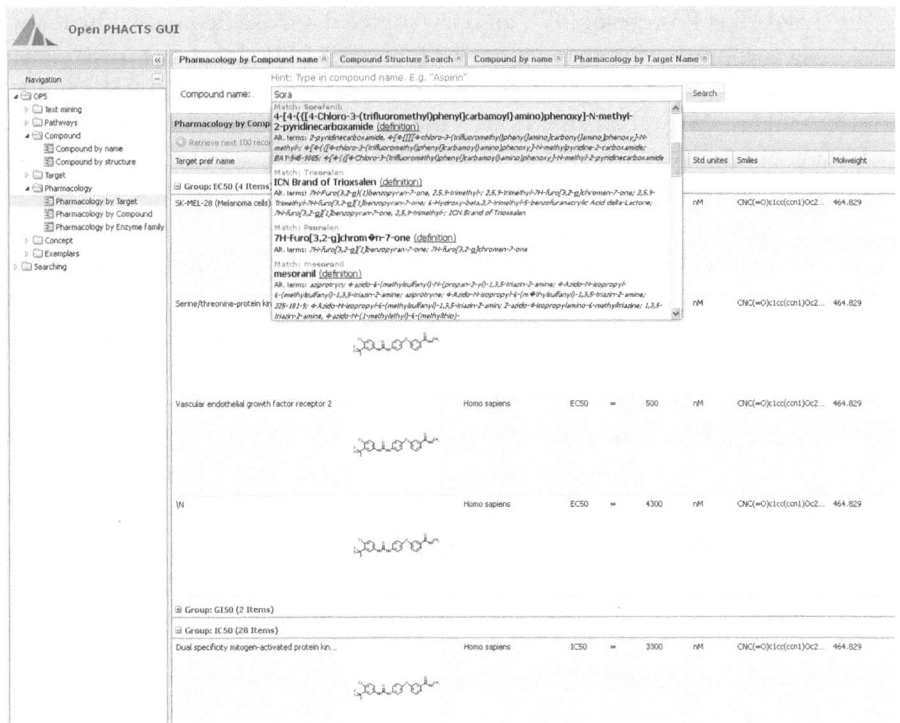

Figure 1.6 IMI OpenPhacts GUI based on the LSP4All frame

To see wide adoption and combined future development would be truly great. In order to encourage as wide an adoption of LSP as possible, we will make the LSP4All port based solely on FLOSS tools. That means moving away from the Oracle database and the Symyx/Accelrys chemical structure tools as we think these licence costs would deter many groups from using LSP4All. The move to LSP4All, or its initiation, has already started as we decided that porting LSP to LSP4All would be a great opportunity to create the scaled-down version of LSP needed to deliver to the OpenPhacts project, while at the same time re-building LSP using the new versions of our technologies in the preferred way for our internal use and also start an open source project. This was basically the window of opportunity we were looking for to start the LSP upgrade project.

Later, when sufficient LSP functionality has been moved, we expect LSP4All to be the version that will run here at Lundbeck. It follows our chosen design principles and it will save us some extra licence costs as it will be based fully on open source tools. The database currently in use for

LSP4All is Postgresql [47], an open source database being used by many big corporations in many different industries [48] and therefore likely to support our database needs. Another, perhaps more natural, option would be MySQL [49]; however, following the acquisition of MySQL by Oracle there seems to be a tendency in the open source community to move to Postgresql.

The chemical structure needs of LSP4All will be handled by something like Ketcher [50], a pure open source Javascript-based editor, for structure depiction and like RDKit [51] for the structure searches. RDKit supports Postgrsql and therefore fits with the choice of open source database. Several new well-functioning open source chemistry packages and cartridges have appeared lately, so in the future there will be more options from which to choose.

The LSP4All project is hosted at Github and we expect to open the repository to the public during 2012. Then all university labs, small biotechs and other relevant organisations will have the opportunity to download and run a full data management package with chemistry support for free. It will be a big step towards more open source collaboration in the biopharmaceutical informatics space.

We hope you will all join in!

1.12 References

[1]   Wikipedia Data warehouse. *http://en.wikipedia.org/wiki/Data_warehouse*. Updated 16 August 2011. Accessed 18 August 2011.

[2]   Homepage of H. Lundbeck A/S. *http://www.lundbeck.com*. Accessed 18 August 2011.

[3]   Agrafiotis, DK, et al. Advanced Biological and Chemical Discovery (ABCD): centralizing discovery knowledge in an inherently decentralized world. *Journal of Chemical Information and Modeling* 2007;47:1999–2014.

[4]   Pfizer Rgate – *http://www.slideshare.net/bengardner135/stratergies-for-the-intergration-of-information-ipiconfex*. Accessed 28 September 2011.

[5]   Homepage of FDA – GLP Regulations. *http://www.fda.gov/ICECI/EnforcementActions/BioresearchMonitoring/NonclinicalLaboratories InspectedunderGoodLaboratoryPractices/ucm072706.htm*. Updated 8 May 2009. Accessed 28 September 2011.

[6]   Homepage of SAP. *http://www.sap.com/uk/index.epx#*. Accessed 18 August 2011.

[7]   Wikipedia MDL Information Systems. *http://en.wikipedia.org/wiki/MDL_Information_Systems*. Updated 3 December 2010. Accessed 18 August 2011.

[8]   Wikipedia Symyx Technologies. *http://en.wikipedia.org/wiki/Symyx_Technologies*. Updated 2 December 2010. Accessed 18 August 2011.

[9] Homepage of Accelrys. *http://accelrys.com/*. Accessed 18 August 2011.

[10] Homepage of IDBS. *http://www.idbs.com/activitybase/*. Accessed 30 August 2011.

[11] Press release from H. Lundbeck A/S. *http://lundbeck.com/investor/releases/ReleaseDetails/Release_89_EN.asp*. Accessed 30 August 2011.

[12] Brafman O, Brafman R. *Sway: The Irresistible Pull of Irrational Behavior.* 1st ed.: Crown Business; 2008.

[13] Wikipedia List of system quality attributes. *http://en.wikipedia.org/wiki/List_of_system_quality_attributes*. Updated 31 July 2011. Accessed 30 August 2011.

[14] The Linux Foundation. *www.linuxfoundation.org*. Accessed 30 August 2011.

[15] TOP500 Super Computer Sites. *http://www.top500.org/stats/list/36/osfam*. Accessed 30 August 2011.

[16] LAMP: The Open Source Web Platform. *http://onlamp.com/pub/a/onlamp/2001/01/25/lamp.html*. Updated 26 January 2001. Accessed 30 August 2011.

[17] Homepage of Debian. *www.debian.org*. Updated 30 August 2011. Accessed 30 August 2011.

[18] Homepage of RedHat Enterprise Linux. *http://www.redhat.com/rhel/*. Accessed 30 August 2011.

[19] Apache http server project. *http://httpd.apache.org/*. Accessed 30 August 2011.

[20] Web server survey. *http://news.netcraft.com/archives/category/web-server-survey/*. Accessed 18 July 2011.

[21] Ruby, A programmer's best friend. *http://www.ruby-lang.org/en/*. Accessed 30 August 2011.

[22] Kernighan BW, Ritchie DM. *C Programming Language.* 2nd ed.: Prentice Hall; 1988.

[23] Gartner Hype Cycle. *http://www.gartner.com/technology/research/methodologies/hype-cycle.jsp*. Accessed 30 August 2011.

[24] About Ruby. *http://www.ruby-lang.org/en/about/*. Accessed 30 August 2011.

[25] Architectural Styles and the Design of Network-based Software architectures. *http://www.ics.uci.edu/~fielding/pubs/dissertation/top.htm*. Accessed 30 August 2011.

[26] About 37Signals. *http://37signals.com/about*. Accessed 30 August 2011.

[27] Homepage of Basecamp. *http://basecamphq.com/*. Accessed 30 August 2011.

[28] Sencha Extjs product site. *http://www.sencha.com/products/extjs/*. Accessed 30 August 2011.

[29] GNU General Public License. *http://www.gnu.org/copyleft/gpl.html*. Accessed 30 August 2011.

[30] Sencha Extjs Licensing Options. *http://www.sencha.com/products/extjs/license/*. Accessed 30 August 2011.

[31] GSL – GNU Scientific Library. *http://www.gnu.org/software/gsl/*. Updated 17 June 2011. Accessed 30 August 2011.

[32] Homepage of flot. *http://code.google.com/p/flot/*. Accessed 30 August 2011.

[33] Homepage of GitHub. *www.github.com*. Accessed 30 August 2011.

[34] Homepage of Git. *http://git-scm.com*. Accessed 30 August 2011.

[35] Torvalds L, Diamond D. *Just for Fun: The Story of an Accidental Revolutionary*. 1st ed.: HarperCollins, 2001.

[36] Chemical computing group MOE product page. *http://www.chemcomp. com/software.htm*. Accessed 30 August 2011.

[37] Getting Started with Rails. *http://guides.rubyonrails.org/getting_started. html*. Accessed 30 August 2011.

[38] Applications Programming in Smalltalk-80™: How to use Model-View-Controller (MVC) *http://st-www.cs.illinois.edu/users/smarch/st-docs/mvc. html* – Updated 4 March 1997. Accessed 30 August 2011.

[39] JavaScript: The World's Most Misunderstood Programming Language. *http://www.crockford.com/javascript/javascript.html*. Accessed 30 August 2011.

[40] IDEO CEO Tim Brown: T-Shaped Stars: The Backbone of IDEO's Collaborative Culture. *http://chiefexecutive.net/ideo-ceo-tim-brown-t-shaped-stars-the-backbone-of-ideoae%E2%84%A2s-collaborative-culture*. Accessed 30 August 2011.

[41] Kelley T, Littman J. *The Art of Innovation: Lessons in Creativity from IDEO, America's Leading Design Firm*. Crown Business; 2001.

[42] Fried J, Hansson DH. *Rework*. Crown Publishing Group; 2010.

[43] Bizer C, Heath T, Berners-Lee T. Linked Data – The Story So Far. *International Journal on Semantic Web and Information Systems (IJSWIS)* 2009;5: 3, accessed 13 September 2011, doi:10.4018/jswis.2009081901

[44] Samwald M, Jentzsch A, Bouton C, et al. Linked open drug data for pharmaceutical research and development. *Journal of Cheminformatics* 2011;3:19. doi:10.1186/1758-2946-3-19.

[45] Pistoia Alliance Sequence Services Project. *http://www.pistoiaalliance.org/ workinggroups/sequence-services.html*. Accessed 14 September 2011.

[46] OpenPhacts Open Pharmacological Space. *http://www.openphacts.org/*. Accessed 30 August 2011.

[47] Homepage of PostgreSQL. *http://www.postgresql.org/*. Accessed 30 August 2011.

[48] PostgreSQL featured users. *http://www.postgresql.org/about/users*. Accessed 30 August 2011.

[49] Homepage of Mysql. *http://mysql.com/*. Accessed 30 August 2011.

[50] GGA Open-Source Initiative: Products, Ketcher. *http://ggasoftware.com/ opensource/ketcher*. Accessed 30 August 2011.

[51] RDKit: Cheminformatics and Machine Learning Software. *http://rdkit.org/*. Accessed 30 August 2011.

## 2

Interactive predictive toxicology with Bioclipse and OpenTox

Egon Willighagen, Roman Affentranger, Roland Grafström, Barry Hardy, Nina Jeliazkova and Ola Spjuth

Abstract: Computational predictive toxicology draws knowledge from many independent sources, providing a rich support tool to assess a wide variety of toxicological properties. A key example would be for it to complement alternative testing methods. The integration of Bioclipse and OpenTox permits toxicity prediction based on the analysis of chemical structures, and visualization of the substructure contributions to the toxicity prediction. In analogy of the decision support that is already in use in the pharmaceutical industry for designing new drug leads, we use this approach in two case studies in malaria research, using a combination of local and remote predictive models. This way, we find drug leads without predicted toxicity.

Key words: toxicity; structure–activity relationship; malaria; Bioclipse; OpenTox.

2.1 Introduction

Much is already known about the toxicity of small molecules, but yet it is a challenge to predict the toxicology of new compounds. At the same time, society demands insight into the toxicity of compounds, such as that outlined by the REACH regulations in the European Union [1].

Where and when experimental testing of toxicity is expensive or unnecessary, experimental data may under certain conditions be complemented by alternative methods. Among these alternative methods are computational tools, such as read-across and (Q)SAR, that can predict a wide variety of toxicological properties. A good coverage of predictions with different alternative models can help provide reasonable estimations for the toxicity of a compound.

Such predictive toxicology draws knowledge from many independent sources, which need to be integrated to provide a weight of evidence on the toxicity of untested chemical compounds. Typical sources include *in vivo* and *in vitro* experimental databases such as ToxCast [2] and SuperToxic [3], literature-derived databases, such as SIDER summarizing adverse reactions [4], and also computational resources based on toxicity data for other compounds including DSSTox [5]. And somehow all this information should be aggregated and presented to the user in such a way that the various, potentially contradictory, pieces of information can help reach a weighted decision on the toxicity of the compound.

Visualization of this information is, therefore, an important tool, and should preferably be linked to the chemical structure of the compound itself. Further visualization that will be important is that of relevant life science data, such as gene, protein and biological pathway information [6–8] or metabolic reactions [9]. Bioclipse was designed to provide such interactive data analysis for the life sciences, although the resources are not yet as tightly integrated as other sources.

Underlying this knowledge integration, there must be a platform that allows researchers to use multiple prediction services. This is exactly what the recently introduced Bioclipse–OpenTox platform is providing [10]. This chapter will describe how this platform can be used to interactively study the toxicity of chemical structures using computational toxicology tools. Its ability to interactively predict toxicity and the ability to dynamically discover computational services, allows users to get the latest insights into toxicity predictions while hiding technicalities. Other tools that provide similar functionality include the OECD QSAR ToolBox [11] and ToxTree [12–13]. Bioclipse, however, is not a platform targeted at toxicity alone, and has been used for other scientific fields too, taking advantage of the ease by which other functionality, both local and remote, can be integrated [14–16].

This chapter will not go into too much technical detail, but instead provide two example use cases of the platform. Detailed descriptions can be found in the papers describing the OpenTox Application Programming Interface (API) [17] and the AMBIT implementation of this API [18], the

Published by Woodhead Publishing Limited, 2012

second Bioclipse paper describing the scriptability [19], a recently published description of the integration of these tools [10], and the paper describing the technologies used to glue together the components [20]. For the user of the platform it suffices to know that all functionality of the Bioclipse–OpenTox interoperability is accessible both graphically and by means of scripting. Table 2.1 shows an overview of all currently available functionality.

Table 2.1 Bioclipse–OpenTox functionality from the Graphical User Interface is also available from the scripting environment. Scriptable commands (left column) and descriptions (right column) for various groups of functionality are provided in this table

Authentication	
login(accountname, password)	Authenticate the user with OpenSSO and login on the OpenTox network.
logout()	Logout from the OpenTox network.
getToken()	Returns a security token when Bioclipse is logged in on the OpenTox network.
Computation	
calculateDescriptor(service, descriptor, molecules)	Calculates a descriptor value for a set of molecules.
calculateDescriptor(service, descriptor, molecule)	Calculates a descriptor value for a single molecule.
predictWithModel(service, model, molecules)	Predicts modeled properties for the given list of molecules.
predictWithModel(service, model, molecule)	Predicts modeled properties for the given molecule.
Data exchange	
createDataset(service)	Creates a new data set on an OpenTox server.
createDataset(service, molecules)	Creates a new data set on an OpenTox server and adds the given molecules.
createDataset(service, molecule)	Creates a new data set on an OpenTox server and adds a single molecule.
addMolecule(dataset, mol)	Adds a molecule to an existing data set.

(Continued)

Table 2.1	Bioclipse–OpenTox functionality from the Graphical User Interface is also available from the scripting environment (*Continued*)

Data exchange (*Continued*)	
addMolecules(dataset, molecules)	Adds a list of molecules to an existing data set.
deleteDataset(dataset)	Deletes a data set.
downloadCompoundAsMDLMolfile (service, dataset, molecule)	Downloads a molecule from a data set as a MDL molfile.
downloadDataSetAsMDLSDfile (service, dataset, filename)	Downloads a complete data set as MDL SD file and save it to a local file in the Bioclipse workspace.
listCompounds(service, dataset)	Lists the molecules in a data set.
Querying	
listModels(service)	Lists the predictive models available from the given service.
getFeatureInfo(ontologyServer, feature)	Returns information about a particular molecular feature (property).
getFeatureInfo(ontologyServer, features)	Returns information about a set of molecular features.
getModelInfo(ontologyServer, model)	Returns information for a computational model.
getModelInfo(ontologyServer, models)	Returns information for a list of computational models.
getAlgorithmInfo(ontologyServer, algorithm)	Returns information for a computational algorithm.
getAlgorithmInfo(ontologyServer, algorithms)	Returns information for a list of computational algorithms.
listAlgorithms(ontologyServer)	Returns a list of algorithms.
listDescriptors(ontologyServer)	Returns a list of descriptor algorithms.
listDataSets(service)	Returns the data sets available at the given OpenTox server.
searchDataSets(ontologyServer, query)	Returns matching data sets using a free text search.
search(service, inchi)	Returns matching structures based on the InChI given.
search(service, molecule)	Returns matching structures based on the molecule given.

Published by Woodhead Publishing Limited, 2012

This chapter will first outline a few basic tasks one can perform with the Bioclipse–OpenTox platform, including the calculation of molecular descriptors (which can also be used to calculate properties important to toxicology, including logP and pK_a), the sharing of data on toxins and toxicants online, and how Bioclipse supports authorization and authentication on the OpenTox network. After that, two more elaborate use cases will be presented, which will make use of the Decision Support extension. Bioclipse Decision Support was developed by the Department of Pharmaceutical Biosciences, Uppsala University, Sweden, and AstraZeneca R&D, Mölndal, Sweden, with the goal of building an extensible platform for integrating multiple predictive models with a responsive user interface. Bioclipse Decision Support is now being used for ADME-T predictions within AstraZeneca R&D.

2.2 Basic Bioclipse–OpenTox interaction examples

As an introduction to the Bioclipse–OpenTox interoperability, this chapter will first introduce a few short examples. Of course, to reproduce these examples yourself, you may have to familiarize yourself with the Bioclipse software, which takes, like any other software, some time to learn. However, the examples given in this chapter will not be that basic. For more tutorials on how to use Bioclipse, we refer the reader to *http:// www.opentox.org/tutorials/bioclipse*.

Figure 2.1 shows the first Bioclipse–OpenTox integration and highlights the Bioclipse QSAR environment for calculating molecular properties and theoretical descriptors [21]. This environment has been extended to discover OpenTox services on the internet. Using this approach, Bioclipse has access to the most recent descriptors relevant to toxicity predictions. The figure shows the equivalence of a number of descriptors provided by both local services (CDK and CDK REST, provided by the Chemistry Development Kit library [22]) and OpenTox, as well as the Ionization Potential descriptor provided only via an OpenTox computational service. Practically, these online computational services are found by querying a registry of OpenTox services: this registry makes use of the OpenTox ontology [17], which Bioclipse queries, as outlined in [10]. The ability to compute descriptors using various local and remote providers creates a flexible application for integration of numerical inputs for statistical modeling of toxicologically relevant endpoints, as well as

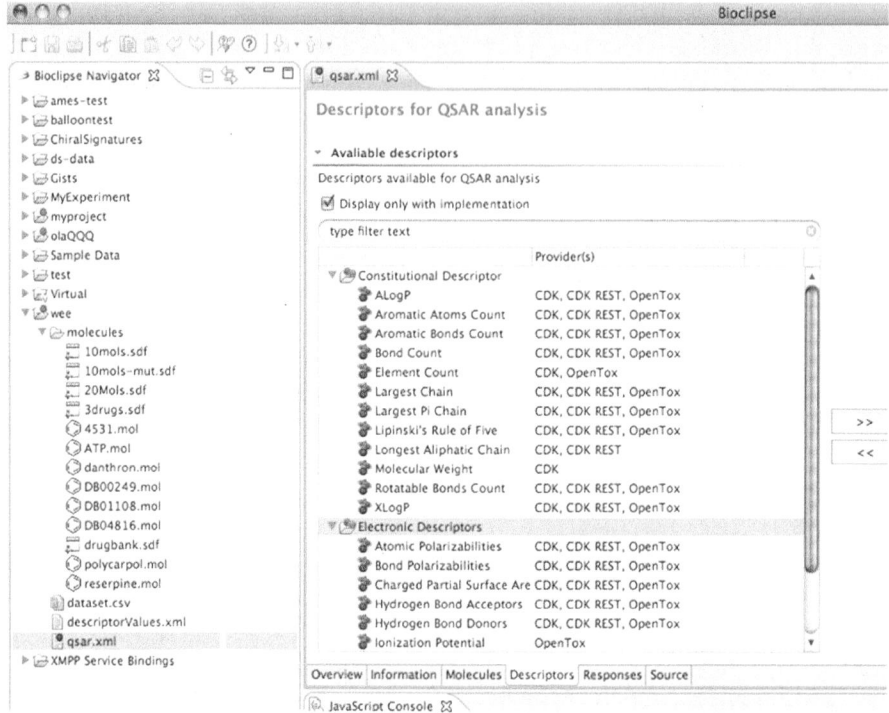

Figure 2.1 Integration of online OpenTox descriptor calculation services in the Bioclipse QSAR environment. Molecular descriptors are frequently used in computational toxicology models. This screenshot from Bioclipse QSAR shows descriptors discovered on the Internet (providers: OpenTox and CDK REST) in combination with local software (provider: CDK)

comparison of various predictive models for a more balanced property analysis.

A second basic example is the sharing of data on the OpenTox network, which demonstrates how all Graphical User Interface (GUI) interaction in Bioclipse can also be performed using a scripting language. Bioclipse extends scripting languages like JavaScript and Groovy with life sciences-specific extensions, defining the Bioclipse Scripting Language (BSL). These extensions are available for much open source life sciences software, including Jmol, the Chemistry Development Kit, but also the OpenTox API. The latter is used in, for example, Figure 2.2, which shows

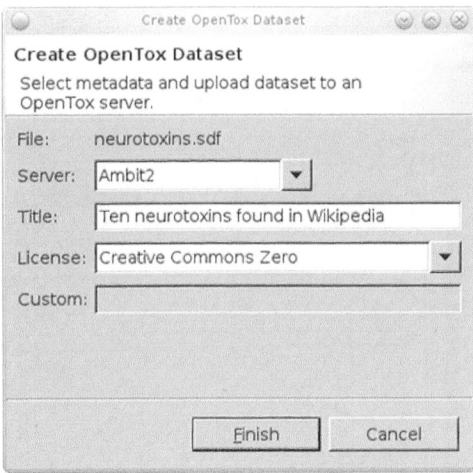

Figure 2.2 The Bioclipse Graphical User Interface for uploading data to OpenTox. Sharing new toxicological data about molecular structures can be done by uploading the data to an OpenTox server. This Bioclipse dialog shows a selected MDL SD file with ten neurotoxins (neurotoxins.sdf) being shared on the Ambit2 server, the OpenTox server to upload to, providing a title for the data set, and the CCO waiver (see main text). Clicking the Finish button will upload the structures and open a web browser window in Bioclipse with the resulting online data set (see Figure 2.3)

the Bioclipse dialog for uploading a small data set with ten neurotoxin structures to an OpenTox server. This dialog asks to which OpenTox server the structures should be uploaded (the Ambit2 server is selected, *http://apps.ideaconsult.net:8080/ambit2/*), a title under which this data set will be available (*'Ten neurotoxins found in Wikipedia'*), and the data license under which the data will be available to others (the Creative Commons Zero license in this case [23]).

At a scripting level, this dialog makes use of the createDataSet(service, molecules), setDatasetLicense(datasetURI, licenseURI), and set DatasetTitle(datasetURI, title) commands, as listed in Table 2.1. The latter two methods use the data set's Universal Resource Identifier (URI) returned by the first method. The dialog, however, opens the data set's web page in a browser window in Bioclipse (see Figure 2.3)

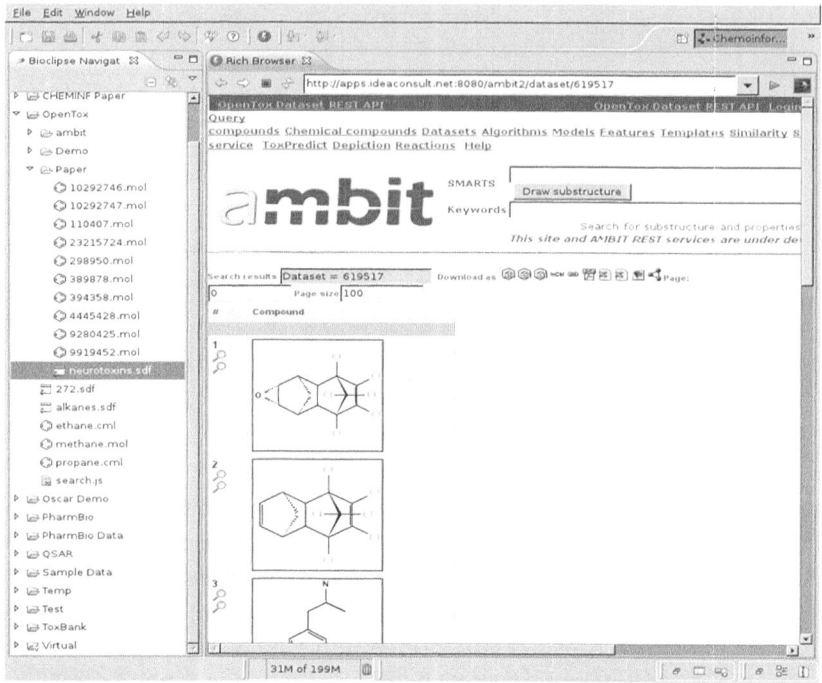

Figure 2.3 OpenTox web page showing uploaded data. Screenshot of Bioclipse showing a web browser window with the neurotoxins data hosted on the Ambit2 OpenTox server after the upload, as shown in Figure 2.2 (see http://apps.ideaconsult.net:8080/ambit2/dataset/619517)

when the upload has finished. This is currently not possible using the scripting language.

Most of the other Bioclipse–OpenTox integration takes advantage of the fact that Bioclipse has all GUI functionality matched by a scripted equivalent too. The use of the BSL directly allows interaction with the OpenTox network to be automated, combined with other Bioclipse functionality into larger workflows, and makes it easier to share procedures with others using social scientific sites like MyExperiment [24].

For example, a BSL script equivalent for the earlier descriptor calculation GUI would look as follows:

Published by Woodhead Publishing Limited, 2012

```
service = 'http://apps.ideaconsult.net:8080/ambit2/';
serviceSPARQL = 'http://apps.ideaconsult.net:8080/ontology/';
stringMat = opentox.listDescriptors(serviceSPARQL);
descriptor = stringMat.get(1, 'algo');
molecules = cdk.createMoleculeList();
molecules.add(
  cdk.fromSMILES('CC(=O)C1=CC=C(C=C1)N')
);
molecules.add(
  cdk.fromSMILES('C1=CC=C(C(=C1)CC(=O)O)NC2=C(C=CC=C2C1)C1')
);
js.say(
  descriptor + ' - ' +
  opentox.calculateDescriptor(service, descriptor, molecules)
);
```

This script combines functionality from the Chemistry Development Kit plug-in for Bioclipse via the cdk BSL extension with the OpenTox functionality provided by the opentox extension.

The third basic functionality of the Bioclipse–OpenTox interoperability we highlight here is the support for accessing protected resources within the OpenTox network. Despite preferences of the authors, we acknowledge that not all scientific data will be Open Data in the foreseeable future. As such, authentication and authorization (A&A) are important features of data access. OpenTox implements both aspects, and provides web services for A&A, allowing users to log in and out of OpenTox applications, accompanied by policy-based specification of OpenTox resource access permissions. Additionally, the same mechanism is used to restrict access to calculation procedures, allowing exposure of software with commercial licenses as protected OpenTox resources.

Bioclipse was extended to support the authentication functionality, allowing OpenTox servers to properly authorize user access to particular web services and data sets. The OpenTox account information is registered with Bioclipse's keyring system, centralizing logging in and out onto remote services, providing the GUI for adding a new OpenTox account and to log in and out. The corresponding script commands for the authentication are given in Authentication category in Table 2.1. Support of authentication is important for industrial environments in which confidential data are handled.

With these basics covered, we will now turn to two use cases much more interesting from a toxicology perspective. For these, we will use the Tres Cantos Antimalarial Compound Set, TCAMS – a collection of 13 533 compounds of the Tres Cantos Medicines Development Campus of GlaxoSmithKline deposited in 2010 at the ChEMBL Neglected

Tropical Disease Database (https://www.ebi.ac.uk/chemblntd), making the data publicly available [25]. These compounds will be tested against a number of toxicity-related predictive models, as outlined in Table 2.2, see [16] for a more detailed description. The Decision Support plug-in allows detailed information as to how the decision for a particular endpoint was reached, using a variety of data types outlined in Table 2.3. The OpenTox extension currently does not make use of this, as there is no ontology available at this moment to communicate this information. However, this is under development.

Each of the chemicals in the TCAMS inhibits growth of the 3D7 strain of *Plasmodium falciparum* – the malaria causing parasite – by at least 80% at a concentration of 2 μM. Many also show a similar effect against the multidrug-resistant *P. falciparum* strain DD2. Evidence for liver toxicity is provided by growth inhibition data against human hepatoma HepG2 cells. Of all TCAMS compounds, 857 are annotated with a so-called target hypothesis. This target hypothesis had been obtained by comparing each compound with public and GSK-internal data of compound-target relationships, accepting a target hypothesis if a homologue to the identified target exists in *P. falciparum* and unless many compound-target relationships had been identified for a given chemical. Of this subset, 233 are annotated as potential Ser/Thr kinase inhibitors.

Table 2.2	Description of the local endpoints provided by the default Bioclipse Decision Support extension. The OpenTox integrates further tests, which are not described in this table. An up-to-date overview of services available on the OpenTox network is provided at http://apps.ideaconsult.net:8080/ToxPredict#Models
Ames mutagenicity	The Ames *Salmonella* microsome mutagenicity assay (AMES test) indicates if a compound can induce mutations to DNA.
CPDB	Carcinogenic Potency Database (CPDB) contains data on compounds known to lead to cancer.
AhR	Aryl hydrocarbon receptor (AhR) is a transcription factor involved in the regulation of xenobiotic-metabolizing enzymes, such as cytochrome P450.

Table 2.3	Various data types are used by the various predictive models described in Table 2.2 to provide detailed information about what aspects of the molecules contributed to the decision on the toxicity
Structural alerts	A substructure that has been associated with an alert, in our case a chemical liability. Can be implemented in many ways, most common is using SMARTS patterns. Also referred to as toxicophore if alerting for toxicity.
Signature alerts	A type of structural alert implemented using Signatures [26].
Signature significance	A QSAR model which is capable of, apart from a prediction, to return the most significant signature in the prediction [27].
Exact match	An identical chemical structure was found in the training data. Can be implemented by InChI and Molecular Signature.
Near Neighbor match	A similar chemical structure was found in the training data. Commonly implemented by binary fingerprint and within a certain Tanimoto distance.

2.3 Use Case 1: Removing toxicity without interfering with pharmacology

From the TCAMS data set, we select the compound TCMDC-135308 for further investigation. TCMDC-135308 is similar (Tanimoto=0.915) to quinazoline 3d (both shown in Table 2.4), a potent human-TGF-β1 inhibitor (Transforming Growth Factor-β1). TCMDC-135308 inhibits growth of the *P. falciparum* strain 3D7 by 98% at a concentration of 2 μM (XC50=700 nM). It is also active against the multidrug-resistant *P. falciparum* strain DD2, inhibiting its growth by 63% at a concentration of 2 μM. The compound has not shown significant growth inhibition of human HepG2 cells (5% at a concentration of 10 μM).

We import TCMDC-135308 using its SMILES to Bioclipse, log in to OpenTox and move to the Decision Support Perspective, where we run all models. It turns out TCMDC-135308 contains a CPDB Signature Alert for Carcinogenicity (see Figure 2.4). Signature Alerts in Bioclipse Decision Support are discriminative signatures identified according to

[26–8], which can be visualized as substructures in the chemical structure of the query molecule.

The carcinogenicity prediction is supported by some positive predictions with OpenTox models 'IST DSSTox Carcinogenic Potency DBS Mouse', 'IST DSSTox Carcinogenic Potency DBS SingleCellCall' and 'ToxTree: Benigni/Bossa rules for carcinogenicity and mutagenicity'

Table 2.4 Structures created from **SMILES** representations with the Bioclipse New from SMILES wizard for various structures discussed in the use cases

TCMDC-135308 **SMILES** CN(C)c1ccc2nc(nc(Nc3ccncc3)c2c1)c4cccc(C)n4	
quinazoline 3d **SMILES** Cc1cccc(n1)c2nc(Nc3ccnc(C)c3)c4ccccc4n2	
TCMDC-134670 **SMILES** n1ccc(cc1)Nc3nc(nc2ccc(OC)cc23)c4nc(ccc4)C	

TCMDC-135174 **SMILES** COc1cc2c(Nc3cccc(NC(=O) c4ccccc4)c3)- ncnc2cc1OCCCN5CCOCC5	
TCMDC-134695 **SMILES** Cc1cc(Nc2nc(Nc3cccc(Cl)c3) nc4ccccc24)n[nH]1	
TCMDC-133364 **SMILES** NCCNS(=O)(=O)c1ccc(cc1) c1ccnc2[nH]c(cc12)-c1ccncc1	

(Continued)

Published by Woodhead Publishing Limited, 2012

Table 2.4	Structures created from SMILES representations with the Bioclipse New from SMILES wizard for various structures discussed in the use cases (*Continued*)

TCMDC-133807

SMILES
CCn1c(nc2c(ncc(OCCCN)c12)c1ccc(CO)o1)-c1nonc1N

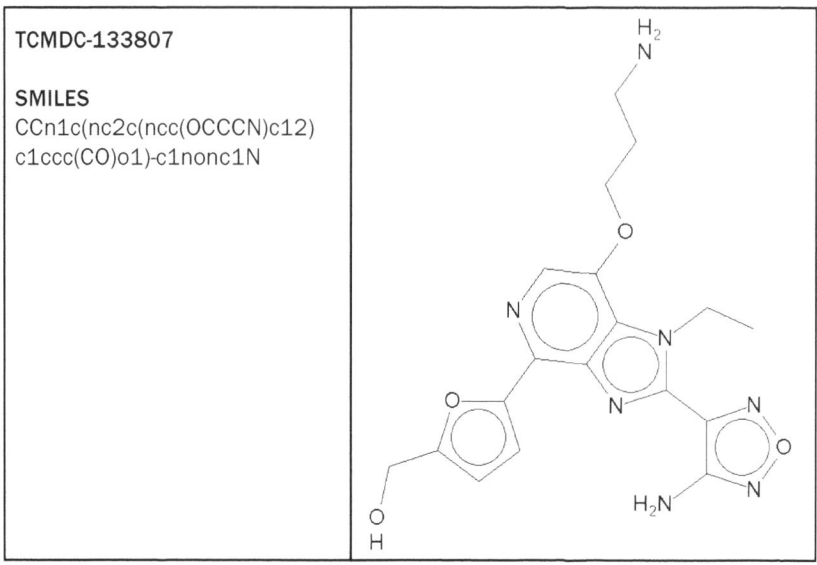

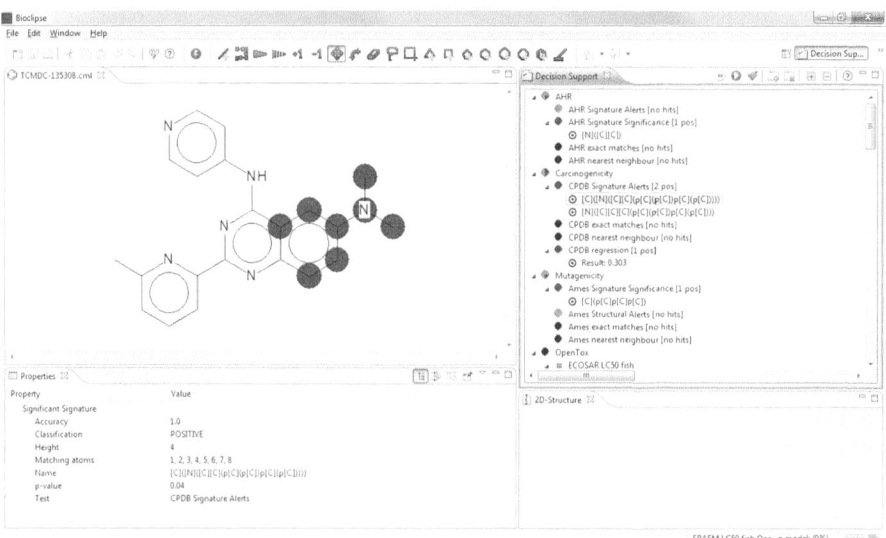

Figure 2.4	CPDB Signature Alert for Carcinogenicity for TCMDC-135308

Published by Woodhead Publishing Limited, 2012

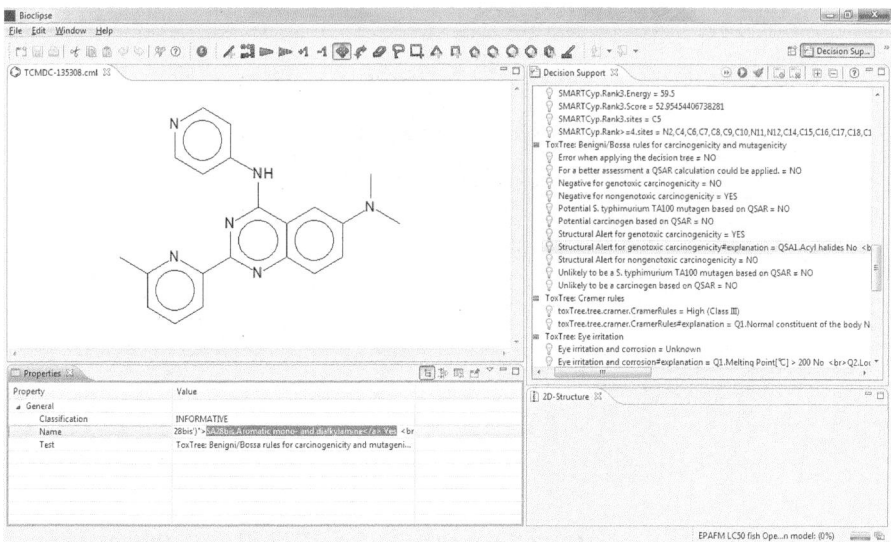

Figure 2.5 Identification of the structural alert in the ToxTree Benigni/Bossa model for carcinogenicity and mutagenicity, available via OpenTax

(structural alert for genotoxic carcinogenicity) as shown in Figure 2.5. Also, the OpenTox models 'IST DSSTox Carcinogenic Potency DBS Mutagenicity', 'IST Kazius-Bursi Salmonella Mutagenicity' and 'ToxTree: Structure Alerts for the *in vivo* micronucleus assay in rodents' are positive.

To identify the structural alerts, we click on the 'Structural Alert for genotoxic carcinogenicity#explanation' line in the Decision Support window. In the property tab, the property 'Name' lists the contents of the #explanation obtained from OpenTox. One can either scroll along the explanation in the value field of the 'Name' property, or copy and paste its content to a text editor to identify which structural alert was fired. For our compound of interest, it is the presence of a dialkylamine, which overlaps with the CPDB Signature Alert for Carcinogenicity. The presence of a dialkylamine is also a structural alert for the '*in vivo* micronucleus assay in rodents' model.

Analyzing the crystal structure of the inhibitor quinazoline 3d bound to human-TGF-β1 (PDB entry 3HMM [29]), we find that the dimethylamine is not deeply buried in the binding pocket, but rather exposed, without making crucial interactions. This is visualized with the Jmol viewer in Bioclipse in Figure 2.6. The dimethylamino group might therefore be easily replaceable without markedly affecting the activity of the compound.

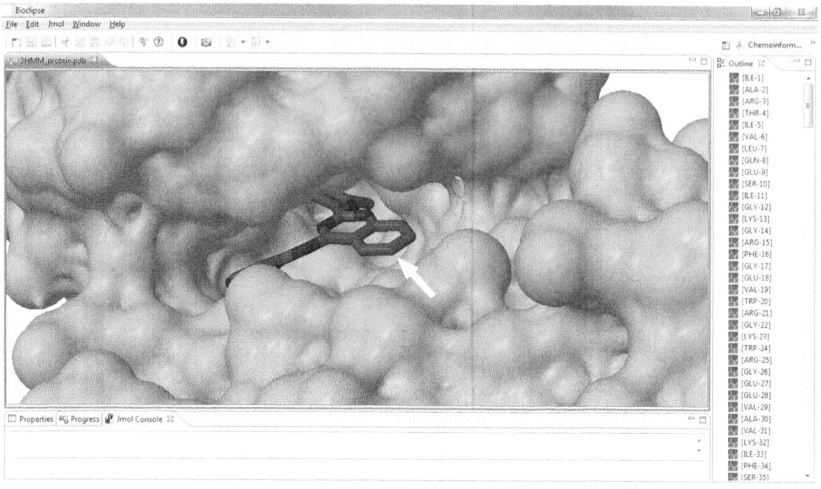

Figure 2.6

Crystal structure of human TGF-β1 with the inhibitor quinazoline 3d bound (PDB-entry 3HMM). The dimethylamino group of TCMDC-135308 is bound to the carbon atom of quinazoline 3d highlighted by the white arrow, thus pointing toward the water-exposed side of the binding pocket. The image was generated from the 3HMM crystal structure, hiding all water molecules, using the Jmol editor in Bioclipse

However, as we do not know the exact target protein in *P. falciparum*, we cannot exclude that the dimethylamino group is important for the activity or selectivity of the compound. Therefore, we try to replace it with a bioisosteric group, the simplest one being a methoxy group.

In the Bioclipse compound window, we modify the structure of TCMDC-135308 by renaming the nitrogen atom of the dimethyl amino group to oxygen, and deleting one of the methyl groups of the former dimethyl amino group.

Re-running all predictions, we find that there are no CPDP signature alerts anymore, and that the OpenTox Benigni/Bossa rules predict the modified compound to be negative for both genotoxic and non-genotoxic carcinogenicity (Figure 2.7). Also the 'IST DSSTox Carcinogenic Potency DBS Mutagenicity' and 'IST DSSTox Carcinogenic Potency DBS Mouse' models are now negative. The OpenTox model for the *in vivo* micronucleus assay is still positive because of the presence of two hydrogen-bond acceptors that are three bonds apart. Also the OpenTox 'IST DSSTox Carcinogenic Potency DBS SingleCellCall' and 'IST Kazius-

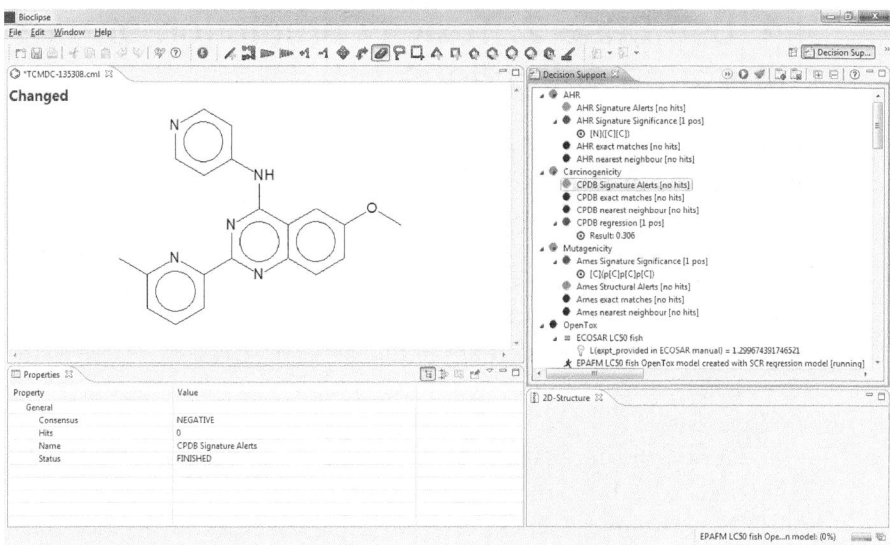

Figure 2.7 Replacing the dimethylamino group of TCMDC-135308 with a methoxy group resolves the CPDB signature alert as well as the ToxTree Benigni/Bossa Structure Alerts for carcinogenicity and mutagenicity as provided by OpenTox. The image illustrates the absence of CPDB signature alerts

Bursi Salmonella mutagenicity' models are still positive. In addition, the 'IST DSSTox Carcinogenic Potency DBS MultiCellCall' model is now positive. As expected, physico-chemical properties such as pK_a or XLogP are not, or only marginally, affected (no change to pK_a, XLogP changes from 0.845 to 0.551), and also, no new positive toxicity predictions are obtained due to the change.

Thus, our bioisosteric replacement of the dimethylamino group has resolved several toxicity predictions, while hopefully only marginally affecting the compound's activity. As it turns out, the methoxy variant we constructed based on TCMDC-135308 exists in TCAMS as well. It is TCMDC-134670. Its activity against *P. falciparum* 3D7 is almost identical to that of TCMDC-135308 (101% growth inhibition at 2 µM, XC50 = 670 nM), whereas its activity against *P. falciparum* DD2 is slightly decreased (from 63% to 42% growth inhibition at 2 µM). TCMDC-134670 is also less active in the *in vitro* hepatotoxicity assay reported (-1% growth inhibition of human HepG2 cells at a concentration of 10 µM, compared to 5% obtained with TCMDC-135308).

2.4 Use Case 2: Toxicity prediction on compound collections

As a second, illustrative example, we look at the US FDA's Adverse Events Reporting System database (AERS) [30]. This database is a unique source of *in vivo* data on observations of the adverse outcomes of human toxicities of drugs. Pharmatrope [31] has processed the AERS data according to statistical considerations, and created the Titanium Adverse Events Database and Models. For all compounds in the TCAMS we predicted association with groups of human adverse events related to the hepatobiliary tract and classified the compounds' association with these adverse events.

All data on the annotated kinase inhibitors in TCAMS – namely the compounds' identifiers and SMILES, the inhibition of *P. falciparum* 3D7 and DD2, of human HepG2 cells, and the degree of association with human adverse events – was combined into an SDF file and imported into Bioclipse. Double-clicking the imported file in the Bioclipse Navigator opens the table, which allows the rearrangement of the columns (Figure 2.8).

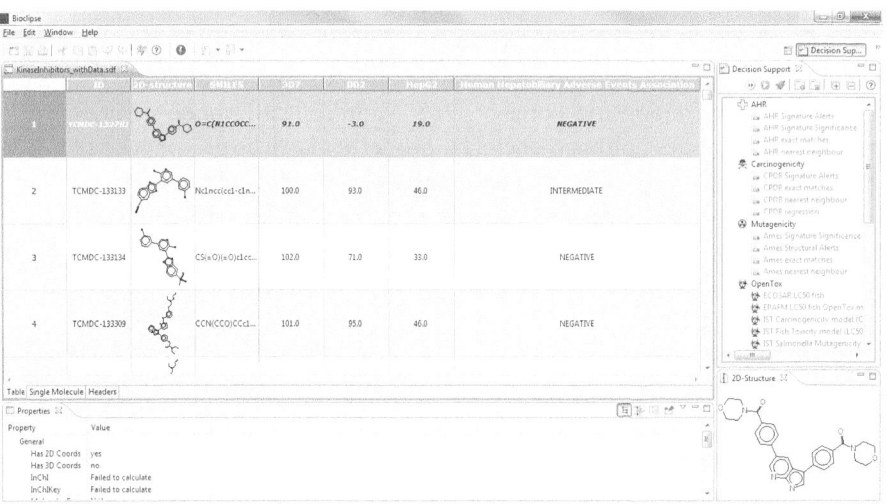

Figure 2.8 Annotated kinase inhibitors of the TCAMS, imported into Bioclipse as SDF together with data on the association with human adverse events

The imported SDF file contained a field for the compounds' SMILES. However, we will not be using the SMILES here, so we hide the column by right-clicking on its header and selecting 'Hide Column'. Clicking again on a column header, all compounds are selected. Right-clicking and selecting 'Decision Support' in the menu that opens gives the list of Bioclipse toxicity predictions that can be run on the compounds (Figure 2.9). We select the CPDB Signature Alerts.

While running the calculations, Bioclipse displays a progress bar, so that users can monitor how far the calculation has proceeded. In a similar manner we calculate the Ames Signature Significance and the AHR Signature Significance. Once the computations are completed, we can save the table under a new name.

Browsing the table, we can identify some interesting compounds, for example TCMDC-133364 is very active against both *P. falciparum* strains, without inhibiting the growth of the HepG2 cells. The compound's association with human adverse events is negative, and three out of four

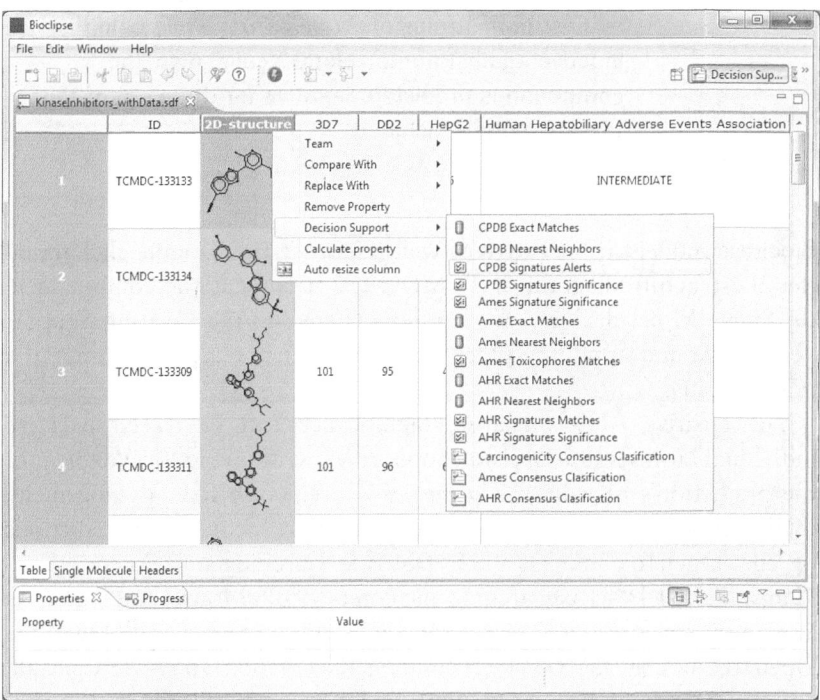

Figure 2.9 Applying toxicity models to sets of compounds from within the Bioclipse Molecule TableEditor

Published by Woodhead Publishing Limited, 2012

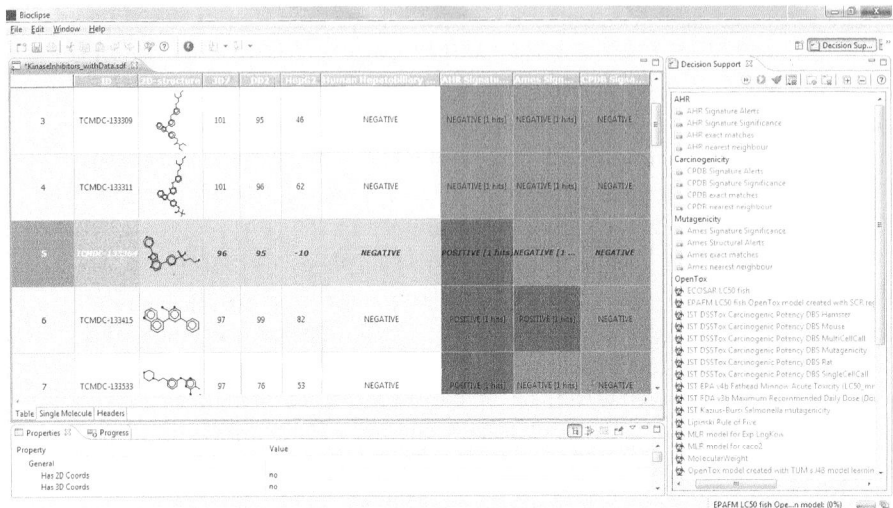

Figure 2.10 Adding Decision Support columns to the molecule table. The highlighted compound – TCMDC-133364 – is an interesting candidate as it is highly active against both strains of *P. falciparum* while being inactive against human HepG2 cells. Also, the compound is predicted negative for three out of four toxicity models included

Bioclipse models are negative as well (Figure 2.10). Double-clicking on the 2D-structure diagram of TCMDC-133364 opens the compound in the Single Molecule View. We can now change to the Decision Support perspective and apply the OpenTox models on the selected compound (Figure 2.11).

The positive AHR Signature Significance can be traced back to individual atoms in the Decision Support View, allowing for judgment of the prediction's relevance. The negative prediction for carcinogenicity by the CPDB model in Bioclipse is confirmed by the negative predictions of the OpenTox models 'IST DSSTox Carcinogenic Potency DBS Hamster', 'IST DSSTox Carcinogenic Potency DBS Mouse' and 'ToxTree: Benigni/Bossa rules for carcinogenicity and mutagenicity'. Only the OpenTox models 'IST DSSTox Carcinogenic Potency DBS MultiCellCall' and 'IST DSSTox Carcinogenic Potency DBS SingleCellCall' are positive. For mutagenicity we have the negative prediction by Bioclipse's Ames model, the negative prediction with the OpenTox Benigni/Bossa model

Published by Woodhead Publishing Limited, 2012

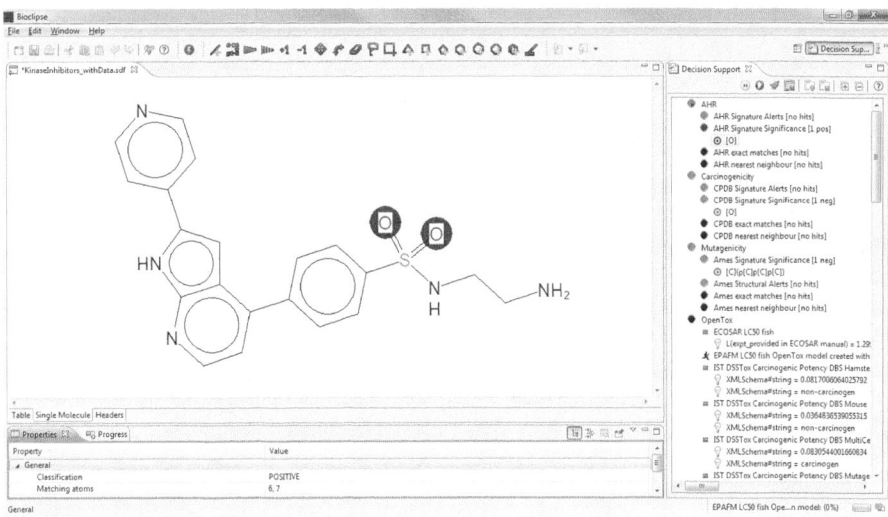

Figure 2.11 Opening a single compound from a table in the Decision Support perspective

and the 'IST Kazius-Bursi Salmonella mutagenicity' model. However, we obtain a positive prediction with the OpenTox 'ToxTree: Structure Alerts for the *in vivo* micronucleus assay' model. Inspecting the details of that prediction, we see that it is related to hydrogen-bond acceptors separated by three covalent bonds in the sulfonamide part of the molecule.

TCMDC-135174 is also very active against both *P. falciparum* strains, without inhibiting the growth of the HepG2 cells, and its association with human adverse events, as well as the Ames Toxicophore Match and the AHR Signature Match, are also negative. However, the Bioclipse CPDB Signature Match is positive, as is visible in Figure 2.12.

This positive carcinogenicity prediction is not confirmed in the DS perspective with any of the OpenTox 'IST DSSTox' carcinogenicity or mutagenicity models, or the 'ToxTree: Benigni/Bossa rules for carcinogenicity and mutagenicity' model. The DS view offers additional information, though: the 'ToxTree: Structure Alerts for the *in vivo* micronucleus assay in rodents' is positive, and two 'ToxTree: Skin sensitization alerts' are triggered. Combining these pieces of information, we might decide against selecting this compound as a drug development candidate, even though we may reject the initial carcinogenicity prediction.

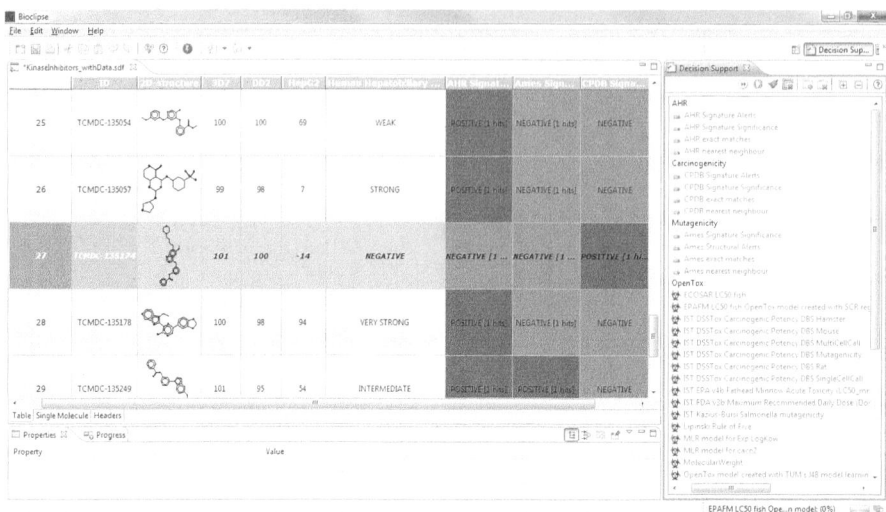

Figure 2.12 The highlighted compound – TCMDC-135174 (row 27) – is an interesting candidate as it is highly active against both strains of *P. falciparum* while being inactive against human HepG2 cells. Also, the compound is predicted negative for three of the four toxicity models included. However, it triggered a CPDB Signature alert for carcinogenicity

In terms of its cytotoxic experimental data, TCMDC-134695 is also a promising candidate, as shown in Figure 2.13. In addition, again three out of four Bioclipse Decision Support models are negative. However, the compound has a weak association predicted with human adverse events related to the hepatobiliary tract.

Running the OpenTox models in the DS perspective on this compound adds to the concerns raised with the predicted association with human adverse events. The compound is predicted to be carcinogenic and mutagenic with several OpenTox models ('ToxTree: Structure alerts for the *in vivo* micronucleus assay in rodents', 'ToxTree: Benigni/Bossa rules for carcinogenicity and mutagenicity', 'IST DSSTox Carcinogenic Potency DBS Mouse', 'IST DSSTox Carcinogenic Potency DBS MultiCellCall' and 'IST DSSTox Carcinogenic Potency DBS Mutagenicity').

TCMDC-133807 (see Figure 2.14) appears to be a rather toxic compound, being predicted to be strongly associated with human adverse events and triggering a structure alert with Bioclipse's CPDB carcinogenicity model.

Published by Woodhead Publishing Limited, 2012

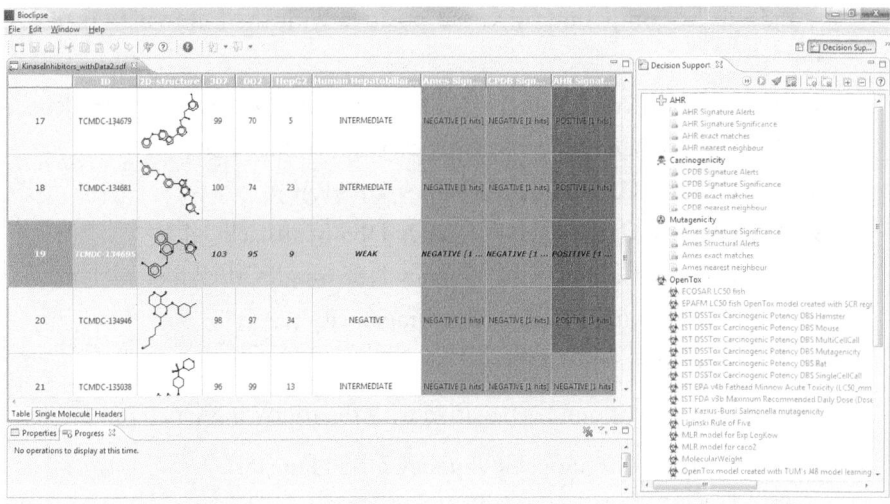

Figure 2.13 Molecule Table view shows TCMDC-134695 in row 19. This compound is promising for most of its properties represented in the table, except for the AHR Signature Significance. However, it is weakly associated with human adverse events

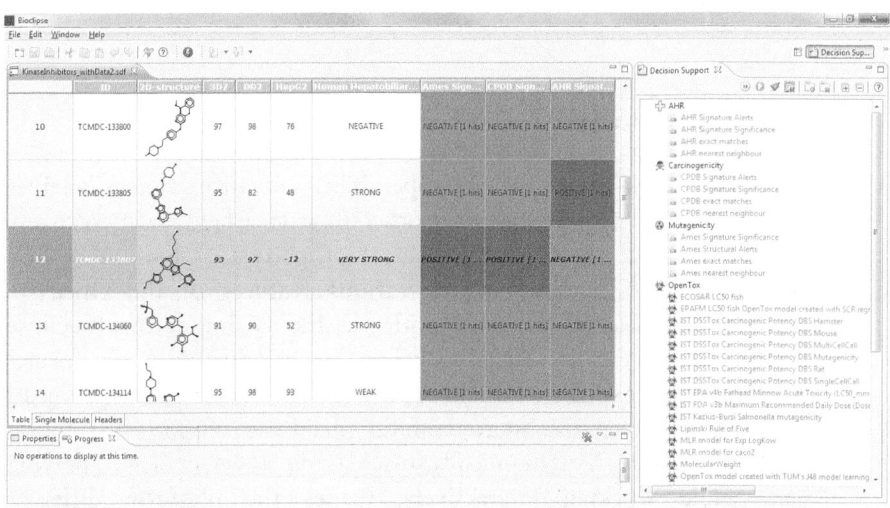

Figure 2.14 The compound TCMDC-133807 is predicted to be strongly associated with human adverse events, and yields signature alerts with Bioclipse's CPDB and Ames models

Indeed the compound turns out positive with the following OpenTox models:

- 'ToxTree: Structure alerts for the in vivo micronucleus assay in rodents';
- 'IST DSSTox Carcinogenic Potency DBS Mutagenicity';
- 'IST DSSTox Carcinogenic Potency DBS Rat';
- 'IST DSSTox Carcinogenic Potency DBS MultiCellCall';
- 'IST DSSTox Carcinogenic Potency DBS SingleCellCall';
- 'IST Kazius-Bursi Salmonella Mutagenicity';
- 'ToxTree: Skin sensitization'.

However, the compound is negative with the following OpenTox models:

- 'IST DSSTox Carcinogenic Potency DBS Hamster';
- 'IST DSSTox Carcinogenic Potency DBS Mouse';
- 'IST DSSTox Carcinogenic Potency DBS Hamster';
- 'ToxTree: Benigni/Bossa rules for carcinogenicity and mutagenicity'.

The considerably higher number of positive than negative predictions confirms the initial toxicity estimation by the human adverse events model and Bioclipse's Decision Support Models.

2.5 Discussion

These use cases show an interoperability advance, which enables toxicologists and people working in other life science fields to interactively explore and evaluate the toxicological properties of molecules. The integration into Bioclipse makes various features of the OpenTox platform available to the user, both via the GUI and the Bioclipse Scripting Language, the latter focusing on reproducibility, the former on interactive exploration and optimization. This dual nature of the Bioclipse–OpenTox interoperability makes it unique. A solution which is capable of dynamically discovering new services, applying them while working interactively, makes the platform different from other software like ToxTree [13], ToxPredict [32], and the OECD QSAR ToolBox [11], or more general tools like Taverna [33] and KNIME [34].

A further difference to some of these alternative tools is that the Bioclipse–OpenTox integration relies on semantic web technologies, which are seeing significant adoption in other areas of the life sciences too, including drug discovery, text mining, and neurosciences [35–37]. Several other chapters in this book describe applications making use of semantic web technologies. In particular David Wild's work in creating the Chem2Bio2RDF framework (Chapter 18) and the chapter by Bildtsen and co-authors in the development of the 'triple map' application (Chapter 19). In addition, the chapters by Alquier (Chapter 16) and by Harland and co-authors (Chapter 17) describe application of semantic MediaWiki in life science knowledge management. The OpenTox platform has demonstrated the provision of a simple but well-defined and consistent ontology for the interaction with their services, providing functionality for both service discovery and service invocation. The SADI framework is the only known semantic alternative [38], but does not currently provide the same level of computational toxicology services as does OpenTox.

The combination of these two unique features makes it possible for Bioclipse–OpenTox to follow the evolution of computational toxicology closely and promptly. We are therefore delighted that this book chapter is available under a Creative Commons ShareAlike-Attribution license, allowing us to update it for changes in the Bioclipse–OpenTox environment.

2.6 Availability

Bioclipse with the Decision Support and OpenTox extensions can be freely downloaded for various platforms from *http://www.bioclipse.net/opentox*.

2.7 References

[1] European Parliament – European Commission. Regulation (EC) No. 1907/2006 of the European Parliament and of the Council. *Official Journal of the European Union* 2006.

[2] Knudsen TB, Houck KA, Sipes NS, et al. Activity profiles of 309 ToxCast™ chemicals evaluated across 292 biochemical targets. *Toxicology* 2011;282(1–2):1–15.

[3] Schmidt U, Struck S, Gruening B, et al. SuperToxic: a comprehensive database of toxic compounds. *Nucleic Acids Research* 2009;37(Database issue):D295–9.

[4] Kuhn M, Campillos M, Letunic I, Jensen LJ, Bork P. A side effect resource to capture phenotypic effects of drugs. *Molecular Systems Biology* 2010;6:343.

[5] Williams-DeVane CR, Wolf MA, Richard AM. DSSTox chemical-index files for exposure-related experiments in ArrayExpress and Gene Expression Omnibus: enabling toxico-chemogenomics data linkages. *Bioinformatics (Oxford, England)* 2009;25(5):692–4.

[6] Huang DW, Sherman BT, Lempicki RA. Systematic and integrative analysis of large gene lists using DAVID bioinformatics resources. *Nature Protocols* 2009;4(1):44–57.

[7] Kelder T, Pico AR, Hanspers K, et al. Mining biological pathways using WikiPathways web services. *PloS One* 2009;4(7):e6447.

[8] Kanehisa M, Goto S. KEGG: Kyoto Encyclopedia of Genes and Genomes. *Nucleic Acids Research* 2000;28(1):27–30.

[9] Rydberg P, Gloriam DE, Olsen L. The SMARTCyp cytochrome P450 metabolism prediction server. *Bioinformatics* 2010; 26(23):2988–9.

[10] Willighagen EL, Jeliazkova N, Hardy B, Grafström RC, Spjuth O. Computational toxicology using the OpenTox application programming interface and Bioclipse. *BMC Research Notes* 2011;4(487).

[11] Diderichs R. Tools for Category Formation and Read-Across: Overview of the OECD (Q)SAR Application Toolbox. In: Cronin M. MJ, ed. *In Silico Toxicology*. RSC Publishing; 2010:385–407.

[12] Patlewicz G, Jeliazkova N, Safford RJ, Worth AP, Aleksiev B. An evaluation of the implementation of the Cramer classification scheme in the Toxtree software. *SAR and QSAR in Environmental Research* 2008;19(5–6):495–524.

[13] ToxTree. 2011. Available at: *http://toxtree.sourceforge.net*.

[14] Spjuth O, Helmus T, Willighagen EL, et al. Bioclipse: an open source workbench for chemo- and bioinformatics. *BMC Bioinformatics* 2007;8(1):59.

[15] Spjuth O, Alvarsson J, Berg A, et al. Bioclipse 2: a scriptable integration platform for the life sciences. *BMC Bioinformatics* 2009;10(1):397.

[16] Spjuth O, Eklund M, Ahlberg Helgee E, Boyer S, Carlsson L. Integrated decision support for assessing chemical liabilities. *Journal of Chemical Information and Modeling* 2011;51(8):1840–7.

[17] Hardy B, Douglas N, Helma C, et al. Collaborative Development of Predictive Toxicology Applications. *Journal of Cheminformatics* 2010;2(1):7.

[18] Jeliazkova N, Jeliazkov V. AMBIT RESTful web services: an implementation of the OpenTox application programming interface. *Journal of Cheminformatics* 2011;3:18.

[19] Spjuth O, Alvarsson J, Berg A, et al. Bioclipse 2: a scriptable integration platform for the life sciences. *BMC Bioinformatics* 2009;10(1):397.

[20] Willighagen EL, Alvarsson J, Andersson A, et al. Linking the Resource Description Framework to cheminformatics and proteochemometrics. *Journal of Biomedical Semantics* 2011;2(Suppl 1):S6.

[21] Spjuth O, Willighagen EL, Guha R, Eklund M, Wikberg JES. Towards interoperable and reproducible QSAR analyses: Exchange of datasets. *Journal of Cheminformatics* 2010;2(1).

[22] Steinbeck C, Hoppe C, Kuhn S, et al. Recent developments of the chemistry development kit (CDK) – an open-source java library for chemo- and bioinformatics. *Current Pharmaceutical Design* 2006;12(17):2111–20.

[23] Anon. CC0 1.0 Universal Public Domain Dedication. Available at: *http://creativecommons.org/publicdomain/zero/1.0/*.

[24] Goble CA, Bhagat J, Aleksejevs S, et al. myExperiment: a repository and social network for the sharing of bioinformatics workflows. *Nucleic Acids Research* 2010;38(suppl 2):W677–W682.

[25] Gamo F-J, Sanz LM, Vidal J, et al. Thousands of chemical starting points for antimalarial lead identification. *Nature* 2010;465(7296):305–10.

[26] Faulon J-L, Visco DPJ, Pophale RS. The signature molecular descriptor. 1. using extended valence sequences in QSAR and QSPR studies. *Journal of Chemical Information and Computer Sciences* 2003;43(3):707–20.

[27] Carlsson L, Helgee EA, Boyer S. Interpretation of nonlinear QSAR models applied to Ames mutagenicity data. *Journal of Chemical Information and Modeling* 2009;49(11):2551–8.

[28] Helgee EA, Carlsson L, Boyer S. A method for automated molecular optimization applied to Ames mutagenicity data. *Journal of Chemical Information and Modeling* 2009;49(11):2559–63.

[29] Gellibert F, Fouchet M-H, Nguyen V-L, et al. Design of novel quinazoline derivatives and related analogues as potent and selective ALK5 inhibitors. *Bioorganic & Medicinal Chemistry Letters* 2009;19(8):2277–81.

[30] Adverse Event Reporting System (AERS). Available at: *http://www.fda.gov/Drugs/GuidanceComplianceRegulatoryInformation/Suveillance/AdverseDrugEffects/default.htm*

[31] Pharmatrope. 2011. Available at: *http:/pharmatrope.com/*.

[32] OpenTox ToxPredict application. 2011. Available at: *http://toxpredict.org*.

[33] Oinn T, Addis M, Ferris J, et al. Taverna: a tool for the composition and enactment of bioinformatics workflows. *Bioinformatics* 2004;20(17):3045–54.

[34] Berthold MR, Cebron N, Dill F, et al. KNIME: The konstanz information miner. In: Preisach C, Burkhardt H, Schmidt-Thieme L, Decker R eds, *Data Analysis, Machine Learning and Applications*. Springer; 2008:319–26.

[35] Ruttenberg A, Clark T, Bug W, et al. Advancing translational research with the Semantic Web. *BMC Bioinformatics* 2007;8(Suppl 3).

[36] Splendiani A, Burger A, Paschke A, Romano P, Marshall MS. Biomedical semantics in the Semantic Web. *Journal of Biomedical Semantics* 2011;2 Suppl 1:S1.

[37] Willighagen EL, Brändle MP. Resource description framework technologies in chemistry. *Journal of Cheminformatics* 2011;3(1):15.

[38] Chepelev LL, Dumontier M. Semantic Web integration of Cheminformatics resources with the SADI framework. *Journal of Cheminformatics* 2011;3:16.

## 3

Utilizing open source software to facilitate communication of chemistry at RSC

Aileen Day, Antony Williams,
Colin Batchelor, Richard Kidd and
Valery Tkachenko

Abstract: The Royal Society of Chemistry is one of the world's premier chemistry publishers and has an established reputation for the development of award-winning platforms such as Prospect and ChemSpider. Using a small but agile in-house development team, we have combined commercial and open source software tools to develop the platforms necessary to deliver capabilities to our community of users. This book chapter will review the systems that have been developed in-house, what they deliver to the community, the challenges encountered in developing our systems and utilizing open source code, and how we have extended available code to make it fit-for-purpose.

Key words: ChemSpider; cheminformatics; wikis; ontologies; OSCAR; OpenBabel; JSpecView.

3.1 Introduction

The Royal Society of Chemistry (RSC) is the largest organization in Europe with the specific mission of advancing the chemical sciences. Supported by a worldwide network of 47 000 members and an international publishing business, our activities span education, conferences, science policy and the promotion of chemistry to the public. The information-handling requirements of the publishing division have always consumed the largest proportion of the available software development resources, traditionally dedicated to enterprise systems to develop robust and well-defined systems to deliver published content to customers. Internal adoption of open source solutions was initiated with the development of Project Prospect [1], and then extended with the acquisition of ChemSpider [2]. ChemSpider delivered both a platform incorporating much open source software, staff expertise in cheminformatics, as well as new and innovative functionality. The small but agile in-house development team have combined commercial and free/open source software tools to develop the platforms necessary to deliver capabilities to the user community. This book chapter will review the systems that have been developed in-house, what they will deliver to the community, the challenges encountered in utilizing these tools and how they have been extended to make them fit-for-purpose.

3.2 Project Prospect and open ontologies

RSC began exploring the semantic markup of chemistry articles, together with a number of other publishers in 2002, providing support for a number of summer student projects at the Unilever Centre in Cambridge University. This work led to an open source Experimental Data Checker [3], which parsed the text of experimental data paragraphs and performed validation checks on the extracted and formatted results. This collaboration led to RSC involvement, as well as collaboration with Nature Publishing Group [4] and the International Union of Crystallography [5], in the SciBorg project [6]. The resulting development of OSCAR [7] (Open Source Chemistry Analysis Routines) as a means of marking up chemical text and linking concepts and chemicals with other resources, was then explored and was ultimately used as the text mining service underpinning the award-winning 'Project Prospect [1]' (see Figure 3.1).

It was essential to develop both a flexible and cost-effective solution during this project. Software development was started from scratch, using standards where possible, but still facing numerous unknowns. Licensing a commercial product for semantic markup would have been difficult to justify and also risked both inflexibility and potential limitations in terms of rapid development. As a result, it was decided to

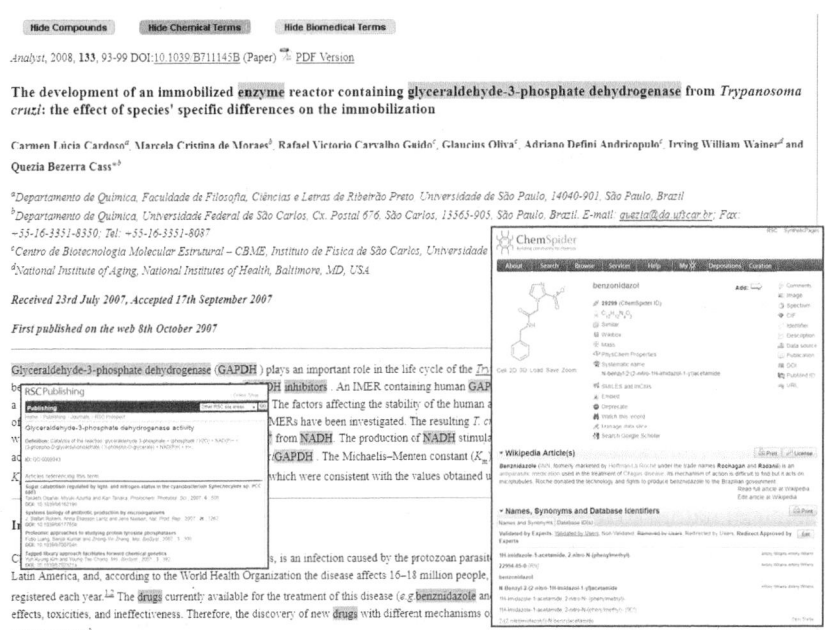

Figure 3.1 A 'prospected' article from RSC. Chemical terms link out to ontology definitions and related articles as appropriate and chemical names link out to ChemSpider (*vide infra*)

work alongside members of the OSCAR development team, contributing back to the open source end product and providing a real business case to drive improvements in OSCAR. This enabled the creation of a parallel live production system, and RSC became the first publisher to semantically markup journal articles, this resulted in the ALPSP Publishing Innovation award in 2007 [8]. More importantly, the chosen path provided a springboard to innovation within the fields of publishing chemistry and the chemical sciences. What follows is a summary of the technical approach that was taken to deliver Project Prospect.

If the architecture for a software project is designed in the right way then in many cases it is possible to set up the system using open source software such as scaffolding [9] and migrate parts of it, as necessary, to more advanced modules, either open or closed source software, as it is determined what the requirements are. For Project Prospect the following needs were identified:

1. a means of extracting chemical names from text and converting them to electronic structure formats;

2. a means of displaying the resulting electronic structure diagrams in an interface for the users;

3. a means of storing those chemicals separately from the article XML;

4. a means of finding non-structural chemical and biomedical terms in the text.

Chemical names commonly contain punctuation, for example [2-({4-[(4-fluorobenzyl)oxy]phenyl}sulfonyl)-1,2,3,4-tetrahydroisoquinolin-3-yl](oxo)acetic acid, or spaces, like diethyl methyl bismuth, or both, and hence cause significant problems for natural language processing code that has been written to handle newswire text or biomedical articles. For this reason, the Sciborg project [6] required code that would identify chemical names so that they would not interfere with further downstream processing of text. Fortunately, a method for extracting chemical structures out of text was already available. The OSCAR software provided a collection of open source code components to meet the explicitly chemical requirements of the Sciborg project. It delivered components that determined whether text was chemical or not, RESTful web services for the Chemistry Development Kit (CDK) [10], routines for training language models, and, importantly, the OPSIN parser [11], which lexes candidate strings of text and generates the corresponding chemical structures. The original version of OPSIN produced in 2006 had numerous gaps but was still powerful enough to identify many chemicals. We also used the ChEBI database [12] as the basis of a chemical dictionary.

In order to display extracted chemical structures, the CDK was used via OSCAR. Although the relevant routines were not entirely reliable and, specifically, did not handle stereochemistry, they were good enough to demonstrate the principle. Following the introduction of the International Chemical Identifier (InChI) [13], and clear interest by various members of the publishing industry and software vendors in supporting the new standard, it was decided to store the connection tables as InChIs. The InChI code is controlled open source, open but presently only developed as a single trunk of code by one development team. The structures were stored as InChIs both in the article XML and in a SQL Server database. For the non-structural chemical and biomedical terms contained within the text, resources that were accessible to the casual reader were identified as being most appropriate. This application was launched with the IUPAC Gold Book [14], which had recently been converted to XML, and with the Gene Ontology [15]. A more detailed account of the integration is given elsewhere [16]. Unfortunately, the Gold Book turned out to have too broad a scope and too narrow coverage to be particularly useful for this work. Initially attempts were made to markup text with all of the entries in the Gold Book, but too many of them, such as *cis-* and *trans-*, were about nomenclature and parts of names, so only hand-picked selections from the resource were used.

Named reactions and analytical methods were additional obvious areas to select for the further development of ontologies applicable to chemistry publishing. RXNO [17] represents named reactions, for example the Diels–Alder reaction, which are particularly easy to identify and a method was established to determine under which classification a reaction falls. The Chemical Methods Ontology (CMO) [18] was initially based on the 600 or so terms contained within the IUPAC Orange Book [19] and then extended based on our experience with text-mining. It now contains well over 2500 terms and covers physical chemistry as well as analytical chemistry. RXNO and CMO have been provided as open source and are available from Google Code where we have trackers and mailing lists [17] [18]. When ChemSpider was acquired in 2009 (*vide infra*), a number of processes were changed to make use of many of the tools and interfaces available within the system. The processes associated with (1), (2) and (3) listed above have been changed but (4), the method by which non-structural chemical and biomedical terms are found in the text, remains the same. At present we utilize OSCAR3, a greatly improved version of OPSIN [11] and ACD/Labs' commercial name to structure software [20]. The large assortment of batch scripts and XSLT transforms has now been replaced by a single program, written in C#, for bulk markup of documents, and it memorizes the results of the name to structure transformations. The key difference for structure rendering is that ChemSpider stores the 2D layouts in addition to the InChIs in the database and as a result the rendering process is now a lot simpler. The ChemSpider image renderer is also used in place of the original CDK, providing significant improvements in structure handling and aesthetics. The original RSC Publishing SQL Server database serving the Prospect Project has now been replaced by integration with ChemSpider, meaning that substructure searching is now available as well as cross-referencing to journal information from other publishers that has been deposited in ChemSpider.

New approaches are being investigated to enhance the semantic markup of RSC publications and to roll out new capabilities as appropriate. These now include the delivery of our semantically enriched articles to the Utopia platform [21], see Chapter 15 by Pettifer and collegues for more details on this innovative scientific literature tool. The development of the RSC semantic markup platform owes much of its success to the availability of the open source software components, developed by a team of innovative scientists and software developers, and these are now used in parallel with both in-house and commercial closed source software to deliver the best capabilities.

3.3 ChemSpider

ChemSpider [22] was initially developed on a shoestring budget as a hobby project, by a small team, simply to contribute a free resource to the chemistry community. Released at the American Chemical Society (ACS) Spring meeting in Chicago in March 2007, it was seeded with just over 10 million chemicals sourced from the PubChem database [23]. Following a two-year period of expanding the database content to over 20 million chemicals, adding new functionality to the system to facilitate database curation and crowd-sourced depositions of data, as well as the development of a series of related projects, ChemSpider was acquired by the Royal Society of Chemistry [2]. The original strategic vision of providing a structure-centric community for chemistry was expanded to become the world's foremost free access chemistry database and to make subsets of the data available as open data.

The database content in ChemSpider (see Figure 3.2), now over 26 million structures aggregated from over 400 data sources, has been developed as a result of contributions and depositions from chemical vendors, commercial database vendors, government databases, publishers, members of the Open Notebook Science community and individual scientists. The database can be queried using structure/ substructure searching and alphanumeric text searching of chemical names and both intrinsic, as well as predicted, molecular properties. Various searches have been added to the system to cater to various user personae including, for example, mass spectrometrists and medicinal chemists. ChemSpider is very flexible in its applications and nature of available searches.

The primary ChemSpider architecture is built on commercial software using a Microsoft technology platform of ASP.NET and SQL Server 2005/2008 as at inception, it allowed for ease of implementation, projected longevity and made best use of available skill sets. Early

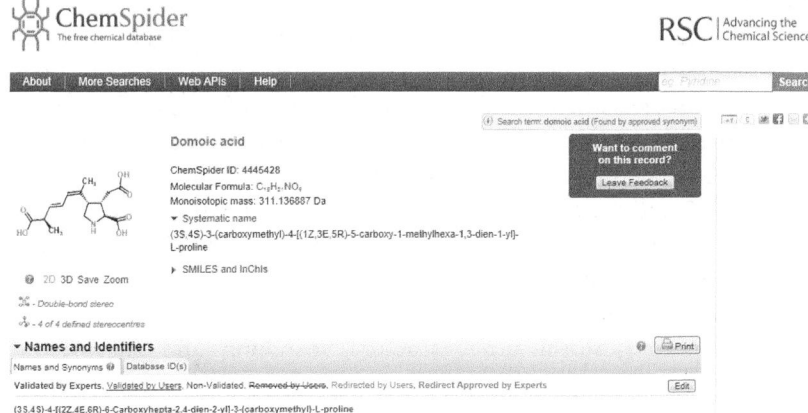

Figure 3.2 The header of the chemical record for domoic acid (*http://www.chemspider.com/4445428*) in ChemSpider. The entire record spans multiple pages including links to patents and publications, pre-calculated and experimental properties and links to many data external data sources and informational web sites

attempts to use SQLite as the database were limited by performance issues. The structure databasing model was completely developed in-house. The InChI library is the basis of the ChemSpider registration system as is the exact searching in ChemSpider (which uses InChI layer separation and comparison). As a result, ChemSpider is highly dependent on the availability of the open source InChI library for InChI generation. The choice between using InChI identifiers versus alternative chemical structure hashing algorithms (e.g. CACTVS hash codes [24]) was largely based on community adoption. Attempts were made to develop our own version of hash codes early on but were abandoned quickly as the standard InChI library was already out of beta and increasingly used in the chemical community. No modifications to the InChI source code have been made except for small changes to the libraries allowing multiple versions of the InChI code to coexist in one process address space.

The GGA Bingo toolkit (SQL Server version) [25] is used for substructure and similarity searches in ChemSpider. The open source library GGA software is developed by a small team of geographically co-located developers and, to the best of our knowledge, they do not allow the source code to be modified outside of their organization. This platform was chosen over other possible solutions for ChemSpider as the team was knowledgeable, professional and agile. The original version of Bingo was made available only on SQL Server 2008 while ChemSpider was running on SQL Server 2005 at that time. As the software was available as open source, we recompiled the source code to work on SQL Server 2005 and the GGA team fixed version discrepancies quickly and provided a working version of Bingo for SQL Server 2005 within a one-day turnaround. This is a testament to the skills of the software team supporting this open source product.

In order to perform both structure searching and substructure searching, a manner by which to introduce a chemical structure drawing is required. We provide access to a number of structure drawing tools, both Java and JavaScript. Two of these structure editors are open source (GGA's Ketcher [26] and JChemPaint [27]) and we implemented them without modification.

There are various needs on the ChemSpider system for the conversion of chemical file formats and we utilize the open source OpenBabel package for this purpose [28]. Although the software code was functional, we have identified a number of general issues with the code including inversion of stereo centers and loss of other chemical information. We believe that OpenBabel is a significant contribution to the cheminformatics community that will continue to improve in quality.

We generate 3D conformers on the fly from the 2D layouts in the database. We chose the freely available Balloon optimizer [29], primarily because it is free and fast; and it is a command line tool and was relatively easy to integrate. Balloon is not, however, open source, and cannot be modified. We use Jmol [30] to visualize the resulting optimized 3D molecular structures as well as crystallographic files (CIFs) where available. We do not use the Java-based JMol to visualize regular 2D images as it can add a significant load to browsers.

Literature linking from ChemSpider to open internet services has been established in an automated fashion taking advantage of freely available application programming interfaces to such web sites as PubMed [31], Google Scholar [32] and Google Patents [33]. Validated chemical names are used as the basis of a search against the PubMed database searching *only* against the title and the abstract. In this way, a search on cholesterol, for example, would only retrieve those articles with cholesterol in the title and abstract rather than the many tens of thousands of articles likely to mention cholesterol in the body of the article. A similar approach has been taken to integrate to Google Scholar and Google Patents. It should

Published by Woodhead Publishing Limited, 2012

be noted that the Application Programming Interfaces (APIs) are free to access but are not open source. PubMed (through Entrez) [34] has both SOAP and RESTful APIs. The Entrez API is both extensive and robust, providing access to most of the NCBI/NLM [35] electronic databases. Google now provides RESTful APIs and has deprecated the SOAP services that it once supported. This probably reflects the trend to support only lightweight protocols for modern web applications. All of these APIs are called in a similar way: a list of approved synonyms associated with a particular ChemSpider record is listed, sorted by 'relevance' (which is calculated based on the length of the synonym as well as its clarity), then used to call against the API. The result (whether SOAP, XML or HTML) is then processed by an 'adapter', transformed into an intermediate XML representation and passed through XSLT to produce the final HTML shown in the ChemSpider records.

The value of analytical data is as reference data for comparing against other lab-generated data. Acquisition of a spectrum and comparison against a validated reference spectrum speeds up the process of sample verification without the arduous process of full data analysis. As a result of this general utility, ChemSpider has provided the ability to upload spectral data of various forms against a chemical record such that an individual chemical can have an aggregated set of analytical data to assist in structure verification. As a result of contributions from scientists supporting the vision of ChemSpider as a valuable centralizing community-based resource for chemical data for chemists, over 2000 spectra have been added to ChemSpider in the past 2 years with additional data being added regularly. These data include infrared, Raman, mass spectrometric and NMR spectra, with the majority being 1H and 13C spectra. Spectral data can be submitted in JCAMP format [36] and displayed in an open source interactive applet, JSpecView [37], allowing zooming and expansion. JSpecView is open source but the code seems to lack a clearly defined architecture and boundaries. JSpecView has been modified to visualize range selection (inverting a region's color while dragging a mouse cursor). One of the main problems faced with supporting JSpecView is that it understands only one of the many flavors of JCAMP produced by spectroscopy vendors. This is not the fault of JSpecView but rather the poor adherence to the official JCAMP standard by the spectroscopy vendors. An alternative spectral display interface is the ChemDoodle spectral web component [38], which is a 'Spectrum Canvas' and renders a JCAMP spectrum in a web page along with controls to interact with it – for example to zoom in on a particular area of interest. However, it relies on HTML5, which limits its usage to modern standards compliant browsers that support HTML5 (for example Google Chrome and Firefox) and, as described earlier, limits its use in most versions of Internet Explorer unless the Google Chrome Frame plug-in is installed. This form of spectral display has not yet been implemented in the ChemSpider web interface but has been installed to support the SpectralGame [39] [40] on mobile devices.

Although ChemSpider is not an open source project per se, depending for its delivery on a Microsoft ASP.NET platform and SQL Server database, it should be clear that the project does take advantage of many open source components to deliver much of the functionality including file format conversion and visualization. In particular the InChI identifier, a fully open source project, has been a pivotal technology in the foundation of ChemSpider and has become essential in linking the platform out to other databases on the internet using InChIs.

3.4 ChemDraw Digester

The ChemDraw Digester is an informatics project bridging the previous two topics discussed – it is a tool that uses the structure manipulation programs contained in ChemSpider's code to help enhance RSC articles. In the first section we saw that if the most important chemical compounds in a paper can be identified and deposited to ChemSpider then the article can be enhanced with links to provide readers with more compound information. These compounds were generated by using name to structure algorithms after extraction of the chemical identifier. However, more often than defining a compound by name, chemistry authors refer to and define compounds in their paper by figures in the manuscript where the molecular structures are being discussed (see Figure 3.3) [41].

Where the images accompanying a manuscript have been generated using the structure drawing package ChemDraw [42], the RSC requests that authors supply these images not only as image files, but also in their original ChemDraw format (with the file extension '.cdx'), as these files preserve the chemical information of the structures within them. As the ChemDraw file format can also incorporate graphical objects and text, the files often contain labels (reference numbers or text) that correspond to the references of the compound in the corresponding manuscript, as in the example figure. Therefore, by 'digesting' a ChemDraw file we could potentially decorate these occurrences of the compounds' identifying

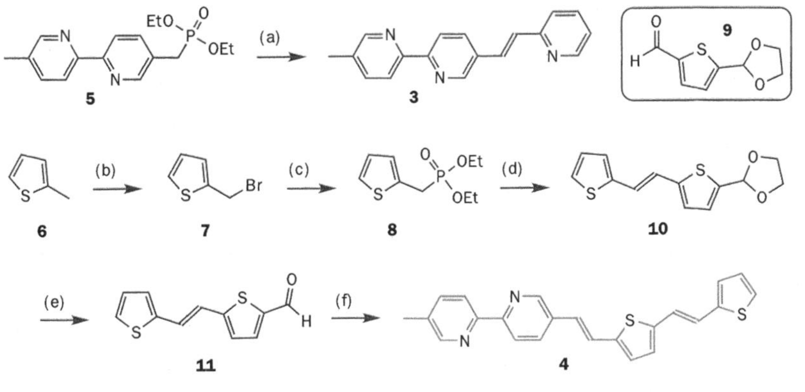

Figure 3.3 Example of figure in article (Reproduced by permission of The Royal Society of Chemistry) defining compounds

labels in the manuscript with a link to its chemical structure in ChemSpider – this is the basic aim of the ChemDraw Digester.

The most crucial part of this digestion process is to find each compound in the original ChemDraw file, match it up with its corresponding label, and then convert its 2D molecular structure into the MDL MOLfile format [43] (with extension.mol). The conversion from ChemDraw to mol format is required so that the files can be concatenated to make a MDL SDF file (with extension .sdf) suitable for deposition to ChemSpider. This SDF file [44] is also supplemented with article publication details in its associated data fields, which are used during deposition to create links from the new and existing compound pages in ChemSpider back to the source RSC article. Once deposited to ChemSpider, the related IDs of each compound can be retrieved and used to markup their names and references in the source article with reverse links to the ChemSpider compounds. The ChemDraw format is, unfortunately, not an open standard and it is not straightforward to digest in order to extract and convert the chemical structures and their associated labels. It is a binary file format, and although there is good documentation [45], deciphering it is a painstaking process and this would require considerable effort.

Fortunately, as discussed previously there is an existing routine to convert ChemDraw files to SDF using the 'convert' function of OpenBabel [46]. The ChemDraw digester was written using a Visual Studio, and .NET framework as a C# service with an ASPX/C# web front-end so that ultimately it can be reintegrated with the main ChemSpider web site. As a result, it could reference the native C++ OpenBabel library in the same way as the main ChemSpider code – via a wrapper managed C++ assembly (OBNET), which only exposed functionality required for the Digester and ChemSpider. The real advantage of OpenBabel being open source is that source code can be adjusted and the assembly recompiled, allowing adjustments required to deal with the real ChemDraw files from authors. These adjustments primarily involved adding new functionality, such as a new 'splitter' function to split ChemDraw files that contain multiple 'fragment' objects (molecules) into separate ChemDraw files so that each could be processed separately. Another issue is that the ChemDraw format supports more features than MOL, so some information is lost in this conversion. As a result, the ChemDraw reader had to be adjusted to read in this information and store it in the associated data fields in the SDF file generated – for example special bond types are represented by the PubChem notation. The other more important example of data lost from the original ChemDraw file is that of text labels associated with molecules. The difficulty in this case is to define how to match up a structure with its label. As a first step, OpenBabel was adjusted to recognize text labels that had been specifically grouped with a particular structure by the author. However, it became clear that in practice authors rarely used this grouping feature for this purpose, so that the vast majority of labels in the figures would be lost.

The ChemDraw Digester incorporates a review step where the digested information can be reviewed in an editable web page as shown in Figure 3.4. If a label is wrong or absent it can be amended but this is a time-consuming process, and the ultimate aim for the ChemDraw Digester is that it could be run as a fully automated process that does not require human intervention.

As an alternative to manual correction, the OpenBabel source code was modified to return labels for structures based on proximity, as well as grouping. A function was added which was called when a 'fragment' object (molecule or atom) was found. The function calculates the distance between the fragment object and all of the 'text' objects in the file (based on their 2D coordinates), so that the closest label to it could be identified. If the distance between the fragment and its closest text was less than the distance between that same text and any other fragment, then the value of the 'text' property of the text object (the text in the label) was associated with the structure and returned in the SDF file produced. Certain checks were also built in to ignore labels that do not contain any alphanumeric characters (e.g. '+').

The ChemDraw Digester is presently in its final stages of development and testing, all of the processed structures in the SDF file will be reviewed

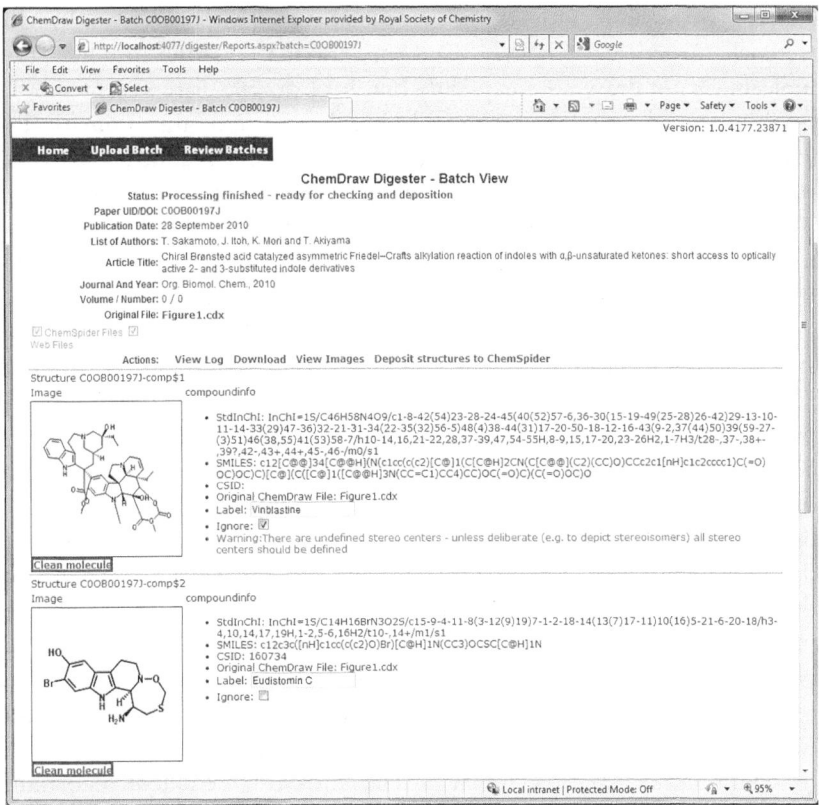

Figure 3.4 A review page of digested information

for mistakes and compared with those in the figures of the article so that all discrepancies are identified. We are already aware of some areas that will need attention. Some can be dealt with by post-processing the structures in the SDF file after digestion – for example, it is common for authors to draw boxes in the ChemDraw files for aesthetic reasons, and to draw these molecules by simply drawing four straight line bonds. As a result, we have added a filter which by default ignores these rogue cyclobutane molecules, and similarly ethane molecules, which are commonly used to draw straight lines. When a structure is flagged to be ignored for any reason it will not be deposited into ChemSpider or marked up in the original article, but at the reviewing stage any automatically assigned 'ignore' flags can be overridden (see the Ignore checkbox in Figure 3.4). Molecules are also flagged to be ignored based on some basic checks of their chemistry such as having a non-zero overall charge, possessing atoms with unusual atom valences, undefined stereochemistry, etc. Another issue encountered was one in which the molecule is not drawn out explicitly but, for example, is represented by a single 'node' which is labeled, e.g. 'FMoc'. These groups are automatically expanded but the placement of atoms in these expanded groups is sometimes peculiar and leads to ugly 2D depictions of the molecules. This can be addressed by allowing the ability to apply a cleaning algorithm to the relevant MOL structure in the SDF file to tidy up and standardize the bond lengths and angles to prevent atom overlap and very long bonds.

These are examples that can be dealt with by post-processing the structures in the SDF file. However, as OpenBabel source code availability allows customization, then issues that cannot be fixed with post-processing can be dealt with during the initial ChemDraw to SDF conversion. One such issue is that authors may use artistic license to overlay another ChemDraw object onto a molecule – for example to only draw part of a larger structure. The objects can be lines to indicate dangling bonds (even more problems are caused when these are not drawn as graphical objects but instead as various variations of ethane molecules as in Figure 3.5(a)), graphical pictures (e.g. circles to indicate beads as in Figure 3.5(b)) or brackets (e.g. commonly used to indicate repeat units polymers as in Figure 3.5(c)). The objects are usually overlaid onto an unlabeled carbon to give the appearance of a bond from the drawn molecule to these objects. The current OpenBabel algorithm would interpret the ChemDraw by simply identifying a carbon atom, and treating any objects overlaid on it as separate entities rather than bonded in any way, and it would not be possible to detect any error in the final molecule that was output. Although it is difficult to envisage any way that we could fully interpret such molecules, we could modify the OpenBabel convert function to return a warning when a chemical structure overlaps any other ChemDraw object so that these can be ignored by default, rather than processed incorrectly. Another very common problem which is difficult to find a solution for, is dealing with Markush structures – see Figure 3.5(d). Authors commonly save valuable space in the figures of their articles by representing multiple, similar compounds by defining part of the structure with a place holder, for example the label 'R' and supplying a label (usually elsewhere in the ChemDraw file) defining the different groups that could be substituted for R. This would require quite an extensive alteration to OpenBabel to deal with it correctly, but it is at least conceivable.

The long-term aim of the ChemDraw Digester is for it to process all ChemDraw files supplied with RSC articles automatically. In fact, with

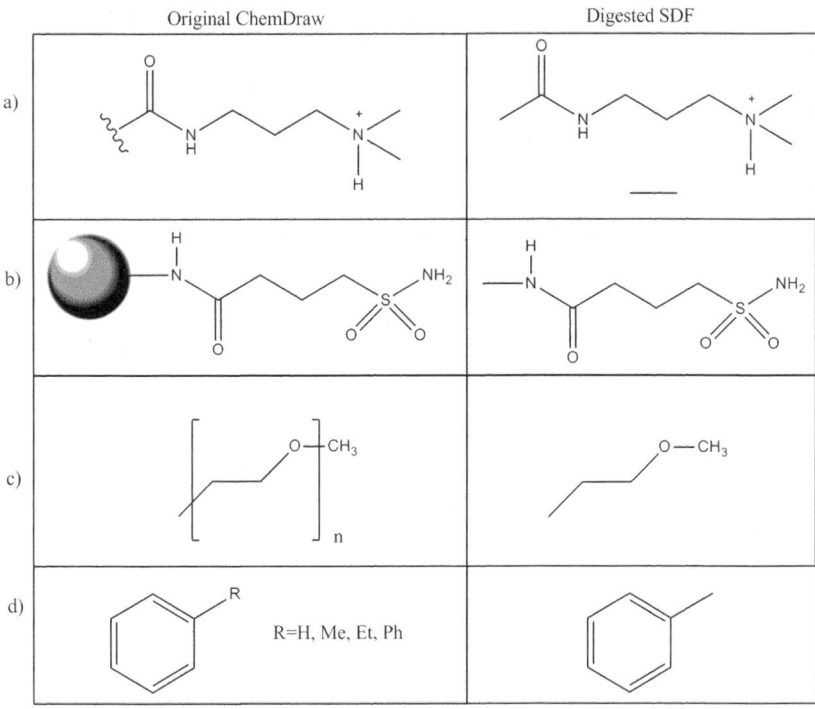

	Original ChemDraw	Digested SDF

Figure 3.5 Examples of ChemDraw molecules which are not converted correctly to MOL files by OpenBabel

some extra effort, it may be possible to extract embedded ChemDraw file objects from Microsoft Word files and digest them, and this may allow even more structures to be identified, even when authors do not send ChemDraw files. However, rather than simply using the various checks on molecules to filter out and omit problem structures automatically, it would be more useful to be able to feed these warnings back to the authors to give them the opportunity to revise their ChemDraw files and images, so that they could be used as intended. For this reason, another long-term aim of the ChemDraw Digester is for it to be made available as part of the ChemSpider web site as an author tool. If this were possible then authors could upload their own ChemDraw files when writing their articles, and view the Review web page (as in Figure 3.4) and see a clear indicator of whether the structures drawn in their figures do indeed adequately define the molecules to which they are referring, and if not what issues need to be fixed. Poorly drawn structures are more common than one would expect in academic papers so the ability to raise the standard of chemistry in RSC articles would benefit both authors and readers. The Digester is presently in a testing phase with the RSC editors and we expect to roll it out to general usage within the organization shortly.

3.5 Learn Chemistry Wiki

The RSC's objective is to advance the chemical sciences, not only at a research level but also to provide tools to train the next generation of chemists. The RSC's LearnChemistry platform [47] is currently being developed to provide a central access point and search facility to make it easier to access the various different chemistry resources that it provides. ChemSpider contains a lot of useful information for students learning chemistry but there is also a lot of information which is not relevant to their studies, which might be confusing and distracting. As a result, the RSC is developing a teaching resource, which will belong to LearnChemistry, for students in their last years of school, and first years of university (ages 16–19), which restricts the compounds and the properties, spectra and links displayed for each, to those relevant to their studies. However, students do not just need compound information in isolation – it is most useful when linked to and from study handouts and laboratory exercises. In addition, this resource is not just intended to be read and browsed, but interactive – allowing students to answer a variety of quiz questions, and allowing chemical educators to contribute to the content – so this resource is to be called the Learn Chemistry Wiki (see Figure 3.6 [48]). The platform on which this community web site is built is required to support the ability for multiple users to contribute collaboratively and to be easily customizable both in terms of appearance and functionality.

As such, an obvious start point was MediaWiki open source software [49]. It is easy to administrate and customize in terms of initial setup, managing user logins, tracking who is making changes and reverting these when necessary. It is also easy for untrained users to add pages and edit existing ones as many people are familiar with the same editing interface as Wikipedia [50]. It is also easy to program enhancements to the basic functionality as not only is all of the PHP code itself open source, but it has also been designed to allow extensions (programs called when the wiki pages are loaded) to be built into it. There are many extensions already available [51] and as these are also open source, it is straightforward to learn how they work and

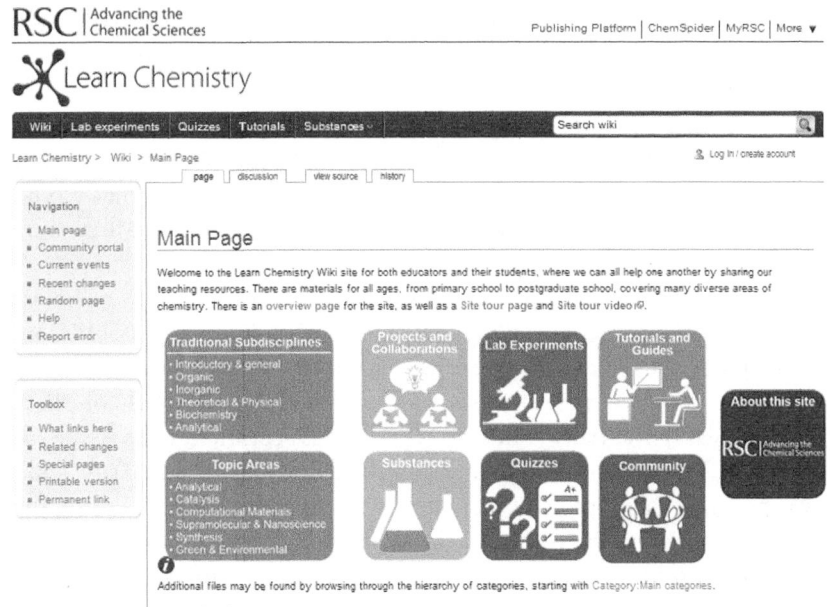

Figure 3.6 The Learn Chemistry Wiki

write new extensions. For all of these levels of MediaWiki use, there is plenty of documentation readily available by performing internet searches. An Industrial application of Mediawiki for knowledge management and collaboration is described in Chapter 13 by Gardner and Revell.

The basic setup of the wiki is straightforward – anyone can view the web site, but a login is required to edit or add pages. Anyone can register for a login and make changes, but this functionality is primarily aimed at teachers. Changes to the site will be monitored by administrators, and can be crowd-curated by other teachers. The content of the web site is separated into different namespaces. The 'Lab', 'TeacherExpt' and 'Expt' sections contain traditional HTML content – a mixture of formatted text, pictures and links which describe experiments (an overview, teachers' notes and students' notes, respectively).

Each page in the 'Substance' section contains compound information. Most of this information is dynamically retrieved from a corresponding ChemSpider compound page – images of its structure and a summary of its properties (molecular formula, mass, IUPAC name, appearance, melting and boiling points, solubility, etc.), and links to view safety sheets and spectra. Where there is a link to a Wikipedia page from the linked ChemSpider compound, the lead section of the Wikipedia article is shown in the page with a link to it. It is also possible to add extra information, for example references to the pages using the regular MediaWiki editing interface. There are currently approximately 2000 of these substance pages that correspond to simple compounds that would commonly be encountered during the last years of school and first years of university.

Creating these substance pages posed a variety of technical challenges. The decision to retrieve as much information as possible from ChemSpider and Wikipedia rather than to store it in the Learn Chemistry Wiki was taken in the interests of maintainability, as both of these sources will potentially undergo continuous curation. The issue of retrieving compound images from ChemSpider's web site and incorporating them into the wiki pages was easily addressed by using the 'EnableImageWhitelist' option [52] in the local settings configuration. However, to retrieve text from the ChemSpider server required installation of the 'web service' extension [53]. This is very easy to install in the same way as the majority of MediaWiki extensions – simply by copying the source PHP file into the extensions directory of the MediaWiki setup files, and referencing this file in the main local settings file of the MediaWiki installation.

In the LearnChemistry wiki, the main identifier for each compound is its name, which appears in the title of the compound page. However, in ChemSpider the main identifier for each compound with a particular structure is its ChemSpider ID and it is more future-proof to call information from its web services by querying this rather than its name, which is potentially subject to curation. This mapping between the wiki page name and ChemSpider ID is crucial, and was painstakingly curated before the LearnChemistry pages were created. This also needs to be easily maintainable in case a change is needed for whatever reason in the future. The 'Data' extension [54] is used to manage this mapping. Each substance page contains hidden text, which uses the extension to set a mapping of the compound name to the ChemSpider ID. The extension can then be used to retrieve this mapped information whenever the ChemSpider ID needs to be used (e.g. in a web service call), in the substance page but also any other in the wiki, rather than 'hardcoding' the ChemSpider ID into many places.

The substance pages themselves were created by writing and running a MediaWiki 'bot' – a PHP script that accesses the MediaWiki API to login to the wiki, read information from it, or edit pages in it. There is a lot of information on the internet describing the MediaWiki API [55], and examples of bot scripts to get started [56]. In two chapters in this book, Alquier (Chapter 16) and also Harland and co-workers (Chapter 17) describe the benefits and application of semantic MediaWiki. For each batch of substance pages to be created, an input file was made containing the basic inputs required to populate the page. The bot script firstly retrieves the login page of the wiki and supplies it with user credentials, then logs in and retrieves a token to be used when accessing other pages on the wiki. The script then recurses through the input file, constructs a URL for each new substance page in edit mode, posts the new content, and then saves the changes made. The Snoopy open source PHP class [57] played the crucial role of effectively simulating a web browser in this process – it was very well documented and straightforward to implement.

Discoverability is also important for these substance pages. An important objective was to make the substance pages searchable by structure (as ChemSpider is). An easy way to do this from outside the site is to use the 'Add HTML Meta And Title' extension [58], which was used to set the meta-data keywords and description on each substance page for search engines to use, and making sure that the InChI key was included in the meta-data. Structure searching within the wiki (not just from internet search engines) is also necessary, so that students or teachers can draw a molecule using a chemical drawing package embedded within a wiki page. When they click on a Search button in the page, the InChI key of the drawn structure is compared with that of all the substance pages in the wiki, and any matches are returned. This is rather a specialized requirement and required the development of a new extension. Developing a new extension was made easier by investigating the range of extensions that are already available for MediaWiki and reviewing the code behind them which is enabled by the fact that the MediaWiki hooks and handlers are all open source, well documented, and transparent. The functionality of this structure search was split so as to create two new MediaWiki extensions rather than one: the first embedded a structure drawer into a wiki page and the second added a Search button, which when clicked would display the search results. The reason for splitting the functionality into two separate extensions was that various other applications of the structure drawer had been suggested (which will be described shortly) and by this design the first extension could be used for various other applications without duplicating code.

Although a new MediaWiki extension needed to be developed to add a structure editor to a wiki page, it was not necessary to start from scratch as various open source structure editors already exist that can be embedded into web pages. The GGA Ketcher [26] structure editor available in ChemSpider was chosen as the structure editor of choice and implemented in the structure drawing extension because it is easy to use and is based on Javascript so does not require any extra additional add-ins or Flash support to be installed (which could be a problem in a school environment). It was also very easy to integrate into a MediaWiki extension. To incorporate a Ketcher drawing frame into a HTML page, it was simply necessary to download the Javascript and CSS files that comprise the Ketcher code, reference these in the head section of the HTML of a wiki, add the Ketcher frame, table and buttons to the body of the HTML, and add an onload attribute to the page to initialize the Ketcher frame. The only part of these steps which was not immediately straightforward for a version 1.16.0 MediaWiki extension to add to the web page in which it was called, was the step of adding an onload attribute to the page, but a workaround was used that involved adding a Javascript function to the HTML head, which was called at the window's onload event. The resulting extension was called the KetcherDrawer extension.

The accompanying extension would add a Search button and would need to perform several actions when clicked. The first action is to take the MOL depiction of the molecule that has been drawn (which is easily retrieved via a call to the Ketcher Javascript functions) and convert it into an InChI key so that this can be searched on. This conversion is done using the IUPAC InChI code [59], and any warnings that are returned are displayed in the wiki page, for example if stereochemistry is undefined or any atom has an unusual valence. The next action is to post this InChIKey to a search of the wiki – this was done by using the MediaWiki API to silently retrieve the results of this search. If one matching substance page was found then the page would redirect to view it. If no match for the full InChIKey was found, then a second search was submitted to the MediaWiki API to find any matches for just the first half of the InChIKey. This roughly equates to broadening the search to find matches for the molecule's skeleton. Any results from this search are listed in the wiki page itself, with a warning that no exact match could be found for the molecule but that these are similar molecules. After these two extensions had been written, it was then possible to add the functionality to perform a structure search within the wiki just by calling the KetcherDrawer and KetcherQuizAnswer extensions in the page.

A DisplaySpectrum extension was also written to add an interactive spectrum to a wiki page. As explained earlier, we use two possible display tools for spectra: JspecView and the ChemDoodle spectral display. Approximately two-thirds of current viewers of the RSC educational web sites use a web browser that does not support canvases, and in most school environments the installation of plug-ins is not an option. To make the best of both worlds, the DisplaySpectrum extension automatically tests the browser being used: if it supports canvases then it displays the spectrum using the ChemDoodle spectral viewer [38], and if not it uses the JspecView applet [37].

This chapter has demonstrated how a simplified version of both the information in, and functionality of ChemSpider has been integrated into the LearnChemistry educational web site, using the collaborative aspects of MediaWiki to allow these and other related pages, such as quizzes and descriptions of experiments to then be built up. The system was pieced together from many different open source programs and libraries, which would not have been possible without the flexibility of the MediaWiki platform on which the platform is based.

3.6 Conclusion

RSC has embraced the use of free/open source cheminformatics and Wiki tools in order to deliver multiple systems to the chemistry community that facilitate learning, data sharing and access to data and information of various types. By utilizing open source code where appropriate, and by integrating with other commercial platforms, we have been able to deliver a rich tapestry of functionality that could not otherwise have been achieved without significantly higher investment. In choosing our commercial vendor for our substructure search engine, we also opted for an open source platform with the GGA software.

Our experiences of using free/open source software are generally very positive. In a number of cases we have been able to take the software components as are and drop them into our applications to be used without any recoding and using the existing software interfaces as delivered. In most cases, our involvement with the code developers has either been negligible or has required significant dialog to resolve issues. In the cheminformatics domain of open source software, we have found commercial open source software to be of excellent quality and rigorously tested and well supported. For open source software of a more academic nature, we have found that small teams (where the software is supported by one group, for example) are highly responsive and effective in addressing identified issues, whereas applications with a broad development base are less so. In certain cases we have had to invest significant resources in optimizing the software for our purposes and knitting it into our applications. We generally find that documentation suffices for our needs, or that our development staff can understand the code even without complete documentation.

The true collaborative benefits of platforms such as ChemSpider will be felt as the multitude of online resources are integrated into federated searches and semantic web linking in a manner that single queries can be distributed across the myriad of resources to provide answers through a single interface. There is a clear trend in life sciences towards more open access to chemistry data. In the near future this may provide additional pre-competitive data allowing the development of federated systems such as the Open PHACTS [60], using ChemSpider as an integral part of the chemistry database and search engine. The Open PHACTS platform will allow pharmaceutical companies to link data across the abundance of life science databases that are already and will increasingly become available. ChemSpider is likely to become one of the foundations of the semantic web for chemistry and, with an ongoing focus for enabling collaboration and integration for life sciences, will be an essential resource for future generations.

3.7 Acknowledgments

ChemSpider is the result of the aggregate work of many contributors. All core ChemSpider development is led by Valery Tkachenko (Chief Technology Officer) and we are indebted to our colleagues involved in the development of much of the software discussed in this chapter. These include Sergey Shevelev, Jonathan Steele and Alexey Pshenichnov. Our RSC platforms are supported by a dedicated team of IT specialists that is second to none. The authors acknowledge the support of the open source community, the commercial software vendors (specifically Accelrys, ACD/Labs, GGA Software Inc., OpenEye Software Inc., Dotmatics Limited), many data providers, curators and users for their contributions to the development of the data content in terms of breadth and quality.

3.8 References

1. Project Prospect. [Accessed September 2011]; Available from: *http://www.rsc.org/Publishing/Journals/ProjectProspect/FAQ.asp*
2. Royal Society of Chemistry acquires ChemSpider. [Accessed September 22nd 2011]; Available from: *http://www.rsc.org/AboutUs/News/PressReleases/2009/ChemSpider.asp*

3. Adams, S.E., et al., Experimental data checker: better information for organic chemists. *Org Biomol Chem*, 2004. 2(21): 3067–70.

4. Nature Publishing Group. [Accessed September 2011]; Available from: *http://www.nature.com/npg_/company_info/index.html*

5. International Union of Crystallography. Available from: *http://www.iucr.org/*

6. Sciborg Project. [Accessed September 2011]; Available from: *http://www.cl.cam.ac.uk/research/nl/sciborg/www/*

7. OSCAR on Sourceforge. [Accessed September 2011]; Available from: *http://sourceforge.net/projects/oscar3-chem/*

8. Project Prospect wins ALPSP award. [Accessed September 2011]; Available from: *http://www.rsc.org/Publishing/Journals/News/ALPSP_2007_award.asp.*

9. Scaffolding. [Accessed September 2011]; Available from: *http://depth-first.com/articles/2006/12/21/scaffolding/*

10. Steinbeck, C., et al., The Chemistry Development Kit (CDK): an open-source Java library for Chemo- and Bioinformatics. *J Chem Inf Comput Sci*, 2003. 43(2): 493–500.

11. Lowe, D.M., et al., Chemical name to structure: OPSIN, an open source solution. *J Chem Inf Model*, 2011. 51(3): 739–53.

12. de Matos, P., et al., Chemical Entities of Biological Interest: an update. *Nucleic Acids Res*, 2010. 38(Database issue): p. D249–54.

13. The IUPAC International Chemical Identifier (InChI). [Accessed September 2011]; Available from: *http://www.iupac.org/inchi/*

14. IUPAC Gold Book. [Accessed September 2011]; Available from: *http://goldbook.iupac.org/*

15. The Gene Ontology. [Accessed September 2011]; Available from: *http://www.geneontology.org/*

16. Batchelor C.R., and Corbett, P.T., Semantic enrichment of journal articles using chemical named entity recognition. ACL '07 Proceedings of the 45th Annual Meeting of the ACL on Interactive Poster and Demonstration Sessions 2007: 45–48.

17. The RXNO Reaction Ontology. [Accessed September 2011]; Available from: *http://code.google.com/p/rxno/*

18. The Chemical Methods Ontology. [Accessed September 2011]; Available from: *http://code.google.com/p/rsc-cmo/*

19. Inczedy, J., Lengyel, T., and Ure, A.M., Compendium of Analytical Nomenclature (definitive rules 1997) – *The Orange Book*, 3rd edition 1998.

20. ACD/Labs Name to Structure batch. Available from: *http://www.acdlabs.com/products/draw_nom/nom/name/*

21. Hull, D., Pettifer, S.R., and Kell, D.B., Defrosting the digital library: bibliographic tools for the next generation web. *PLoS Comput Biol*, 2008. 4(10): e1000204.

22. ChemSpider. [Accessed September 2011]; Available from: *http://www.chemspider.com*

23. The PubChem Database. [Accessed September 2011]; Available from: *http://pubchem.ncbi.nlm.nih.gov/*

24. Gregori-Puigjane, E., Garriga-Sust, R. and Mestres, J. Indexing molecules with chemical graph identifiers. *J Comput Chem*, 2011. 32(12): 2638–46.

25. The GGA Software Bingo Toolkit. [Accessed September 2011]; Available from: *http://ggasoftware.com/opensource/bingo*

26. The GGA Ketcher Structure Drawer. [Accessed September 2011]; Available from: *http://ggasoftware.com/opensource/ketcher*

27. JChemPaint Sourceforge Page. [Accessed September 2011]; Available from: *http://sourceforge.net/apps/mediawiki/cdk/index.php?title=JChemPaint*

28. OpenBabel Wiki Page. [Accessed September 2011]; Available from: *http://openbabel.org/wiki/Main_Page*

29. The Balloon 3D Optimizer. [Accessed September 2011]; Available from: *http://users.abo.fi/mivainio/balloon/.*

30. Jmol: An Open Source Java viewer for chemical structures in 3D. [Accessed September 2011]; Available from: *http://jmol.sourceforge.net/*

31. PubMed. [Accessed September 2011]; Available from: *http://www.ncbi.nlm.nih.gov/pubmed/*

32. Google Scholar. [Accessed September 2011]; Available from: *http://scholar.google.com/*

33. Google Patents. [Accessed September 2011]; Available from: *http://www.google.com/patents*

34. Entrez, the Life Sciences Search Engine. [Accessed September 2011]; Available from: *http://www.ncbi.nlm.nih.gov/sites/gquery*

35. NCBI, the national Center for Biotechnology Information. [Accessed September 2011]; Available from: *http://www.ncbi.nlm.nih.gov/*

36. Published JCAMP-DX Protocols. [Accessed September 2011]; Available from: *http://www.jcamp-dx.org/protocols.html*

37. Lancashire, R.J., The JSpecView Project: an Open Source Java viewer and converter for JCAMP-DX, and XML spectral data files. *Chem Cent J*, 2007. 1: 31.

38. ChemDoodle web components [Accessed September 2011]; Available from: *http://web.chemdoodle.com/*

39. Bradley, J.C., et al., The Spectral Game: leveraging Open Data and crowdsourcing for education. *J Cheminform*, 2009. 1(1): 9.

40. The SpectralGame. [Accessed September 2011]; Available from: *http://www.spectralgame.com/*

41. Younes, A.H., et al., Electronic structural dependence of the photophysical properties of fluorescent heteroditopic ligands – implications in designing molecular fluorescent indicators. *Org Biomol Chem*, 2010. 8(23): 5431–41.

42. Cambridgesoft ChemDraw. [Accessed September 2011]; Available from: *http://www.cambridgesoft.com/software/chemdraw/*

43. The Molfile Format. [Accessed September 2011]; Available from: *http://goldbook.iupac.org/MT06966.html*

44. The SDF file format. [Accessed September 2011]; Available from: *http://www.epa.gov/ncct/dsstox/MoreonSDF.html#Details*

45. CDX File format specification. [Accessed September 2011]; Available from: *http://www.cambridgesoft.com/services/documentation/sdk/chemdraw/cdx/index.htm*

46. Guha, R., et al., The Blue Obelisk-interoperability in chemical informatics. *J Chem Inf Model*, 2006. 46(3): 991–8.

47. Learn Chemistry. [Accessed September 2011]; Available from: *http://www. rsc.org/learn-chemistry*

48. Learn Chemistry Wiki. [Accessed September 2011]; Available from: *http:// www.rsc.org/learn-chemistry/wiki*

49. MediaWiki. [Accessed September 2011]; Available from: *http://www. mediawiki.org/wiki/MediaWiki*

50. Wikipedia. [Accessed September 2011]; Available from: *http://www. wikipedia.org/*

51. Mediawiki Extensions. [Accessed September 2011]; Available from: *http:// www.mediawiki.org/wiki/Category:All_extensions*

52. MediaWiki EnableImageWhitelist extension. [Accessed September 2011]; Available from: *http://www.mediawiki.org/wiki/Manual:$wgEnableImage Whitelist*

53. MediaWiki Webservice extension. [Accessed September 2011]; Available from: *http://www.mediawiki.org/wiki/Extension:Webservice*

54. Mediawiki Data extension. [Accessed September 2011]; Available from: *http://www.mediawiki.org/wiki/Extension:Data*

55. MediaWiki API. [Accessed September 2011]; Available from: *http://www. mediawiki.org/wiki/API:Main_page*

56. MediaWiki Bot to make pages. [Accessed September 2011]; Available from: *http://meta.wikimedia.org/wiki/MediaWiki_Bulk_Page_Creator*

57. Snoopy PHP class. [Accessed September 2011]; Available from: *http:// sourceforge.net/projects/snoopy/*

58. MediaWiki Add HTML Meta and title extension. [Accessed September 2011]; Available from: *http://www.mediawiki.org/wiki/Extension:Add_ HTML_Meta_and_Title*

59. IUPAC InChI v1.03. [Accessed September 2011]; Available from: *http:// www.iupac.org/inchi/release103.html*

60. Open PHACTS. [Accessed September 2011]; Available from: *http://www. openphacts.org/*

## 4

Open source software for mass spectrometry and metabolomics

Mark Earll

Abstract: This chapter introduces open source tools for mass spectrometry (MS)-based metabolomics. The tools used include the R language for statistical computing, mzMine and KNIME. A typical MS-metabolomics data set is used for illustration and the tasks of visualisation, peak detection, peak identification, peak collation and basic multivariate data analysis are covered. A discussion of vendor formats and the role of open source software in metabolomics is also included. Example R scripts and KNIME R-nodes are shown, which should enable the reader to apply the principles to their own data.

Key words: data visualisation; KNIME; mass spectrometry; metabolomics; mzMine; R language.

4.1 Introduction

Mass spectrometry has advanced enormously since the invention of the first mass spectrometer built in 1919 by Francis Aston at Cambridge University [1, 2]. This first instrument is reported to have had a mass resolving power of 130. Less than a century later, many laboratories have at their disposal instruments capable of resolving powers in excess of 60 000 and a few specialist instruments may reach as high as 200 000. Technologies such as Time of Flight, Fourier Transform Ion Cyclotron resonance (FTICR) mass spectrometry and the new generation of ion traps (Orbitrap, Qtrap) have provided the researcher with unprecedented resolving power and potential.

In some ways it could be argued that the technology has moved faster than our ability to deal with the data. Despite such high-resolution instruments being available for several years, even instrument vendors' software has struggled to keep pace with the potential abilities these instruments provide. In addition, the modern spectrometer places a huge strain on the computing hardware both in terms of storage and processing power. Our ability to process and make sense of such large data sets relies on cutting edge data visualisation and multivariate statistical methods.

Possibly for these reasons there have been a large number of open source projects started by researchers in mass spectrometry, chemometrics, metabolomics and proteomics. It is a rapidly developing area with many new innovations still to be made and one where collaboration is essential. We are moving into an age of research where no one person can possibly keep up with all the skills required. Instead communication and networking skills are becoming as important as scientific knowledge. Fortunately, the open source software community is an excellent forum for such collaborations.

In this chapter a few of these open source tools will be demonstrated.

4.2 A short mass spectrometry primer

As some readers of this book may be unfamiliar with mass spectrometry, here is a short explanation. (Mass spectrometry experts may wish to skip this section.)

A mass spectrometer can be described simply as a device that separates ions based on their mass to charge ratio. It is frequently used as a detector as part of a chromatography system such as high pressure liquid chromatography (HPLC), gas chromatography or used alone with samples being introduced from surfaces using laser desorption (MALDI) or directly from air (DART). Chemical samples are therefore introduced to the mass spectrometer as solutions, gases or vapours. The introduced substances are then ionised so that they may be deflected by electrostatic or magnetic fields within the spectrometer. Ions with different masses are deflected to different extents, so, by varying the deflection strength, a range of mass/charge ratios may be determined. The resulting plot of the molecular weight versus ion abundance is known as a mass spectrum.

From the pattern of masses detected, or from the exact mass, important clues to the structure of the original molecule may be determined. Molecules may travel through the instrument relatively intact or may fragment into smaller parts. A skilled mass spectrometrist is often able to infer the chemical structure of the original molecule by studying the fragments produced.

Some methods of mass spectrometry produce highly fragmented spectra, which represent 'fingerprints' characteristic of a particular molecule so searching against a stored library is possible. Modern multistage instruments are also able to deliberately fragment individually selected molecules by collision with low-pressure gas. This is known as MS/MS or sometimes MS^n.

Rather surprisingly to an outsider, mass spectrometers often give poorly reproducible results between instruments, so spectra acquired from one instrument may not exactly match those from another. It is common practice to run large numbers of known molecules on a particular instrument to build a library specific to that instrument. The reason for the poor reproducibility is that ionisation of compounds is dependent on many factors and subtle changes between instruments can result in drastic differences in fragmentation and instrument response. Indeed ionisation of different molecules is unpredictable so that there is no direct relationship between the composition of a mixture and the response of the spectrometer. For quantitative work careful calibration of the mass spectrometer is required to determine the response of each chemical component.

With the newest high-resolution spectrometers, the precise mass of the compound can yield molecular formula information [3]. This is commonly termed 'accurate mass measurement'. How can the mass of a compound lead to its molecular formula? To understand this, first a little theory is required.

4.2.1 Spectrometer resolution and accuracy

The resolution of a spectrometer is defined as the mass number of the observed mass divided by the difference between two masses that can be separated. For instance a resolution of 100 would allow a m/z of 100 to be distinguished from a m/z of 101.

The mass accuracy is defined in ppm (parts per million) units, and expresses the difference between an observed m/z and theoretical m/z. Today sub-ppm accuracy is not uncommon.

mass accuracy = [m/z (observed − exact)/exact × 1 000 000].

4.2.2 Definitions of molecular mass

In mass spectrometry, it is important to realise that the spectrometer will be measuring individual isotopes, so that, for instance, elements with two or more commonly occurring isotopes will be seen as multiple peaks (e.g. chlorine ^{35}Cl, ^{37}Cl).

The average mass of a molecule is the sum of the average atomic masses of the constituent elements. The average atomic mass is determined by the naturally occurring abundances of isotopes. For instance, carbon has an average mass of 12.0107(8) but is made up of 98.93% ^{12}C (mass = 12 exactly) and 1.07% ^{13}C (mass = 13.0033548378). The average mass is used in general chemistry but not mass spectrometry.

The monoisotopic mass is the sum of the most abundant isotopes in each molecule. For most typical organic molecules, this means the lightest of the naturally occurring isotopes. The monoisotopic mass will therefore represent the biggest peaks detected in the mass spectrometer. (However, for some heavier atoms this does not hold, for example iron (Fe) the lightest isotope is not the most abundant.)

The fact that the mass of an isotope is not exactly the sum of its neutrons and protons is due to an effect called the 'mass defect' and is due to the binding energy that holds the nuclei together. Some of the atomic mass is converted to energy according to the principle of relativity $(E = mc^2)$. Each isotope has a characteristic mass defect and this can be used to calculate an exact mass (for instance, see Table 4.1).

Thus, by measuring the mass accurately on the spectrometer and matching that to a theoretical calculated mass, a molecular formula may be determined. Of course, many molecular structures are possible from a single formula and depending on the precision of the measurement there may be more than one molecular formula that fits the accurate mass. As molecular weight increases, the number of formulae that will fit a measured mass will increase for any given mass accuracy.

Table 4.1

Isotope	Mass	Example
^{12}C	12.00000	C6H5NO2
^{1}H	1.007825	
^{14}N	14.003074	Monoisotopic mass =123.032029
^{16}O	15.994915	

The nominal mass is the integer part of the molecular mass and was commonly used with older low-resolution spectrometers. Mass spectrometry is a complex subject. Reference [4] gives a good overview of the field.

4.3 Metabolomics and metabonomics

In this chapter the focus will be on the application of mass spectrometry to metabolomics and metabonomics. Both terms refer to the study of naturally occurring metabolites in living systems. 'Metabo*l*omics' originally referred to the study of metabolites in cellular systems, whereas 'metabo*n*omics' [5] referred to the metabolic response of entire living systems. Subsequently, confusion has arisen by the inconsistent adoption of either term [6]. These terms are often used interchangeably, so the term 'metabolomics' will be used in the text that follows.

In animal and human studies the study of metabolites is often by analysis of fluids such as urine, saliva, blood plasma or cerebral-spinal fluid. In this way the sampling may be non-invasive (urine/saliva) or at least non-lethal, and therefore allows the repeated collection of samples from the same individual. In plant metabolomics, a portion of the plant (or whole plant) is used. As removal of plant tissue is a disruptive event, it is rare that any repeated sampling will come from the same part of the plant tissue or even the same plant. Instead groups of plants are grown under carefully controlled conditions, with the assumption that the variation between individuals will be smaller than any treatment effects.

Extracts of the biological material are then presented to the analysis system where a 'snapshot' of the metabolites present in the sample is determined. This kind of global metabolite profiling is often called 'untargeted analysis' where the hope is that many compounds may be identified either by comparison with known, previously run compounds or that tentative identification by the mass spectrum or accurate mass may be made. This is in comparison to 'targeted analysis' where standard dilutions of known compounds are run in advance to determine the quantitative response of the instrument and to identify metabolites.

The consequence of untargeted analysis is that the quantitative information is relative between treatment groups. Within any particular sample the amounts of analytes detected may not represent their true concentration but a comparison of a control and treated sample for a particular peak will give a relative measure of that metabolite's concentration between samples. Moreover, the data revealed by metabolomics represents a 'snapshot' at one point in time and gives no information about the dynamic flux of metabolites. For example, a metabolite showing a small concentration may in fact be transported in large quantities or one that shows high quantity may be produced very slowly but accumulates over time.

Identification of metabolites from the mass spectrum is one of the biggest challenges in metabolomics. Accurate mass and fragmentation patterns may assist in the determination of structure but for absolute certainty, preparative isolation of the substance followed by other methods such as NMR may need to be employed. Therefore in many studies it must be realised that identifications are somewhat tentative. This is not necessarily a major problem in hypothesis-generating research, but could potentially limit the use of untargeted methods in critical areas such as clinical diagnosis. Fortunately, robust multivariate methods may be used to fingerprint the combination of many metabolite signals in order to produce classifications with high levels of accuracy (low false positives and negatives) so the precise identification of individual components may be unnecessary for some applications.

4.4 Data types

Occasionally mass spectrometry data may be in the form of a mixed spectrum of all components in the case of 'direct infusion' where a sample is infused into an instrument. This is a simple 2D mass spectrum of mass/charge ratio (m/z) and ion abundance (intensity).

More commonly data come from LC-MS, GC-MS or CE-MS where the complex mixtures are separated before being introduced to the mass spectrometer. The data which result are three-dimensional, having axes of mass/charge ratio (m/z), retention time and ion abundance (intensity).

4.4.1 Vendor formats versus open formats

The data from mass spectrometry, in common with many areas of analytical chemistry, are often saved in proprietary formats dictated by instrument vendors. There are several disadvantages with this situation, the difficulties of using a common software platform in labs with different manufacturers' equipment, the hindrance of free interchange of data in the scientific community and the lack of ability to read archived data in the future. Although it is unlikely that instrument vendors will accept a common file format any time soon, many have the ability to export data into common open formats. There are also a number of converters available [7, 8], but care must be taken to preserve the integrity of the data.

MzXML [9], mzData and mzML [10, 11] are open, XML (eXtensible Markup Language)-based formats originally designed for proteomics mass spectrometric data. mzData was developed by the HUPO Proteomics Standards Initiative (PSI), whereas mzXML was developed at the Seattle Proteome Center. In an attempt to unify these two rival formats, a new format called mzML has been developed [11]. A further open format is JCAMP-DX [12], an ASCII-based representation originally developed for infra-red spectroscopy. It is seldom used for MS due to the size of the data sets. Finally, ANDI-MS is also open, and based on netCDF, a data interchange format used in a wide variety of scientific areas, particularly geospatial modelling. It is specified under the ASTM E1947 [13] standard. These are complemented by the wide variety of vendor-led formats in routine use in the field, the most common of which are .raw (Thermo Xcalibur, Waters/ Micromass MassLynx or Perkin Elmer); .D (Agilent); .BAF, .YEP and .FID(Bruker); .WIFF (ABI/Sciex) and .PKL (Waters/Micromass MassLynx).

The remainder of this chapter will demonstrate the use of open source tools to typical mass spectrometry situations.

4.4.2 Analysing mass spectroscopy data using R

A number of tools for mass spectrometry have been written in the R language for statistical computing [14]. Versions are available for Linux, Mac or Windows, making it compatible with a broad range of computing environments. One of the most convenient ways to run R is to use RStudio (*http://rstudio.org/*), which serves as integrated development environment. This software allows the editing of scripts, running commands, viewing graphical output and accessing help in an integrated system. Another very useful feature of R is the availability of 'packages', which are pre-written functions that cover almost every area of mathematics and statistics.

4.4.3 Obtaining a formula from a given mass

Our first example uses a Bioconductor package for the R language written by Sebastian Böcker at the University of Jena [15, 16]. The package uses various chemical intelligence rules to infer possible formula from a given mass. With this library, users are able to generate possible formulas from an input mass value and output this data via the command *decomposeMass* as shown below.

```
#-----------------------
# Install Package
#-----------------------
install.packages('Rdisop')
library (Rdisop)

molecules <- decomposeMass(89.99531, ppm = 10)

#-------------------------------------------------------------------------
# The output shows the formula, exact mass, score probability, charge,
# validity (adherence to the Nitrogen rule), and a list of isotopes and
# abundances. (Note only the first 5 isotopes are listed for clarity)
#-------------------------------------------------------------------------
molecules

$formula
[1] 'C2H2O4'
$score
[1] 1
$exactmass
[1] 89.99529
$charge
[1] 0
$parity
[1] 'o'
$valid
[1] 'valid'
$DBE
[1] 2
$isotopes
$isotopes[[1]]
      [,1]      [,2]         [,3]         [,4]        [,5]
[1,] 89.99529  90.99873554  91.99959798  9.30E+01    9.40E+01
[2,] 0.968356  0.02350477   0.007928286  1.86E-04    2.46E-05
```

Subsequently, a formula can be selected and the isotopic distribution calculated using the getMolecule command. Plotting as a graph produces the output shown in Figure 4.1, representing the composition cystine with two sulphur atoms that have four isotopes ^{32}S (95.02%), ^{33}S (0.75%), ^{34}S (4.21%) and ^{36}S (0.02%).

```
cystine<-getMolecule('C6H12N2O4S2')
spectrum<-matrix(unlist(cystine$isotopes), nr=2)

plot(spectrum[1,], spectrum[2,], type='h', main='Isotope Pattern',
  ylab='Abundence', xlab='m/z')
```

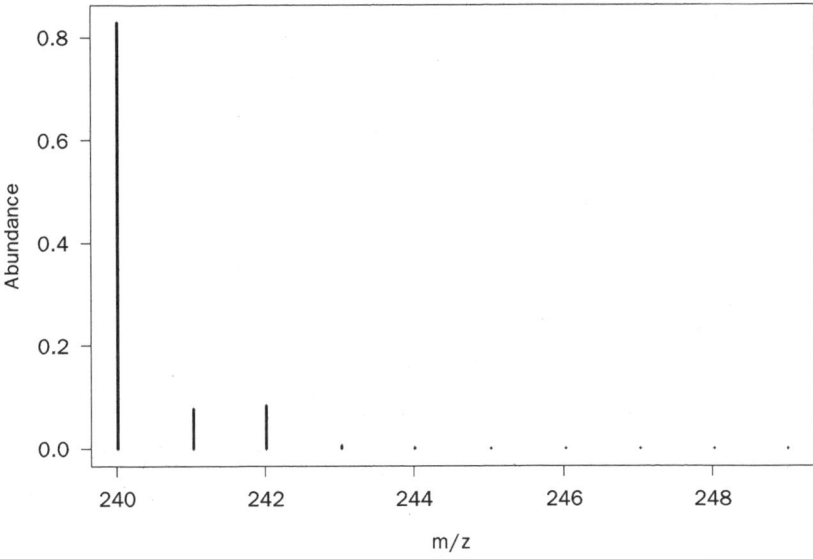

Isotope Pattern

Figure 4.1 Isotope pattern for cystine

4.4.4 Data visualisation

A vital part of data analysis is the visualisation of data, it is good practice to check the data visually for instrumental problems such as drift or baseline shifts. The XCMS package may be used for this purpose.

Examining a LC-MS spectrum in XCMS

The XCMS [16–20] package was written at the Scripps Center for Metabolomics and Mass Spectrometry and is now provided as part of the Bioconductor R package [15]. XCMS can be used to display the results of an LC-MS scan using a few straightforward steps. In the example, data from a single LCMS run are loaded into an '*xcmsRaw*' object designated x1. The content and structure of this R object are viewed in order to extract data for plotting.

```
#------------------------
# Install Package
#------------------------
source('http://bioconductor.org/biocLite.R')
biocLite('xcms')
library(xcms)


#-----------------------------------------------------
# set the working directory of R from the command
# line or alternatively from R Studio
#-----------------------------------------------------
setwd('C:/Your_Directory_Name')


x1=xcmsRaw(filename='QC1_917_01.mzData', includeMSn=F)
x1
An 'xcmsRaw' object with 1000 mass spectra

Time range: 0.4-839.7 seconds (0-14 minutes)
Mass range: 85.0023-899.9206 m/z
Intensity range: 127.735-49679200

MSn data on 0 mass(es)
with 0 MSn spectra
Profile method: bin
Profile step: 1 m/z (816 grid points from 85 to 900 m/z)

Memory usage: 28 MB


#-----------------------------------
#list the structure of the x1 object
#-----------------------------------
str(x1)


Formal class 'xcmsRaw' [package 'xcms'] with 20 slots
  ..@ env        : <environment: 0x0000000009376678>
  ..@ tic        : num [1:1000] 0 0 0 0 0 0 0 0 0 ...
  ..@ scantime   : num [1:1000] 0.411 1.554 2.835 4.144 5.464 ...
  ..@ scanindex  : int [1:1000] 0 1613 3309 4997 6774 8534 10273 11981
    13691 15411 ...
  ..@ polarity   : Factor w/3 levels 'negative','positive',..: 3 3 3 3
    3 3 3 3 3 ...
  ..@ acquisitionNum   : int [1:1000] 1 3 5 7 9 11 13 15 17 19 ...
  ..@ profmethod   : chr 'bin'
  ..@ profparam    : list()
  ..@ mzrange      : num [1:2] 85 900
  ..@ gradient     : logi[0, 0 ]
  ..@ msnScanindex     : int(0)
  ..@ msnAcquisitionNum    : int(0)
  ..@ msnPrecursorScan     : int(0)
  ..@ msnLevel     : int(0)
  ..@ msnRt    : num(0)
  ..@ msnPrecursorMz   : num(0)
```

```
..@ msnPrecursorIntensity    : num(0)
..@ msnPrecursorCharge    : num(0)
..@ msnCollisionEnergy    : num(0)
..@ filepath    : chr 'QC1_917_01.mzData'
```

We can then use various commands to determine the start and stop times, mzrange and generate the TIC (total ion chromatogram) plot, shown in Figure 4.2. It is also possible to output the raw data as a retention time versus m/z versus intensity table as a CSV file.

```
RTstart=min(x1@scantime)

RTstop=max(x1@scantime)

RTstart
[1] 0.41058

RTstop
[1] 839.6642

x1@mzrange
[1] 85 900

#------------------------------------------------------------------
# To plot the TIC (Total Ion Chromatogram) use the getEIC method
#------------------------------------------------------------------
x2 <- getEIC(x1, mzrange=cbind(85,900),
     rtrange = cbind(RTstart,RTstop),step = 0.1)
plot(x2)

#-------------------------------------------------------------------------
# To extract a one Dalton wide Ion Chromatogram just change the mzrange
#-------------------------------------------------------------------------
x2 <- getEIC(x1, mzrange=cbind(200.0,201.0), rtrange = cbind(1,800),
     step = 0.1)
plot(x2)

#------------------------------------------
# To output the data as RT vs m/z table
#------------------------------------------
object <- rawMat(x1, mzrange = cbind(100.0,300.0),
          rtrange = cbind(1,800),log=FALSE);

write.csv(object, file='MStest.csv')

"",'time','mz','intensity'
'1',1.55418,100.039428710938,10355.81640625
'2',1.55418,100.047500610352,1614.08435058594
'3',1.55418,100.050613403320,1042.66650390625
'4',1.55418,100.058731079102,487.58349609375
```

```
'5',1.55418,100.075805664062,15026.720703125
'6',1.55418,100.083862304688,591.471557617188
'7',1.55418,100.112190246582,3342.90063476562
'8',1.55418,100.932266235352,585.759033203125
'9',1.55418,100.943664550781,240.230697631836 . . .

#---------------------------------------
# The TIC may be exported to file
#---------------------------------------
>TIC <- rawEIC(x1, mzrange=cbind(85,900),
    rtrange = cbind(RTstart,RTstop))
>write.csv(TIC, file='MS_TIC.csv')

",'scan','intensity'
'1',1,5438430.45281982
'2',2,4019233.98780823
'3',3,3866972.27947998
'4',4,3805243.32279968
'5',5,3760837.48051453
'6',6,3709146.05143738
'7',7,3794325.40415955
'8',8,3768514.95440674
'9',9,3697039.30259705 . . .
```

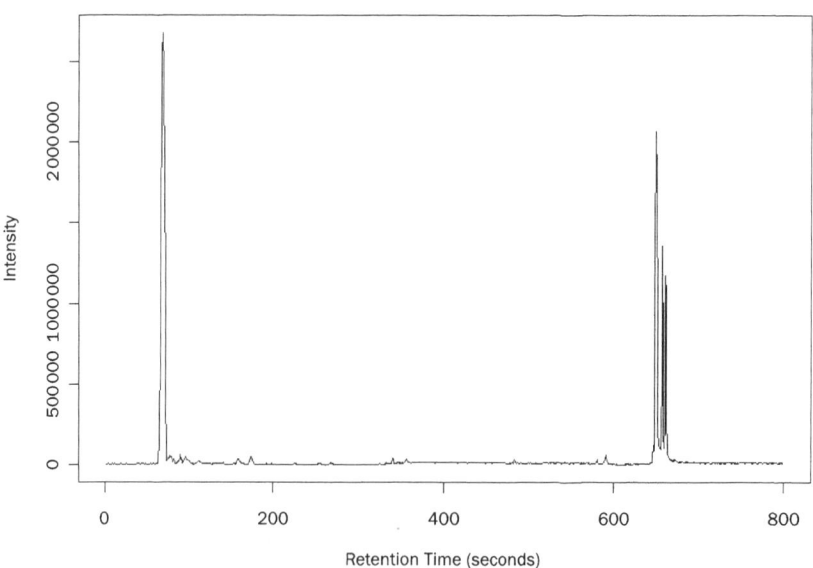

Extracted Ion Chromatogram: 200-201 m/z

| **Figure 4.2** | Ion chromatogram produced in R (xcms) |

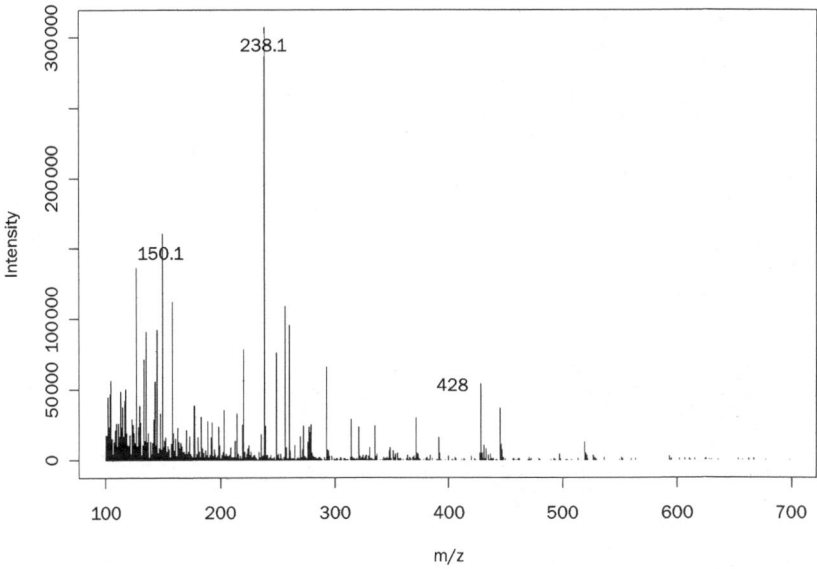

Mass Spectrum: 177.6 seconds (scan 200)

| **Figure 4.3** | A mass spectrum produced from R (xcms) |

The mass spectrum may also be plotted using the *plotScan* function:

```
plotScan(object, scan, mzrange = numeric(), ident = FALSE)
```

where scan is the scan number, ident allows annotation of the peaks interactively with the mouse, see Figure 4.3 as an example.

```
plotScan(x1,200,c(100,700),ident=TRUE) # plot spectrum
```

For a global overview of the LC-MS scan, a rotatable 3D image can be generated via the '*plotSurf*' command within the RGL package [22], as shown in Figure 4.4. Although only a few elementary features of XCMS have been shown here, XCMS is a comprehensive metabolomics processing package [19–21] and there are many good tutorials [23].

Visualising an LC-MS run in R is useful but lacks a certain degree of interactivity. Another open source MS package is mzMine [24–26] developed at Okinawa Institute of Science and Technology, Japan and VTT Finland. It is a Java-based program and is therefore platform-independent. mzMine works with a rich set of file types including Net CDF, mzData, mzML, mzXML, Xcalibur Raw files and Agilent CSV

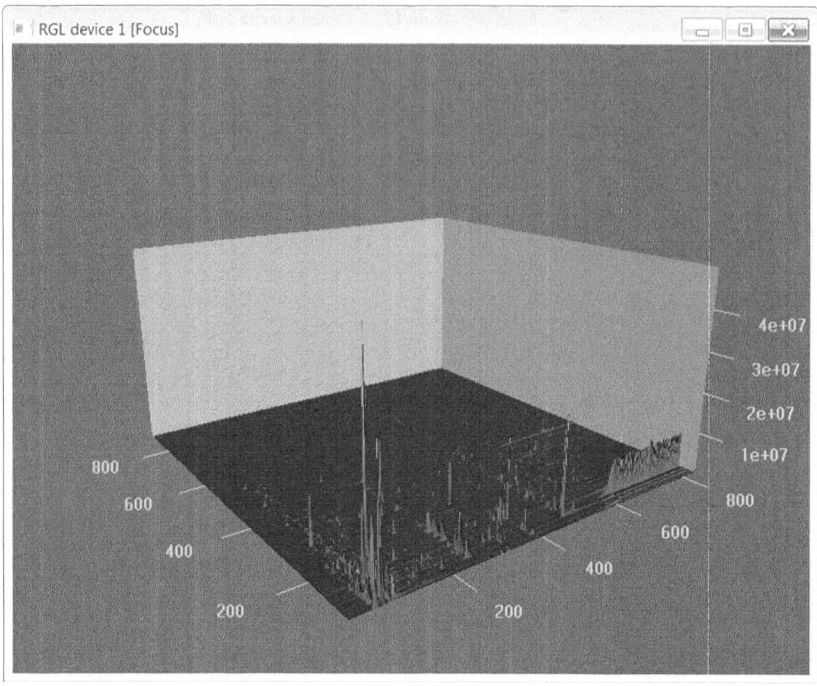

Figure 4.4 3D Image of a LC-MS scan using the plot surf command from the RGL R-package

files. (For the Thermo Xcalibur files it is necessary either to have the Thermo Xcalibur software installed on the same machine or to have downloaded and installed the free ThermoMSFilereader software [27].)

One of the great advantages of mzMine is its interactivity. Importing and processing the data is achieved using standard graphical file dialogs. For instance, the TIC visualiser produces a high-quality spectrum plot shown in Figure 4.5. The plot is fully zoomable and interactive, and double clicking a peak leads to its mass spectrum and associated data.

Peak detection is enabled using the chromatogram builder that has several different peak detection methods. Figure 4.5 demonstrates the use of the peak-list Wavelet method where the threshold for peak detection in the mass dimension and other options can be interactively adjusted. The ultimate aim is to create a set of parameters able to distinguish between true peaks and irrelevant/noisy features. Other options for peak detection in the mass dimension are Centroid (for previously centroided data), and Exact Mass Detector (which defines the peak centre at half maximum height), Local Maximum (simple local maxima) and Recursive Threshold (for noisy data).

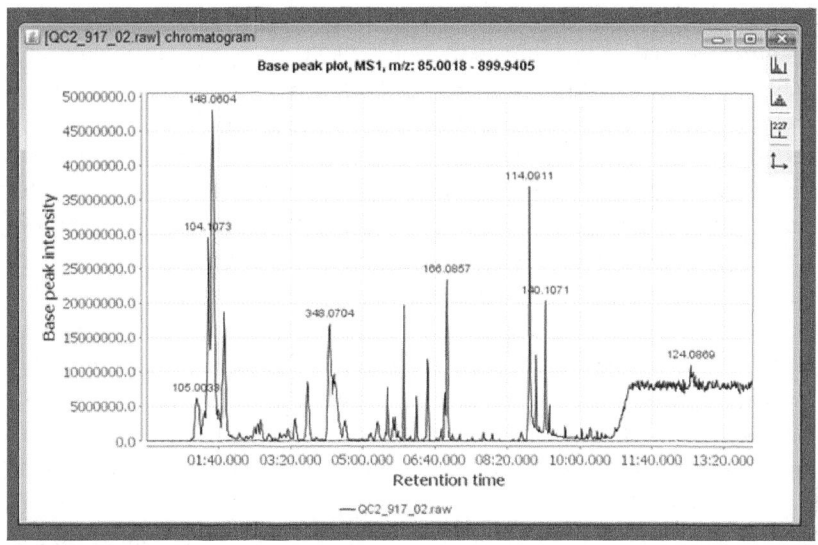

Figure 4.5 A total ion chromatogram (TIC) plot from mzMine

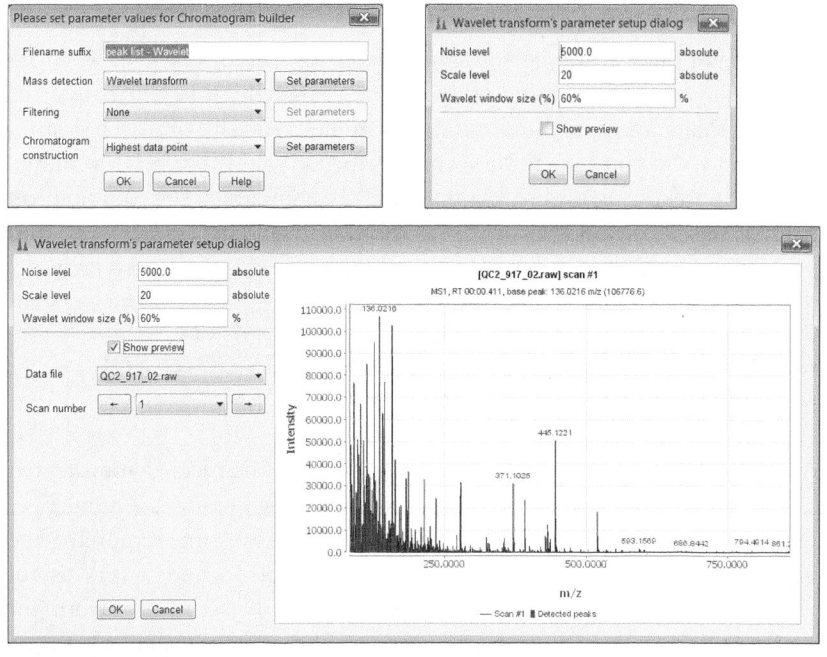

Figure 4.6 Configuring peak detection

The peak list comprises a series of ion chromatograms taken at each mass channel detected in the chromatogram builder. Some ion chromatograms may contain more than one peak so a second peak deconvolution stage is required. Chromatogram deconvolution has several different methods to choose from, here local minimum search was used. This method searches for local minima in partially overlapping peaks and works well for chromatograms with well-defined peaks and low noise. Alternative methods are simple Baseline Cut-Off (threshold), Noise Amplitude (detects baseline noise amplitude) and Savitsky-Golay (standard peak detection method using second derivatives). After this step, mzMine produces a resolved peak list with one peak per row as shown in Figure 4.7. These data can be further explored by visualisation as a 3D plot (Figure 4.8) or 2D-Gel view. The 3D view is particularly useful as the detected peaks from a peak list are shown on the plot, which enables a visual check that peaks have been found correctly. Alternatively, several peak lists may be obtained by varying the detection parameters and visualised in the 3D view, the aim being to recognise the main peaks without excessive detection of baseline noise.

4.5 Metabolomics data processing

In metabolomics a number of samples are measured, resulting in a 3D LC-MS or GC-MS scan for each sample. Typically, this will consist of control, treated and repeated standard samples. The aim of metabolomics processing is to combine these scans together so that the relative amounts of metabolites occurring in all samples may be compared. The combination of data has to be done in a consistent way over all the data sets, and problems have to be accounted for such as small retention time drifts.

4.5.1 Processing a metabolomics data set in mzMine

Aside from the peak detection and data display features demonstrated above, mzMine is primarily a tool for metabolomics and has a number of useful features to support the extensive data processing required. The batch mode tool allows a chain of processes to be created; a very useful feature with large data sets as the processing can be set to run overnight in unattended operation. Furthermore, once the parameters for a particular operation have been configured, mzMine remembers the last used settings such that they may be applied across all samples in the study.

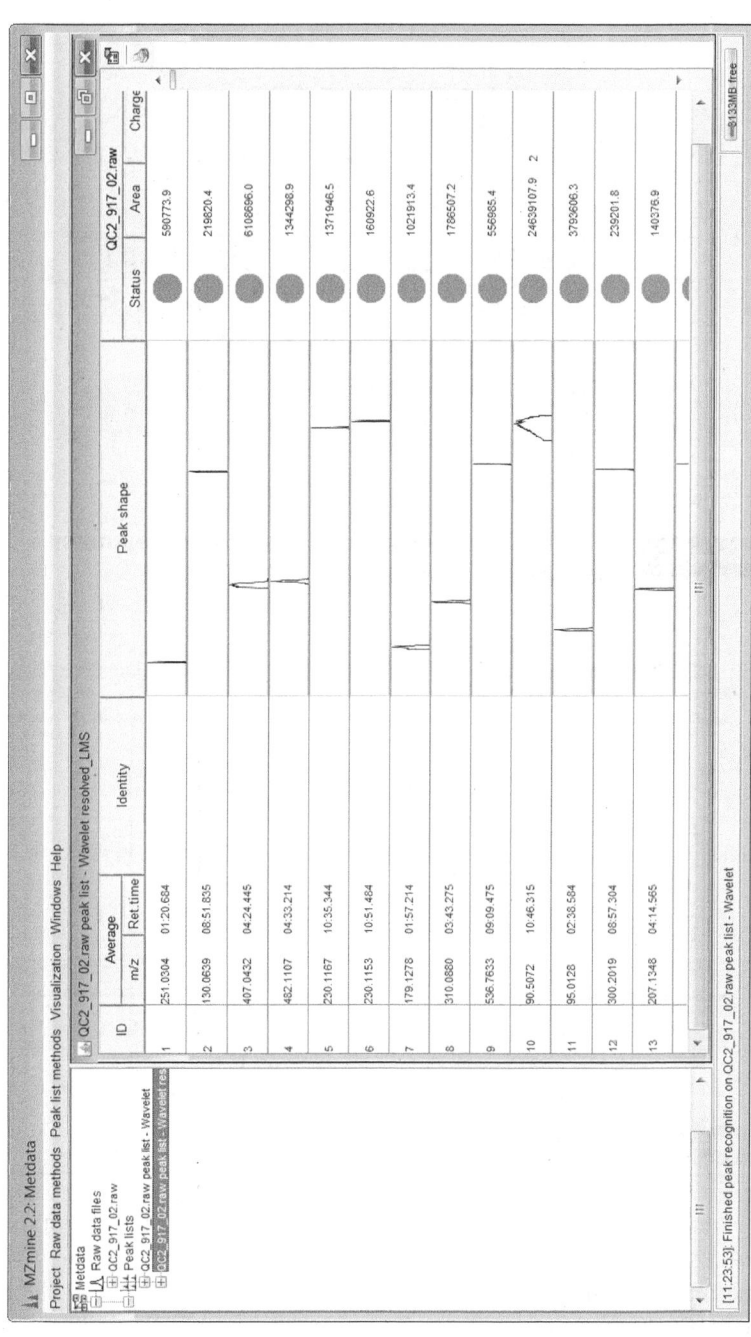

Figure 4.7 Deconvoluted peak list

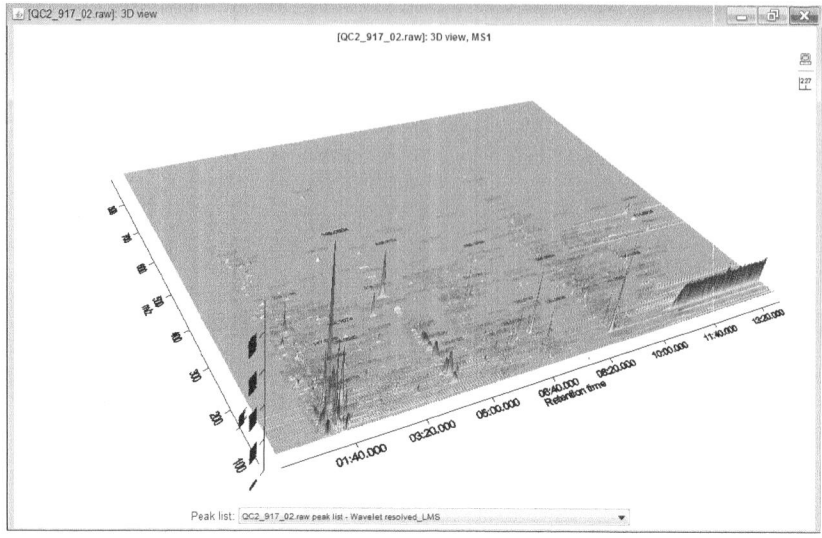

Figure 4.8 3D view of an LC-MS scan. The plot may be rotated and zoomed interactively

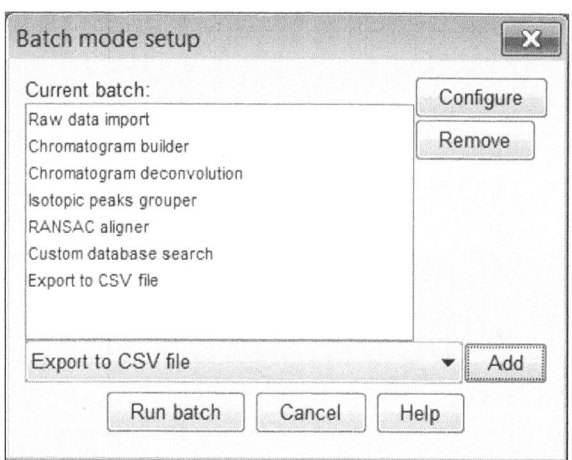

Figure 4.9 Example of a Batch mode workflow

An example of a complete metabolomics workflow is shown in Figure 4.9. Each item in the list is associated with an operation from the mzMine menu, each with its own parameter settings dialog box. First (Figure 4.10(a)), the Isotopic Peaks Grouper is configured to find all ^{13}C peaks and groups the area with the main ^{12}C peak. Next (Figure 4.10(b)), the peaks are aligned using the RANSAC [28] Peak aligner, which is a method to join all the separate peak lists into one master list, accounting for both linear and non-linear deviations in retention time. An alternative is the simple Join aligner, which uses mz and RT windows. Because many small, possibly spurious, peaks may be detected in single runs, the combined table can be constrained to entries where there are a minimum of, say, 20 occurrences of that peak. Figure 4.10(c) shows how this is configured in mzMine. Finally (Figure 4.10(d)), in order to identify the peaks, a custom database of accurate mass/retention times measured on standard compounds is used. This library is simply a comma separated value file (.CSV) listing the mz, RT, molecular formula and name of each metabolite. The retention times are determined by the previous injection of standard samples onto the system. There are also options to search online databases such as ChemSpider, KEGG, METLIN, etc., but the hits are often rather promiscuous returning many research chemicals, drugs and mammalian metabolites. These may be irrelevant and misleading when the experiment concerns a limited, defined space, such as plant metabolites for example.

Once all the stages are configured satisfactorily it is possible to run the operations in batch mode. This can take some time and having a multicore processor is useful as mzMine is multithreaded. For the small example data set illustrated, this operation took approximately 5 minutes (PC = HP Zeon Z600 8- core 2.4 GHz workstation with 8 GB RAM running Windows 7, 64 bit). It is not uncommon in our laboratory to run analyses that take many hours of overnight operation for a typical metabolomic study. The final step in the workflow is an Export to CSV option that allows the export of the final spreadsheet for downstream analysis.

The end result of the data processing workflow is shown in Figure 4.11 'RANSAC Aligned min 20 peaks'. Peaks that are missing are shown as red spots in the table (shown boxed in the figure). As missing data is undesirable, mzMine can be configured to fill missing peaks using the regions defined in the peak table. This ensures a reading of real data which is preferable for later statistical analysis. mzMine has two main gap-filling options; 'Peak finder' and 'm/z and RT range gap filler'. The former looks for undetected peaks in the same region as other scans, whereas the mz and RT gap filler simply finds the highest data point within the defined range.

At this stage it is most likely that the data will be processed further in a commercial data analysis package, but there are a few basic data visualisation tools included in mzMine. Analysis options include: coefficient of variation (CV) analysis, log ratio analysis, principal component analysis, curvilinear distance analysis, Sammon's projection

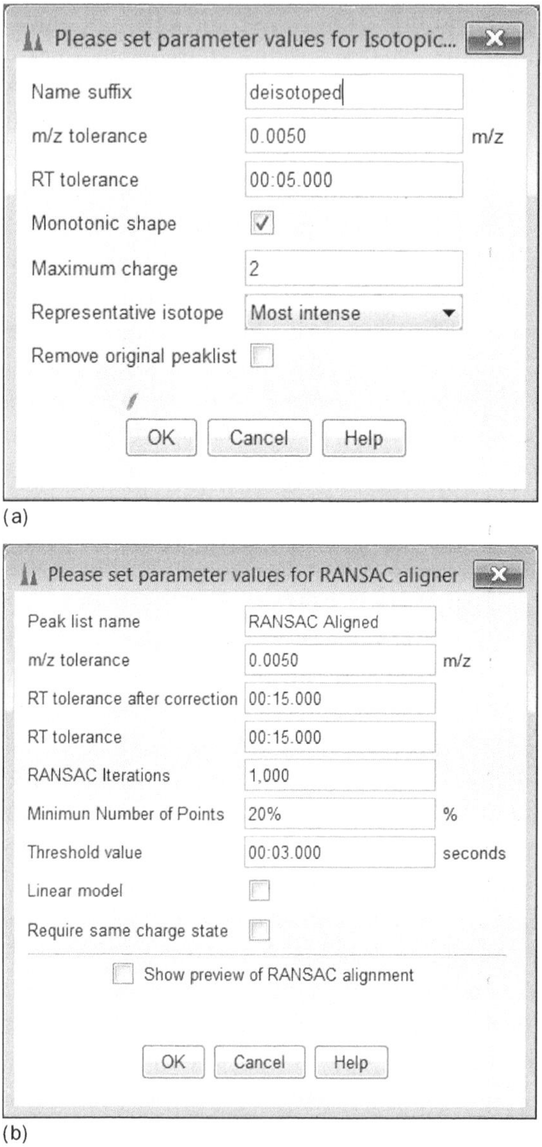

(a)

(b)

Figure 4.10 Configuring mzMine for metabolomics processing

and clustering. Use of these functions is illustrated below. (The data are taken from a metabolomic study on the ripening of fruits, comparing wild-type and non-ripening mutants, plus controls.)

Coefficient of variation analysis (Figure 4.12(a)) calculates the coefficient of variation of each peak and displays the result as a colour-coded plot.

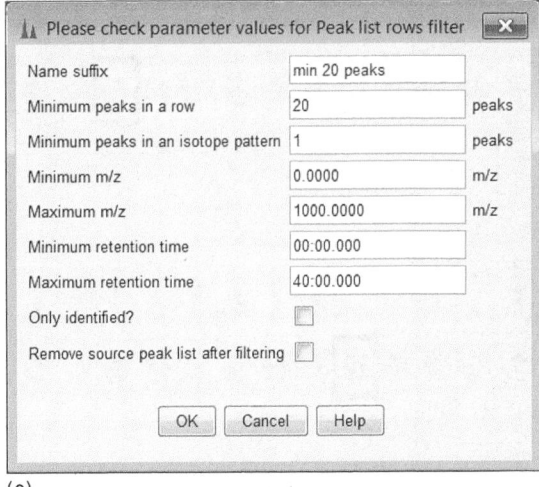

(c)

(d)

Figure 4.10 Continued

Log ratio analysis (Figure 4.12(b)) looks at the difference between two groups. It is the ratio of the natural logarithm of the ratio of each group average to the natural logarithm of 2. Principal component analysis (Figure 4.12(c)) is a multivariate visualisation method. The first principal component shows a separation of the control samples from the wild type and mutant. The second principal component shows a vertical trend. This, at first glance, appears to be related to ripening, but on closer inspection

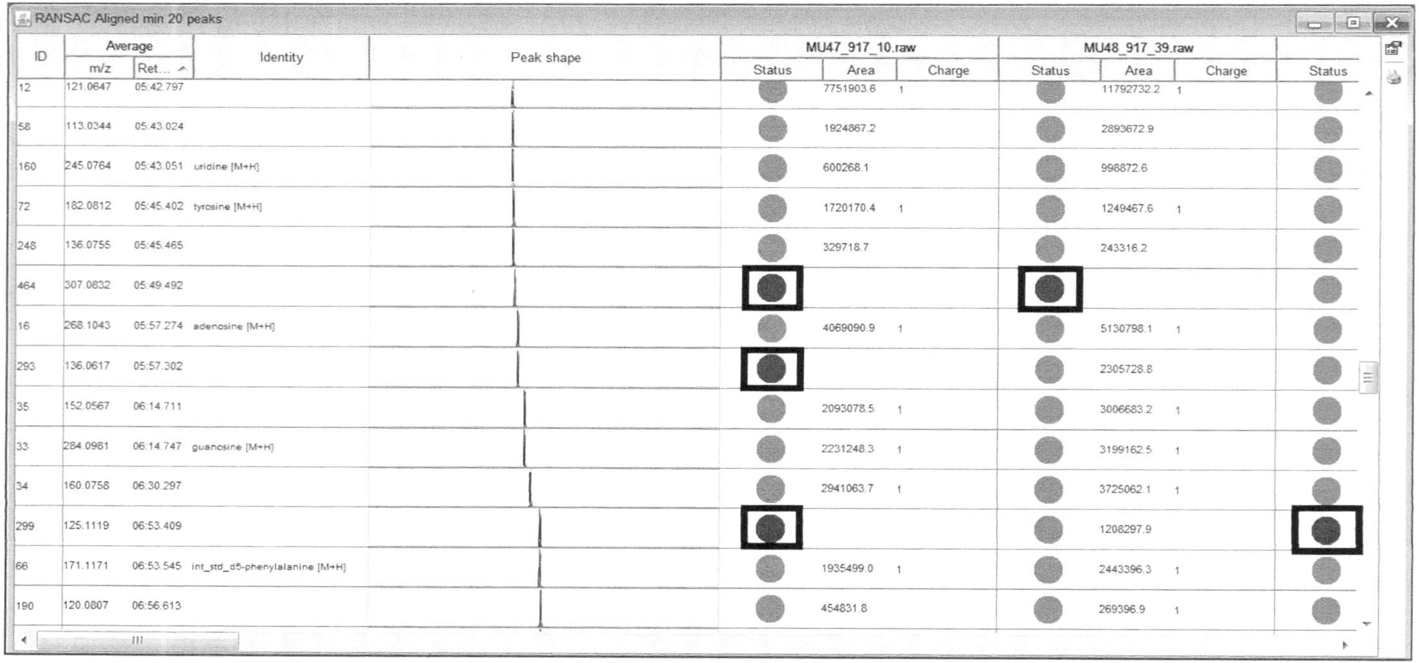

RANSAC Aligned min 20 peaks

ID	Average m/z	Average Ret...	Identity	Peak shape	MU47_917_10.raw Status	MU47_917_10.raw Area	MU47_917_10.raw Charge	MU48_917_39.raw Status	MU48_917_39.raw Area	MU48_917_39.raw Charge	Status
12	121.0647	05:42.797				7751903.6	1		11792732.2	1	
58	113.0344	05:43.024				1924867.2			2893672.9		
160	245.0764	05:43.051	uridine [M+H]			600268.1			998872.6		
72	182.0812	05:45.402	tyrosine [M+H]			1720170.4	1		1249467.6	1	
248	136.0755	05:45.465				329718.7			243316.2		
464	307.0832	05:49.492									
16	268.1043	05:57.274	adenosine [M+H]			4069090.9	1		5130798.1	1	
293	136.0617	05:57.302							2305728.8		
35	152.0567	06:14.711				2093078.5	1		3006683.2	1	
33	284.0981	06:14.747	guanosine [M+H]			2231248.3	1		3199162.5	1	
34	160.0758	06:30.297				2941063.7	1		3725062.1	1	
299	125.1119	06:53.409							1208297.9		
66	171.1171	06:53.545	int_std_d5-phenylalanine [M+H]			1935499.0	1		2443396.3	1	
190	120.0807	06:56.613				454831.8			269396.9	1	

Figure 4.11 mzMine results

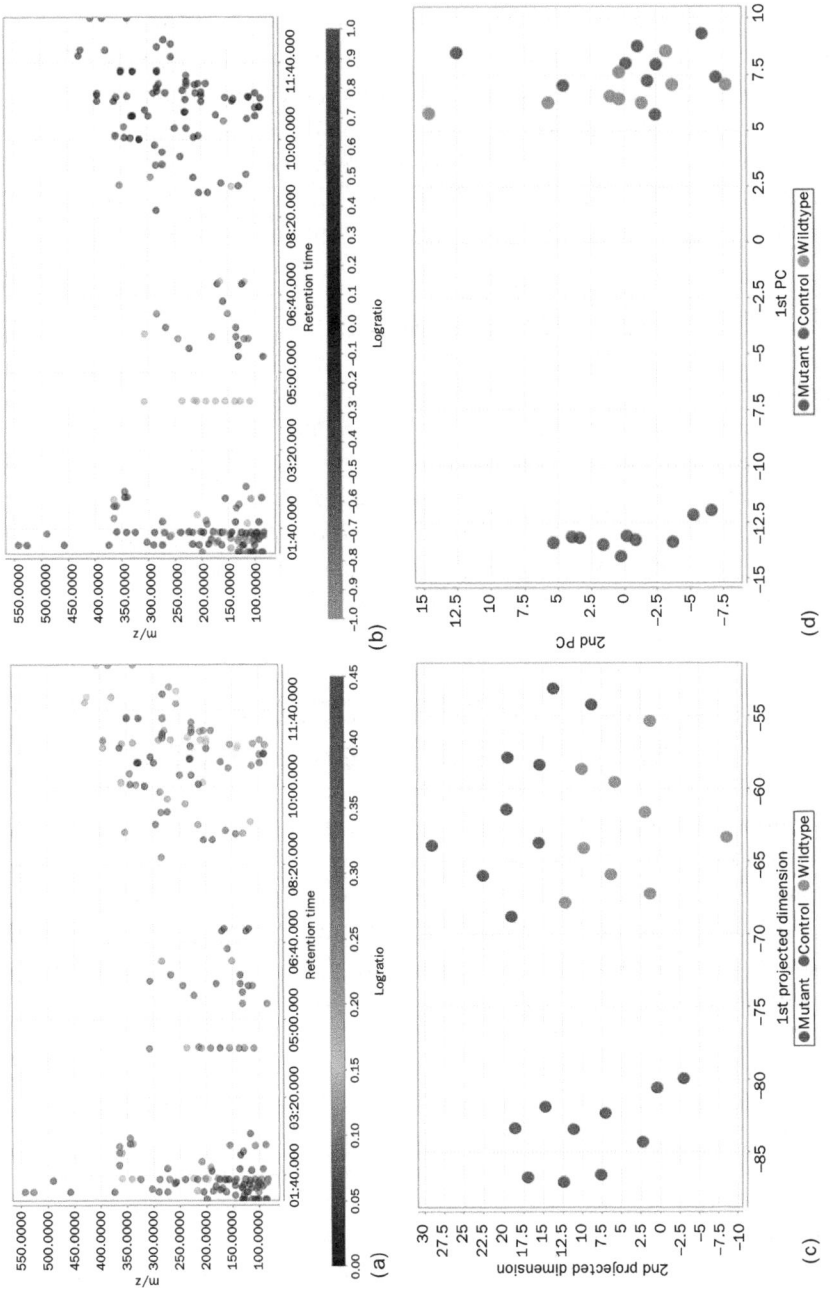

Figure 4.12 Data analysis in mzMine. (a) Coefficient of variation. (b) Log ratio analysis. (c) Principal components analysis. (d) Sammon's projection

the control samples also exhibit the same effect, showing that the variation is almost certainly due to spectrometer drift. The data have not yet been normalised to the internal standards which were spiked into the samples, an operation for the moment that has to be done outside mzMine. By removal of the control samples some separation of wild type from mutant was observed (not shown). Sammon's projection (Figure 4.12(d)) is a non-linear multidimensional scaling method which projects multidimensional data down to just two dimensions. It is a useful as a method to examine approximate clustering in data but offers no useful interpretability.

4.5.2 Other features in mzMine

There are many more features in mzMine, including some support for ms/ms data and extensive searching of external internet databases. More features are being added all the time, including several experimental baseline correction algorithms. Development in this area includes a raw data baseline correction module based on asymmetric least squares from the R package ptw: parametric time-warping [29] and an interface to the NIST MS Search [30] program to allow the use of mzMine for GC-MS data.

4.6 Metabolomics data processing using the open source workflow engine, KNIME

Data output from metabolomics software almost always requires further formatting before the data analysis stage. Procedures such as normalising the data to internal standards, or to total signal and subtotalling adducts are possible using spreadsheets and manual manipulation. The disadvantage of this approach is the ability to make unintentional errors and also the lack of transparency as to what was done to the data.

An alternative to this approach is to use a workflow tool which uses small data processing nodes linked together and therefore inherently documents the data set operations such as the open source project KNIME [31, 32]. Chapter 6 by Meinl, Jagla and Berthold describes applications of KNIME in chemical and bioinformatics. KNIME was developed by the Centre for Bioinformatics and Information Mining at the University of Konstanz and is based on the Eclipse platform for Java [33]. The KNIME software comes with a number of standard nodes but also there is a growing community of both commercial and open source developers writing new nodes for many data processing tasks. Included in KNIME are a number of statistical and data mining tools as well as data manipulation nodes. One of the really useful features is the support for writing code snippets in R, Java, Perl or Python, so if a KNIME node is not available for a particular application it is possible to write workarounds. Two examples of using KNIME and R for common metabolomics workflows are shown below.

4.6.1 Componentisation using mzMine

The first workflow will process the output of mzMine and componentise the identified metabolites. In other words, all identified adducts of a metabolite will be summed together in order to produce a component

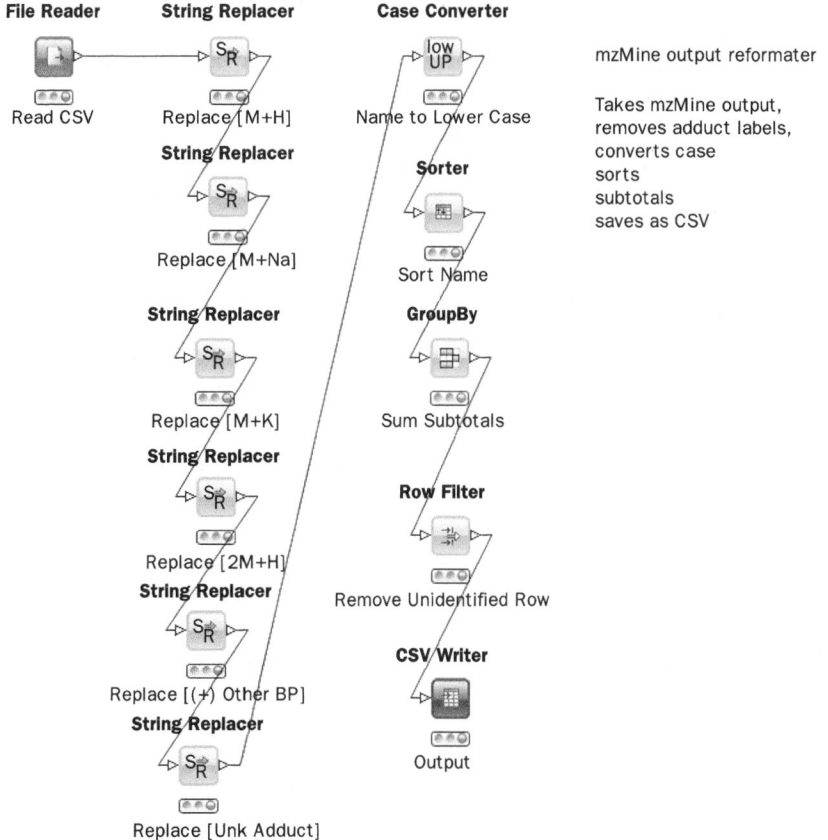

File Reader
Read CSV

String Replacer
Replace [M+H]

String Replacer
Replace [M+Na]

String Replacer
Replace [M+K]

String Replacer
Replace [2M+H]

String Replacer
Replace [(+) Other BP]

String Replacer
Replace [Unk Adduct]

Case Converter
Name to Lower Case

Sorter
Sort Name

GroupBy
Sum Subtotals

Row Filter
Remove Unidentified Row

CSV Writer
Output

mzMine output reformater

Takes mzMine output,
removes adduct labels,
converts case
sorts
subtotals
saves as CSV

Figure 4.13 A metabolomics componentisation workflow in KNIME

table. If, for instance, citric acid occurs as both [M+H], [M+Na] adducts then the workflow will add those peaks together. The actual workflow as seen in the KNIME GUI is shown in Figure 4.13. This workflow reads a CSV file, replaces adduct strings, converts all names to lowercase then subtotals the spreadsheet. In fact, in a spreadsheet such as Excel, string replacement and subtotalling are quite involved procedures and with large data sets the subtotalling may take a long time. In contrast, the KNIME workflow takes seconds to run and will run in a consistent manner eliminating manual processing errors. The workflow is constructed using various nodes. For example, the String Replacer node replaces any occurrence of a text (such as [M+H]) in the Name column. The Group By node groups rows by the Name column, here aggregations such as mean and sum of row subsets can be performed. The output is a CSV file with only the summed, identified metabolites in alphabetical order.

4.6.2 Internal standard normalisation using KNIME and R

In a typical metabolomics experiment, it is standard practice to include isotopic internal standards (such as d4-alanine, d5-phenylalanine) which

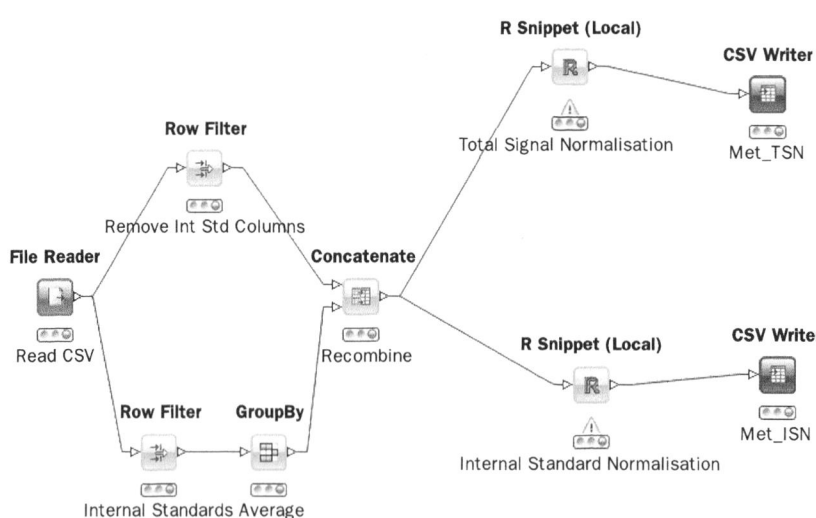

Figure 4.14 Workflow to normalise to internal standard or total signal

are spiked into the solvent used to dissolve freeze-dried plant material. The reason for this is to ensure there is a constant amount of standard in each sample so that instrumental response may be normalised. This avoids amplitude-based errors such as instrument drift, sample dilution or concentration. A KNIME workflow (Figure 4.14) can be created that identifies internal standards, averages the values and then divides each row in the original data by the internal standard. Although KNIME has many nodes for data manipulation, as yet there are none that allow mathematical functions to be applied to rows or columns within a data set so a custom R node (Labelled 'R-Snippet') can be used in order to do the division.

In cases where internal standards are not available several other methods are possible, one of the most common being total signal normalisation where each observation is divided by the total signal for that observation. In this way dilution effects may be eliminated. As with all normalisation methods, it is helpful to study replicate or pooled samples to see the effect of normalisation. If correctly normalised these samples should cluster into a tight group.

The R code for the Internal Standard node is shown below.

```
>intstd<-R[28,8:35]   # get int_std row
>mdata<-R[,8:35]   # get numerical part of data frame
>normalised<-sweep(as.matrix(mdata),2,as.matrix(intstd),'/')
>R<-cbind(R[1:7],normalised) # recombine ID's with data and output
```

4.7 Open source software for multivariate analysis

Metabolomics data consist of very large numbers of variables and relatively few observations. Such data are inherently co-linear, which leads to the use of chemometric techniques that can handle highly correlated data by using latent variable methods [34]. These methods [35] include principal components analysis (PCA), principal components regression (PCR), Projection to latent structures (PLS), PLS discriminant analysis (PLS-DA), orthogonal PLS (OPLS®) [36, 43, 44], orthogonal PLS discriminant analysis (OPLS-DA®) [37] and kernel OPLS (K-OPLS) [38].

Once the data have been formatted and normalised, it is commonly analysed interactively in a commercial multivariate analysis package. However, the world of open source does offer some multivariate tools, mainly in the R language. There are several chemometrics packages for R.

- Chemometrics with R from the book *Chemometrics with R – Multivariate Data Analysis in the Natural Sciences and Life Sciences* by R Wehrens [39]. This package contains PCA and MCR routines.

- Chemometrics. This package is the R companion to the book *Introduction to Multivariate Statistical Analysis in Chemometrics* by K Varmuza and P Filzmoser (2009) [40]. This includes PCA, PLS, clustering, self-organising maps and support vector machines.

- pls by R Wehrens and B-H Mevik [41]. Contains both PLS and PCR methods. This package is easily adapted for PLS-DA using a categorical Y variable denoting class membership (i.e. 0=control 1= treated).

- pcaMethods [42], initiated at the Max-Planck Institute for Molecular Plant Physiology, Golm, Germany. Now developed at CAS-MPG Partner Institute for Computational Biology (PICB) Shanghai, P.R. China and RIKEN Plant Science Center, Yokohama, Japan. pcaMethods has a number of alternative PCA methods for missing data including NIPALS and support for cross-validation.

- Kopls [38]. An implementation of the kernel-based orthogonal projections to latent structures (K-OPLS) method for MATLAB and R. The package includes cross-validation, kernel parameter optimisation, model diagnostics and plot tools.

4.7.1 Important considerations with multivariate analysis

The most critical aspect of multivariate analysis is the ability to estimate the predictive power, or model stability. This is usually implemented using cross-validation [45] where some data are sequentially left out of the model and the model re-calculated. The left-out data are then estimated from the model and the differences are summarised in a parameter called Q2, the predictive variance. Without an estimate of predictivity, there is no objective way to estimate the optimum number of components or even if any components are actually predictive at all. The variance explained or R2 of a model will keep increasing with every component and so there is a great danger of overfitting the model if this is the only criterion used to judge the model.

The ability to estimate predictivity becomes of paramount importance when using supervised methods such as PLS-DA. Without the measure of Q2 it may be possible to get discriminant models which are effectively worthless, for example getting separations with random data [45]. In addition to cross-validation, permutation testing is also a highly effective method of detecting overfit [46]. A second critical feature of multivariate methods is the use of the NIPALS [47] and related algorithms [35], which are able to handle small amounts of missing data without resorting to possibly misleading imputation (guessing) methods. Models are built with the data that are present, effectively ignoring the missing parts. Only some of the R packages implement cross-validation and missing value tolerance. Two of these are pcaMethods [42] and kopls [38].

4.8 Performing PCA on metabolomics data in R/KNIME

The pcaMethods package contains all required functionality in order to perform a simple PCA analysis. The input is an internal standard normalised data set as described in the previous section. A KNIME workflow to perform this analysis is shown schematically in Figure 4.15 and serves as a useful container for a number of small R scripts.

The PAR Scale node is a wrapper around R package pcaMethods, in this example, using Pareto scaling. (Pareto scaling is commonly applied to 'omics data sets and is the division of mean centred variable columns by the square root of the standard deviation. It up-weights medium scale features without excessive increases in baseline noise.)

```
require(pcaMethods)
R<-prep(R, scale = 'pareto',centre = TRUE)
```

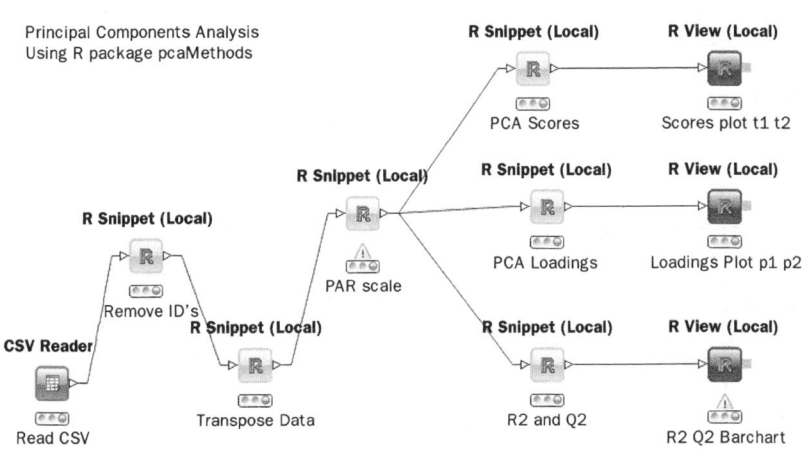

Figure 4.15 PCA analysis using KNIME and R

Published by Woodhead Publishing Limited, 2012

Alternatively, if scaling to unit variance is required the following code may be used.

```
R<-scale(R, center = TRUE,scale = TRUE)
```

Next, PCA scores are calculated using the NIPALS algorithm using pcaMethods.

```
#--------------------------------------
# PCA Scores Node (Snippet Node)
#--------------------------------------
require(pcaMethods)
results=pca(R, method='nipals', centre=FALSE, nPcs=3)
t1=results@scores[,1]
t2=results@scores[,2]
t3=results@scores[,3]
R<-cbind(t1,t2,t3)
R<-R
```

The R View node provides the ability to produce R plots. The code below sets up a graph, colours the points according to class and plots the first two scores from the PCA node output.

```
#---------------------------------------------
# Scores Plot (RView Node)
#---------------------------------------------

#calculate axis limits for X
mnX=min(R$'t1')
mxX=max(R$'t1')
plotrangeX=1.2*(max(abs(mnX),abs(mxX))) # add 20%

#calculate axis limits for Y
mnY=min(R$'t2')
mxY=max(R$'t2')
plotrangeY=1.2*(max(abs(mnY),abs(mxY))) # add 20%

A=substr(row.names(R), 1, 2) # extract first 2 characters
B=factor(A)     # change to a factor

# Plot the scores
plot(R$t1, R$t2,
  main= 'Scores Plot', sub='First two components',
  xlab='t1',
  ylab='t2',
  xlim = c(-plotrangeX, +plotrangeX),
  ylim = c(-plotrangeY, +plotrangeY),
```

```
  col=as.integer(B),
  pch=(as.integer(B)),
  cex=3, lwd=3)

segments(-plotrangeX, 0, plotrangeX, 0, col='grey')
segments(0, -plotrangeY, 0, plotrangeY, col='grey')

text (R$'t1', R$'t2',
  labels = substring(rownames(R),1,4),
  pos = 4,
  offset = 0.5,
  cex=1)
```

To produce the loadings plots, the PCA model is recalculated to obtain the loadings and R View is used to plot the loadings.

```
#------------------------------------------------
# PCA Loadings Node (Snippet Node)
#------------------------------------------------
require(pcaMethods)
results=pca(R, method='nipals', centre=FALSE, nPcs=3)
p1=results@loadings[,1]
p2=results@loadings[,2]
p3=results@loadings[,3]
R<-cbind(p1,p2,p3)
R<-R


#------------------------------------------------
# Loadings Plot (RView Node)
#------------------------------------------------
#calculate axis limits for X
mnX=min(R$'p1')
mxX=max(R$'p1')
plotrangeX=1.2*(max(abs(mnX),abs(mxX))) # add 20%

#calculate axis limits for Y
mnY=min(R$'p2')
mxY=max(R$'p2')
plotrangeY=1.2*(max(abs(mnY),abs(mxY))) # add 20%

# Plot the loadings
plot(R$p1, R$p2,
  main= 'Loadings Plot', sub='First two components',
  xlab='p1',
  ylab='p2',
  xlim = c(-plotrangeX, +plotrangeX),
  ylim = c(-plotrangeY, +plotrangeY),
  col='red',
  pch= 16,
  cex=2)
```

```
segments(-plotrangeX, 0, plotrangeX, 0, col='grey')
segments(0, -plotrangeY, 0, plotrangeY, col='grey')
text (R$'p1', R$'p2', labels = rownames(R),
  pos = 4,
  offset = 0.5,
  cex=1)
```

Finally, the R2 (Variance explained) and the Q2 (Predictive variance) are calculated via the following R code.

```
#-----------------------------------------------
# PCA R2 Q2 Node (Snippet Node)
#-----------------------------------------------
require(pcaMethods)
X=as.matrix(R)
results=pca(X, method='nipals', centre=FALSE, nPcs=3)
Q=Q2(results,X,fold=7,nruncv=1) @lculate Q2
R_2=c(results@R2cum[1],results@R2cum[2],results@R2cum[2])
Q_2= c(Q[1],Q[2],Q[3])
model_stats=cbind(R_2,Q_2) #R2 and Q2 together - concatenate
R<-model_stats


#-----------------------------------------------
# R2 and Q2 barchart (RView Node)
#-----------------------------------------------
R = as.matrix(R)

model_stats= c(R[1,1],R[1,2],R[2,1],R[2,2],R[3,1],R[3,2])
barplot(model_stats,
    ylab= 'Value',
    xlab='Component',
    beside=TRUE,
    main='R2 (fit) and Q2 (prediction)',
    col=c('lightgreen','lightblue'),
    ylim = c(0,1),
    space=0)
axis(1, at=c(1.0,3.0,5.0), lab=c('1','2','3'))

legend(0.2, 1, c('R2','Q2'),
    cex=1.5,
    col=c('lightgreen','lightblue'),
    pch=15)
```

The output of the R View nodes may be seen by right clicking the node and selecting View: R View (Figure 4.16(a)). The PCA scores plot (Figure 4.16(B)) is shown in the output from the first RView node. The plot shows the improved clustering of the quality control samples after the internal standard normalisation (the cluster of triangles on the right hand side). Some, but not complete, separation in the two plant genotypes is seen. A new model containing just the plant samples would be the logical next step in the analysis. Figure 4.16(c) shows the loadings plot, which shows the pattern of the metabolites corresponding to the patterns observed in the scores plot. It appears that glutamine, gaba and citric acid are higher for the WT fruits, and glutamic acid is higher in the controls. Figure 4.16(d) shows the model is predictively sound as both the variance explained (R2) and predictive variance(Q2) are high.

Further possibilities for analysis may entail the use of the pls package for PLS-DA replacing the Y variable with class membership (0 or 1) for the genotypes or the more sophisticated kopls. This exercise is left for the reader to complete using the general principles outlined above.

4.9 Other open source packages

There are many more useful packages being developed for mass spectrometry and 'omics data processing. For example, TOPPView is a viewer from the OpenMS proteomics pipeline for the analysis of HPLC/ MS data [48]. A full tutorial is available [49]. Cytoscape [50, 51] is an open source platform for network visualisation with good connectivity to bioinformatics resources. Network features such as node and edge parameters may be visualised in many ways, for instance levels of detected metabolites may be superimposed on a biological network. There is also an easy to follow tutorial available [52]. Ondex [53] is another data integration platform, which is designed to integrate diverse data sets and has a number of graph visualisation methods.

Finally, no discussion of metabolomics and mass spectrometry would be complete without mentioning the vast amount of data that studies are capable of producing. Although original data are often stored in instruments or dedicated LIMS systems, data modelling produces many types of intermediate data files. Many laboratories are resorting to backing up data onto high-capacity USB drives for storage or transfer. Although Linux users are in the fortunate position of having good operating system support for backups, Cobain Backup 8 [54] is an excellent open source program for automating backup of data in Windows. It is able to run either as a background service or an application and can schedule backup jobs to be started at daily or weekly intervals. The software supports FTP transfers and contains several methods of compression and strong encryption.

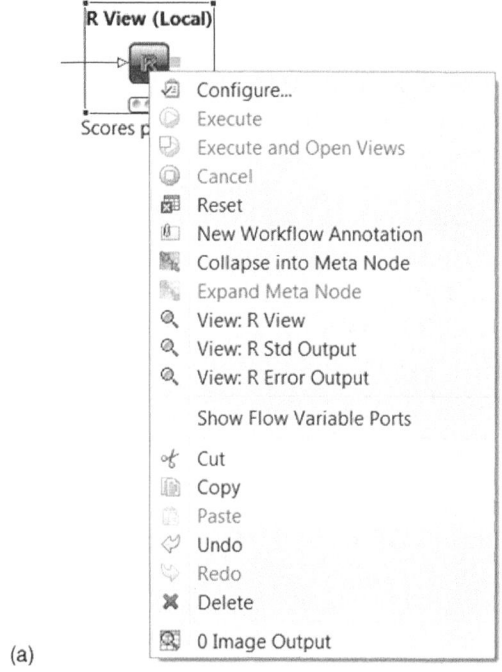

(a)

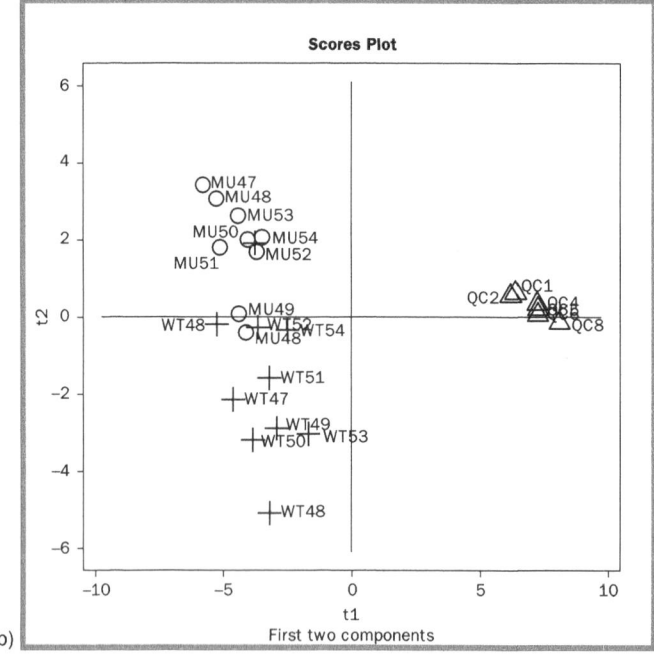

(b)

Figure 4.16 The plots from the R PCA nodes

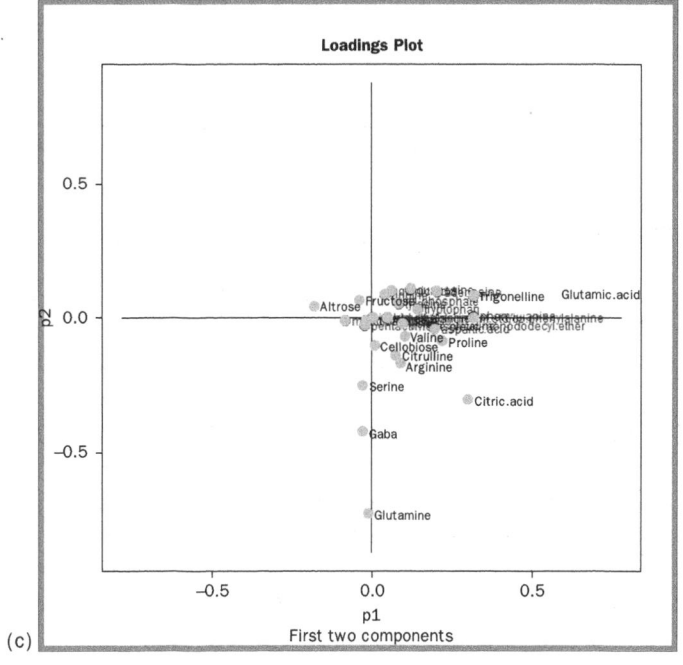

(c)

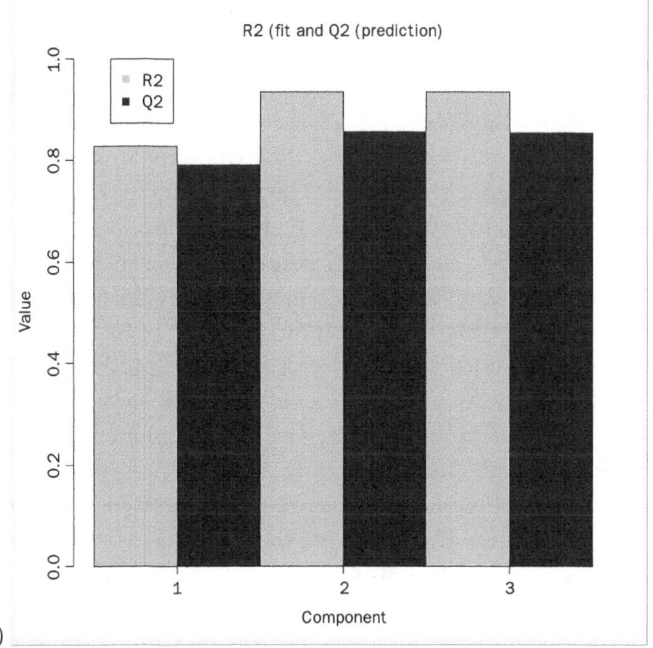

(d)

Figure 4.16 Continued

4.10 Perspective

As previously mentioned, the 'omics technologies are developing at a rapid pace and the open source world is playing a valuable role in plugging the gap between newly developed academic methods and commercial vendors' software. Many innovative and imaginative software packages are being developed, indeed since first writing this chapter the author has become aware of three more metabolomics packages namely MAVEN [55], R 'Metabonomic' Package [56] and Automics [57], the former for LC-MS and the latter two for NMR metabolomics.

Some issues with open source are difficult to use interfaces, experimental or buggy features, fragmentation of effort and poor availability of training materials. However, the growth of video-based training is beginning to improve the ease of use.

Another concern is the abandonment of projects. However, unlike commercial software, the source code is available so it is always possible to inspect, revive and adapt code which is no longer developed. Security is another concern, but as the code is under open scrutiny there should be less reason to fear open source than closed software.

One area that is frequently talked about, especially by commercial vendors, is the 'lack of technical support' with open source software. In our experience we have found completely the opposite. The open source community is very responsive to queries and more often than not bugs are fixed within days rather than the often months or years taken by commercial organisations. Active contribution to the projects also generates a lot of goodwill and need not be limited to programming skills as documentation and tutorials are highly valued.

Quite where the rise of open source in science leaves commercial software vendors is an interesting question. Having worked both in the software business and as an end user in industry, I can see advantages and disadvantages to both open and closed source products. The value of easy-to-use, well documented, validated and tested products that commercial vendors offer may often be underestimated. High licence fees may not look so high after considering the extra time needed to implement and learn less well-documented products (discussed in more detail by Thornber in Chapter 22). However, these costs are frequently hidden so that free open source packages may look very attractive from a budget holder's point of view. Also, the ability to customise and tailor software exactly to your own needs is a big benefit that is seldom available with commercial products.

One area in which commercial vendors still have a key role is the of use of data analysis tools in regulated environments. Extensive software validation is a costly exercise and currently commercial software vendors take this responsibility and worry away from the customer, naturally in exchange for licence fees. For those interested in this area, Chapter 21 by Stokes discusses this in much more detail.

In more research-based areas, open source solutions may have an edge over commercial software due to the openness of algorithms. There have been recent moves to make data analysis more transparent by embedding algorithms inside publications using tools such as R and Sweave [58]. There has been much discussion about 'reproducible research' [59, 60], where not only the data but also the methods used for analysis are freely available. This could be a valuable contribution to the aim of being able to reproduce complex 'omics analyses.

Open source is certainly a challenge to the commercial software world and we have seen the rise of many new business models, such as open community development with support, consultancy and training being run as commercial services. One area of relevance to Metabolomics is the provision of online services [61]. How open and free these type of services are likely to remain if massive transfers of data occur, remains to be seen. Commercial cloud computing solutions may emerge; but data security is of great concern to commercial organisations. I believe most commercial researchers would be very uneasy about transferring sensitive metabolomics data, particularly in highly commercial areas such as disease diagnosis or regulatory studies.

Lastly in the data analysis area, it appears so far only commercial vendors have been able to produce truly interactive data analysis tools. Products such as SIMCA-P, Unscrambler and JMP allow interactive plotting, data exclusion, filtering, transformation and many other manipulations. These operations are essential to any large-scale data analysis task where several rounds of quality control and data cleanup must be performed. To date, most of the data analysis tools in the open source are script-based and offer little opportunity for truly interactive analysis. Two packages which are attempting to make R more user-friendly and interactive are Rattle [62, 63] and GGobi [64, 65] but to date these have focussed more on 'machine learning' for business applications rather than chemometrics. (In general terms, machine learning algorithms require many more observations and fewer variables than are encountered in metabolomics, hence the preferred use of Chemometric methods, which are specifically designed for 'wide' data sets.) It will be interesting to see if any open source interactive tools will be developed in the future.

4.11 Acknowledgements

The author would like to thank the following people for help past and present who have guided the way through difficult and challenging territory. No single person can possibly know everything in today's cross-disciplinary world of science but knowing who to ask is a great help!

Dave Portwood, Mark Seymour, Mansoor Saeed, Mark Forster, Charlie Baxter, Stuart Dunbar at Syngenta, Tomas Pluskal, Chris Pudney and all the mzMine developers, Tim Ebbels, Richard Barton, Elaine Holmes, Hector Keun at Imperial College, Svante Wold, Lennart Eriksson, Erik Johansson, Johan Trygg, Thomas Jonsson, Mattias Rantalainen, Oliver Whelehan and all my friends from Umeå and Umetrics, Charlie Hodgman, Graham Seymour at Nottingham University, Madalina Oppermann at Thermo Instruments, Michael Berthold at Knime, Stephan Neumann at IPB-Halle and Stephan Biesken at the EBI.

4.12 References

[1] Griffiths J. A Brief History of Mass Spectrometry. *Analytical Chemistry* 2008;80(15).

[2] *http://en.wikipedia.org/wiki/History_of_mass_spectrometry*

[3] Webb K, Bristow T, Sargent M, Stein B and ESPRC National Mass Spectrometry Centre. *Methodology for Accurate Mass Measurement of Small Molecules Best Practice Guide*. Swansea UK. ISBN 0-948926-22-8.

[4] de Hoffmann E, Stroobant V. *Mass Spectrometry: Principles and Applications*, 3rd Edition September 2007 Wiley. ISBN: 978-0-470-03310-4

[5] Nicholson JK. *Molecular Systems Biology 2:52 Global systems biology, personalized medicine and molecular epidemiology*. Department of Biomolecular Medicine, Faculty of Medicine, Imperial College London, South Kensington, London, UK doi:10.1038/msb4100095

[6] Robertson DG. Metabonomics in Toxicology: A Review 1. *Toxicological Sciences* 2005;85:809–22. doi:10.1093/toxsci/kfi102.

[7] *http://tools.proteomecenter.org/wiki/index.php?title=Formats:mzXML*

[8] *http://en.wikipedia.org/wiki/Mass-spectrometry_data_format*

[9] Pedrioli PG, Eng JK, Hubley R, et al. A common open representation of mass spectrometry data and its application to proteomics research. *Nat Biotechnol* 2004;22(11):1459–66.

[10] Deutsch EW. mzML: A single, unifying data format for mass spectrometer output. *Proteomics* 2008;8(14):2776–7.

[11] *http://www.psidev.info/index.php?q=node/257*

[12] Hau J, Lampen P, Lancashire RJ, et al. *JCAMP-DX V.6.00 for chromatography and mass spectrometry hyphenated methods* (IUPAC Technical Note 2005).

Published by Woodhead Publishing Limited, 2012

[13] ASTM E1947 – 98(2009) *Standard Specification for Analytical Data Interchange Protocol for Chromatographic Data* DOI: 10.1520/E1947-98R09

[14] R Development Core Team. *R: A Language and Environment for Statistical Computing.* R Foundation for Statistical Computing Vienna, Austria 2011 ISBN 3-900051-07-0 *http://www.R-project.org*

[15] Gentleman R, Carey VJ, Bates DM, et al. Bioconductor: Open software development for computational biology and bioinformatics. *Genome Biology* 2004;5:R80.

[16] Neumann S, Pervukhin A, Böcker S. *Mass decomposition with the Rdisop package.* Leibniz Institute of Plant Biochemistry, Department of Stress and Developmental Biology, Bioinformatics, Friedrich-Schiller-University Jena, April 22, 2010.

[17] Böcker S, Letzel M, Lipták Z, and Pervukhin A. SIRIUS: Decomposing isotope patterns for metabolite identification. *Bioinformatics* 2009;25(2):218–24.

[18] Tautenhahn R, Böttcher C, Neumann S. Highly sensitive feature detection for high resolution LC/MS. *BMC Bioinformatics* 2008.

[19] Smith CA, Want EJ, Tong GC, Abagyan R and Siuzdak G. XCMS: Processing Mass Spectrometry Data for Metabolite Profiling Using Nonlinear Peak Alignment, Matching, and Identification. *Analytical Chemistry* 2006.

[20] Smith CA, Want EJ, Tong GC, et al. *Metlin XCMS: Global metabolite profiling incorporating LC/MS filtering, peak detection, and novel nonlinear retention time alignment with open-source software.* 53rd ASMS Conference on Mass Spectrometry, June 2005, San Antonio, Texas.

[21] Benton HP, Wong DM, Trager SA, Siuzdak G. XCMS2: Processing Tandem Mass Spectrometry Data for Metabolite Identification and Structural Characterization. *Analytical Chemistry* 2008.

[22] *www.neoscientists.org*

[23] *http://metlin.scripps.edu/xcms/faq.php*

[24] Pluskal T, Castillo S, Villar-Briones A, Orešič M, MZmine 2: Modular framework for processing, visualizing, and analyzing mass spectrometry-based molecular profile data, *BMC Bioinformatics* 2010;11:395.

[25] Katajamaa M, Miettinen J and Orešič M. MZmine: Toolbox for processing and visualization of mass spectrometry based molecular profile data, *Bioinformatics* 2006;22:634–6.

[26] Katajamaa M and Orešič M Processing methods for differential analysis of LC/MS profile data, *BMC Bioinformatics* 2005;6:179.

[27] *http://sjsupport.thermofinnigan.com/public/detail.asp?id=586*

[28] Fischler MA and Bolles RC. 1981. Random sample consensus: a paradigm for model fitting with applications to image analysis and automated cartography. Commun ACM 1981;24(6):381–95. DOI=10.1145/358669.358692 http://doi.acm.org/10.1145/358669.358692.

[29] Boelens HFM, Eilers PHC, Hankemeier T. Sign constraints improve the detection of differences between complex spectral data sets: LC-IR as an example. *Analytical Chemistry* 2005;77:7998–8007.

[30] *http://chemdata.nist.gov/mass-spc/ms-search/*

[31] *http://www.inf.uni-konstanz.de/bioml2/publications/Papers2007/BCDG+07_knime_gfkl.pdf*

Published by Woodhead Publishing Limited, 2012

[32] http://www.knime.org/

[33] http://www.eclipse.org/

[34] Eriksson L, Antti H, Gottfries J et al. Using chemometrics for navigating in the large data sets of genomics, proteomics, and metabonomics (gpm). *Analytical and Bioanalytical Chemistry* 2004;380(3).

[35] Fonville JM, Richards SE, Barton RH, et al. The evolution of partial least squares models and related chemometric approaches in metabonomics and metabolic phenotyping. *J Chemometrics* 2010;24:636–49.

[36] Trygg J and Wold S. Orthogonal Projections to Latent structures (O-PLS). *J Chemometrics* 2002;16:119–28.

[37] Bylesjö M, Rantalainen M, Cloarec O, et al. OPLS discriminant analysis: combining the strengths of PLS-DA and SIMCA classification. *J Chemometrics* 2006;20:341–51.

[38] Bylesjö M, Rantalainen M, Nicholson JK, et al. K-OPLS package: Kernel-based orthogonal projections to latent structures for prediction and interpretation in feature space (2008) *BMC Bioinformatics* 2008;9:106.

[39] Wehrens R. Chemometrics with R: Multivariate Data Analysis in the Natural Sciences and Life Sciences (Use R). Springer, 2011. ISBN-13: 978-3642178405.

[40] Varmuza K and Filzmoser P. *Introduction to Multivariate Statistical Analysis in Chemometrics.* Taylor & Francis – CRC Press, Boca Raton, FL, 2009. ISBN: 978–1420059472

[41] Mevik B-H, Wehrens R. The pls Package: Principal Component and Partial Least Squares Regression in R. *Journal of Statistical Software* 2007;18(2): 1–24.

[42] Stacklies W, Redestig H, Scholz M, Walther D, Selbig J. pcaMethods – a bioconductor package providing PCA methods for incomplete data. Bioinformatics. 2007;23(9):1164–7. Epub 2007 Mar 7.

[43] Trygg J, Wold S. Umetrics AB No. 10204646 filed on 22 February 2001.

[44] OPLS Trademark 78448988 Registration Number: 3301655 Filing Date: 12 July 2004

[45] Westerhuis JA, Huub CJ, Hoefsloot S. et al. Assessment of PLSDA cross validation. *Metabolomics* 2008;4:81–9.

[46] Eriksson L, Trygg J, Wold S. SSC10 CV-ANOVA for significance testing of PLS and OPLS® models. Special Issue: Proceedings of the 10th Scandinavian Symposium on Chemometrics. *Journal of Chemometrics* 22(11–12):594–600.

[47] Wold H, Lyttkens E. Nonlinear Iterative Partial Least Squares (NIPALS) Estimation Procedures. 1969 Bull ISI PLS.

[48] Sturm M and Kohlbacher O. TOPPView: An Open-Source Viewer for Mass Spectrometry Data. *J Proteome Res* 2009;8(7):3760–3.

[49] http://www-bs2.informatik.uni-tuebingen.de/services/OpenMS/OpenMS-release/TOPP_tutorial.pdf

[50] Smoot ME, Ono K, Ruscheinski J, Wang PL, Ideker T. Cytoscape 2.8: new features for data integration and network visualization. *Bioinformatics* 2011;27(3):431–2. Epub 2010 Dec 12.

[51] http://www.cytoscape.org/.

[52] http://cytoscape.wodaklab.org/wiki/CytoscapeRetreat2008/UserTutorial.

[53] *http://www.ondex.org/*

[54] *http://www.educ.umu.se/~cobian/cobianbackup.htm.*

[55] Melamud E, Vastag L and Rabinowitz JD. Metabolomic Analysis and Visualization Engine for LC?MS Data. *Analytical Chemistry* 2010;82(23): 9818–26.

[56] Izquierdo-García JL, Rodríguez I, Kyriazis A, et al. A novel R-package graphic user interface for the analysis of metabonomic profiles. *BMC Bioinformatics* 2009, 10:363 doi:10.1186/1471-2105-10-363.

[57] Wang T, Shao K, Chu Q, et al. Automics: an integrated platform for NMR-based metabonomics spectral processing and data analysis. *BMC Bioinformatics* 2009,10:83 doi:10.1186/1471-2105-10-83

[58] *http://www.stat.uni-muenchen.de/~leisch/Sweave/*

[59] *http://reproducibleresearch.net/index.php/RR_links*

[60] *http://journal.r-project.org/archive/2010-1/RJournal_2010-1_Thioulouse~et~al.pdf*

[61] *https://xcmsonline.scripps.edu/*

[62] *http://rattle.togaware.com/*

[63] Williams GJ. Rattle: A Data Mining GUI for R. *The R Journal* 2009;1/2. ISSN 2073-4859.

[64] *http://www.ggobi.org/*

[65] Temple Lang D, Swayne DF. DGGobi meets R: an extensible environment for interactive dynamic data Visualization. Proceedings of the 2nd International Workshop on Distributed Statistical Computing 15–17 March, 2001. Vienna, Austria *http://www.ci.tuwien.ac.at/Conferences/DSC-2001*

## 5

Open source software for image processing and analysis: picture this with ImageJ

Rob Lind

Abstract: Image processing and analysis is fundamental to extract useful information from images. To achieve this end, open source image analysis software, exemplified by the Java application ImageJ, can be used very flexibility to create workflows and is open to customisation due to its open source architecture. ImageJ has a strong academic community with many macros, Java scripts and plug-ins available online, a help forum, regular updates and face to face conferences. Furthermore, running Java ensures that ImageJ is platform-independent so that executable code can be shared easily between researchers using different operating systems. Lastly, ImageJ can be integrated into workflows of other open source applications such as KNIME.

Key words: image analysis processing, ImageJ.

Computers are useless. They can only give you answers.

Pablo Picasso (1881–1973)

5.1 Introduction

Pablo Picasso's words ring true with imaging software in that all you get out is the answers to the questions you ask. Asking the right question is something the computer cannot do. Only with a scientist asking the right question coupled with a programmer who can translate that question into code does the computer become useful rather than useless!

In fact the computer becomes invaluable when performing imaging tasks in a non-subjective manner and at a speed that might take a human operator a lifetime to complete. Microprocessor performance has increased according to the predictions of Moore's law [1]. Around 20 years ago the desktop PC would have been stretched to perform image analysis on a 20 megapixel, 24 bit image, having just 32 MB RAM and a 25 MHz CPU. Today's PCs can cope with this high demand, having increased data-handling speeds, multicore processors in combination with GBs of RAM and clock speeds in the GHz range.

The aim of this chapter is to provide users with an introduction to the capabilities and potential applications of open source image processing tools, and to help find solutions to problems. The predecessor of ImageJ first appeared in the 1990s, when it was known as NIH image and was available to run under the Macintosh operating system. The author and collaborator Iain Couzin created a custom imaging suite of bespoke tools to address the challenging visual research questions, the software produced was known as ICBiovision [2]. Such a bespoke piece of software had some advantages in that it fully met the requirements for which it was designed. However, when an analysis outside the defined scope was required, there was no way to achieve this goal without the further effort of low-level coding – the software was not easily extensible.

In the 1990s, desktop PCs increased in power and digital camera specification followed a similar upwards curve. The need for, and utility of, image analysis as a way to non-subjectively and quantitatively assess many different biological questions therefore increased. The utility of ICbiovision made clear the potential for image analysis in many different scientific arenas and a new tool box was needed to provide imaging solutions. Another imaging software package utilised was Aphelion [3]. This piece of software provides the user with access to a large number of tools that could be threaded together into macros. One downside with Aphelion is that it is a commercial piece of software, with a restrictive licence. Also, the application was dependent on a particular operating system. As the company upgraded its operating system then Aphelion was left behind. There were many other pieces of commercial image analysis software available but these came shackled with licence fees. It was at this point that ImageJ was rediscovered, after its metamorphosis from NIH image, to an open source Java language program on the PC. Java enabled ImageJ to be platform-independent and its free distribution made it suitable to roll out as an imaging platform across a wide network of imaging enthusiasts without restrictive licensing issues.

5.2 ImageJ

ImageJ (J for Java) evolved from NIH image, was developed at the National Institutes of Health by Wayne Rasband [4] and has developed a large global following since being released in 1997. It was conceived with an open architecture that could be extended through Java scripts, plug-ins and recordable macros. Custom acquisition, analysis and processing can be created with ImageJ's built-in editor and Java compiler. ImageJ is platform-independent and only requires Java 5 or a later virtual machine. It can run on Microsoft Windows, Mac OS, Mac OS X and Linux, and its source code is freely available [4].

As ImageJ is open source, an ImageJ user has the four essential freedoms defined by Richard Stallman in 1986 [5]. First, the freedom to run the program, for any purpose. Second, the freedom to study how the program works, and change it to make it do what you wish. Third, the freedom to redistribute copies so you can help your neighbour. Lastly, the freedom to improve the program, and release your improvements to the public, so that the whole community benefits.

In addition to the many freely available macros, plug-ins and scripts, commercially available plug-ins have been developed for which a licence needs to be purchased. For example, Image-Pro Plus from Media Cybernetics is a commercially available ImageJ plug-in [6].

5.2.1 The many faces of ImageJ

The customisable nature of ImageJ has allowed users to design custom user interfaces, such as NeuronJ shown in Figure 5.1. Note that the menu titles remain the same but the bottoms can have custom icons that link to particular functionality imparted by linked scripts or macros.

Alternatively, ImageJ can be customised by the production of custom folders in the plug-in menu, such as used for the astronomy plug-ins shown in Figure 5.2 [8].

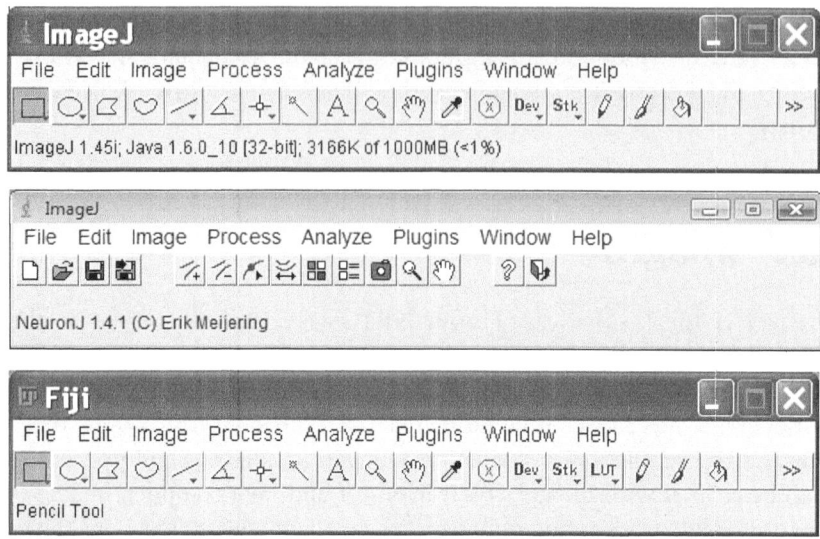

Figure 5.1 The many faces of ImageJ. The open source nature of ImageJ (native version shown top running on Windows Vista) allows for the main window to be fully customised to a particular function as has been done in the case of NeuronJ [7] (middle running on Windows XP) and for biological imaging Fiji (bottom running on Windows Vista)

5.2.2 Workflow

Before getting to grips to write a solution to an imaging problem in ImageJ, a clear workflow must be conceived incorporating any preprocessing of the images, the functions required, combining the functions into the programmable elements of ImageJ, internal calculations and finally the outputs. Once this has been clearly stated, a workflow within imageJ can be transcribed onto the plan and coding can begin. Customer input is invaluable to understand the exact nature of the input images and what the output should contain. Input images can be discrete image documents, such as lossless formats such as TIFFs, or more commonly the lossy formats such as JPGs, which are typical outputs from digital capture equipment. Alternatively, images can be in the form of an image sequence as in an AVI movie file, which can be read into ImageJ as a sequence of images. ImageJ can display multiple spatially or

Figure 5.2 ImageJ can be customised by defining the contents of the various menus. Custom plug-ins can be added under the plug-in menu in a subfolder as shown here for the astronomy suite of functionality

temporally related images in a single window. These image sets are called stacks and are often generated in biological imaging either by timelapse images or by Z stacks used in microscopy. The images that make up a stack are called slices. In stacks, a pixel (which represents 2D image data in a bitmap image) becomes a voxel (volumetric pixel), that is an intensity value on a regular grid in a 3D space. Furthermore, ImageJ can create and handle hyperstacks of images that are composed of multicolour channel images having multiple slices and frames.

The image itself may have data attached to it in the form of an EXIF data file and this too can be read by ImageJ and contains useful information for outputs such as the date the image was captured and many other image parameters. The image name may also contain useful information required in the output, which can be easily captured by ImageJ using the 'getTitle()' command. Images come in various bit depths from one bit binary to 24 bit colour. ImageJ can handle various bit depths but errors will occur if these images are processed by functions that require images at certain bit depths so it is important to appreciate this requirement. For example, the 'Analyse particles' function can only work on binary images in which objects have been segmented. As well as bit depth, the images can be of any physical pixel dimensions providing enough RAM has been allocated to ImageJ, which can be set in the menu Edit/Options/Memory and Threads. When running ImageJ under a 64 bit operating system, more RAM can be allocated than is possible with a 32 bit system. This is useful when working with large images or arrays of data. When opening large image sequences, the problem of running out of RAM memory can be averted by opening them in a virtual sequence. An alternative way of handling large sets of images is to handle images one at a time within a macro using the 'openNext' command.

The colour space of an image describes the gamut of colours and because human vision is based on trichromatic perception, most of the colour models use three values, others use more [9]. The colour models all form a 3D space where the colour components describe a location (i.e. colour) in that space. Images are opened as 24 bit images with 8 bits per red, green, blue (RGB) channel. In other words, each channel is assigned 8 bits per pixel, which is 256 possible shades of that colour channel. A colour space has to be described by a minimum of three different parameters and many different models exist. Within ImageJ it is possible to split an image up by either the RGB or hue, saturation, brightness (HSB) channels into a stack of three images.

The questions as to why you need to be able to work in different colour spaces and when is one more useful than another are often asked. The need to be able to do this becomes evident in particular imaging tasks where a particular parameter becomes critical to Segment your image. For example, when trying to separate yellow objects from purple objects using the RGB colour space this may result in a blurred segmentation; however, by switching to HSB and using the hue channel then objects of different colours can easily be segmented using this channel independently of their brightness or saturation. Besides 24 bit colour, ImageJ can also handle images with larger bit depths per pixel, for example 32 bit greyscale images. Look up tables (LUTs) can also be applied to 8 bit images to quickly make false colour images and ImageJ provides many LUT pallets of 256 colours ready to apply to 8 bit greyscale images.

These can be particularly useful to apply to processed images to provide the user with a visual check of the numerical results.

In addition to opening and closing of images, the ImageJ macro language offers a good deal of file and folder handling to smooth your workflow. For example, result images, plots and data can be saved to particular locations and temporary images stored and then deleted at the end of the workflow. Making a plan of what you intend to do in terms of image processing on your images is the next important step in the workflow. It is a good working knowledge of all the ImageJ functions that allows the user to describe in his or her own words what they want to do and then translate this into the ImageJ commands. The macro language of ImageJ is focused in its functionality towards its core function of analysing images and is constantly being added to in new versions. Generally, macros containing these latest commands are not back compatible with older versions of ImageJ so using the ImageJ updater regularly is a good habit to get into. This command simply checks the web for a newer version and if found can download the update automatically and after a restart of ImageJ it will be running the latest release. To prevent older versions running macros that include functionality only available in later versions, a function call such as 'requires ("1.45e") can be added to the macro to prevent trying to run a macro using an outdated version. There are excellent guides available that give a full account of the ImageJ functions, some of these have links to examples of macros which use them and are included when downloading ImageJ. Further macro examples can be found on the ImageJ website [10] where, 300+ macros, 500+ plug-ins and 20+ scripts are available.

5.2.3 *ImageJ community*

ImageJ has a large following in academia and beyond due to its open source nature, being freely available for download and use without restriction. This has led to actual and virtual meetings of ImageJ users to share experience and, of course, knowhow, macros, scripts and more. Indeed, posting a request on how to do a task on the ImageJ forum often meets with a quick and successful response from within the community. Users can also search the archives to see if a question has already been answered [11]. This global network of ImageJ users is a great source of knowhow and advice to help get a novice user up and running. In addition to the ImageJ forum, the ImageJ web site hosts a large selection of macros that not only provide useful functionality but also demonstrate how the different ImageJ command functions work. It is beneficial to spend time becoming familiar with ImageJ, trawling through these macros to understand the capabilities these functions bring by example and how these could be applied to the users own imaging problems. The size of the ImageJ user community is so large that it even has its own conference series [12].

5.2.4 Programming ImageJ with macros, plug-ins and Javascript

There are several methods of coding within ImageJ, namely with macros, scripts and Javascript. Each has their own advantages and disadvantages, which are now discussed. Macros are simple programs and are the easiest way to automatically execute a series of ImageJ commands. The native ImageJ macro language is Java-like and contains a set of control structures, operators and built-in functions, and can be used to call built-in commands and other macros, which are saved simply as text files.

Plug-ins are much more powerful, flexible and faster than macros (most of ImageJ's built-in menu commands are actually plug-ins), but harder to write and debug. Plug-ins are written in the Java programming language (.java source files) and compiled to .class files.

Scripts within ImageJ use the Mozilla Rhino interpreter to run Javascripts. Similar to plug-ins, scripts have full access to all ImageJ and Java application programming interfaces (APIs) but do not need to be compiled (scripts and macros run interpretively). On the other hand, scripts lack the simplicity of the macro language and feel less integrated in ImageJ. They do provide a much richer language together with documentation and standardisation. Support for other languages is possible in ImageJ using Fiji and its powerful editor. Fiji adds extra support for the Java scripting and programming languages of BeanShell, Clojure, Python and Ruby. Several packages exist that allow ImageJ to interact with other applications/environments. R is a free software environment for statistical computing and graphics and RImageJ provides the linkage between the two software packages [13]. MATLAB is a commercial package and MIJ provides a bi-directional communication data exchange between MATLAB and ImageJ. MIJ also allows MATLAB to access all built-in functions of ImageJ as well as third-party ImageJ plug-ins. Fiji features MIJ, which makes it even more convenient to use the libraries and functions provided by Fiji's components from within MATLAB.

Published by Woodhead Publishing Limited, 2012

A couple of disadvantages exist in that scripts cannot access most of the ImageJ built-in macro functions and also they do not support 'batch mode'. Batch mode is a function within the macro language and is used to effectively speed up ImageJ processing times by turning off the image display allowing the macro to run up to 20 times faster.

An example of an executable Javascript is SmartRoot [14], an ImageJ plug-in that is a semi-automated image analysis software which streamlines the quantification of root growth and architecture for complex root systems measuring parameters such as root angle, area, branching, growth, length, topology and volume. Because this is a Javascript plug-in, the greater flexibility in software expression is evident over the macro language with a tailored graphical user interface (Figure 5.3). Furthermore, SmartRoot is an operating system-independent freeware based on ImageJ and uses cross-platform standards (XML, SQL, Java) for communication with a database and data analysis software.

ImageJ can also be a node in integration, processing, analysis and exploration platforms such as KNIME (Konstanz Information Miner (see Chapter 6)) [15], which is a user-friendly and comprehensive open source workflow software toolkit. This is an attractive proposition for ImageJ

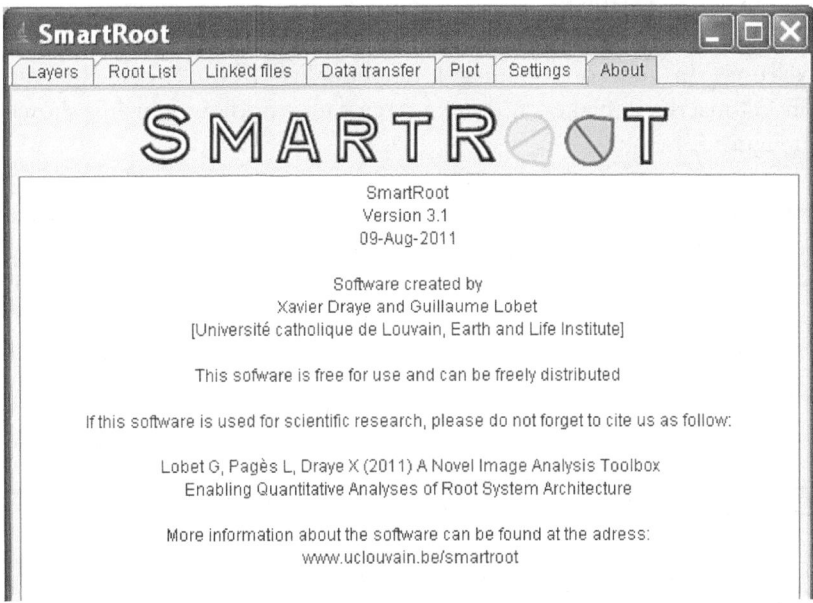

Figure 5.3 Smartroot displays a graphical user interface that only Javascript can deliver within ImageJ

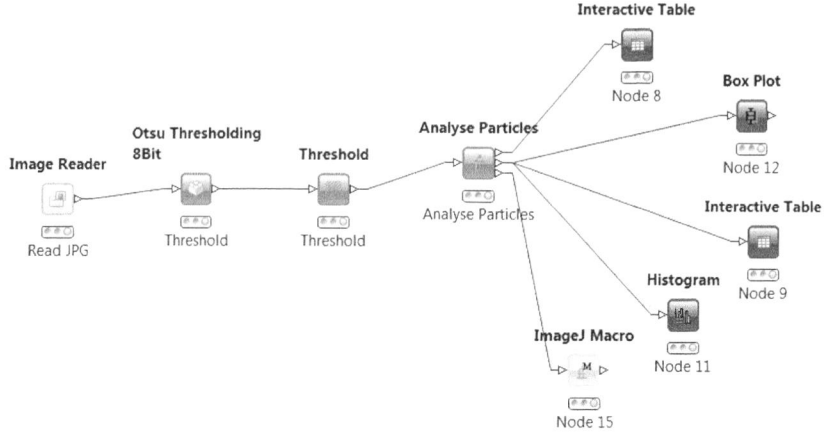

Figure 5.4 A KNIME workflow that integrates ImageJ functions in nodes as well as custom macros and the outputted data can be taken on by nodes to display the data in sophisticated plots or interactive tables

users who effectively take images and extract the useful data from them into a results table. Although a certain amount of calculations and graphs can be plotted within ImageJ, a specialist package to interrogate and visualise large data sets that can directly interface with ImageJ is welcome. In addition to ImageJ functionality, KNIME can also run ImageJ macros to make a node in a large data workflow, which is shown in Figure 5.4.

The use of quick response (QR) codes is ever increasing due to their popularity among mobile phone users who take an image of the code which, say, encodes a URL, which can then be used to direct the user to a web site (Figure 5.5). QR codes can be generated online [16], and, if placed in an image, then there is a Java plug-in available for ImageJ that can read the codes and this information can then be used to rename the image or used in a results table, for example [17].

5.3 ImageJ macros: an overview

The most basic form of macro creation is to simply use the ImageJ recorder to record actions, and then use the write macro function to output to a text file. This can then be saved. It is important that the suffix '.txt' is applied, otherwise ImageJ will not recognise the file as a text file

Figure 5.5	Example of a QR code that can be read by a plug-in based on ZXing (pronounced 'zebra crossing'), an open source, multiformat 1D/2D barcode image processing library implemented in Java for ImageJ, which will decipher this QR code as 'Open source software for image processing: Picture this with ImageJ'

which contains a macro. To install a recorded macro, the plug-ins menu has the ability to install it under plug-ins/macros/install, after which it will appear in the plug-ins/macros menu. However, a far easier method to access macros is to make a new subfolder in the ImageJ plug-ins folder and place personal macros there, as earlier shown for the astronomy plug-ins. It is important that the macro name contains an underscore for ImageJ to recognise it and for it to appear in the plug-ins menu. Alternatively, macros can be assigned to buttons on the menu bar and, of course, being an open source piece of software even a custom icon can be applied to the button! It is in this fashion that the ImageJ interface can be customised such as in the case of NeuronJ, for example, as shown in Figure 5.1. This makes it a much more intuitive environment with which users can interact.

The Java-like macro language within ImageJ has function commands as a cornerstone, but much more is required to put some cement between these stones to define the workflow from start to finish, and make it interactive with the user where necessary. Inputs from the user can be captured and assigned as variables to create a semi-automatic macro, whereas a fully automatic macro needs no input other than an open image, for example.

Published by Woodhead Publishing Limited, 2012

5.3.1 Macro language in ImageJ

The macro language used by ImageJ is much like Java, so some knowledge of similar languages will help get a user up and running. There are some basic elements within the language, which are now discussed, which can be blended together to provide a sophisticated and fairly rich functionality dedicated to image processing.

Variables

Variables can be assigned within a macro which are case-sensitive. For example, counter=0, where 'counter' is the variable and 0 is its initial value. Variable names must not contain only numbers or any spaces. Variable values can also be strings, for example, day='Tuesday', in this example the value is not a number so must be encased in speech marks. Without speech marks ImageJ would assume that 'Tuesday' was already a variable and if it has not been defined an error message would ensue! Alphanumeric strings can be constructed in ImageJ as in the following example.

```
// macro start
year = 2011;
weekday = 'Tuesday';
month = 'August';
day = '20th';
date = 'weekday';
print(weekday + ' ' + day + ' ' + month + ' ' + year);
// macro end
```

In this case, the variable 'date' is built from a string containing other variables. When such a macro is run, the result is printed in the log window as the 'print' command is used. However, the variable can be used for many other functions such as giving a file a name or appearing in a custom results table. Note in this example it is a true macro and each line is followed by a semi colon to evaluate this statement. Comments are an important part of writing a macro to help users understand what the code is doing and are denoted using '//', whereby the macro language will ignore any following code in that line. It is good practice to have as many comments as there is code in a macro, which helps when it comes to debugging!

For loops

For loops are a long-established programming statement, having been available in the late 1970s with the Basic language on the Commodore

PET, enabling much more compact code. They are as invaluable now in ImageJ as they were back then and are more often than not the main driving element of macros. ImageJ does not offer the luxury of 2D arrays that may negate for loops, so they are an essential way of either running through images within a folder, or pixels within an image, for example. Indeed, to run through every pixel in an image, then a for loop nested within a for loop may be required to scan through every pixel in the x direction first, before starting scanning through a new y row of pixels. Multidimensional arrays can, however, be implemented in ImageJ using Java coding by the more experienced programmer. Additionally, another take on the for loop can be executed using a 'do . . . while' loop.

Arrays

Arrays are essentially sequences of numbers or alphanumeric values, which can be defined up front or captured by the macro to later be used as output. When they are numeric they can be processed by ImageJ, for example they can be sorted. An example of an array within a for loop is:

```
// Demonstration of for loop containing an array
number = newArray(10);
for (x=0; x<number.length; x++) {
number[x] = (x);
print(number[x]);
}
// end of macro.
```

Arrays are particularly useful to capture long strings of data coming from a macro, which can be stored in a results table on which further analysis can be performed. Arrays may also be used to input pixel values from an image, which can then be modified and the new array saved back as an image. There are built-in functions within ImageJ to handle arrays, such as to trim, sort and invert them as well as output statistics. When used in collaboration with for loops to populate and read them, they provide an alternative to using the central results table used by ImageJ to output measured parameters. This is useful when it is inconvenient to clear the results table to make way for a new set of measurements. Arrays can be large in size and are only limited by the RAM allocation to ImageJ.

Results and custom tables

ImageJ delivers all measured results into a central results table. In addition to ImageJ adding results to the table, the user can also add columns and rows of results with the 'setResults' command. These can be new results or calculations. However, in macros the user often wants to capture many different sets of results throughout the macro by clearing the results table. To bring all the results back together, several options exist. First, copies of the results table can be made with the 'IJ.renameResults(name)' function, which can be saved out to compile with others in a different application such as Microsoft Excel. Second, all results can be saved in a single 1D array, which can be read back into the results table using the 'setResults' command using offset for loops to turn the 1D array back into a 2D array of rows and columns.

Other elements in ImageJ macro language

Lastly, the conditional statements of 'if' and 'else' are implemented within ImageJ to provide a dichotomous selection of which piece of code to execute.

The statements in ImageJ are supported by standard Java boolean operators, but with fewer precedence levels. Care must taken to use these correctly, for example '=' is used to assign a value to a variable and '==' is used as an operator to check if variables are equal.

The macro language can also call on plug-ins that are installed within the plug-in folder. This allows the user to blend bespoke code and discrete functionality of a plug-in within the workflow. The syntax to call on a plug-in is straightforward using the run command shown below, which in this case calls on a plug-in called 'colour transformer' and provides an input variable for the plug-in, in this case the colour space to convert to.

```
run('Color Transformer Corrected', 'colour=LCHLab');
```

5.4 Graphical user interface

The main tool and menu bar in ImageJ (Figure 5.1) allows the user to interact with all the functionality and tools, and, in addition, to access links to service ImageJ such as updates and help, and to get started with macros and Javascript. The macro language allows the display of splash screens, input boxes and dialog boxes. If more complex GUIs are needed, then this requires involvement of Javascripts as in the example of SmartRoot (Figure 5.3). The following macro is given as an example of a macro that comes ready installed with ImageJ to demonstrate dialog

Published by Woodhead Publishing Limited, 2012

boxes and how to use them to capture user information that can then be subsequently used by the macro.

```
// This macro demonstrates how a macro can display a
// data input dialog box. The dialog it creates contains
// one string field, one popup menu, two numeric fields,
// and one check box.
title = 'Untitled';
width=512; height=512;
Dialog.create('New Image');
Dialog.addString('Title:', title);
Dialog.addChoice('Type:', newArray('8-bit', '16-bit', '32-bit', 'RGB'));
Dialog.addNumber('Width:', 512);
Dialog.addNumber('Height:', 512);
Dialog.addCheckbox('Ramp', true);
Dialog.show();
title = Dialog.getString();
width = Dialog.getNumber();
height = Dialog.getNumber();;
type = Dialog.getChoice();
ramp = Dialog.getCheckbox();
if (ramp==true) type = type + ' ramp';
newImage(title, type, width, height, 1);
// End of macro
```

When executed, this macro provides the following GUI, which incorporates free text fields, drop-down options and tick boxes with options to confirm these or cancel the macro (Figure 5.6).

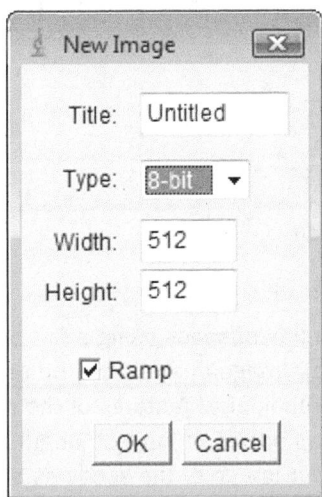

Figure 5.6 An example of a GUI that can be generated within the ImageJ macro language to capture user inputs

5.5 Industrial applications of image analysis

In the field of plant sciences, image analysis provides a way to non-subjectively measure many parameters specific to plants. Two examples are discussed here that have different objectives and associated challenges, to phenotype seeds and whole plants. The phenotype is the physical appearance of the plant and is linked both to its genome and environment, so measuring its parameters accurately is a core competency within plant sciences. Before image processing or analysis can be done, the image has first to be captured consistently and this is as important as the subsequent *in silico* steps.

The first example used a flat bed scanner, which gave the advantage of consistent lighting across the field of view, and a known image scale often expressed in dots per inch (DPI). The disadvantage of using the scanner is that only small or 2D subjects are suitable for scanning. The starting and fully processed and analysed image is shown in Figure 5.7.

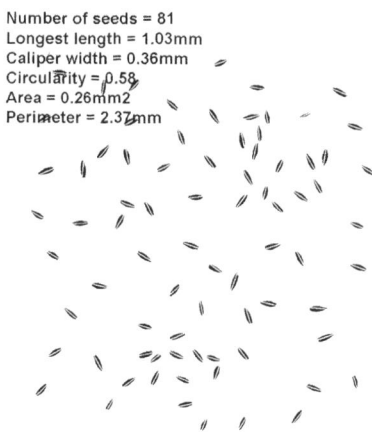

Number of seeds = 81
Longest length = 1.03mm
Caliper width = 0.36mm
Circularity = 0.58
Area = 0.26mm2
Perimeter = 2.37mm

Figure 5.7 Imaging of seeds using a flat bed scanner (left-hand side). Image analysis can be used to measure many morphological features of the seeds and return the mean scores of longest length (shown on the false colour image of the seeds as a white line in the right-hand side), calliper width, circularity (in which a value of 1 is a perfect circle and values are from 0–1), area and, finally, perimeter

When imaging at industrial scales, automation is crucial so the macro written within ImageJ was fully automatic and only had to be directed towards the folder of starting images. Using the control structures of for loops, if and else statements and arrays together with the ImageJ commands and macro functions, a creative workflow was built to fully automate image analysis tasks. Output was in the form of a comma delimited text file, which can be further manipulated in applications such as Microsoft Excel. The output of false colour images of the seeds showed exactly which objects were classified as seeds, with a white line drawn on them that represented the feret distance (also known as the maximum calliper), which was the longest dimension of the seed.

The next example is a type of analysis which is challenging to the human observer, and measures both colour and sizes. Plant leaves can take on a variety of hues and the goal was to classify them into different categories. Furthermore, the plant spread, represented by measuring the convex hull, was also needed. The starting image and the processed and analysed image are shown in Figure 5.8.

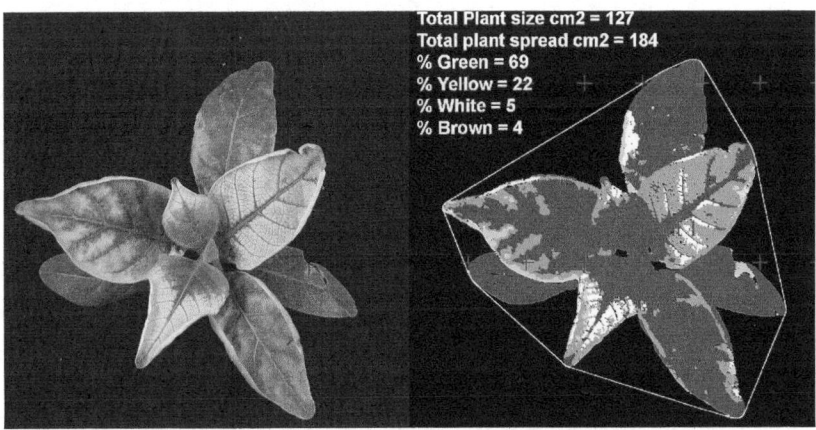

Total Plant size cm2 = 127
Total plant spread cm2 = 184
% Green = 69
% Yellow = 22
% White = 5
% Brown = 4

Figure 5.8 Plant phenotyping to non-subjectively quantify the areas of different colour classifications. The starting image is shown on the left-hand side and the segmented plant in classified false colour is shown on the right-hand side (converted here to a greyscale image). Note that the convex hull area to measure plant spread is shown surrounding the plant. False colour images are as important as the actual text results as they demonstrate to the user that the parameters have been measured correctly

The starting point to achieve accurate measurement of parameters is capturing consistent images in terms of lighting, focus, exposure and quality. Setting a digital SLR camera to full manual settings ensures a consistent focus, exposure and quality. The lighting is crucial, and consistency in this department is made by the use of grey cards and colour checking cards, such as those made by xrite (*http://www.xrite.com*). During imaging, inclusion of a QR code could be made to automatically assign the results in the image against a treatment identifier.

As with the seed images, automation is key within industry and in this case a semi-automatic approach is taken as the image scale was undefined. The user must first ascribe this to the images either by using a calibration object in the image of a known size, to allow a fully automatic solution (it could be the QR code for example), or by drawing a line of a known length just once if all images are to the same scale. Once this has been completed, the ImageJ macro is pointed at a folder hosting all the images, which are then automatically processed into false colour images and a results table produced. Importantly, false colour images can be used to allow the user to check that the parameters have been measured correctly, and to check whether any erroneous results may have been recorded. A key advantage of image analysis is that the original images are stored and can be reprocessed should novel parameters be identified that need to be subsequently measured. In this example, plants could also be imaged at different wavelengths, for example using infrared to examine water content, or using a laser scanner to generate 3D information. This information can be used in conjunction to allow a multiparameterisation of plant features to measure phenotype.

5.6 Summary

ImageJ is a freely distributed, open source solution for image analysis, which affords the user many different levels of interaction ranging from performing simple tasks from the menus through to high levels of automation by its in-built macro language, plug-ins and Javascripts. It benefits greatly from the large number of users in a global community that comes together virtually online in forums and in reality at conferences, which leads to its continual development, support and new version releases.

From a user and company perspective, ImageJ offers some unique characteristics associated with open source software. First, its free distribution means that everyone in a 20 000+ organisation can potentially run it on their PC without licensing issues. Second, it is platform-independent so that macros and plug-ins can be shared easily between researchers using different operating systems. Third, it is robust to use and the author has encountered few error messages apart from the unintentional ones in your own code (!) and running out of memory when trying to open too many images at once! Lastly, the way ImageJ integrates with other open sources packages to provide seamless workflows allows much flexibility to extract the useful data from your images, which is the very essence of image analysis.

5.7 References

[1] Moore GE. Cramming more components onto integrated circuits. *Electronics* 1965;38:8.
[2] ICBiovision (*http://icouzin.princeton.edu/*).
[3] Aphelion (*http://www.adcis.net*)
[4] Rasband WS. *ImageJ*, U. S. National Institutes of Health, Bethesda, Maryland, USA, 1997–2011. *http://imagej.nih.gov/ij/*
[5] Free software definition *http://www.gnu.org/philosophy/free-sw.html*
[6] Media cybernetics. *http://www.mediacy.com*
[7] *http://www.imagescience.org/meijering/software/neuronj/*
[8] *http://www.astro.physik.uni-goettingen.de/~hessman/ImageJ/Book/The%20Astronomy%20Plugin%20Tools/index.html#intro*
[9] Colour Models *http://en.wikipedia.org/wiki/Colour_models*
[10] ImageJ Macro functions *http://rsb.info.nih.gov/ij/developer/macro/functions.html*
[11] ImageJ interest group *https://list.nih.gov/cgi-bin/wa.exe?A0=IMAGEJ*
[12] ImageJ user and developer conference *http://imagejconf.tudor.lu/archive/imagej-user-and-developer-conference-2010*
[13] ImageJ package for R *http://cran.r-project.org/web/packages/RImageJ/RImageJ.pdf*
[14] Draye X, Yangmin K, Lobet G, Javaux M. Model-assisted integration of physiological and environmental constraints affecting the dynamic and spatial patterns of root water uptake from soils, *Journal of Experimental Botany* 2010;61(8): 2145–55.
[15] KNIME workflow tool *http://www.knime.org/*
[16] Online QR code generator *http://qrcode.kaywa.com/*
[17] QR code usage example *http://elliottslaughter.com/2011/07/qr-decoder-imagej*

## 6

Integrated data analysis with KNIME

Thorsten Meinl, Bernd Jagla and Michael R. Berthold

Abstract: In this chapter the open source data analysis platform KNIME is presented. KNIME allows the user to create workflows for processing and analyzing almost any kind of data. There is a short introduction to KNIME and its key concepts such as nodes and views, followed by a short review of KNIME's history, and the origin and key aspects of its success today. Finally, several real-world applications of KNIME are presented. These examples include workflows from the fields of chemoinformatics and bioinformatics, and image processing. All these workflows apply freely available extensions to the basic KNIME distribution contributed by community members.

Key words: interactive data analysis, professional open source, life science, workflow, chemoinformatics, bioinformatics, image processing.

6.1 The KNIME platform

The Konstanz Information Miner (KNIME) has been developed by the Nycomed Chair for Bioinformatics and Information Mining at the University of Konstanz since 2004. KNIME is open source and since 2010 it has been licensed under GPLv3. Therefore, for researchers, at universities or in industry, it is a free yet powerful tool to perform all kinds of data analysis. Based on the Eclipse platform (and thus Java) it runs on every major operating system (Windows, Linux, and MacOS X, both 32bit and 64bit) and is very easy to install – all users need to do is download [1] and unpack an archive (which already includes Java).

With KNIME, the user can model workflows, consisting of nodes that process data, which is transported via connections between the nodes. A flow usually starts with a node that reads in data from some data source, usually text files, but databases can also be queried by special nodes. Imported data are stored in an internal table-based format, where columns have a certain data type (integer, string, image, molecule, etc.) and an arbitrary number of rows conforming to the column specifications. These data tables are sent along the connections to other nodes. In a typical workflow, the data will first be pre-processed (handling of missing values, filtering columns or rows, partitioning into training and test data, etc.) and then predictive models are built with machine learning algorithms such as decision trees, naive Bayes classifiers or support vector machines. A number of view nodes are available to inspect the results of analysis workflows, which display the data or the trained models in various ways. Figure 6.1 shows a small workflow with some nodes.

The figure also illustrates how workflows can be documented by use of annotations: in the upper part of the flow classified molecules are read in, properties are calculated, and finally a decision tree is built to distinguish between active and inactive molecules. The lower part reads unclassified molecules and predicts activity by using the decision tree model.

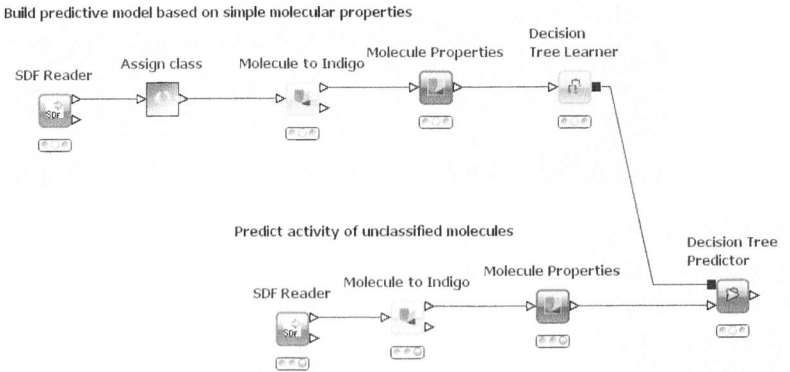

Figure 6.1 Simple KNIME workflow building a decision tree for predicting molecular activity

In contrast to many other workflow or pipelining tools, KNIME nodes first process the entire input table before the results are forwarded to successor nodes. The advantages are that each node stores its results permanently and thus workflow execution can easily be stopped at any node and resumed later on. Intermediate results can be inspected at any time and new nodes can be inserted and may use already created data without preceding nodes having to be re-executed. The data tables are stored together with the workflow structure and the nodes' settings. The small disadvantage of this concept is that preliminary results are not available as soon as possible as in real pipelining (i.e. when single rows are sent along and processed as soon as they are created).

Alternatively, many tasks performed via workflows can also be (often very simply) accomplished by applying simple spreadsheet programs or hand-written scripts. However, a workflow is a much more powerful method. In contrast to spreadsheets, it allows access to intermediate results at any time and can handle many more data types than just numbers or strings – to name only two advantages. In comparison to a set of scripts, a workflow is much more self-documenting, thus even non-programmers can easily understand which tasks are performed. Moreover, KNIME offers many more useful features, some of which we briefly describe in the following sections.

6.1.1 Hiliting

One of KNIME's key features is hiliting. In its simple form, it allows the user to select and highlight several rows in a data table with the result that the same rows are also hilited in all other views that show the same data table (or at least the hilited rows). This type of hiliting is simply accomplished by using the 1:1 correspondence between the tables' unique row IDs. There are, however, several nodes that completely change the input table structure and yet there is still some relation between input and output rows. A nice example is the MoSS node, which searches for frequent fragments in a set of molecules. The node's input are the molecules, the output the discovered frequent fragments. Each of the fragments occurs in several molecules. By hiliting some fragments in the output table, all molecules in which these fragments are contained are hilited in the input table. Figure 6.2 shows this situation in a small workflow. In this flow, public nodes from Indigo (see also below) are used to display the molecular structures.

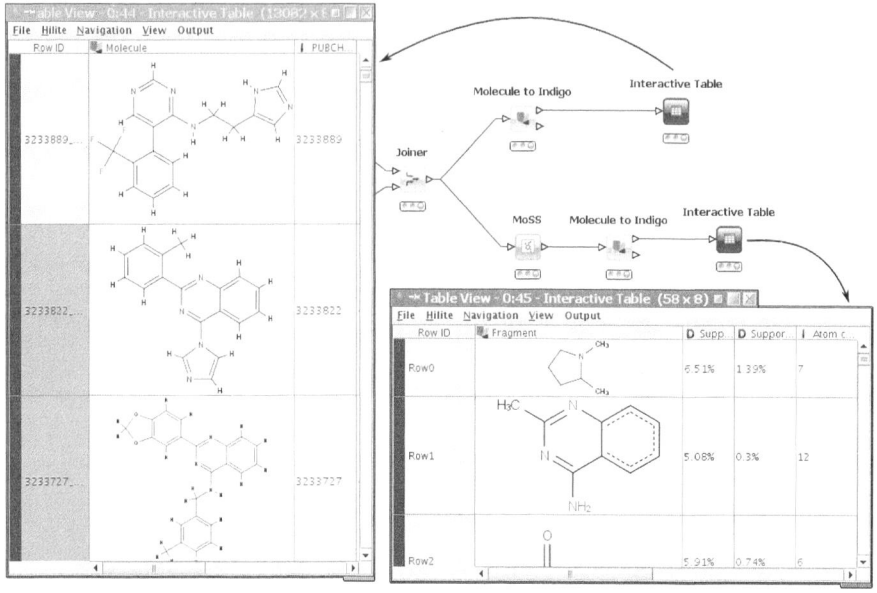

Figure 6.2 Hiliting a frequent fragment also hilites the molecules in which the fragment occurs

6.1.2 Meta-nodes

Workflows for complex tasks tend to increase considerably in size. Meta-nodes are a practical way of structuring them. They encapsulate sub-workflows, which require several nodes to accomplish a certain task in the parent workflow. Meta-nodes can either be created as empty sub-workflows, to which other nodes are added, or a number of nodes in an existing workflow can be selected and collapsed into a meta-node. Meta-nodes can even be nested to arbitrary depths. Moreover, certain common tasks, such as cross-validation or feature elimination, which require more than one node are available as pre-configured meta-nodes in the node repository.

6.1.3 Loops

The workflows' conceptual structure is a directed acyclic graph, that is there are no loops from the output of one node to the input of one of its predecessors. Data flows strictly in one direction. However, there are cases in which the repeated execution of parts of the workflow with changed parameters is desirable. This can range from simple iterations over several input files, to cross-validation where a model is repeatedly trained and evaluated with different distinct parts of data and can include even more complex tasks such as feature elimination. In order to be able to model such scenarios in KNIME, two special node types are available: loop start- and loop end-nodes. In contrast to normal nodes (inside the loop), they are not reset while the loop executes, both nodes have access to its counterpart, and they can directly exchange information. For example, the loop end-node can tell the start-node which column it should remove at the next iteration or the start-node can tell the end-node whether or not the current iteration is last. Figure 6.3 shows a feature elimination loop in which the start- and end-nodes are visually distinguishable from normal nodes. The feature elimination model can subsequently be used by the feature elimination filter to remove attributes from the data table.

The node repository contains several predefined loops encapsulated in meta-nodes, such as a simple 'for' loop, which is executed a certain number of times, cross validation, iterative feature elimination, or looping over a list of files.

6.1.4 Extensibility

KNIME is an Eclipse-based application, which has two major implications. First, it is written in Java meaning that it runs on all major operating systems (Windows, Linux, MacOS X). Second, the Eclipse platform was designed to be highly extensible – as is KNIME. Developers can easily

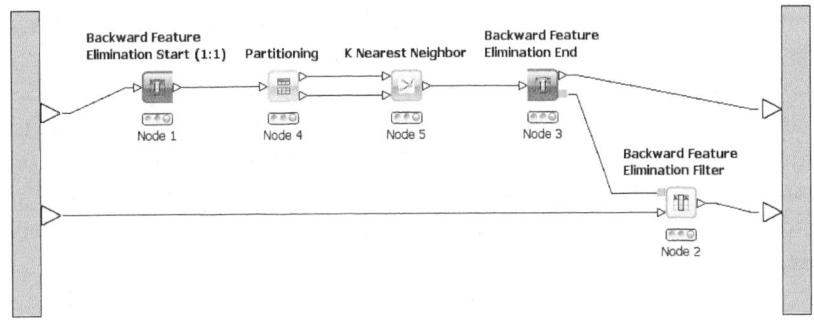

Figure 6.3 Feature elimination is available as a loop inside a meta-node

write their own extensions to KNIME by using the public API. Once packaged appropriately – either as ZIP files or as online update sites – users can quickly install (and uninstall) them into their existing KNIME installation by using the graphical update manager.

6.2 The KNIME success story

As already mentioned, KNIME initially started as a university project with a group of four people. One advantage about this kind of project is that it is relatively easy to find developers in the form of PhD students or student assistants who invest considerable amounts of time on the project. In the case of KNIME, one of the four people had already gained several years' experience as a professional software developer. However, it still took more than two years before the first version was ready for the public (28 July 2006). One big disadvantage of many university projects is the sustainability of the project: when developers leave – because they have graduated – they can only spend little if any time on further development, which often leads to the death of the project. There are two ways out of this dilemma. One way round this problem is to attract a large number of developers outside the group right from the beginning to ensure that when one developer leaves, another developer is there to step in and continue with the work. This, however, usually only works for very wide-spread projects such as the Linux kernel, the Apache web server, or the Secure Shell, for example. Another solution to the problem is to found a commercial company to take care of and continue with further development. A combination of both is possible of course.

For KNIME, the second option, the foundation of KNIME.com [2] in 2008, proved to be strategic to its success. Once the key people from the university group had completed their PhD, they switched to the company. Besides being in charge of further development of the open source KNIME core, the KNIME.com team is currently developing commercial extensions such as a KNIME server and cluster execution components. Another important focus of the company is to provide training courses for users and developers as well as support in case of problems. The university group continues to develop new nodes based on current research topics. Both parties collaborate well and ensure that the basic KNIME functionality is extended and new analysis methods are added.

Even though about 15 people are currently working on KNIME in both groups, it is obvious that they cannot develop solutions for all kinds of applications areas nor do they have the means to develop all the existing methods further in a given scientific area. Fortunately, two other parties jumped in. From the beginning there were commercial vendors (such as Schrödinger or Tripos) who integrated their existing tools into KNIME. The number of such companies has grown tremendously in recent years. A growing number of community members have also started developing extensions, which they provide free of charge to the community. Initially they were scattered over different locations, but since 2011 many of them are centrally hosted on the official KNIME server. This has benefits for the users, who have a central access point to obtain extensions for KNIME and do not need to search and collect them from various sources, while developers also benefit from the fact that they are provided with a source code repository as well as a nightly build service, a web page, and a discussion forum. At the time of writing, seven different projects exist in the KNIME Community Contributions [3] and many others are already in the queue. Remarkably, not all projects are from universities or other governmentally funded groups, but also come from industry.

6.3 Benefits of 'professional open source'

The term 'professional open source' stands for a business model where a vendor provides support and commercial extensions to an otherwise open source program. This concept was successfully pioneered by companies such as Red Hat or MySQL. In the same fashion, KNIME. com offers the 'KNIME Professional' package, which adds personal support and access to emergency patches in addition to the free open source version (see *http://www.knime.org/products* for details). This is especially useful when KNIME is used in production environments. To facilitate collaboration inside or between groups in a company, KNIME. com also provides the KNIME Team Space and the KNIME Server which enable (besides other features) easy sharing of workflows and meta-node templates.

Not only is the 'professional' aspect beneficial to (industry) users, but also the open source 'core' has several advantages for all user groups. First of all, the usage is completely free of charge, no matter how many computers it is installed on. This may not be an issue for some companies with regard to using the graphical user interface on a limited number of desktop computers, but as soon as it comes to deploying KNIME on a large compute cluster or even in a cloud, the benefits of open source and its licenses are obvious. The latter is especially interesting because KNIME can also be run in batch mode without a graphical user interface to execute pre-built workflows. This allows huge amounts of data to be processed on a large scale with no additional software costs.

Another aspect of KNIME's free accessibility is that many more people can easily use it and start to write extensions for their own or other freely available programs. These not only include Indigo [4], RDKit [5], or Imglib [6] as described in the next section, but also connectors to other well-known open source software programs such as the R statistics software and the Weka Data Mining suite, or integration of programming languages such as Perl or Python. Having access to such well-established software packages from inside KNIME is a benefit to all users, both in academia and industry.

6.4 Application examples

KNIME is used in many areas, such as finance, publishing, and life sciences. A private bank in Switzerland, for example, is applying KNIME workflows for customer segmentation based on investment philosophy and purchase behavior. A German publishing company is using KNIME to conduct propensity-to-purchase scoring and to look for early non-renewal indicators. A world-wide B2B-organization applies KNIME to perform needs and value segmentation of their customers to better plan campaigns via their sales force.

These are only a few examples of KNIME's usage in non-life science industries. However, the life sciences such as chemoinformatics and bioinformatics are the fields in which KNIME is currently used the most. Therefore the following example workflows deal with problems relevant to these areas. In order to use these workflows, one first needs to download and unpack KNIME [1]. As all workflows make use of nodes contributed by community members, you need to install the respective extensions from the Community Contributions [3].

All workflows are available on the public workflow server, which can be conveniently accessed directly from any KNIME installation.

6.4.1 Comparing two SD files with Indigo

The following scenario depicts a common use case in chemoinformatics. A supplier delivers an SD file with new structures and the user wants to check whether these structures are already in the corporate database or if

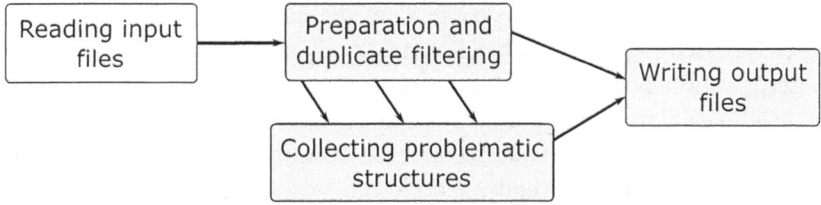

Figure 6.4   Outline of a workflow for comparing two SD files

there are new structures that should be purchased. This task can be accomplished with the following KNIME workflow. As the workflow is too big to be shown on a single page in readable form, we have divided it into four different parts. The overall process is depicted in Figure 6.4.

The first part is shown in Figure 6.5 and is quite simple: first, it reads two SD files (e.g. the first containing the new structures from the vendor and the second containing the corporate database) with the SDF Reader, then adds a column indicating the source of each molecule with a Java Snippet node, and finally concatenates both tables.

This small workflow already shows two interesting concepts in KNIME. First, the SDF Reader has two output ports: the upper port provides a table with all the structures that have been successfully parsed and the second port contains a (possibly empty) table with structures that could not be parsed due to syntactic errors in the SD file. The reader node

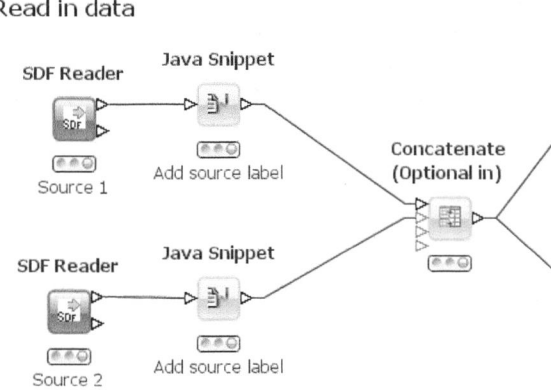

Figure 6.5   Reading and combining two SD files

does not try to interpret the contents of the file but merely extracts the text representing each molecule into a cell of the output table. The interpretation is done by special conversion nodes (e.g. CDK, Indigo, Schrödinger, Tripos, etc.), which are more or less strict about the actual representation of a molecule. Second, the Concatenate node has so-called optional input ports. Usually, a node can only be executed if all input ports are connected, but for optional input ports this is not necessary. This makes perfect sense when concatenating tables because it frees the user from having to build a cascade of two-port concatenation nodes when there are several tables to combine.

The next part of the workflow deals with the preparation of the molecules. The Indigo [4] nodes are used for the whole filtering process. These nodes were recently contributed as open source by GGA Software and build on their Indigo chemical library.

In the first step (Figure 6.6), the SD records read in by the previous step are converted into the internal Indigo format. Molecules that cannot be converted, for example because of wrong stereochemistry or invalid atom types, are transferred to the second output port and added to the list of problematic structures (not shown). All remaining molecules are checked for correct valences and structures further down the pipeline. Structures failing the test are again added to the problematic structures. In the next step very small compounds (having a molecular weight of less than 10) are filtered out as uninteresting. Also, all structures that consist of several fragments are removed. This is necessary because in the next step (see Figure 6.7) canonical SMILES are generated, which cannot handle disconnected structures.

The most important step involves the Group By node, which groups all incoming rows based on the canonical Smiles string. In addition, it adds two columns to each canonical SMILES: a list of IDs that share the same SMILES, and the number of rows with that SMILES. The following row splitter is then used to divide the molecules into unique molecules (first

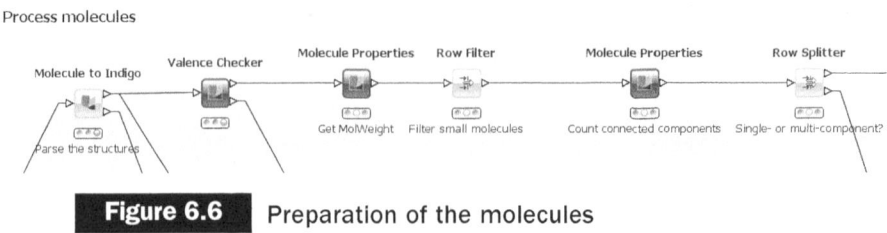

Figure 6.6 Preparation of the molecules

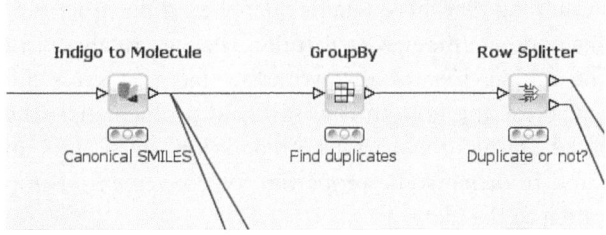

Figure 6.7 Filtering duplicates

output port) and duplicates (second output port) simply based on the number previously mentioned.

The last part of the workflow consists of writing the different sets of molecules to files (Figure 6.8). The upper row takes the unique molecules determined in the previous step and joins them with the table containing the converted molecules. This is necessary because they get lost during the group by operation (although this step can be omitted if one is only interested in the IDs of the SMILES). The converted molecules also provide a good 2D depiction for further inspection. The Row Filter node is important. It removes the unique molecules that come from the second

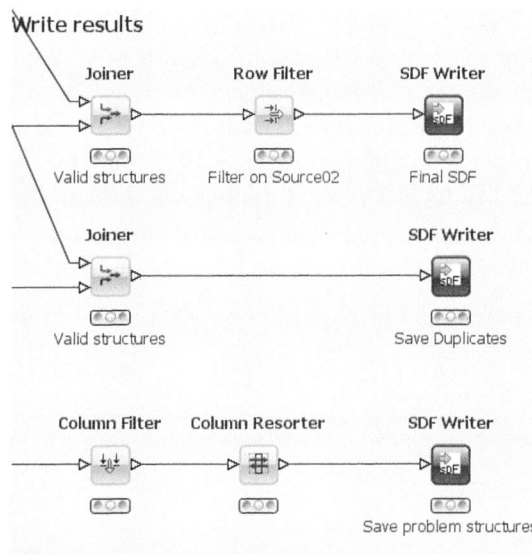

Figure 6.8 Writing out the results

Published by Woodhead Publishing Limited, 2012

input file. Assuming that this contains the pre-existing structures, all that is left after this node are the new structures. They are finally written into an SD file. The middle row of the workflow likewise saves all duplicate structures, whereas the bottom row saves all problematic structures that were gathered during the various preparation steps into another file (removing and resorting some properties for convenience before they are actually written to the file).

6.4.2 Enumerating amides with RDKit

The next workflow deals with reaction simulation, in particular the formation of amides by a two component reaction of amines and acids. Similar workflows are in use at Novartis and other pharmaceutical companies. This workflow makes use of the RDKit [5] nodes. A lot of functionality is provided from this well-known open source toolkit in KNIME. The concepts of quickforms and variables are also introduced by this workflow. The workflow is shown in Figure 6.9.

First, an SD file containing several amines and acids (in any order) is read in. Next, a meta-node encapsulates some preprocessing of the workflow. Specifically, it splits the input file into amines and acids and also reduces their number based on the user selection (e.g. three amines and three acids). Each group is then output to one of the two ports and passed to the Two Component Reaction node. This node simulates pairwise reactions between all molecules from both input ports and outputs a table with the resulting products. As the products are stored in the internal RDKit format, they need to be converted back to SDF before they can be written into a file (upper branch). A PDF file containing 2D depictions of all products is created in the lower branch. In the process, 2D coordinates are generated first and then the column with the 2D depictions is transformed into a grid. This means that all cells in the

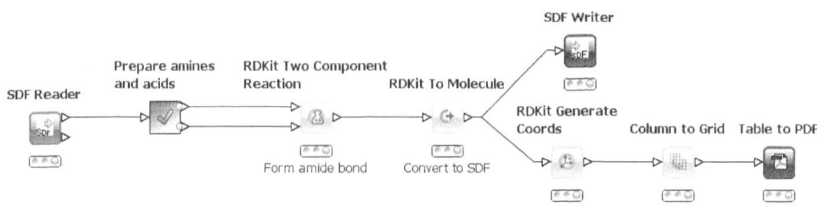

Figure 6.9 Amide enumeration workflow

selected columns are arranged in a data table with, for example, four columns and as many rows as necessary to show all the products. A PDF is subsequently created from this table using the Table to PDF node, which is available on the KNIME Update Site as an extension.

The content of the meta-node that creates two sets of reactants is shown in Figure 6.10. First, the table containing the SDF records is processed with the Molecule to RDKit node, which converts the molecules into an internal format. Subsequently the converted molecules are passed to two Substructure Filter nodes that filter the molecules based on a SMARTS pattern. In the upper branch only amines pass the filter (SMARTS pattern [NX3;H2,H1;!$(NC=O)]), whereas in the lower branch only acids can pass through (SMARTS pattern [CX3](=O)[OX2H1]). Both molecule sets are then passed through a Row Sampling node. Usually, the user can enter several parameters for sampling (e.g. the number of rows to pass through, if random or stratified sampling should be performed, or if only

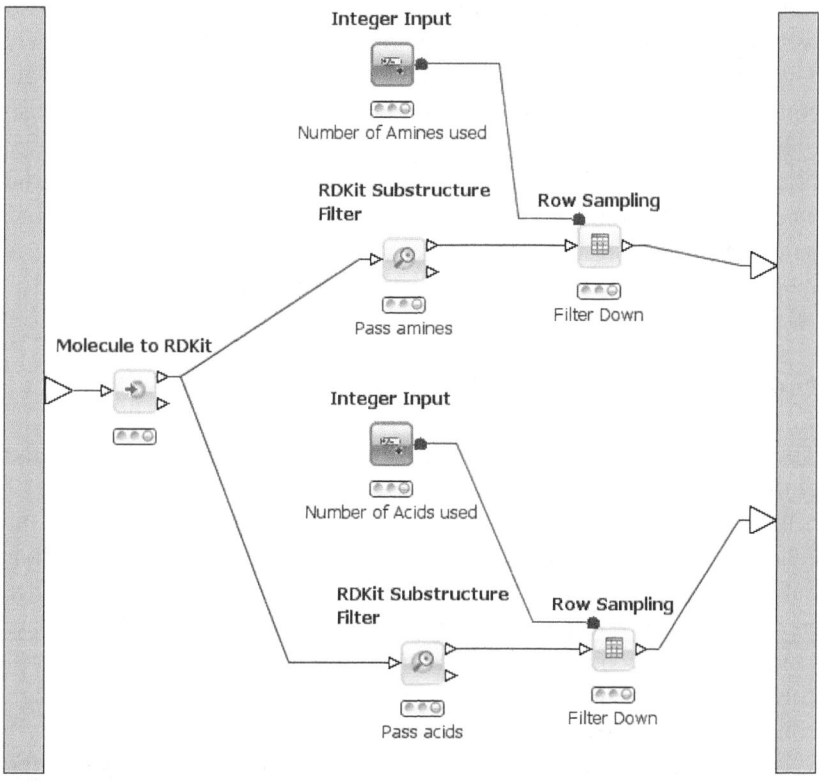

Figure 6.10 Contents of the meta-node

the top-n rows should be passed). However, in this case some of the parameters for the node are passed from a different node. The Integer Input node is part of a collection of so-called quickform nodes that allow injection of certain variables into a workflow from the outside. In the quickform node a description and a default value for the variable is entered, for example the number of amines/acids that should be passed through. When used inside a meta-node all variables defined by quickform nodes are exposed in the dialog of the meta-node, that is the user can configure the meta-node from the outside without opening it. Once values for each variable are available, they are passed into the row sampling nodes via special variable ports (red balls instead of the usual white triangles for data tables). In the sampling nodes' dialog the parameter for the number of rows is subsequently linked to the variable instead of taking the (default) value that was directly entered in the dialog.

In combination with the commercial KNIME Server, quickform nodes can also be used outside meta-nodes. Here, variables are exposed in a web interface allowing the user to easily configure a workflow from a web browser and execute the workflow on the server, see Figure 6.11. The results (files, numbers, . . .) are also available via the browser. This concept allows 'power users' to build elaborate workflows and expose only certain parameters to the outside. Once uploaded to the server (directly from within KNIME), the 'normal' users can access the workflows via the web portal with the browser, provide input files and parameters, execute it on the server, and finally retrieve the results. A workflow can even be executed several times and all results are stored.

6.4.3 Classifying cell images

The third example workflow solves a common task in bioinformatics: the automated classification of cell images. One use case is the treatment of cells with different substances at various concentrations. At some point cells start to die and the task is to ascertain the lethal concentration automatically. For this, cell cultures are grown in 96-well plates and are treated with the substances in different concentrations. On each plate there are always positive and negative controls, for example the cell cultures without additional substances and cultures with very high concentrations of known activity, respectively.

Automated imaging systems take pictures, one for each well. Using the Image Processing nodes (which again rely on the free Imglib [6]), which are also available from the community contributions, these images can be

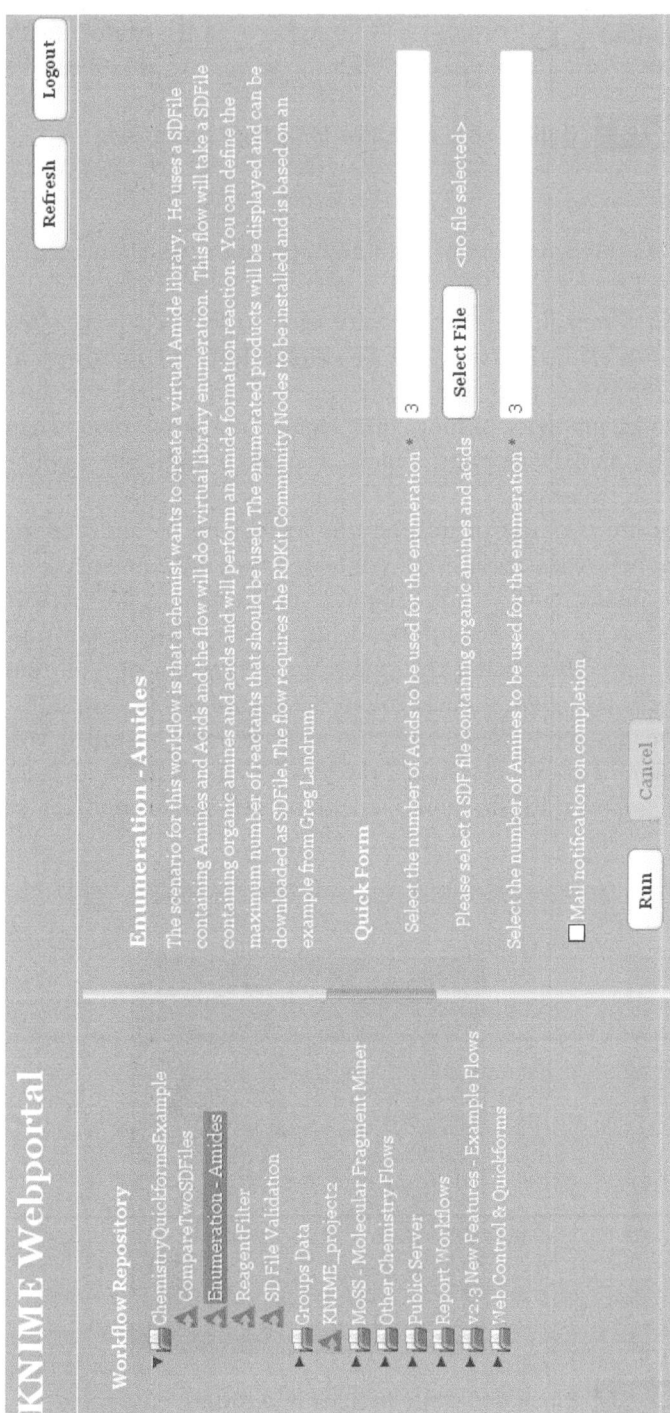

Figure 6.11   KNIME Enterprise Server web portal

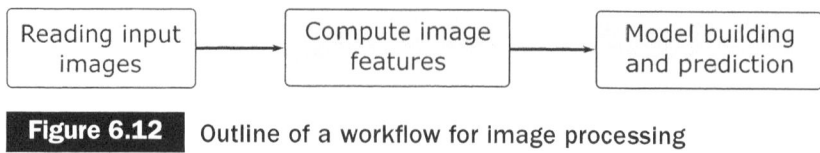

Figure 6.12   Outline of a workflow for image processing

analyzed. As this workflow is again quite complex, we show the outline in Figure 6.12.

In the first step, Image Readers are used to load all images from a directory into KNIME. The Image Processing feature comes with a new data type for images to enable them to be used in data tables in the same way as strings, numbers, or molecules. Figure 6.13 shows one such image in KNIME's table view (the images are taken from the public SBS Bioimage CNT dataset).

As there are two sets of images, one for the nuclei and one for the cytoplasm, before the features are computed they are combined into one table. The image features are computed independently for each single cell. Therefore the cells have to be identified first. This is performed by taking the nuclei images and applying a binary thresholder to distinguish the nuclei clearly from the background, see Figure 6.14.

These images are subsequently used as seeds for a so-called Voronoi segmentation. This process takes the cytoplasm images and segments them into many different non-overlapping regions around the seeds.

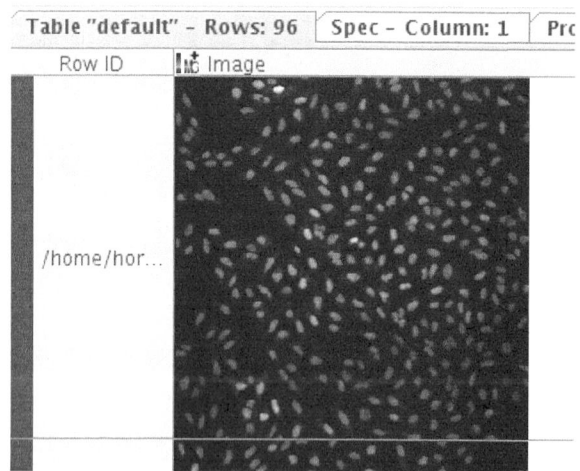

Figure 6.13   Black-and-white images in a KNIME data table

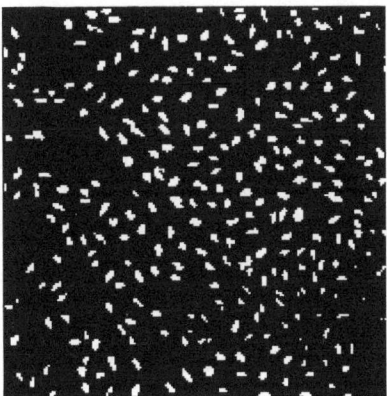

Figure 6.14 Image after binary thresholding has been applied

These segments are used to compute a variety of features such as histogram-based, texture-based, or segment-based features. The workflow part that deals with segmentation and feature generation is shown in Figure 6.15. As KNIME makes use of multicore systems and executes independent branches of a workflow in parallel, the workflow layout shown in the figure ensures that all features are computed in

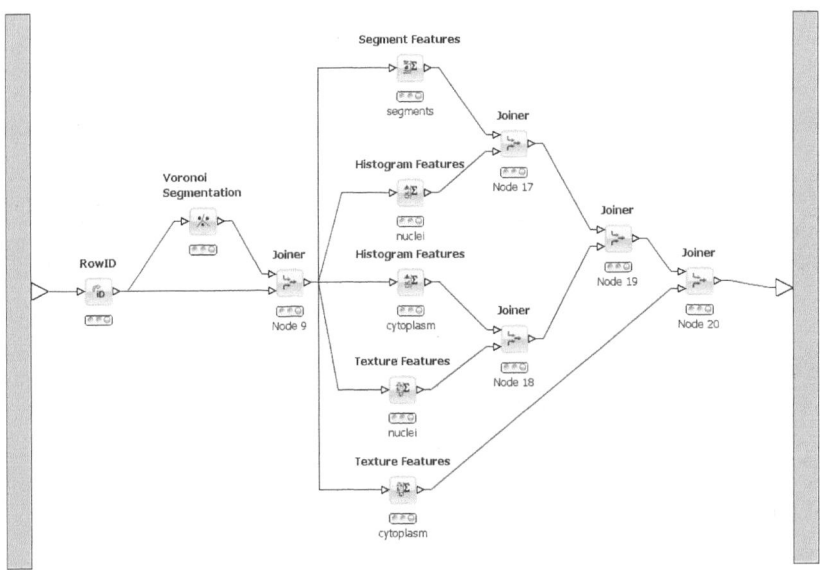

Figure 6.15 Meta-node that computes various features on the cell images

parallel. In the end they need to be re-combined into a single table using a cascade of Joiner nodes.

The computed features are all numeric or nominal, with the result that the remaining parts of the workflow – building a predictive model and applying it to unclassified images – are straightforward (and therefore not shown here; a similar example is given in Figure 6.1). All features originating from the control samples are used to build the model, that is a decision tree and all other segments are subsequently classified using this model.

6.4.4 Next Generation Sequencing

The workflow described here (Figure 6.16) is an example of how to use KNIME for Next Generation Sequencing (NGS) data analysis [7]. In particular, it describes typical parts of a data analysis workflow with regard to RNA-sequencing, DNA-sequencing, and ChIP-sequencing. It does not try to show a complete or perfect workflow but rather to point out some of the features of what can be done and explicit NGS relevant tools that can be used. This and similar workflows are used by Institute Pasteur.

In general, this workflow reads-in FastQ formatted data, which is cleaned and filtered, then aligned to a reference genome (hg19) using bowtie [8] (upper part). The results are read and filtered and then written

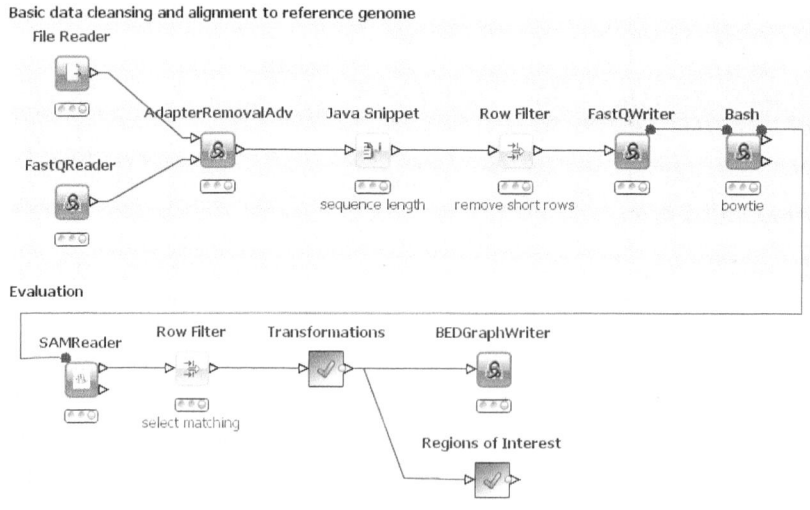

Figure 6.16 A workflow for large-scale analysis of sequencing data

into a BEDGraph [9] file for visualization using, for example, GBrowse [10] or UCSC [11] genome browser (lower part). Regions of interest are also being identified.

As this workflow is also quite complex, we only highlight the most interesting parts. A complete description of an even more elaborate workflow is available in a separate publication [7]. Similar to the previous workflows, this one also makes use of the KNIME Community Contributions. The NGS package offers special nodes for dealing with NGS-related data. One example is the FastQ Reader, which reads the de facto standard FastQ file format using the BioJava library. Its output is a data table containing the cluster ID and the sequence along with the quality information. The File Reader reads parameters for the subsequent Adapter Removal Adv node, such as adapter sequences or other contaminating sequences, similarity threshold, quality threshold, and minimum overlap. This latter node compares each sequence from the FastQ file (target) with all sequences from the parameter file (query) and removes contaminations. The output is the second input table with adapters removed from the input sequences. The following nodes compute the sequence length and filter out very short sequences before writing everything back into a FastQ file. The subsequent 'Bash' node is only connected to its predecessor with a variable port. This ensures that it is not executed before the FastQ file has been written. The node executes a bash-script, which calls the *bowtie* program to align all sequences to the reference genome (hg19). Its output is a SAM formatted file with the information from the alignment. This is read in with the SAM Reader (again connected with a variable port to its predecessor). The next nodes select only sequences that align to the reference genome (Row Filter) and apply various transformations that result in a data table holding the original sequences, the chromosomes from the reference genomes and the positions in the sequences where there has been a mismatch in the alignment process (Meta Node). This information is subsequently written out into a BEDGraph file.

In the second branch, the ROI meta-node (see Figure 6.17) extracts the regions of interest (i.e. consecutive regions of coverage between the sequences and the reference chromosome). This is especially interesting when analyzing small RNA. The input is a list of positions from the reference genome with associated coverage. This list is already sorted by chromosome and position. The *GetRegions* node identifies regions of interest (ROIs). A ROI is defined as having entries in the input table with the same chromosome name and increasing (by one) positions, that is consecutive regions of coverage. Values are stored in a string column and concatenated using a space as a separator. Next, a Java Snippet retrieves

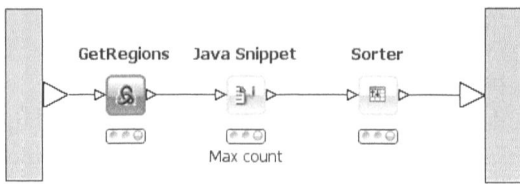

Figure 6.17 Identification of regions of interest

the maximum count value from the count string whereupon all rows are sorted by that value to retrieve the regions with the highest coverage as the top entries in the output table.

Further open source tools for genomic data analysis are described in Chapter 8 by Tsirigos and co-authors.

6.5 Conclusion and outlook

It is possible to perform sophisticated data analysis tasks in KNIME using completely free software. The many contributors from the KNIME community ensure that the number of freely available extensions will continue to grow in the future, providing even more application areas in which KNIME can be used at no cost. There are, naturally, some areas in which no free methods are available and situations where guaranteed support is necessary when reliability is crucial for the success of projects. This is where commercial vendors come into play. KNIME's modularity and licensing model facilitates the usage of a mixture of free and non-free components. Another important aspect is the existence of the KNIME.com company which not only sells training courses, provides professional support, and develops enterprise extensions but also plays a key role in the further development of KNIME's free open source core.

6.6 Acknowledgments

We would like to thank the authors of the extensions described here for making their nodes available to KNIME. Special acknowledgments also go to Greg Landrum, Dmitry Pavlov, Frank Schaffer, and Martin Horn for providing some of the example workflows shown in this chapter.

6.7 References

[1] KNIME. *http://www.knime.org/*. Accessed 18 August 2011.

[2] KNIME.com. *http://www.knime.com/products*. Accessed 18 August 2011.

[3] KNIME Community Contributions. *http://tech.knime.org/community*. Accessed 18 August 2011.

[4] GGA Software Services LLC. Indigo. *http://ggasoftware.com/opensource/indigo*. Accessed 18 August 2011.

[5] RDKit. *http://www.rdkit.org/*. Accessed 18 August 2011.

[6] Imglib. *http://pacific.mpi-cbg.de/wiki/index.php/Imglib*. Accessed 18 August 2011.

[7] Jagla B, Wiswesdel B and Coppee JY. Extending KNIME for Next Generation Sequencing data analysis. *Bioinformatics*, 2011;20:2907–9.

[8] Langmead B, Trapnell C, Pop M, Salzberg SL. Ultrafast and memory-efficient alignment of short DNA sequences to the human genome. *Genome Biol* 10:R25.

[9] BEDGraph. *http://genome.ucsc.edu/goldenPath/help/bedgraph.html*. Accessed 17 August 2011.

[10] Gbrowse. *http://gmod.org/wiki/GBrowse*. Accessed 27 December 2011.

[11] Kent WJ et al. The human genome browser at UCSC. *Genome Res* 2002;12(6):996–1006.

## 7

Investigation-Study-Assay, a toolkit for standardizing data capture and sharing

Philippe Rocca-Serra,
Eamonn Maguire, Chris Taylor,
Dawn Field, Timo Wittenberger,
Annapaola Santarsiero, Alejandra Gonzalez-
Beltran and Susanna-Assunta Sansone

Abstract: This chapter introduces the problems experimentalists in all sectors face in utilizing third party data sets given the unhelpful wealth of formats and terminologies and the consequent mountain of technical frameworks needed to achieve data interoperability. We argue on the importance of a complementary set of open standards, the challenges we must overcome and the role the BioSharing effort is set to play. As an example of progress, we present the open source ISA software solution in action during the curation of the InnoMed PredTox data set, along with its growing active developer and user community, including academia and industrial sectors, such as The Novartis Institutes for BioMedical Research and Janssen Research & Development.

Key words: data management; biocuration; community standards; ontology; open source software; InnoMed PredTox; BioSharing, ISA.

The successful integration of heterogeneous data from multiple providers and scientific domains is already a major challenge within industry. This issue is exacerbated by the absence of agreed standards that unambiguously identify the entities, processes and observations within experimental results.

> Empowering Industrial Research With Shared Biomedical
> Vocabularies, *Drug Discovery Today* (2011)

7.1 The growing need for content curation in industry

High-quality public bioscience research data should be readily available for re-use in private sector research and development. At present, public resources, diverse in implementation, provide data whose formatting and annotation vary widely, requiring extensive manipulation to open their content to integrative analyses. Such hindrances have motivated community standardization initiatives to develop minimum information checklists, terminologies and file formats, which are increasingly used in the structuring, description, formatting and curation of data sets, although in most cases only within the originating community. These standards aim to ensure that descriptions of entities of interest (e.g. genes, receptors) and related assays contain sufficient contextual information ('experimental meta-data' – e.g. provenance of study materials, technology and measurement types, sample-to-data relationships) to be comprehensible and (in principle) reproducible; without such context, data are of little value.

The process of utilizing shared, publicly available data using community-sourced standards can still test the resolve of even the most ardent advocate [1]. The focus of most community standardization efforts on their own interests or technologies has led to development of equivalent, yet (largely arbitrarily) different localized standards and esoteric repositories, hindering data integration. Whether searching for the scattered files from the various assays in a broadly based study, or assembling the available data on a species or feature of interest, fragmented data sets can only be re-assembled by those equipped to navigate the various terminologies and formats used to represent and annotate their parts (assuming their annotation is sufficient even to reliably identify them), impacting the ability of the R&D community to utilize such data. And, of course, the dearth of accepted standards extends to commercial knowledge providers, whose information also comes in many forms, magnifying the challenge.

Although several integration workflows are routinely run internally, the person-hours cost of 'deciphering' the heterogeneous experimental meta-data still remains significant. Companies must invest significant effort to integrate public bioscience data with their own data; or outsource such activity [2]. The mountain of technical frameworks needed to achieve interoperability between community standards has also hindered the development of general tools. The diversity of standards – and the consequent lack of general tools – hinders discovery, because only a very willing few can even discover, never mind integrate, information scattered across several standalone resources.

7.2 The BioSharing initiative: cooperating standards needed

Left unresolved, or separately and therefore inefficiently addressed by individual companies, the lack of agreed standards will continue to limit the utility of public data. The solution lies in an open collaborative approach between the public and private sectors, lowering individual risk and costs [3, 4]. To establish the lay of the standards landscape, and to build graphs of relationships and complementarities in scope and functionality, the BioSharing community catalogues available standards [5], extending the work started with the Minimum Information for Biological and Biomedical Investigations (MIBBI) portal [6]. These standards often exhibit different levels of maturity and inevitably duplication of effort. Lack of overall coordination also ensures that significant gaps in coverage remain. Although individual communities cannot be corralled into collaboration, the BioSharing initiative is intended to promote those that already exist, discouraging redundant if unintentional competition. In time and after consultation, a set of criteria for assessing the usability and popularity of the standards listed will be implemented, along with links to tools that use them, or data resources annotated with them.

If a common or at least complementary set of standards existed and was widely used by the academic and commercial sectors, routine tasks in the exploitation of both public and proprietary data such as text mining, re-annotation and integration would be greatly facilitated [7]. There are also other benefits accruing to the development and acceptance of general data and reporting standards. For example, by limiting the range and variability of standards, the development and maintenance costs of commercial and academic software come down, resulting in more appropriate resources for the public and private scientific community. In turn, this makes the job of capturing, annotating, integrating, sharing and exploiting data simpler, increasing the prima facie value of the data to others (secondary users) and increasing the return on the investment that supported their generation.

Many challenges lie ahead. Unequivocal 'rules of engagement' must be defined, extensive community liaison managed and rewards and incentives identified for all contributors, whether from the commercial or public sector. In particular, ownership of standards can be problematic in broad collaborations. The appropriate legal framework is still in embryonic form, yet IP rights and licenses must be established to (1) define the boundaries for commercial exploitation by creators and contributors; (2) enable commercial entities to freely contribute time, use cases and requirements; and (3) manage revenues arising from the commercial exploitation of the IP inherent in a resource.

Another critical issue that remains to be addressed is the development of a strategy for the long-term sustainability of this endeavor. Industry funding cannot be the sole source of support, as budgets and priorities fluctuate year to year. Robust relationships among appropriate commercial stakeholders, coupled with participation by governments and research funding organizations will make for a more diversified funding portfolio, buffering the project against fluctuations in the ability of any one partner to contribute. Overall, the cost of implementing this vision is significant and requires BioSharing to continue enlarging its community; particularly, in close partnership with the industry-driven Pistoia Alliance [4].

7.3 The ISA framework – principles for progress

1. Standards initiatives should be more like rafts than cruise liners: simple and unsinkable.

2. As with any evolutionary change, each step should bring reward, not depend on belief.

3. Existing initiatives should be leveraged where possible, for efficiency and buy-in.

4. Solutions should be forward-looking with respect to further, deeper integration.

In combination, these principles require that any solution be simple, immediately beneficial, and respectful of existing work. The 'Investigation/Study/Assay' (ISA) framework, the product of an ongoing collaboration between various research and service groups actively involved in the development of community standards [8], offers such a solution. By providing a generic backbone for structured descriptions of bioscience research – the Investigation, Study, Assay hierarchy around which all else is built – the ISA framework simultaneously offers an interim solution that respects existing data formats and a workable scaffold around which to build new, integrated standards. The basic ISA backbone, extended in an appropriately generic manner, has been implemented as 'ISA-Tab' – a simple format supported by several projects, not least the ISA software suite whose component modules constitute the core elements of a collaborative framework [9].

Using the shared, meta-data-focused ISA framework it is now possible to (1) aggregate investigations of biological systems – where source material has been subject to several kinds of analyses (e.g. genomic sequencing, protein-protein interaction assays and the measurement of metabolite concentrations) – as coherent units of research; (2) perform meta–analyses; and (3) more straightforwardly submit to public repositories, where required. The latter, unfortunately, are still designed for specific analyses types, necessitating the fragmentation of data sets because of the diversity of reporting standards with which the parts must be formally represented.

7.3 The ISA-Tab format

ISA-Tab is the result of a painstaking exercise to map a number of repository submission formats onto one structure for experimental meta-data, to facilitate bidirectional conversion; leveraging common elements while intentionally keeping data files external in their native or community formats to side-step interoperability issues. ISA-Tab, illustrated in Figure 7.1, has become parent to a variety of spreadsheet-based formats for data sharing [10]. The Investigation file contains all the information needed to understand the overall goals and means used in an experiment; experimental steps (or sequences of events) are described in the Study and in the Assay file(s). For each Investigation file there may be one or more Study file; for each Study file there may be one or more Assay files.

The Investigation file is intended to meet three needs: (1) to define key entities, such as factors, protocols, parameters, which may be referenced

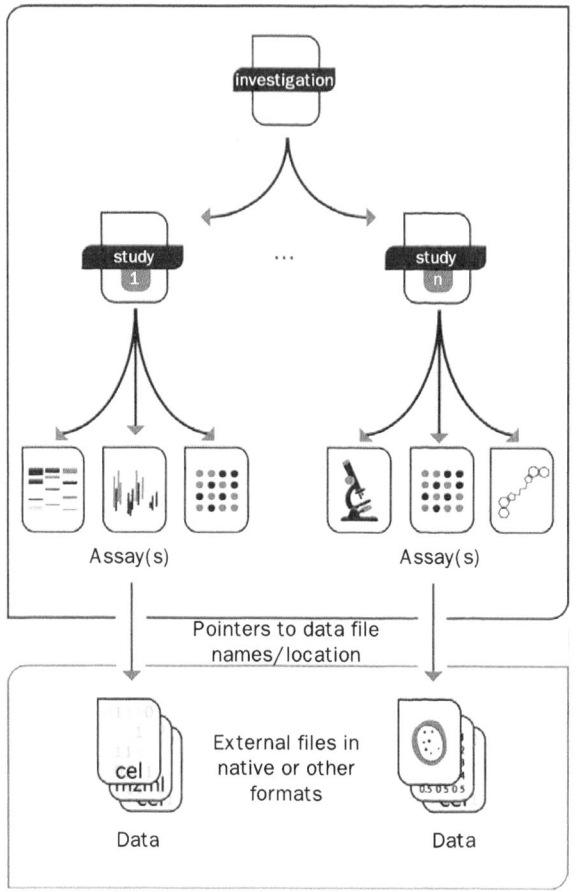

Figure 7.1 Overview of the ISA-Tab format, a general purpose framework with which to collect and communicate complex meta-data (i.e. sample characteristics, technologies used, type of measurements made) from experiments employing a combination of technologies

in the other files; (2) to relate Assay files to Study files; and, optionally, (3) to relate each Study file to an Investigation (this only becomes necessary when two or more Study files need to be grouped). The declarative sections cover general information such as contacts, protocols and equipment, and also – where applicable – the description of terminologies and other annotation resources that were used. The Study file contains contextualizing information for one or more assays, for example the subjects studied; their source(s); the sampling methodology; their characteristics; and any treatments or manipulations performed to prepare the specimens. Note that 'subject' as used above could to refer inter alia to an organism, or tissue, or an environmental sample. The Assay file represents a portion of the experimental graph (i.e. one part of the overall structure of the workflow); each Assay file must contain assays of the same type, defined by the type of measurement (i.e. gene expression) and the technology employed (i.e. DNA microarray). Assay-related information includes protocols, additional information relating to the execution of those protocols and references to data files (whether raw or derived).

For example, in a study looking at the effect of a compound inducing liver damage in rats by characterizing the metabolic profile of urine (by NMR spectroscopy) and measuring protein and gene expression in the liver (by mass spectrometry and DNA microarrays, respectively), there will be one Study file and three Assay files, in addition to the Investigation file. The Study file will contain information on the rats (the subjects studied) their source(s) and characteristics, the description of their treatment with the compound and the steps undertaken to take urine and liver (samples) from the treated rats. The Assay file for the urine metabolic profile (measurement) by NMR spectroscopy (technology) will contain the (stepwise) description of the methods by which the urine was processed for the assay, subsequent steps and protocols, and the link to the resultant raw and derived data files. The Assay file for the gene expression profile (measurement) by DNA microarray (technology) will contain the (stepwise) description of how the RNA extract was prepared from the liver (or a section), how the extract was labeled, how the hybridization was performed and so on, and will also contain the links to the resultant raw and derived data files. The Assay file for the protein expression profile (measurement) by mass spectrometry (technology), will contain the (stepwise) description of how the protein extract was prepared from the liver (or a section), how the extract was labeled, how the hybridization was performed and so on, and will also contain the links to the resultant raw and derived data files.

7.3.2 The ISA software suite

The modular ISA software suite, which implements the ISA-Tab format, offers tools to regularize the local management of experimental meta-data, facilitate curation and support conformance to community-defined reporting standards [9]. The modular nature of the ISA software suite separates conformance to standards from the reporting of experimental details (or meta-data), and provides the ability to convert experimental meta-data into various formats (e.g. for submission to some public databases) or to retain it in local storage. The suite's components are variously intended either for experimentalists or those supporting them. The editor tool (ISAcreator) offers familiar spreadsheet-based data entry and facilitates ontology-based curation at source via the BioPortal [11] and the Ontology Lookup Service [12]. Support for conformance to relevant minimum requirements, and the use of specified terminologies, is configurable across the suite (via the support-person-focused ISAconfigurator tool); it can be rigid or flexible to meet in-house requirements. The BioInvestigation Index is a searchable repository through which experimental meta-data and the associated data files can be managed and shared among the users granted access to them (including publicly). Conversion to and from any of a growing number of acceptable formats is enabled by a further module (the ISAconverter tool) [13–15]. The ISA software suite is available for download from the project web site, including the component's technical documentation and users' guide [16].

New collaborative activities continue to move the ISA community along the path to knowledge discovery through broad-scope data integration. For example, work is in progress to (1) augment the ISA code base with Application Programming Interfaces (APIs) to support further collaborative development; (2) facilitate visualization, manipulation and analysis data analysis, informed by the experimental context (the rISA, a R-package for ISA-Tab formatted files) using existing open source analysis platforms [17, 18]; (3) explore cloud-based resource management systems through the ISA suite being deployed on a Bio-Linux platform [19]; and (4) use semantic web approaches to make existing knowledge available for linking, querying, and reasoning in collaboration with the World Wide Web Consortium (W3C) Semantic Web for Health Care and Life Sciences Interest Group (HCLSIG)'s Scientific Discourse task force [20].

7.3.3 The ISA commons

As the collaboration continues to grow and new groups join in, we are on the path to building the ISA commons [21], a growing, exemplar ecosystem of data curation and sharing solutions built on the ISA framework. Rooted in real case studies, this framework is already used by many communities in domains as diverse as environmental health, stem cell discovery, toxicogenomics, environmental genomics, plant metabolomics and metagenomics while maintaining cross-domain compatibility. For example, the Novartis Institutes for BioMedical Research (NIBR [22]) conducts research aimed at drug discovery and development and is developing an instance of selected ISA software components integrated as part of an extended workflow for a microarray gene expression resource. Janssen Research & Development, LLC (*http://www.janssenrnd.com*) discovers and develops innovative medicines in several therapeutic areas including immunology where the ISA framework is being used to collect, annotate, and search relevant data sets. The use of these software components is aimed at enhancing curation efforts for data integration and analysis of in-house and public data sets. Furthermore, the endorsement of the ISA software by public systems, continued community engagement and growing list of project contributors have engendered a bioscience 'commons' of interoperable tools and data sets [21, 23]. The collaborative ISA framework offers a novel approach to the unsettled status quo by restricting itself to the harmonization of the structure of experimental meta-data only, allowing users of (parts of) data sets to 'connect the meta-data dots'. Harmonization of experimental meta-data is important for many resource providers, as our case studies illustrate, for more efficient and better-informed comparison of studies across assays and domains.

To better understand the use of the ISA framework and related developments, we offer the example of the curation process for the European PredTox data sets – a collaborative, distributed study where source material is subject to several kinds of assay in parallel in the search for markers to predict the toxicity of drug candidates [24–27]. The ISA framework was used to enable standards-compliant harmonization and curation prior to the release of study to the public in 2011.

7.3.4 The InnoMed PredTox case study

PredTox [28] was part of the InnoMed Integrated Project, coordinated by the European Federation of Pharmaceutical Industries and Associations (EFPIA [29]), a body representing the research-based pharmaceutical industries and biotech SMEs operating in Europe. InnoMed is a precursor to the Innovative Medicines Initiative, IMI; Europe's largest current public-private initiative [30] with 19 partners (14 pharma companies, three universities, two technology providers). The goal of PredTox was to assess the value of combining results from 'omics technologies with the results from more conventional toxicology methods to support more informed decision-making in pre-clinical safety evaluation. An overview of the PredTox data sets is given in Figure 7.2. The depth and breadth of

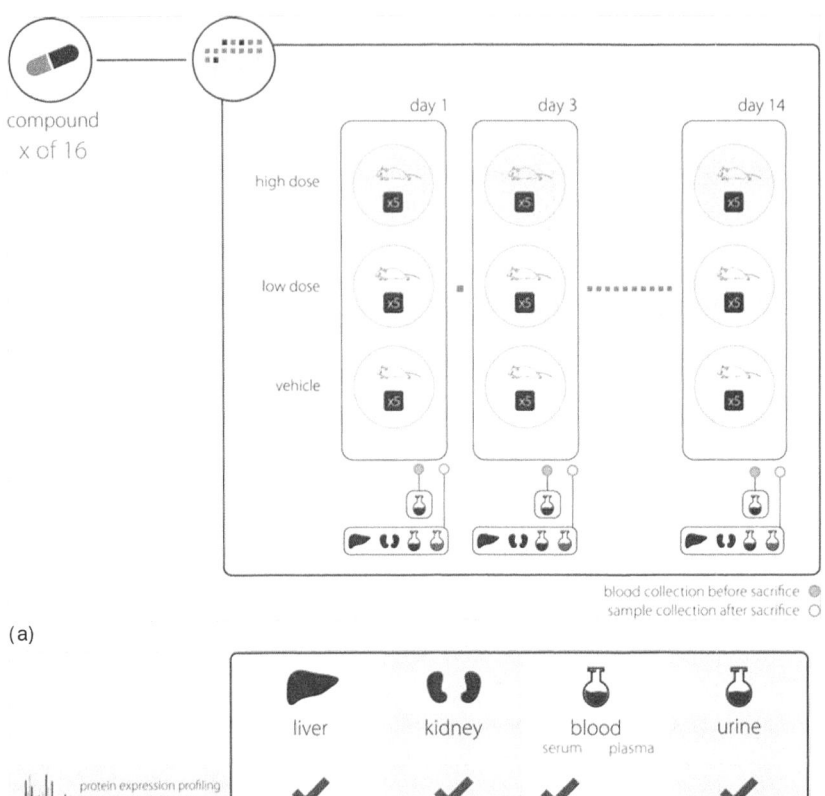

(a)

(b)

Figure 7.2 An overview of the depth and breadth of the PredTox experimental design (a), and content assay types (b)

the PredTox data sets define a landmark for data curation, highlighting the difficulties of structuring multi-assays studies according to common standards.

Raw and processed data, as well as experimental descriptions, protocols, compounds and sample collection procedures were provided by GeneData [31], a member of the PredTox consortium, in tab-delimited free text, with audit information detailing the person by whom data files were produced or sample processing performed, along with the date. As a first step, the identification schemas devised by the data producers to model animals, samples and aliquots and to link those to data files were disambiguated. Tabular information was then mapped to ISA-Tab syntactic elements, and the data imported by ISAcreator using that mapping, allowing for quick upload.

Sixteen studies were created, one per drug candidate. Experimental descriptions were harmonized to meet the requirement of relevant community-standards, and semantic tagging was performed using OBO foundry ontologies [32] accessed via the ISAcreator ontology widget (Figure 7.3). The Ontology of Biomedical Investigations (OBI [33]) was used for many elements of experimental steps; the vocabulary from the HUPO Proteomics Standards Initiative (PSI) [34] was used to identify and describe mass spectrometers and analytical columns. The same terms were also used for metabolite-focused applications of that analytical technique. Biological annotations relating to organ or tissue were sourced from the Foundational Model of Anatomy (FMA [35]). The drug candidates were submitted to the Chemical entities of biological interest (CHEBI [36]) database and ontology for inclusion. For targeted metabolomics applications, internal standards were used (lactate and trimethylsilyl propionate); those were also tagged using CHEBI as well as all identified metabolites identified and used in the data matrices, whereas units were expressed using Unit ontology [37].

Completing the semantic tagging, raw and processed data files were linked to the detailed description of the studies. Then ISA-Tab formatted files were run through the ISAvalidator, which checked and ensured that all annotation followed the community-defined minimum information requirements set for each of the omics components: MIAME (transcriptomics [38]), MIAPE-MS (proteomics [39]) and CIMR (metabolomics [40]). The completed PredTox data set was then uploaded to a public instance of the BioInvestigation Index hosted at The European Bioinformatics Institute (accession numbers: BII-S-8 to BII-S-23).

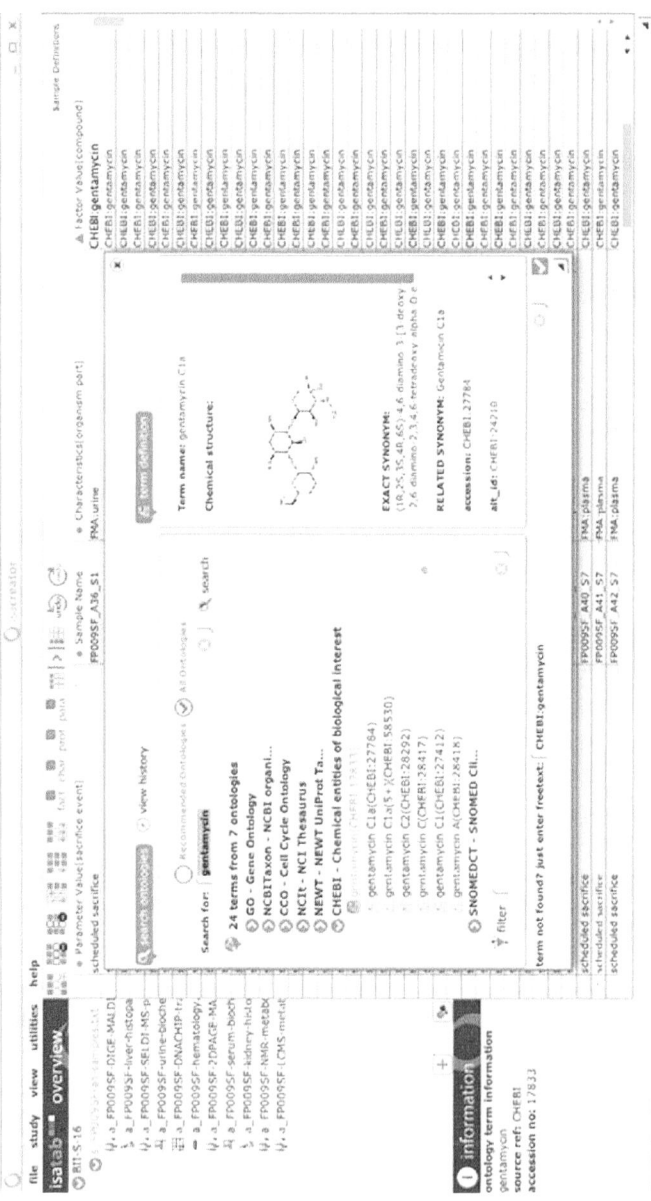

Figure 7.3 The ontology widget illustrates here how CHEBI and other ontologies can be browsed and searched for term selection. Along with the selected term, the internal ontology manager displays the information on each ontology term and records the ID and parent ontology, enabling provenance and tracking

7.4 Lessons learned

7.4.1 Open source, collaborative environment

The ISA software solution is a suite of open source and freely available modules, with an active community of users (researchers and curators) and contributing developers. GitHub [41] is the social repository selected to support an environment for code sharing and collaborative development, with the ultimate goal to achieve long-term sustainability and foster self-reliance. Feature requests and bug reports are tracked, discussed and assigned to specific developers.

Due to the modular nature of the ISA software, contributors can take responsibility for the development of specialized modules, contributions to and extensions of the core code. In turn, these are reviewed, vetted or rejected thus making it possible to implement agile practice in an open framework. The main requirement in such open environment is compliance to the open source licensing contract, which is essential to ensure that contributions are duly acknowledged while limiting aggressive open source 'free loading'.

Augmenting the ISA code base with APIs will also support further collaborative development and connections to other widely used data management systems (e.g. [42]). Dealing with feature requests, code forking and subsequent reviews, however, is always a challenging and demanding task, particularly when the ecosystem of users, contributors, and collaborators continues to grow [22, 23].

7.4.2 Curation practices

Generally, even if some good data management practices are in place when data sets are produced, without community standards any subsequent curation or re-use is neither simple nor straightforward. The current crop of minimum information checklists, terminologies and file formats is still growing, lacking control, creating integration headaches especially when new technologies or combinations of technologies are employed. The harmonization of standards remains patchy at best and far from crossing all life science and biomedical domains. Minimum information checklists are often seen as burdensome and over-prescriptive; ontologies too rich and complex; formats intractable. Although the ISA framework can assist in the process of selecting and using the appropriate standards, manual intervention is seldom completely avoidable; our job is to minimize the effort required.

For the IMI InnoMed PredTox, the necessity for well-annotated data and unambiguous meta-data was especially apparent during data analysis. Data sets from the same experimental series were generated in different laboratories, applying diverse technologies, and then had to be integrated for cross-platform and cross-study analyses. Therefore, both the methods and procedures in the laboratories, and the collection and reporting of meta-data had to be highly standardized. The integrated data analysis was performed on three levels. First, profiling data (e.g. transcriptomics) was integrated with conventional data, for example histopathology and serum chemistry, to identify expression markers through phenotypic anchors. Already, this level of integration required reliable identification of animals, treatment regimen, and derived samples that were further processed and used in the different laboratories to yield the said data sets. Second, data from multiple profiling technologies together with conventional data were analyzed across the different tissue types, to identify, for example, how transcriptional changes in the target organ relate to metabolome changes in serum in dependence of histopathological outcome. Third, data from multiple compounds (studies) were integrated to allow for the identification of common mechanisms between compounds causing similar phenotypic (i.e. histopathological) endpoints.

In summary, the IMI InnoMed PredTox generated a rich data set that will be of value to the general public in the evaluations of those highly parallel profiling techniques and in the curation and annotation practice applied using the ISA software suite.

7.5 Acknowledgments

The authors would like to thank all the collaborators who have contributed to the development of the ISA software suite. Special acknowledgement goes to the InnoMed PredTox consortium, for the data sets, and Stephen Marshall, Dorothy Reilly and Stephen Cleaver (of NIBR's Quantitative Biology Unit, Developmental and Molecular Pathways Platform) for the NIBR's case study. The ISA software suite and the BioSharing initiative are currently funded by grants from the Biotechnology and Biological Sciences Research Council and the Natural Environment Research Council's Environmental Bioinformatics Centre.

Published by Woodhead Publishing Limited, 2012

7.6 References

[1] The Value of Outsourcing Bioinformatics, Bio-IT: *http://www.bio-itworld. com/2011/04/21value-outsourcing-bioinformatics.html*

[2] Dealing with data. *Science* 2011;331:692.

[3] Barnes M et al. Lowering industry firewalls: pre-competitive informatics initiatives in drug discovery. *Nature Reviews Drug Discovery* 2009;8:701.

[4] Pistoia Alliance: *http://www.pistoiaalliance.org/*

[5] BioSharing: *http://www.biosharing.org/*

[6] Taylor CF et al. MIBBI: A Minimum Information Checklist Resource, *Nature Biotechnology* 2008;26:889.

[7] Harland L, et al. Empowering Industrial Research With Shared Biomedical Vocabularies. *Drug Discovery Today* 2011;16:940.

[8] Field D, et al. 'Omics Data Sharing. *Science* 2009;326:234.

[9] Rocca-Serra P, et al. ISA software suite: supporting standards-compliant experimental annotation and enabling curation at the community level. *Bioinformatics* 2010;26:2354.

[10] Baker N: 'ISA-TAB has become a parent standard for a variety of spreadsheet-based formats for data sharing . . .' Evaluation of: [Rocca-Serra P et al. ISA software suite: supporting standards-compliant experimental annotation and enabling curation at the community level *Bioinformatics* 2010 Sep 15;26(18):2354–6; doi: 10.1093/bioinformatics/btq415]. Faculty of 1000, 19 August 2010. F1000.com/4839956.

[11] Noy NP, et al. BioPortal: ontologies and integrated data resources at the click of a mouse, *Nucleic Acids Research* 2009;37:W170–W173.

[12] Côté R, et al. The Ontology Lookup Service: bigger and better, *Nucleic Acids Research* 2010;38(Web Server issue):W155–60.

[13] Parkinson H, et al. ArrayExpress update-from an archive of functional genomics experiments to the atlas of gene expression, *Nucleic Acids Research* 2009;37:868.

[14] Vizcaíno JA, et al. The Proteomics Identifications database: 2010 update, *Nucleic Acids Research* 2010;38:736.

[15] Shumway M, et al. Archiving next generation sequencing data, *Nucleic Acids Research* 2010;38:870.

[16] ISA software suite: *http://isa-tools.org*

[17] BioConductor: *http://bioconductor.org*

[18] Galaxy: *http://galaxy.psu.edu*

[19] Cloud Bio-Linux: *http://cloudbiolinux.com*

[20] W3C HCLSIG, Scientific Discourse task force, Discourse, Data and Experiment sub-task: *http://www.w3.org/wiki/HCLSIG/SWANSIOC*

[21] NIBR: *http://www.nibr.com/*

[22] Sansone SA et al. Towards interoperable bioscience data. *Nat Genet* (accepted).

[23] ISA commons: *http://isacommons.org*

[24] Ellinger-Ziegelbauer H et al. The enhanced value of combining conventional and 'omics' analyses in early assessment of drug-induced hepatobiliary injury. *Toxicology and Applied Pharmacology* 2011;252:97–111.

[25] Boitier E, et al. A comparative integrated transcript analysis and functional characterization of differential mechanisms for induction of liver hypertrophy in the rat. *Toxicology and Applied Pharmacology* 2011;252: 85–96.

[26] Hoffmann D, et al. Performance of novel kidney biomarkers in preclinical toxicity studies. *Toxicological Sciences* 2010;116:8–22.

[27] Collins BC, et al. Use of SELDI MS to discover and identify potential biomarkers of toxicity in InnoMed PredTox: a multi-site, multi-compound study. *Proteomics* 2010;10:1592–608.

[28] InnoMed PredTox: *www.innomed-predtox.com*

[29] EFPIA: *http://www.efpia.org*

[30] IMI: *http://www.imi.europa.eu*

[31] GeneData: *http://www.genedata.com*

[32] Smith B, et al. The OBO Foundry: coordinated evolution of ontologies to support biomedical data integration. *Nature Biotechnology* 2007;25:1251.

[33] OBI: *http://purl.obolibrary.org/obo/obi*

[34] Martens L, et al. Data standards and controlled vocabularies for proteomics. *Methods Molecular Biology* 2008;484:279–86.

[35] Rosse C and Mejino JLV. A reference ontology for biomedical informatics: the Foundational Model of Anatomy. *Journal of Biomedical Informatics* 2003;36:478–500.

[36] Degtyarenko K, et al. ChEBI: a database and ontology for chemical entities of biological interest. *Nucleic Acids Research* 2008;36:D344–350.

[37] Unit Ontology: *http://www.berkeleybop.org/ontologies/owl/UO*

[38] Brazma A, et al. Minimum information about a microarray experiment (MIAME)-toward standards for microarray data. Nature Genetics 2001;29:365–71.

[39] Taylor C, et al. The minimum information about a proteomics experiment (MIAPE). Nature Biotechnology 2007;25(8):887–93.

[40] Sansone SA, et al. The metabolomics standards initiative. Nature Biotechnology 2007;25(8):846–8.

[41] ISA on GitHub: *https://github.com/ISA-tools*

[42] Szalma S, et al. Effective knowledge management in translational medicine. *Journal of Translational Medicine* 2010;19(8):68.

## 8

GenomicTools: an open source platform for developing high-throughput analytics in genomics

Aristotelis Tsirigos, Niina Haiminen, Erhan Bilal and Filippo Utro

Abstract: Following the dramatic reduction of sequencing cost, research laboratories have been producing huge amounts of data, measuring DNA variations, RNA abundances, protein–DNA interactions, DNA methylation levels, and even chromosomal conformations. Making sense of terabytes of data requires reliable data management, computational resources, and, eventually, efficient computational methods for pre-processing, quality control, analysis and meta-analysis. In this work, we present a flexible computational platform for facilitating the development of pipelines to accomplish such computational tasks. For example, the user can easily create average read profiles across transcriptional start sites or enhancer sites, quickly prototype customized peak discovery methods for ChIP-seq experiments, perform genome-wide statistical tests such as enrichment analyses, and design controls via user-designed randomization schemes, among other applications.

Key words: computational genomics; ChIP-seq; RNA-seq; BED format; SAM format.

8.1 Introduction

The advent of high-throughput sequencing techniques initiated by pyrosequencing in 2004 [1] is expected to accelerate the pace of discovery in life sciences. Indeed, the rapidly and inexpensively produced super-exponential amount of data (e.g. short sequence patterns referred to as reads) from various high-throughput sequencing platforms allows the scientific community to study specific biological problems in depth, such as quantification of alternative splicing in tissues [2, 3], human disease [4], discovery of new fusion genes in cancer [5, 6], improvement of genome assembly [7], and transcript identification [8–11].

The common steps in many high-throughput sequencing studies include: (1) alignment of reads directly to a reference transcriptome or genome ('read mapping'); (2) identification of expressed genes, isoforms or binding sites; and (3) differential analysis across samples. An in-depth review of standard steps in RNA-seq and ChIP-seq computational pipelines is published by Pepke and colleagues [12]. It is worth pointing out that genome-wide data, such as transcripts/genes, exons/introns, promoter sites, sequences, multiple sequence alignments, transcription factor binding sites, intergenic regions, repeat elements, microarray probes (expression, SNP, CNV, etc.), sequencing data (RNA-seq, ChIP-seq, DNA-seq, etc.), chromosomal conformations (3C-seq, 4C-seq, etc.), and inter-chromosomal associations can easily be represented as sets of genomic intervals (see Figure 8.1).

Given the huge volume of available data, new efficient computational tools are required in order to efficiently perform analysis tasks such as those outlined above [13]. Currently, freely available computational tools for large-scale data analytics include Bioconductor [14], Galaxy [15], Genomic Regions Enrichment of Annotations Tool (GREAT) [16], USCS genome browser [17] and Integrated Genome Browser (IGB) [18]. For the readers' convenience, we report here the fundamental aspects of each tool. Bioconductor uses the R statistical programming framework to provide tools for the analysis and comprehension of high-throughput genomic data. The functional scope of Bioconductor packages includes the analysis of DNA microarray, sequence, flow, and SNP data. Galaxy is an open web-based platform for genomic research, based around reusable analysis templates that users can manipulate and run repeatedly on different data sets. Galaxy has been used for different types of genomic research, for example investigations of epigenetics, chromatin profiling, transcriptional enhancers, and genome–environment interactions. GREAT is available as a web application that was designed to analyze the

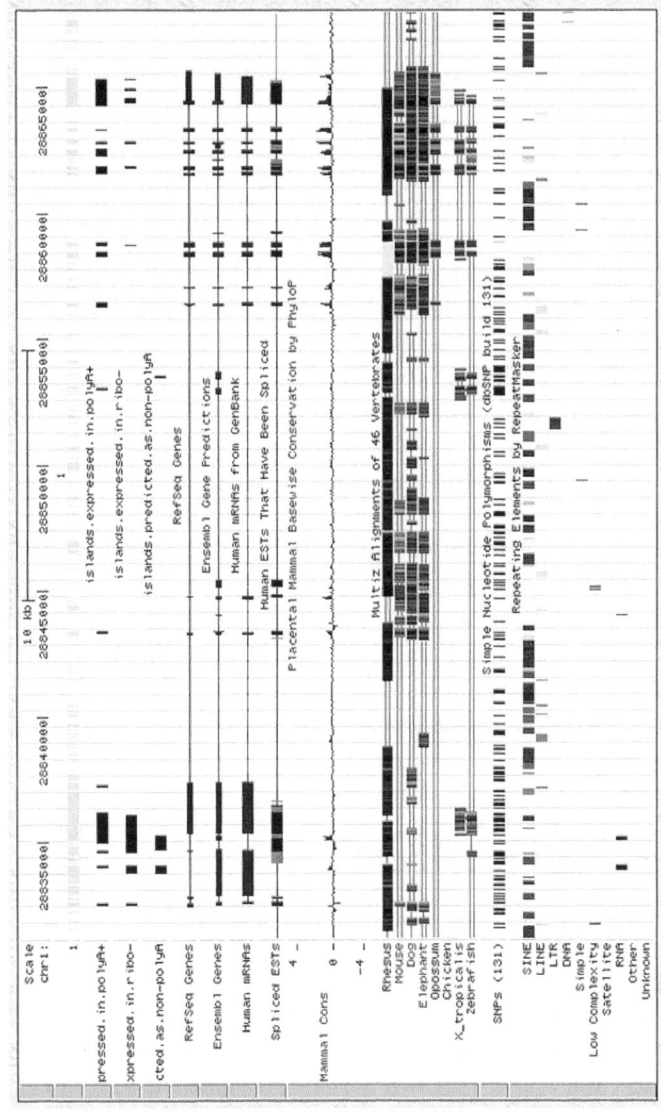

Figure 8.1 Diverse types of genomic data

functional significance of cis-regulatory regions identified by localized measurements of DNA binding events across an entire genome. The USCS genome browser and IGB are web-based visualization platforms that incorporate data from several public databases.

Despite the existence of these sophisticated tools for genome analysis, they are not designed to efficiently process multiple large data sets – such as the ones obtained from high-throughput sequencing – and suffer from poor memory management as essentially all data needs to be loaded into memory and/or sent over the network.

In an attempt to address some of these issues, Quinlan and Hall have developed the BEDTools suite [19]. We predict that the BEDTools initiative will lead into a competition for a new set of tools focused on processing genomic data as streams. This new set of tools will provide the means to efficiently handle large genomic data sets, thus providing a computational platform that facilitates the development of bioinformatics applications. These tools may then be integrated with or incorporated into other bioinformatics tools/environments, including those mentioned above.

The motivation behind GenomicTools (first presented in [20] as an 'applications note') was to create a computational platform for developing customized analytics for genomic data sets with minimal memory and intermediate file requirements in order to address the bottleneck caused by the increasing influx of genome-wide data sets. GenomicTools is available both as command-line tools for building applications in a UNIX-like environment and as documented C++ classes for further development. The open source aspect of GenomicTools is important as it allows users to easily incorporate their own analysis methods with the published tools and to modify the tools to suit their specific data analysis needs. Although similar in motivation to BEDTools, it is in several aspects more general than BEDTools and it addresses several issues that BEDTools do not adequately address. We summarize the novelty of GenomicTools below.

- Novel operations; in GenomicTools the focus is not simply on overlap computations as in BEDTools. GenomicTools is designed to perform a variety of simple mathematical operations on set genomic intervals (as a pre-processing step) and then a variety of complex operations can be performed, such as overlap, offset or scanning computations (Figure 8.2), that is a superset of the operations offered by BEDTools.

- Relaxed data set restrictions: GenomicTools allows several of its operations to operate on sets of genomic regions rather than sets of single genomic intervals as in BEDTools, for example it makes full use of all 'exons' in BED entries. BEDTools operates on single-'exon' BED entries.

- Full stream-computing design: in GenomicTools no files are loaded into memory but are processed instead as streams: this minimizes memory requirements and allows the simultaneous processing of several files (e.g. different replicates, patient samples, etc.).

- C++ API: GenomicTools command-line operations are implemented as C++ class methods in a convenient API, which may be used by developers to write new applications entirely in C++, for example novel peak finders.

- Auxiliary tools: GenomicTools offers a set of auxiliary command-line tools (permutation_test, vectors, and matrix) to facilitate the construction of command-line pipelines as they implement basic mathematical and statistical operations on vectors and matrices.

- Performance: GenomicTools improves performance over BEDTools both in terms of time and memory requirements.

Research institutions as well industry sectors in life sciences, such as pharmaceutical and medical research companies, that make extensive use of high-throughput sequencing technologies, are expected to use these tools. Naturally, these tools can also be used for different kinds of genomics studies, and were in fact initially developed for this reason. More specifically, our computational genomics group at IBM Research has used an early version of this tool – before it was released to the public – for computational studies of repeat elements in mammalian genomes [21, 22], analysis of gene expression tiling array data in Drosophila [23], and the study of dynamic changes in human DNA methylation during differentiation [24].

The rest of this chapter presents the GenomicTools platform (version 2.0, released in September 2011) and is organized as follows. The following section provides the necessary definitions as well as fundamental information on the input file formats used in GenomicTools. There is then an overview of the tools, followed by a more in-depth presentation of some aspects of the C++ implementation. Several practical examples are given using GenomicTools for computational genomics analyses in the context of a simple ChIP-seq pipeline case study. Finally, a comparison of the performance of GenomicTools against BEDTools and Bioconductor is provided.

8.2 Data types

In this section, we will introduce the basic definitions of a genomic interval and genomic regions. We will also describe the input file formats supported by GenomicTools.

8.2.1 Definitions

A genomic interval is a tuple: <chromosome, strand, start position, end position>.

A genomic region is an ordered set of genomic intervals. Note that this is a rather broad definition, which allows for the inclusion of genomic intervals from different chromosomes and/or strands, as well as intervals that overlap. In GenomicTools, this definition of genomic regions is implemented in the REG file format, which we introduce in the next section.

Genomic regions are characterized by several properties. A genomic region is compatible if and only if all its intervals are in the same chromosome and (optionally) strand. A genomic region is sorted if and only if its intervals are sorted first by chromosome, then optionally by strand and finally by start position. A genomic region is non-overlapping if and only if its intervals are non-overlapping in all pair-wise combinations. A genomic region is a single-interval region if and only if it contains exactly one interval. Operations on genomic regions may require that certain properties be satisfied before they can be successfully executed.

A genomic region set is an ordered set of genomic regions. A genomic region set is sorted if and only if its regions are single-interval regions and they appear in the sort order described above. As before, operations on genomic region sets may require that certain properties be satisfied before they can be successfully executed.

In GenomicTools, all input files contain a single genomic region set as defined above. Every line of these files corresponds to a single genomic region possibly annotated with additional information, such as labels, depending on the particular file format (see next section).

8.2.2 Supported file formats

The GenomicTools platform supports the standard BED [25], GFF [26], and SAM [27] file formats. Input files can also be converted into WIG format [28]. Additionally, we propose a new simple format, the REG format, as an attempt to distill the minimum common information from the BED/GFF/SAM formats while allowing for the more general definition of a genomic region as defined above. Each line in a REG file represents a labeled genomic region, where the label is separated from the genomic region via a <TAB> character. A simple REG file representing a set of RNA-seq reads is shown below.

```
Read#1<TAB>1 + 100 149
Read#2<TAB>1 + 102 151
. . .
Read#N<TAB>Y - 10001 10050
```

Another example is the following REG file carrying information on gene exons (note that every line is a set of genomic intervals).

```
Gene#1<TAB>1 + 160446 161690 1 + 161314 161525
. . .
Gene#N<TAB>Y - 279704 279708 Y - 279741 279839 Y - 279911 279916
```

Note that this format is a generalization of the BED format because it allows overlapping intervals within a given region (Gene#1 in the above example). This is particularly useful when we need to group exons of a set of transcript isoforms of the same gene. Additionally, it allows intervals from different chromosomes and strands to be grouped in each line, and this helps represent gene fusions and interchromosomal associations.

In terms of C++ implementation, each genomic region (i.e. each line in an input file) is stored as an instance of the GenomicRegion class or its derived classes for BED, GFF, and SAM formats (see C++ *API for developers* for details). The entire file is stored as an instance of the GenomicRegionSet class, although not necessarily fully loaded in memory.

8.3 Tools overview

The GenomicTools platform is built on top of the genomic_intervals C++ library described in the next section. Its functions are bundled in four command-line tools: (1) genomic_regions, for basic genomic regions operations; (2) genomic_overlaps for comparing sets of regions and computing offsets; (3) genomic_scans for window-based operations; and

(4) permutation_test for enrichment analyses. Additionally, in this distribution, we are including auxiliary tools for manipulating vectors and matrices.

The flow-chart in Figure 8.2 summarizes the role of each of the command-line tools in our pipeline model. Briefly, the pipeline inputs in REG/GFF/BED/SAM format represent either sequenced read alignments (e.g. from DNA-seq, RNA-seq, ChIP-seq experiments) or annotations from public databases. The annotations that can be utilized in our pipeline fall into two categories: (1) genomic annotations represented as genomic regions, such as the known genes set in the UCSC Genome Browser; and (2) functional annotations, for example from Gene Ontology [29]. Computations are performed in two levels: (1) basic interval operations (such as unions, intersections, shifting, flanking, etc.) are implemented by the genomic_regions command-line tool; and (2) complex operations (overlaps, offsets, scanning and permutations) implemented by the genomic_overlaps, genomic_scans and permutation_test command-line tools take as input the regions that have undergone basic processing and produce results. Basic and complex operations

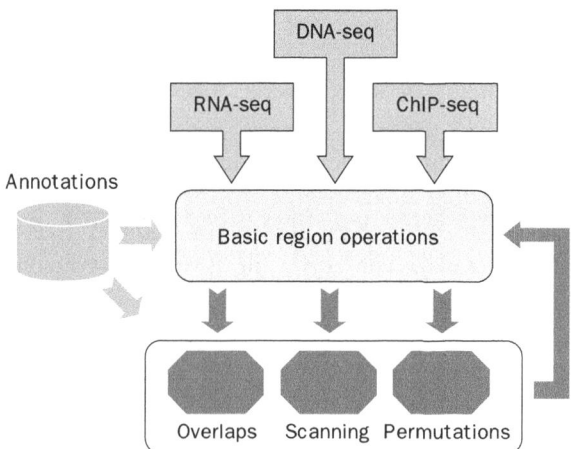

Figure 8.2 Flow-chart describing the various functionalities of the GenomicTools suite: basic region operations are implemented in genomic_regions whereas complex operations are implemented in genomic_overlaps, genomic_scans and permutation_test (source: adapted figure 1 from Tsirigos et al. [20])

can be combined into fairly elaborate scripts that can address a wide range of issues during the course of a bioinformatics project.

In large-scale analyses it is important to avoid biases introduced by the complexity and redundancy of large genome-wide data sets. For example, when computing an average gene TSS profile for ChIP-seq reads, it is important to create a 'non-redundant' set of gene TSSs, which is not a trivial task because gene transcripts with different identifiers and possibly originating in different databases may report slightly different TSSs for transcripts that are in fact the same. The operations implemented in the GenomicTools platform can help correct for these biases as a pre-processing step before the actual computation takes place, or assess the effect of the bias on the result.

8.3.1 The genomic_regions tool

The genomic_regions tool is designed to perform basic operations on genomic region files. These are: (1) line-based operations, such as shifting, shuffling, sorting, and modifying genomic regions; and (2) file-based operations such as inverting or linking genomic regions. Table 8.1 comprises the complete list of operations and each operation is documented in the user's manual, an entry of which is shown in Figure 8.3. To get a list of options without the need to refer to the manual, simply use the '-h' option.

```
$ genomic_regions shuffle -h

USAGE:
  genomic_regions shuffle [OPTIONS] <REGION-SET>

DESCRIPTION:
  Shuffles intervals within given reference region-set.

DETAILS:
  * Input formats: REG, GFF, BED, SAM
  * Operand: interval
  * Region requirements: single-interval
  * Region-set requirements: none

OPTIONS:
  --help     help                    [true]
  -h         help                    [true]
  -v         verbose mode            [false]
  -r         reference region file   []
```

| Table 8.1 | Summary of operations of the genomic_regions tool |

Operation	Description
align	Aligns sequences to reference genome (line-based)
bed	Converts input regions to BED format (line-based)
bounds	Checks interval against chromosome bounds and removes invalid intervals (line-based)
center	Prints center interval (line-based)
connect	Connects intervals from minimum start to maximum stop (line-based)
diff	Computes the difference between successive intervals (line-based)
dist	Computes distances between successive intervals (line-based)
divide	Divides intervals in the middle (line-based)
fix	Removes invalid intervals, i.e. start<1 or start>stop (line-based)
gdist	Computes distances of successive regions (file-based)
int	Computes the intersection of input intervals (line-based)
inv	Inverts regions given the genome chromosomal boundaries (file-based)
link	Links consecutive regions to produce a non-overlapping set (file-based)
n	Computes total interval length, including possible overlaps (line-based)
pos	Modifies interval start/stop positions (line-based)
reg	Converts to REG format (line-based)
rnd	Randomizes region across entire genome (line-based)
select	Selects a subset of intervals according to their relative start positions (line-based)
shift	Shifts interval start/stop positions (line-based)
shiftp	Shifts interval 5'/3' positions (line-based)
shuffle	Shuffles intervals within given reference region (line-based)
sort	Sorts intervals (line-based)
split	Splits regions into their intervals which are printed on separate lines (line-based)
strand	Modifies interval strand information (line-based)

test	Tests whether genomic regions are sorted and non-overlapping (file-based)
union	Computes the interval union (line-based)
wig	Converts to UCSC wiggle format (line-based)
win	Creates new intervals by sliding windows (line-based)
x	Extracts corresponding sequences from DNA (line-based)

```
USAGE: genomic_regions -shuffle [OPTIONS] <REGION-SET>
DESCRIPTION: Shuffle intervals within given reference region.
OPTIONS:
  -v          verbose mode                              [false]
  -help       help                                      [true]
  -h          help                                      [true]
  -r          reference region file                     []
```

```
FUNCTION

Reference         ============          ===============================

Input             AAAAAAAAAA    BBBB          CCC

Output                        BBBB             AAAAAAAAAA    CCC
```

```
EXAMPLES

$ cat inp.reg
geneA  1 + 1 100
geneB  1 + 1000 1100

$ cat reference.reg
chromosome_1   1 + 1 249250621

$ cat inp.reg | genomic_regions -shuffle -r reference.reg
geneA  1 + 889 988
geneB  1 + 7795 7895
```

Figure 8.3 Example entry from the user's manual for the 'shuffle' operation of the genomic_regions tool

8.3.2 The genomic_overlaps tool

The genomic_overlaps tool allows the user to compute various measures of overlaps between sets of regions (see online documentation for the complete list of operations). This is achieved by providing a set of

Table 8.2	Summary of usage and operations of the genomic_overlaps tool

Operation	Description
count	Counts the number of overlapping test regions per reference region
coverage	Calculates the depth coverage (i.e. the total number of overlapping nucleotides) per reference region
density	Computes the density (i.e. the coverage divided by the size of the reference region) of overlaps per reference region
intersect	Computes the intersection between all pairs of test and reference regions
offset	Computes the distances of test regions from their overlapping reference regions
overlap	Finds the overlaps between all pairs of test and reference regions
subset	Picks a subset of test regions depending on their overlap with reference regions

operations and a variety of options for: (1) finding regions that match or partially overlap; (2) counting the number of matches; (3) calculating densities of matched regions; and (4) computing overlap offsets. Applications include computation of gene expression from RNA-seq data (e.g. RPKMs [30]), construction of average read profiles or heatmaps across transcriptional start sites (TSSs), enrichment analyses of virtually any genomic data set, such as genes of specific functional categories, repeat types, single nucleotide polymorphisms (SNPs), or cancer-associated regions.

8.3.3 The genomic_scans tool

The genomic_scans tool can be used for window-based computations such as peak discovery. The command-line version offers several parameters for controlling the window size, statistical tests, etc. Users with basic C/C++ skills can easily modify the source code to perform the statistical test of their choice using the GenomicRegionSetScanner class described below.

Table 8.3 Summary of usage and operations of the genomic_scans tool

Operation	Description
counts	Determines input read counts in sliding windows of reference regions
peaks	Scans input reads to identify peaks

8.3.4 The permutation_test tool

This tool executes row permutations to determine p-values and q-values for all the categories contained in the input file given the statistic chosen by the user (see Table 8.4 for available statistical tests). More specifically, the statistic on the set of rows annotated by a given category is compared against the same statistic on permuted versions of the input on the value column. For an example, see *Identifying enriched Gene Ontology terms*.

8.3.5 Vector and matrix operations

The GenomicTools distribution includes two command-line tools for manipulating labeled vectors and matrices. The input format for the vectors command-line tool is a series of lines, each of which has two TAB-separated fields (label and SPACE-separated vector elements). The label field is optional. Similarly, the input format of the matrix

Table 8.4 Supported statistics for the permutation tests

Statistic	Description
sum	The sum of values in a given category
n	Number of values > 0
sens	Sensitivity
spec	Specificity
ratio	Mean of values divided by mean of values in the background
t	t-test between mean of values against mean of values in the background

command-line tool is an optional header containing column labels followed by a series of labeled vectors (defined above) of the same number of elements. The full list of supported vector and matrix operations can be found in the User's Manual online (see section 8.8).

8.4 C++ API for developers

Users with basic C++ skills can make use of the genomic_intervals library, particularly for window-based computations and overlaps. All implemented classes are fully documented using Doxygen [31] (see Figure 8.4 for a snapshot). Access to full documentation is provided along with the source code distribution. In the following sections, we describe the main classes that are used to represent the genomic data and perform the various operations.

8.4.1 The GenomicInterval and GenomicRegion classes

The GenomicInterval class implements the notion of a genomic interval, that is an interval annotated with chromosome and strand information. The GenomicRegion class implements the notion of a genomic region (in REG format), that is a labeled ordered set of genomic intervals, and corresponds to one single line in the input file. The genomic intervals are stored as C++ STL vectors, but there is also an option of C++ STL lists for developers. This class has a series of constructors, which create genomic regions from an input file (accessed via the FileBuffer class), or from a character array.

The methods of this class are classified into four categories:

- read & print methods: read and print genomic intervals in various formats;
- get & set methods: retrieve and set class variables, such as label, chromosome, etc.;
- check & compare methods: obtain information about region properties (sorted, compatible, etc.), and their relationship with other regions (overlaps, order, etc.);
- operations: execute operations between or within regions, such as union, difference, etc.

This GenomicRegion class contains just enough information for the minimal requirements of the REG format. Most methods described above

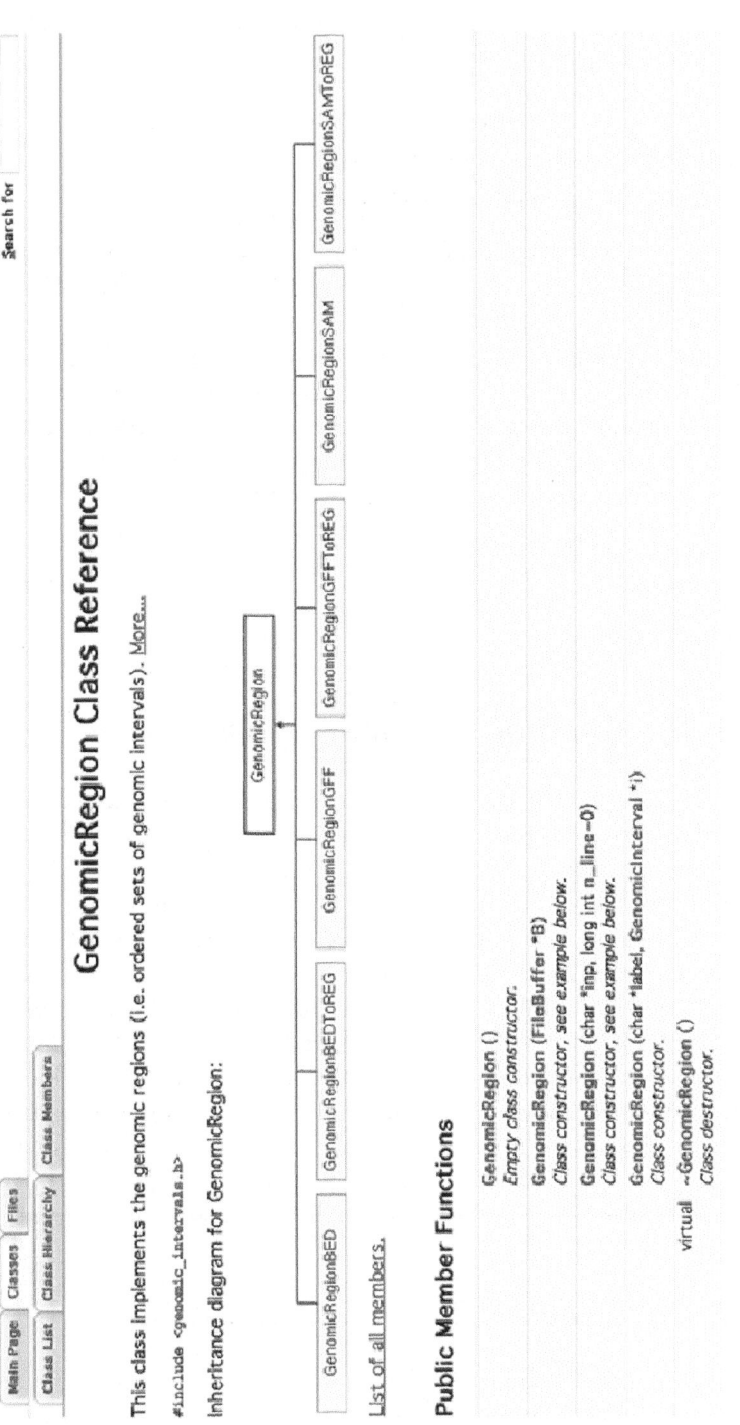

Figure 8.4 Example entry (partial) from the C++ API documentation produced using Doxygen and available online with the source code distribution

are implemented as virtual so as to allow for class extensions, which provide full support for:

- BED format (GenomicRegionBED class);
- GFF format (GenomicRegionGFF class);
- SAM format (GenomicRegionSAM class).

The virtual methods are redefined – when necessary – for each derived class in order to properly read, print, and update the extra variables of each format.

Additionally, for each format, we provide simple classes whose goal is to only read in the corresponding format and immediately convert it into the REG format. These classes are used when the extra variables of the input format are not needed for a particular computation, and can save both time and memory resources. These classes are:

- GenomicRegionBEDToREG;
- GenomicRegionGFFToREG;
- GenomicRegionSAMToREG.

8.4.2 The GenomicRegionSet class

The GenomicRegionSet class implements the notion of a set of genomic regions that corresponds to an entire input file, for example the set of known genes, or aligned reads from a sequencing experiment. The main data element stored in this class is an array of instances of the GenomicRegion class (or any of its derived classes). The input file can be read from the standard input to facilitate pipelined execution, and is not necessarily loaded fully in memory in order to minimize memory requirements: essentially, the input files are read and processed sequentially one line at a time, and the data for each line are discarded when no longer needed for the particular computation. In the next section we show an example of allocating an instance of this class. The operations of the genomic_regions command-line tool summarized in Table 8.1 are implemented as methods in this class (see the User's Manual for details).

8.4.3 The GenomicRegionSetScanner class

This class helps scan by sliding windows an instance of GenomicRegionSet and is used by the genomic_scans command-line tool. The main advantage of this implementation of sliding window computations is that it is done sequentially without the need of storing the entire input intervals in memory. A practical use of this class is to develop customized window-based peak discovery algorithms. As shown in the example below, this class can be used to determine the number of reads in signal and control region sets in sliding windows along the entire genome.

```
# include 'core.h'
# include 'genomic_intervals.h'
// parameters
long int w = 500; // window size
long int d = 25; // window distance
bool verbose = true;
bool load_in_memory = false;

// initialize: create input region sets and associated scanners
map<string,long int> *bounds = ReadBounds(genome_file);
GenomicRegionSet signalReg(signal_file,10000,verbose,load_in_memory);
GenomicRegionSet controlReg(control_file,10000,verbose,load_in_memory);
GenomicRegionSetScanner signal_scanner(&signalReg,bounds,d,w,false,false
,'c');
GenomicRegionSetScanner control_scanner(&controlReg,bounds,d,w,false,
false,'c');

// run: compute read counts in sliding windows for both files while
(true) {
  long int n = signal_scanner.Next();
  long int m = control_scanner.Next();
  if (n==-1) break;
  // ADD your statistical test HERE
}
```

8.4.4 The GenomicRegionSetOverlaps class and its extensions

This class is an abstract class used for determining and manipulating overlaps between two regions sets. It is extended into two classes. SortedGenomicRegionSetOverlaps is used on sorted region sets. The sort order is first by chromosome, then (optionally) by strand, and finally by start position. The algorithm used to compute overlaps in this class is a generalization of the standard merge-sort algorithm modified so as to handle intervals. As before, the main advantage of this implementation is that processing is done sequentially without the need of storing the entire input intervals in memory. The algorithm operates on sorted inputs, scans the files sequentially and computes all overlaps essentially using a merge-sort algorithm adapted to handle intervals. An intermediate buffer keeps all the overlaps of index regions with the current query, as they may also overlap with the next query.

```
INPUTS: query intervals Q and index intervals I.

PSEUDOCODE:
# Q and I are read sequentially as input streams
1.      q = next (Q)                 # read first query
2.      i = next (I)                 # read first interval
3.      B = {   }           # buffer for local overlaps
4.      while (q) {
5.        if (q < i) q = next(Q)     # read next query
6.        else if (q > i) i = next (BUI)    # read next interval
7.        else {
8.          B = {   }
9.          while (overlap(q,i)) {
10.         print q,i                # print overlaps
11.         B = BUI                  # store interval in buffer
12.         i = next (I)
13.         }
14.       }
15.     }
```

UnsortedGenomicRegionSetOverlaps is used on unsorted region sets. The algorithm used here is a modification of the algorithm proposed by Kent *et al.* [17], where we allow the number of levels and the number of bins per level to be chosen arbitrarily.

The example below demonstrates the use of both derived classes (this is taken from the source code file 'genomic_overlaps.cpp').

```
# include 'core.h'
# include 'genomic_intervals.h'

// open region sets
char *REF_REG_FILE = 'exons.bed';
char *TEST_REG_FILE = 'rnaseq.reads.bed';
GenomicRegionSet RefRegSet(REF_REG_FILE,BUFFER_SIZE,VERBOSE,true);
GenomicRegionSet TestRegSet(TEST_REG_FILE,BUFFER_SIZE,VERBOSE,false);

// process overlaps
GenomicRegionSetOverlaps *overlaps;
if (IS_SORTED) overlaps = new SortedGenomicRegionSetOverlaps(&TestRegSet,
&RefRegSet,false);
else overlaps = new UnsortedGenomicRegionSetOverlaps(&TestRegSet,&RefReg
Set);
unsigned long int *coverage;
```

```
coverage = overlaps->CalcIndexCoverage(MATCH_GAPS,IGNORE_STRAND,USE_
VALUES);
Progress PRG('Printing densities . . .',RefRegSet.n_regions);
for (long int k=0; k<RefRegSet.n_regions; k++) {
GenomicRegion *qreg = RefRegSet.R[k];
  long int qreg_size = MATCH_GAPS? (qreg->I.back()->STOP-qreg->I.
front()->START+1): qreg->GetSize();
  double density = (double)coverage[k]/qreg_size;
  if (density>=MIN_DENSITY) printf('%s/t%.4e/n', qreg->LABEL, density);
  PRG.Check();
}
PRG.Done();
delete coverage;
delete overlaps;
```

8.5 Case study: a simple ChIP-seq pipeline

In this chapter, we demonstrate the utility of GenomicTools in constructing a simple pipeline for ChIP-seq analysis. The pipeline helps accomplish the following tasks: (1) produce data for popular plots such as read profiles and read density heatmaps; (2) create genome browser tracks for visualization; (3) identify peaks as potential binding sites; and (4) perform an enrichment analysis. ChIP-seq studies are now widely used to elucidate the molecular function of the cell under normal conditions as well as under stress or disease (see for example [32–34]). As they reveal the genomic positions of protein interactions, such as transcription factors and histone modification, with DNA, they can help create networks of interactions and reveal undiscovered biological mechanisms. Our tools help set up computational pipelines that drive this discovery. The following examples use UNIX command-line functions, but they also run on Cygwin under MS Windows.

8.5.1 Creating ChIP-seq read profiles

ChIP-seq read profiles are heavily used in ChIP-seq studies because they offer an easy method for data validation regarding the relative position of the ChIP-seq peaks (i.e. potential binding sites) with respect to chosen genomic features, such as gene transcriptional start sites (TSSs) or binding sites of other factors, such as enhancers. Additional validation is possible if expression data are available and the transcription factor or histone modification mark under ChIP-seq investigation is activating or repressive. In such a case, its read profile can be computed separately for genes of high versus low expression and its activating or repressive role confirmed.

Creating read profiles using GenomicTools is straightforward, as demonstrated in the following example. First, the user creates the TSS regions using as input the gene transcript chromosomal coordinates in 'genes.bed', which can be downloaded, for example, from the UCSC Genome Browser web site. This is done using the genomic_regions tool 'pos' and 'shift' operations: the former chooses the 5' end of gene transcripts (i.e. the TSS) and the latter performs a 10 kb flanking operation upstream and downstream of the TSS.

```
$ head genes.bed
chr1 3044313 3044814 ENSMUSG00000090025:ENSMUST00000160944 1000 +
chr1 3092096 3092206 ENSMUSG00000064842:ENSMUST00000082908 1000 +
chr1 3456667 3503634 ENSMUSG00000089699:ENSMUST00000161581 1000 +
chr1 3670235 3671871 ENSMUSG00000073742:ENSMUST00000097833 1000 +
. . .

$ cat genes.bed | genomic_regions pos -op 5p | genomic_regions shiftp
-5p -10000 -3p +10000 > TSS.10kb.bed

$ head TSS.10kb.bed
chr1 3034313 3054314 ENSMUSG00000090025:ENSMUST00000160944 1000 +
chr1 3082096 3102097 ENSMUSG00000064842:ENSMUST00000082908 1000 +
chr1 3446667 3466668 ENSMUSG00000089699:ENSMUST00000161581 1000 +
chr1 3660235 3680236 ENSMUSG00000073742:ENSMUST00000097833 1000 +
chr1 4509097 4529098 ENSMUSG00000064376:ENSMUST00000082442 1000 +
chr1 4787868 4807869 ENSMUSG00000025903:ENSMUST00000134384 1000 +
chr1 4787903 4807904 ENSMUSG00000025903:ENSMUST00000027036 1000 +
. . .
```

Next, the distances of the mapped ChIP-seq reads from the TSS regions are computed using the genomic_overlaps tool 'offset' operation. The 'offset' operation allows the user to choose a reference point for the query regions ('-op' option), and to express the computed offset as a fraction of the query region size ('-a' option) instead of an absolute number. Also, in this particular application, the strand information is ignored ('-i' option), because binding occurs both sense and anti-sense of the affected transcript.

```
$ head chipseq.bed
chr1 3001228 3001229
chr1 3001228 3001229
chr1 3001438 3001439
. . .

$ cat chipseq.bed   genomic_overlaps offset -v -i -op 5p -a TSS.10kb.bed
cut -d' ' -f1 > offset.txt
```

Published by Woodhead Publishing Limited, 2012

```
$ head offset.txt
ENSMUSG00000090025:ENSMUST00000160944      0.007850
ENSMUSG00000090025:ENSMUST00000160944      0.007850
ENSMUSG00000090025:ENSMUST00000160944      0.021899
ENSMUSG00000090025:ENSMUST00000160944      0.021899
ENSMUSG00000090025:ENSMUST00000160944      0.030098
ENSMUSG00000090025:ENSMUST00000160944      0.030098
. . .
```

Finally, the computed offsets can be separated in genes of high versus low expression, histogrammed using the vectors tool (operation 'hist') and plotted using R (see Figure 8.5 for a sample plot), Excel or any other similar tool or environment. For example, if the 'offset.txt' file computed above was separated into two files 'offset.high.txt' for the genes of high expression and 'offset.low.txt' for the genes of low expression, then:

```
$ cat offset.high.txt   vectors -hist -n 6 -b 100 > profile.high.txt

$ cat offset.low.txt    vectors -hist -n 6 -b 100 > profile.low.txt

$ head offset.high.txt
#bin-start              bin-freq              bin-counts
0.000000                0.008555              20676
0.010000                0.008522              20596
0.020000                0.008128               19644
. . .
```

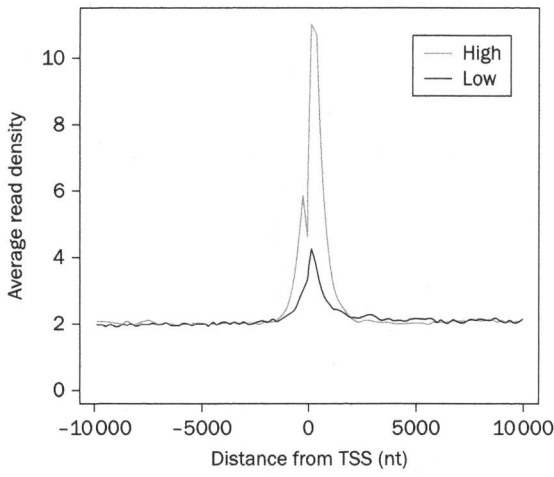

Figure 8.5 Example of TSS read profile for genes of high and low expression

Note that the histogram counts in column #3 need to be normalized by the number of genes in each group. Option '-n 6' sets the number of decimals to 6 and '-b 100' the number of histogram bins to 100.

8.5.2 Creating ChIP-seq read density heatmaps

Although average ChIP-seq profiles are useful for easy visualization and validation, they do not reveal the exact binding site position per gene. This can be achieved by ChIP-seq read density heatmaps around TSSs (Figure 8.6). To produce the data for this type of plot, the user can simply utilize the vectors operations '-merge' and '-bins', so that now the histograms are produced per gene rather than for the entire offset file.

```
$ cat offset.high.txt | sort | vectors -merge | vectors -bins -b 200 -m
10 > heatmap.high.txt

$ head heatmap.high.txt
ENSMUSG00000090025:ENSMUST00000160944    0 0 4 0 4 0 4 0 0 0 0 . . .
ENSMUSG00000064842:ENSMUST00000082908    0 0 0 0 0 0 0 0 0 0 0 . . .
ENSMUSG00000051951:ENSMUST00000159265    0 0 0 0 0 0 0 0 4 0 4 . . .
. . .
```

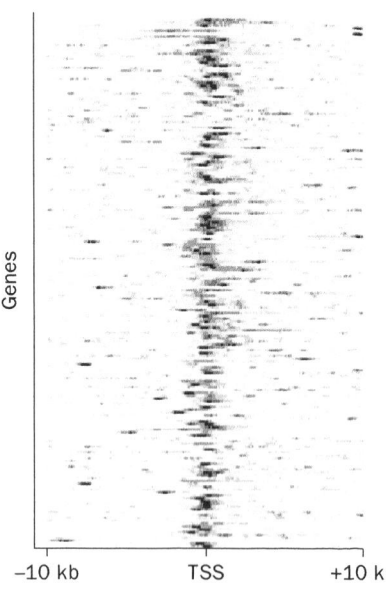

| -10 kb | TSS | +10 kb |

Figure 8.6   Example of TSS read heatmap for select genes

In the example above, we used a total of 200 bins (option '-b 200'), and a smoothing parameter '-m 10', which sums the results in each series of consecutive 10 bins.

8.5.3 Creating window-based read densities

In ChIP-seq studies, researchers are interested in visualizing the densities of their ChIP-seq reads at a genome-wide scale so that they can understand the behavior of the studied protein in genes of interest. GenomicTools can be used to create window-based read densities to be displayed as a wiggle track in the UCSC Genome Browser (Figure 8.7). First, the user needs to decide on the window parameters: (1) the size of the window; (2) the distance between consecutive windows; and (3) minimum number of read allowed in each window. The last two parameters establish a tradeoff between resolution and output file size. Here are some typical values:

```
$ set win_size = 500     # must be a multiple of win_dist
$ set win_dist = 25
$ set min_reads = 20
```

Then, the user needs to create a file describing the chromosomal bounds in REG or BED format:

```
$ head genome.bed
chr1 0 197195432
. . .
```

The genomic_scans tool can be used to compute the counts of reads stored in 'chipseq.bed' in sliding windows across the genome. Finally, the center of each window is computed and the '-wig' operation of genomic_regions converts to the wiggle format for display in the UCSC Genome Browser (Figure 8.7):

```
$ head chipseq.bed
chr1 3001228 3001229
chr1 3001228 3001229
chr1 3001438 3001439
. . .

$ cat chipseq.bed | genomic_scans counts -v -min $min_reads -w $win_size
-d $win_dist -g genome.bed | genomic_regions center | genomic_regions
wig -t 'densities' -s $win_dist -c '0,0,150' > densities.wig

$ head densities.wig
```

```
track type=wiggle_0 name='densities' color=0,0,150
variableStep chrom=chr1 span=25
4775325 20
4775375 22
4775400 22
4775425 26
4775450 26
4775475 28
4775500 28
. . .
```

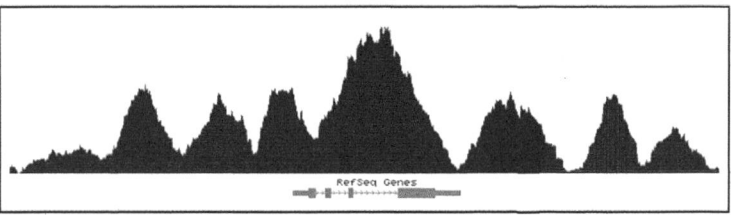

Figure 8.7   Example of window-based read densities in wiggle format

8.5.4 Identifying window-based peaks

GenomicTools can also be used to identify window-based peaks and to display them as a BED track in the UCSC Genome Browser. This is achieved using the operation 'peaks' of the genomic_scans tool. As in the 'counts' operation in the example above, the user needs to determine the window size and distance as well as the minimum number of reads in the window. For each window, p-values are computed using the binomial probability. A p-value cutoff can be enforced using the '-pval' option. The user can specify a file containing control reads ('control.bed' in our example below). If no control reads are specified, the computed p-values are based on a random background. Finally, the 'bed' operation of the genomic_regions tool is used to convert the output to the BED format for visualization in the UCSC Genome browser (Figure 8.8, upper track):

```
$ genomic_scans peaks -v -cmp -w $win_size -d $win_dist -min $min_reads
-pval 1e-05 -g genome.bed chipseq.bed control.bed | genomic_regions -bed
-t 'peaks' -c '0,150,0' > peaks.bed

$ head peaks.bed
track name='peaks' itemRgb=On visibility=1
chr1 4775075 4775575 2.27e-09 1000 + 4775075 4775575 0,150,0
chr1 4775125 4775625 6.90e-11 1000 + 4775125 4775625 0,150,0
chr1 4775150 4775650 6.90e-11 1000 + 4775150 4775650 0,150,0
. . .
```

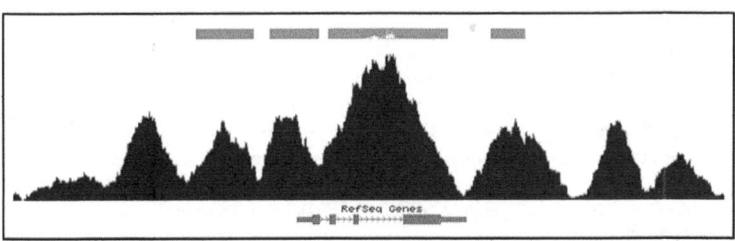

Figure 8.8 Example of window-based peaks in bed format

8.5.5 *Identifying enriched Gene Ontology terms*

In a given biological context, for example a tissue type or disease state, certain proteins (transcription factors or modified histones) tend to bind to genes of specific functional categories. Gene enrichment analysis can identify these categories. In the GenomicTools platform, enrichment analysis is performed using the permutation_test tool, which performs row permutations of inputs comprising measurements and annotations (see below). Using the ChIP-seq peaks computed above, we first calculate their densities across gene TSS regions – flanked by 10 kb – using the 'density' operation of the genomic_overlaps tool:

```
$ cat peaks.bed | genomic_overlaps density -v -i TSS.10kb.bed > tss.val

$ head tss.val
ENSMUSG00000090025:ENSMUST00000160944    0.0000e+00
ENSMUSG00000064842:ENSMUST00000082908    0.0000e+00
ENSMUSG00000051951:ENSMUST00000159265    0.0000e+00
. . .
```

Then, suppose we have a file containing gene annotations in a TAB-separated format where the first column is a gene id and the second column is a SPACE-separated list of annotations for the corresponding gene. The file must be sorted by gene id. For our example, we will use 'gene.go', which contains annotations from the Gene Ontology [29]:

```
$ head gene.go            •
ENSMUSG00000000001        membrane_fusion . . .
ENSMUSG00000000028        DNA-dependent_DNA_replication . . .
ENSMUSG00000000049        acylglycerol_metabolism angiogenesis . . .
ENSMUSG00000000058        M_phase_of_mitotic_cell_cycle . . .
ENSMUSG00000000078        cytokine_and_chemokine_mediated_sign . . .
ENSMUSG00000000085        gene_silencing
ENSMUSG00000000093        aging cardiac_muscle_development . . .
ENSMUSG00000000094        angiogenesis appendage_development . . .
ENSMUSG00000000120        axon_guidance axonogenesis cell_proj . . .
ENSMUSG00000000125        Wnt_receptor_signaling_pathway . . .
```

As an input, permutation_test needs to take a file containing three TAB-separated fields: the first field is an id (e.g. gene id), the second field is a value (e.g. density) and the third field is a SPACE-separated list of annotations. We first group the results in 'tss.val' by gene using 'vector-merge', then choose the maximum density per gene across transcripts using 'vector-max', and finally perform a join operation with 'gene.go' (note that the delimiter used in the join operation specified by option '-t' must be a TAB, i.e. Control-V-I):

```
$ cat tss.val | tr ':' '\t' | cut -f1,3 | sort | uniq | vectors -merge
-n 6 | vectors -max -n 6 | join -a1 -t ' ' – gene.go > tss.val+go

$ head tss.val+go
ENSMUSG00000000544        0.000000
ENSMUSG00000000817        0.000000 I-kappaB_kinase/NF-kappaB_cascade . . .
ENSMUSG00000001138        0.849960
ENSMUSG00000001143        1.549900
ENSMUSG00000001305        0.524970
ENSMUSG00000001674        0.399980
ENSMUSG00000002459        0.000000 negative_regulation_of_signal . . .
ENSMUSG00000002881        0.774960 Schwann_cell_differentiation . . .
ENSMUSG00000003134        0.049998 regulation_of_GTPase_activity . . .
```

Now, we can run the permutation_test tool (for details about the tool's options the reader is referred to the User's Manual):

```
$ permutation_test -v -h -S n -a -p 10000 -q 0.05 tss.val+go > peaks.
enriched.go.in.tss.txt

$ head peaks.enriched.go.in.tss.txt
CATEGORY          CATEGORY-SIZE               Q-VALUE  P-VALUE  STATISTIC
M_phase_of_mitotic_cell_cycle           13       0.00e+00 0.00e+00 13
regulation_of_lymphocyte_activation     12       6.92e-03
```

```
         0.00e+00        12
translation                     12      6.92e-03
         0.00e+00        12
mitosis                         13      6.92e-03
         0.00e+00        13
membrane_lipid_metabolism       15      6.92e-03
         1.38e-05        14
lymphocyte_activation           26      6.92e-03
         1.45e-05        22
ubiquitin_cycle                 13      2.27e-02
         6.19e-05        12
protein_catabolism              13      2.27e-02
         6.19e-05        12
T_cell_activation               20      2.27e-02
         8.46e-05        17
```

The output of the tool ranks each gene category according to the adjusted p-values (i.e. q-values) from the most to the least significant. In this particular scenario, where the input values correspond to binding sites in promoters as identified by our ChIP-seq analysis, the results of the enrichment analysis indicate that the assayed protein binds on genes that play a role in lymphocyte activation. This kind of analysis can be very powerful at generating novel biological hypothesis. Suppose that the assayed protein was never shown to participate in lymphocyte activation. Then, based on the evidence produced by the enrichment analysis above, biologists can design further experiments to prove (or reject) this hypothesis.

8.6 Performance

Typically, during the course of computational analyses of sequencing data, computational biologists experiment with different pre-processing and discovery algorithms. GenomicTools is designed to take advantage of sorted inputs (the sort order is chromosome $\rightarrow$ strand $\rightarrow$ start position) to create efficient pipelines that can handle numerous operations on multiple data sets repeated several times under different parameters. In the GenomicTools platform, the original data sets (e.g. the mapped reads in BED format) need to be sorted only once at the beginning of the project. As we show below, sorted inputs can lead to dramatic improvements in performance.

We evaluated the time and memory usage of GenomicTools and compared its performance to (1) the IRanges Bioconductor package [14]; and (2) the BEDTools suite [19]. The evaluation was run on a RHEL5/ x86-64bit platform with 12 GB of memory on the IBM Research Cloud.

For this evaluation we used sequenced reads obtained from the DREAM project [35], more specifically from challenge #1 of the DREAM6 competition. We downloaded the original FASTQ files (paired-end reads) representing mRNA-seq data from human embryonic stem cells from *http://www.the-dream-project.org/challenges/dream6-alternative-splicing-challenge*.

The FASTQ files were aligned to the reference human genome (version GRCh37, February 2009) using TopHat version 1.3.1 [36]. In total ~86 million reads were aligned and converted from BAM to BED format. In this evaluation, we measured how both CPU time and memory scale with increased input size. The task was to identify all pair-wise overlaps between a 'test' genomic interval file and a 'reference' genomic interval file. The former was obtained from the set of ~86 million sequenced reads using re-sampling without replacement (re-sampling of 1, 2, 4, 8, 16, 32 and 64 million reads), and the latter contained all annotated transcript exons from the ENSEMBL database [37], as well as all annotated repeat elements from the UCSC Genome Browser [17], that is a total of ~6.4 million entries.

As demonstrated in Figure 8.9, GenomicTools improves greatly on time performance (speed-up of up to ~3.8 compared to BEDTools and ~7.0 compared to the IRanges package of Bioconductor) if the inputs are sorted,

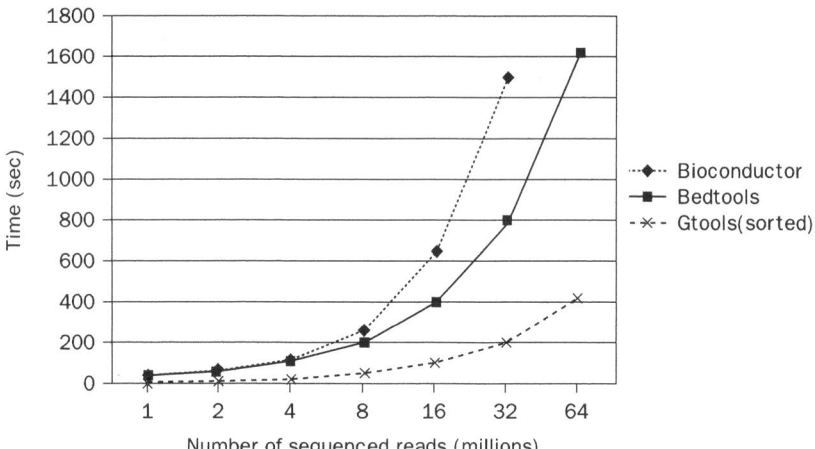

Figure 8.9 Time evaluation of the overlap operation between a set of sequenced reads of variable size (1 through 64 million reads in logarithmic scale) and a reference set comprising annotated exons and repeat elements (~6.4 million entries). Using GenomicTools on sorted input regions yields a speed-up of up to ~3.8 compared to BEDTools and ~7.0 compared to the IRanges package of Bioconductor

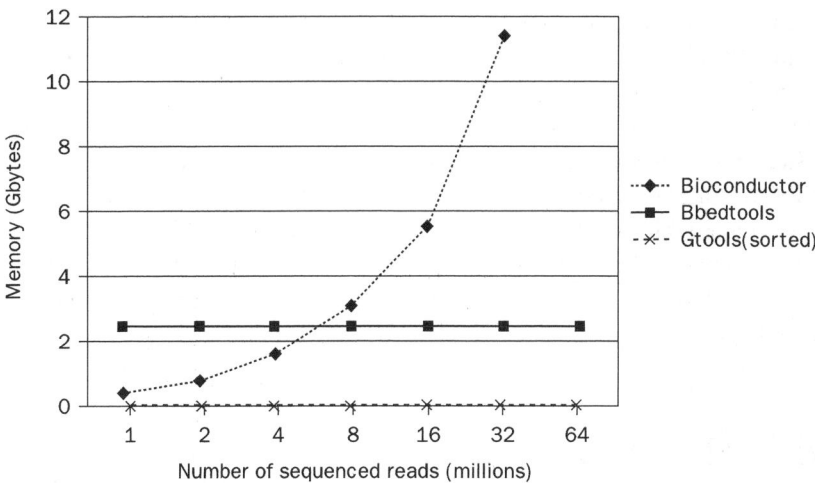

Figure 8.10 Memory evaluation of the overlap operation presented in Figure 8.10. Memory requirements for the IRanges package of Bioconductor increase with input size, which makes it impossible to handle big data sets (in this particular hardware setup of 12 GB memory we could only compute overlaps for up to 32 million reads). BEDTools uses a fixed amount of memory, which depends on the size of the reference input set (i.e. exons and repeat elements). GenomicTools uses no significant amount of memory, as all input files are read sequentially, but, of course, it has to rely on sorted inputs

and is therefore well suited for large-scale Bioinformatics projects. Additionally, GenomicTools makes virtually no use of memory, unlike IRanges and BEDTools, both of which use a significant amount of memory (Figure 8.10), thus limiting the number of such processes that can simultaneously run on the same system.

8.7 Conclusion

It is becoming increasing apparent that humanity in general, and science in particular, can greatly benefit from properly applying the principles of openness, transparency, and sharing of information. The free open source initiatives in recent years have led to new interesting phenomena, such as crowd-sourcing, which bring together entire communities to everyone's benefit in some sort of collaborative competition. This notion of contribution was the single most significant motivation for making this suite of tools publicly and freely available. Soon after we decided to open source our software, we realized that it is one thing to use software internally, and a totally different thing to actually expose it to public usage. For example, we understood that for open source to be meaningful, the developer needs to undertake a variety of tasks, such as documenting the source code, creating a user's manual, tutorials, a web site, and also using some version control and a bug tracking system. Overall, we are convinced that this effort was worthwhile and that the open source trend is only going to get stronger in the next few years.

8.8 Resources

This chapter described version 2.0 of GenomicTools released in September 2011. The source code, documentation, user manual, example data sets and scripts are available online at *http://code.google.com/p/ibm-cbc-genomic-tools/*.

To install GenomicTools, follow these simple instructions:

```
tar xvzf genomic-tools-VERSION-src+doc.tgz
cd genomic-tools
make
sudo cp bin/*    /usr/local/bin
```

There is a dependency on the GNU Scientific Library (GSL), which can be downloaded from *http://www.gnu.org/software/gsl/*.

Examples and associated input files and scripts are part of the source code distribution in the 'examples' subdirectory. The API documentation is stored in the 'doc' subdirectory.

For details about the options for each tool, the reader is referred to the User's Manual, which can be downloaded from *http://code.google.com/p/ibm-cbc-genomic-tools/#Documentation*.

For any questions, suggestions or comments contact Aristotelis Tsirigos at atsirigo@us.ibm.com.

8.9 References

[1] Margulies M, et al. 'Genome sequencing in microfabricated high-density picolitre reactors,' *Nature* 2005;437(7057):376–80.

[2] Wang ET, et al., 'Alternative Isoform Regulation in Human Tissue Transcriptomes,' *Nature Nature*;456(7221)470–6.

[3] Blekhman R, Marioni JC, Zumbo P, Stephens M, and Gilad Y. 'Sex-specific and lineage-specific alternative splicing in primates,' *Genome Research* 2010;20(2):180–9.

[4] Wilhelm BT, et al. 'RNA-seq analysis of 2 closely related leukemia clones that differ in their self-renewal capacity,' *Blood* 2011;117(2):e27–38.

[5] Berger MF et al. 'Integrative analysis of the melanoma transcriptome,' *Genome Research* 2010;20(4):413–27.

[6] Maher CA, et al. 'Transcriptome Sequencing to Detect Gene Fusions in Cancer,' *Nature* 2009;458(7234): 97–101.

[7] Mortazavi A, et al. 'Scaffolding a Caenorhabditis nematode genome with RNA-seq,' *Genome Research* 2010;20(12):1740–7.

[8] Denoeud F, et al. 'Annotating genomes with massive-scale RNA sequencing,' *Genome Biology* 2008;9(12):R175.

[9] Yassour M, et al. 'Ab initio construction of a eukaryotic transcriptome by massively parallel mRNA sequencing,' *Proceedings of the National Academy of Sciences of the United States of America* 2009;106(9):3264–9.

[10] Guttman M, et al. 'Ab initio reconstruction of cell type-specific transcriptomes in mouse reveals the conserved multi-exonic structure of lincRNAs,' *Nature Biotechnology* 2010;28(5):503–10.

[11] Trapnell C, et al. 'Transcript assembly and quantification by RNA-Seq reveals unannotated transcripts and isoform switching during cell differentiation,' *Nature Biotechnology* 2010;28(5):511–15.

[12] Pepke S, Wold B, and Mortazavi A. 'Computation for ChIP-seq and RNA-seq studies,' *Nature Methods* 2009;6(11):S22–32.

[13] Garber M, Grabherr MG, Guttman M and Trapnell C. 'Computational methods for transcriptome annotation and quantification using RNA-seq,' *Nature Methods* 2011;8(6):469–77.

[14] Gentleman RC, et al. 'Bioconductor: open software development for computational biology and bioinformatics,' *Genome Biology* 2004;5(10):R80.

[15] Goecks J, Nekrutenko A and Taylor J. 'Galaxy: a comprehensive approach for supporting accessible, reproducible, and transparent computational research in the life sciences,' *Genome Biology* 2010;11(8):R86.

[16] McLean CY, et al. 'GREAT improves functional interpretation of cis-regulatory regions,' *Nature Biotechnology* 2010;28(5):495–501.

[17] Kent WJ, et al. 'The human genome browser at UCSC,' *Genome Research* 2002;12(6):996–1006.

[18] Nicol JW, Helt GA, Blanchard Jr SG, Raja A and Loraine AE. 'The Integrated Genome Browser: free software for distribution and exploration of genome-scale datasets,' *Bioinformatics (Oxford, England)* 2009;25(20):2730–1.

[19] Quinlan AR and Hall IM. 'BEDTools: a flexible suite of utilities for comparing genomic features,' Bioinformatics 2010;26(6):841–2.

[20] Tsirigos A, Haiminen N, Bilal E and Utro F. 'GenomicTools: a computational platform for developing high-throughput analytics in genomics,' *Bioinformatics (Oxford, England)* November 2011.

[21] Tsirigos A and Rigoutsos I. 'Human and mouse introns are linked to the same processes and functions through each genome's most

frequent non-conserved motifs,' *Nucleic Acids Research* 2008;36(10): 3484–93.

[22] Tsirigos A and Rigoutsos I. 'Alu and b1 repeats have been selectively retained in the upstream and intronic regions of genes of specific functional classes,' *PLoS Computational Biology* 2009;5(12), p. e1000610.

[23] Ochoa-Espinosa A, Yu D, Tsirigos A, Struffi P and Small S. 'Anterior-posterior positional information in the absence of a strong Bicoid gradient,' *Proceedings of the National Academy of Sciences of the United States of America*, 2009;106(10): 3823–8.

[24] Laurent L, et al. 'Dynamic changes in the human methylome during differentiation,' *Genome Research* 2010;20(3):320–331.

[25] 'BED format.' [Online]. Available: *http://genome.ucsc.edu/FAQ/FAQformat.html#format1*.

[26] 'GFF format.' [Online]. Available: *http://www.sanger.ac.uk/resources/software/gff/spec.html*.

[27] 'SAM format.' [Online]. Available: *http://samtools.sourceforge.net/SAM1.pdf*.

[28] 'WIG format.' [Online]. Available: *http://genome.ucsc.edu/goldenPath/help/wiggle.html*.

[29] Ashburner M, et al. 'Gene ontology: tool for the unification of biology. The Gene Ontology Consortium,' *Nature Genetics* 2000;25(1):25–9.

[30] Mortazavi A, Williams BA, McCue K, Schaeffer L and Wold B. 'Mapping and quantifying mammalian transcriptomes by RNA-Seq,' *Nature Methods* 2008;5(7):621–8.

[31] 'Doxygen.' [Online]. Available: *http://www.doxygen.org*.

[32] Rada-Iglesias A, Bajpai R, Swigut T, Brugmann SA, Flynn RA and Wysocka J. 'A unique chromatin signature uncovers early developmental enhancers in humans,' *Nature* 2011;470(7333):279–83.

[33] Ernst J, et al. 'Mapping and analysis of chromatin state dynamics in nine human cell types,' *Nature* 2011;473(7345):43–9.

[34] Wu H, et al. 'Dual functions of Tet1 in transcriptional regulation in mouse embryonic stem cells,' *Nature* 2011;473(7347):389–93.

[35] Stolovitzky G, Monroe D, and Califano A. 'Dialogue on reverse-engineering assessment and methods: the DREAM of high-throughput pathway inference,' *Annals of the New York Academy of Sciences* 2007;1115:1–22.

[36] Trapnell C, Pachter L, and Salzberg SL. 'TopHat: discovering splice junctions with RNA-Seq,' *Bioinformatics (Oxford, England)* 2009;25(9): 1105–11.

[37] Birney E, et al., 'An overview of Ensembl,' *Genome Research*, 2004;14(5):925–8.

Published by Woodhead Publishing Limited, 2012

## 9

Creating an in-house 'omics data portal using EBI Atlas software

Ketan Patel, Misha Kapushesky and David Dean

Abstract: The amount of public gene expression data has dramatically increased in recent years. This has resulted in several efforts to curate and make accessible this huge amount of public data in a useful and easy to use way. One such project, the European Bioinformatics Institute's Gene Expression Atlas, led to the development of a portal that was intuitive to use and yet still incorporated sophisticated queries and statistical analysis results. The authors describe how the public Atlas project was adapted to become a standalone product, which can be installed anywhere. This Atlas allows both private companies and public institutes to develop in-house data portals.

Key words: gene expression; microarray; portal; 'omics; curation; MAGE-TAB; intuitive user interface.

9.1 Introduction

Within the pharmaceutical industry, there is an ever increasing amount of biological data being generated. Much of this data is from the 'omics disciplines such as transcriptomics, proteomics and metabolomics. The most prevalent of these is transcriptomics, with a mountain of public data available and increasing amounts generated within companies. To process, analyze and interpret this data requires sophisticated software, which is available both as open source packages and also several vendor solutions. To date, most of the work in developing software has focused on processing, analytics and storage of the data. This software has a high learning curve and is not easily accessible without training. Within the environment of a large pharmaceutical company, we saw a need to create a portal where users could browse and query 'omics data sets, in an easy to use way using just a web browser. This would enable a whole new category of users to access the data and make sense of it, and hopefully use it in day-to-day scientific decision making. The solution should also complement existing 'omics infrastructure rather than replace it. Here, we describe how we have adapted the European Bioinformatics Institute (EBI) Gene Expression Atlas software (the Atlas), which was developed for a similar purpose, into a standalone product that can be deployed within a company firewall. This in-house installation of the Atlas allows loading of public and internal data sets, and gives scientists access to these data sets in a nice easy-to-use interface. First, we give an overview of how 'omics data is used in the drug industry today, and the motivation for building a portal. Next, we detail the process of adapting Atlas from an existing public web site into an installable standalone piece of software. Finally, an overview of deploying and running the portal within a large company will be described.

9.2 Leveraging 'omics data for drug discovery

9.2.1 'omics data is used throughout the drug discovery and development pipeline

The use of 'omics data within drug discovery has increased in recent years in particular gene expression profiling using transcriptomics [1]. In the past it was used in pre-clinical studies to understand drug action, animal models of disease and also to understand toxicology. Later, 'omics studies utilizing human samples became more prevalent, and this is now done in all phases of the drug discovery and development pipeline (see Figure 9.1). In the early phase of discovery, a typical scenario is to look for differences in expression between normal and diseased tissue (both in humans and in animals), to discover new potential targets for therapeutic intervention [2]. It is also common at this early stage to look at gene expression following compound treatment in

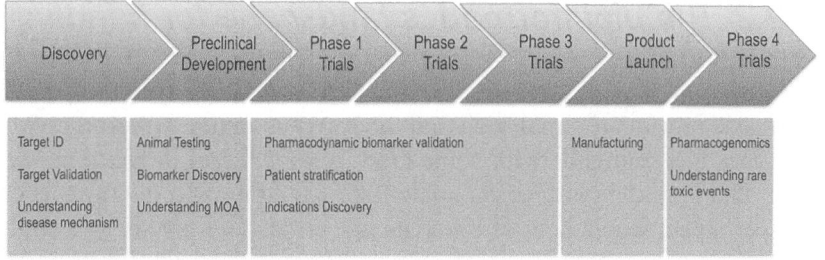

Discovery	Preclinical Development	Phase 1 Trials	Phase 2 Trials	Phase 3 Trials	Product Launch	Phase 4 Trials
Target ID	Animal Testing	Pharmacodynamic biomarker validation			Manufacturing	Pharmacogenomics
Target Validation	Biomarker Discovery	Patient stratification				Understanding rare toxic events
Understanding disease mechanism	Understanding MOA	Indications Discovery				

Figure 9.1 Applications of 'omics data throughout the drug discovery and development process

cell lines to discern patterns and signatures of mode of action. These can later be utilized to discover interesting connections between drugs and disease [3].

In the later phases of drug discovery, 'omics data can be utilized for biomarker discovery and also increasingly to stratify patient populations. A typical scenario is using pre-clinical species or cell line-based studies to identify biomarkers using the lead compound, and then moving into the clinic to validate the best candidate marker(s) to use in later trials. A crucial application of a biomarker is to find the optimum efficacious dose in humans, which in the past, was usually done by estimating downwards from the maximum tolerated dose. This is not always the best approach, as sometimes the maximum tolerated dose is not reached, so the best dose may be much lower resulting in a safer drug. Many different 'omics technologies can be utilized for biomarker discovery (see [4]), and there is an increasing need for software that can integrate data from multiple technologies.

There is also now an acute understanding that to truly treat human disease, it is critical to study the disease as much as we can in humans. Therefore, there is a large effort to do deep molecular profiling of human diseased samples, to gain a better understanding of disease pathophysiology. Furthermore, it is important to also know target expression and distribution in normal tissues to understand how the drug will act on the body and to highlight any potential side effects early on. It is therefore crucial that not only computational experts, but also discovery bench scientists, clinicians, drug metabolism and pharmacokinetics (DMPK) scientists and toxicologists can easily access and interpret this type of data.

9.2.2 The bioinformatics landscape for handling 'omics data

The basic pipeline for analyzing and interpreting 'omics data starts with raw data files and ends with some kind of knowledge. In between, we have data QC, data reduction, analytics, annotation and interpretation and perhaps a whole host of other smaller steps. This results in a large landscape of bioinformatics applications, which cater to the various steps in the data pipeline. We can break down the users of these applications into three crude bins. The first are the computational experts who are familiar with programming and can script highly custom code to process complex data sets. These users usually use open source solutions, although they are also very savvy with the vendor tools out there. Second, there are the power users, who include computational biologists who do not know how to program and also computationally adept scientists who are willing to learn complex software. These users tend to work on this kind of data a lot, and use tools such as Genespring and Expressionist routinely for project data analysis and support. They tend to work on one data set at a time, and are less concerned with mining across a large compendium of data. Hence, the solutions in this space tend to be very project-focused and are great at getting useful knowledge from one data set. The last category of users are the average bench scientists and managers. They do not have time to learn a lot of complex software, and need to know the answers to simple questions. They want to make queries on data that have already been generated and analyzed either inside or outside the company. They want to verify what those data mean, and be presented with the data in a form that can be easily interpreted such as a graphical output. These user communities are summarized in Figure 9.2, along with some typical applications that cater for each community.

9.2.3 Requirements for an 'omics portal

The EBI Gene Expression Atlas was constructed as a large-scale portal for publicly available functional genomics data. Public domain 'omics data sets originate from a variety of sources; the ArrayExpress Archive of Functional Genomics contains data sets performed on more than 5000 different platforms coming from hundreds of different laboratories. Naturally, raw data come in a great variety of formats, whereas meta-data, that is experiment description representation can be highly idiosyncratic, oftentimes incomplete or even missing. Significant efforts

User communities for 'omics data

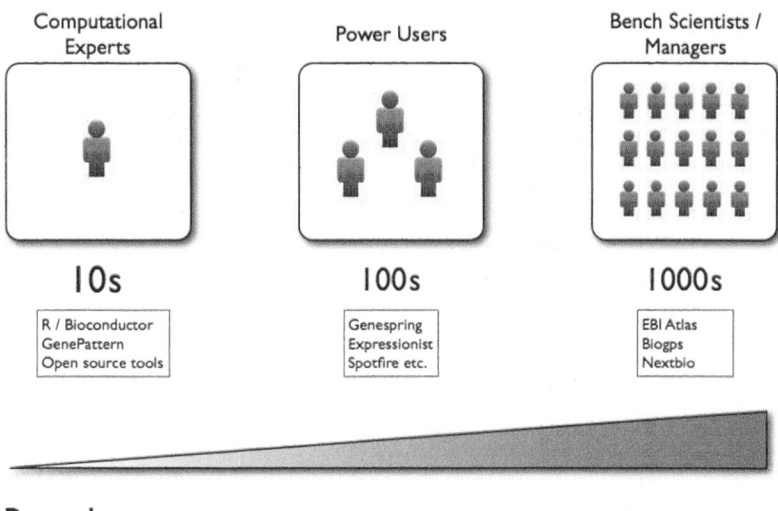

Computational Experts

Power Users

Bench Scientists / Managers

10s	100s	1000s
R / Bioconductor GenePattern Open source tools	Genespring Expressionist Spotfire etc.	EBI Atlas Biogps Nextbio

Raw data   Knowledge

Figure 9.2   User communities for 'omics data

have been undertaken in recent years to bring this situation under control; establishing the Minimal Information About a Microarray Experiment (MIAME) [5] and the Minimum Information about a high-throughput SeQuencing Experiment (MINSEQE) [6] standards for microarrays and NGS experiments and a common data exchange format, MAGE-TAB [7].

In order to create a useful 'omics portal, two additional steps are needed beyond aggregation of standards-compliant data:

1. curation towards a commonality of experimental descriptions across all available data; and

2. statistical analysis enabling queries over the entire curated data collection.

Curation is a part-automated, part-manual process, the aim of which is to prepare data for meaningful downstream analysis, by ensuring data completeness and correctness and mapping free-text descriptions to controlled vocabularies, or ontologies. In bioinformatics, ontologies are used widely to model knowledge of various domains: the OBO Foundry [8] and NCBO BioPortal [9] host between them more than 300 ontologies for describing things as diverse as medical terms for disease and loggerhead turtle nesting behaviors. The Experimental Factor Ontology (EFO) [10] was created at the EBI specifically to describe 'omics experiments and is used widely in the curation and data analysis process. It is described in detail in [11].

The EBI Atlas is constructed primarily for transcriptomics experiments. In these data sets, one of the aspects of curation is to identify and mark experimental variables being tested, for example disease, tissue, cell type or drug response, and map these terms to the EFO hierarchy uniformly. Once the terms are identified, statistical analysis is performed automatically to identify significantly differentially expressed elements in the transcriptome (e.g. genes, their transcripts, microRNA). The statistical framework is described in [12].

In addition to these browsing requirements, which serve the majority of the bench user community described in Figure 9.2, it is also useful to have the ability to process the curated data sets using more sophisticated algorithms, which are not pre-computed automatically for all data and may be thought of at a later date. To serve this purpose, the data in the 'omics portal must be accessible through programmatic APIs in order to automate data processing using other algorithms. The EBI Atlas software again deals with this in an elegant way, by storing the data in self-describing and accessible NetCDF format [13] files, and also providing RESTful APIs to the curated data sets, as described in the next section.

9.3 The EBI Atlas software

The Atlas was constructed to provide a simple, easy to use and understand interface to the computed statistics. As such, the requirements are to be able to query by gene, including any of its attributes such as synonyms, protein domains, pathway and process annotations, or by curated experimental variable, or both. Expecting the user to be unfamiliar with the statistical underpinnings of the analysis, simple summaries of up-/down-expression patterns are provided: for each gene, the number of public data sets is reported where it is over- or under-expressed, color-coded red and blue, respectively. More sophisticated users can create more complex queries and drill down to individual data points underlying the analysis results.

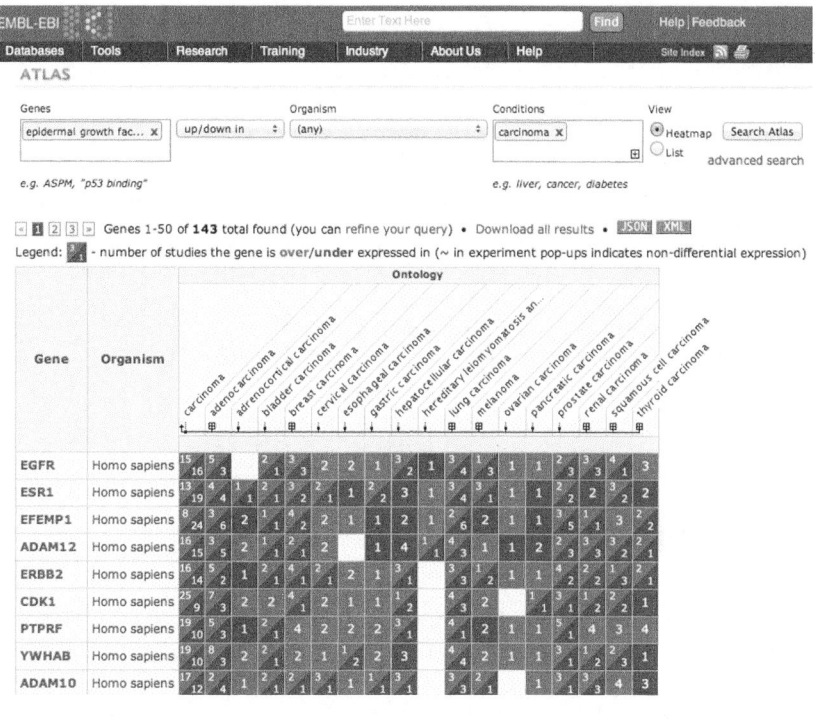

Figure 9.3 Atlas query results screen

Figure 9.3 presents a selection of results for the search for members of the EGFR signaling pathway differentially expressed in carcinomas. Genes are in rows, ranked by the number of found studies, and columns provide a view of EFO carcinoma hierarchy. The numbers in dark/light squares indicate numbers of studies with differential expression.

Bioinformaticians and computational biology researchers often need direct access to the data and analysis results, bypassing the graphical user interface. For such users, extensive application programming interfaces (APIs) are provided, enabling them to extract data from the Atlas as XML or Javascript Object Notation (JSON) feeds. Every result page has a JSON and XML button providing direct access to these data feeds.

9.3.1 Adapting Atlas for the real (internal) world

In order to adapt the EBI Atlas to standalone deployments, we undertook a significant engineering effort. The greatest effort was devoted to

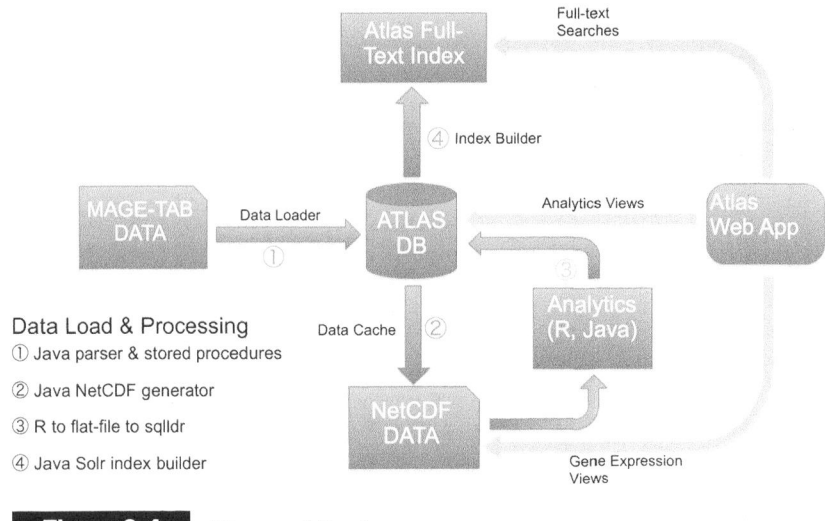

Figure 9.4 Atlas architecture

simplifying the underlying data model and creating a new framework for loading data. The overall architecture is described in Figure 9.4 below.

9.3.2 Simplifying the data model

To simplify maintenance and data administration, we moved from a highly formalized data structure, where every attribute was encoded explicitly as a separate set of tables, into a more lightweight data model, fewer than 30 tables in total. This also allowed us to simplify greatly the data loading procedure, going almost directly from MAGE-TAB parser objects to Atlas data objects and then to the relational database. The database itself does not store the raw data or analytics: these are stored in NetCDF format [13] files and accessed directly by the web application and R-based analytics. These NetCDF files are also available for power users to access via R scripts, if they need to do more sophisticated processing of the raw data.

9.3.3 Building an easy to use installation procedure

The original Atlas was not meant to be installed anywhere except at the EBI. This meant there were a number of dependencies on existing infrastructure. We developed a regular software and data release procedure, together with detailed documentation. To install the standalone Atlas, two files are downloaded – a web application and a data archive. The data archive contains an installer, which unpacks the files and loads them automatically into the specified database. The web application contains an administration interface with configuration options and data loading interface. We can release incremental updates to the data archive, and update an Atlas installation without re-installing the database.

9.3.4 Creating a new administration interface for loading data

A completely new administration interface was developed, as shown in the figure below. This interface allows the Atlas administrator to load and unload experiments, update data and analytics, and manage various configurable options of the software, for example custom branding templates and external database/cross-reference links.

9.3.5 Future enhancements

As Atlas matures into a larger platform for transcriptomics data, we have started to expand its scope from microarrays towards other 'omics data types – first of all, high-throughput sequencing-based, or next-generation sequencing (NGS), transcriptomics. To this end, we developed a pipeline, ArrayExpressHTS [14], for processing NGS data sets from raw reads to quantified transcript isoform expressions, and have been working on integrating it into the Atlas. NGS data, RNA-Seq, and ChIP-Seq will be a key focus for future Atlas developments.

The amount of data in the public Atlas has grown more than sixfold since its start in 2009 and we are seeing that for a growing number of experiments our statistical approach may not be best suited. We must turn our attention now to method development: in particular, we are gradually switching to using raw data wherever possible and reprocessing these data for best standardization and comparability, and we are working with other groups to incorporate advanced gene set-based statistics into our pipeline.

Lower costs and technological barriers have made possible two comparatively new kinds of experiments: large meta-analysis experiments, such as the global map of human gene expression (Atlas experiment ID: E-MTAB-62) and large reference experiments such as the Illumina Body

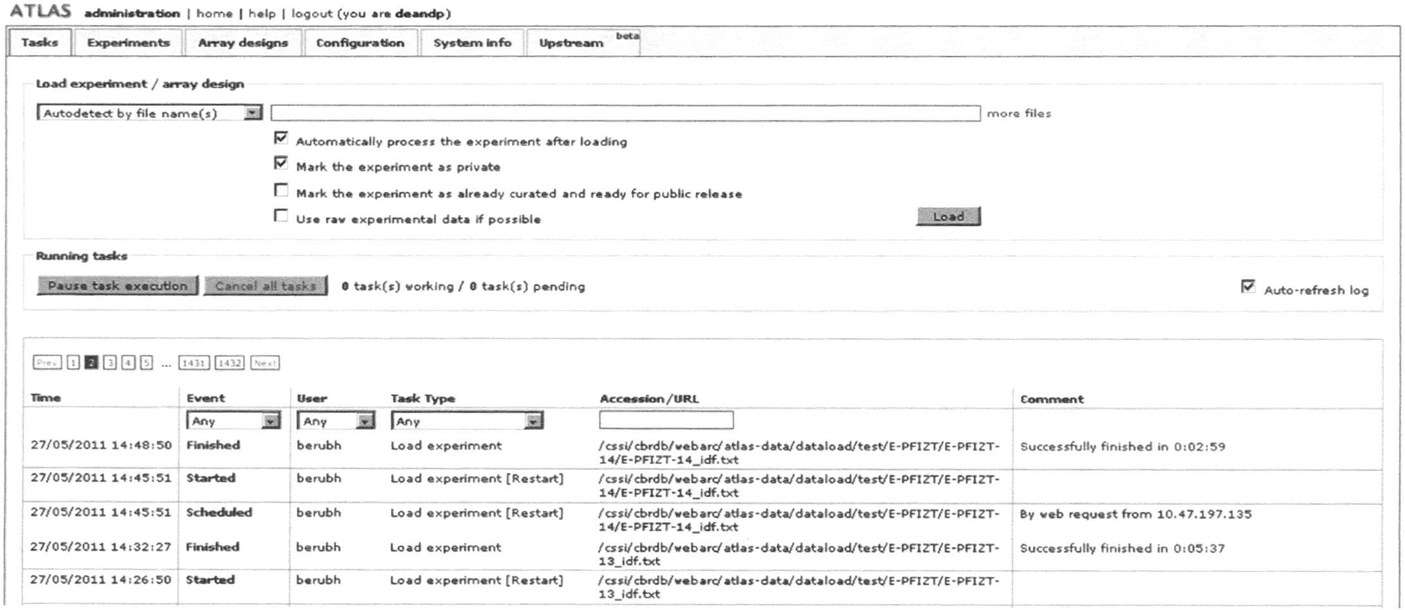

ATLAS administration | home | help | logout (you are deandp)

| Tasks | Experiments | Array designs | Configuration | System info | Upstream beta |

Load experiment / array design

Autodetect by file name(s) ▾ [   ] more files

☑ Automatically process the experiment after loading
☑ Mark the experiment as private
☐ Mark the experiment as already curated and ready for public release
☐ Use raw experimental data if possible

[Load]

Running tasks

[Pause task execution] [Cancel all tasks]   0 task(s) working / 0 task(s) pending

Prev 1 2 3 4 5 ... 1431 1432 Next

☑ Auto-refresh log

Time	Event	User	Task Type	Accession/URL	Comment
	Any ▾	Any ▾	Any ▾	[ ]	
27/05/2011 14:48:50	Finished	berubh	Load experiment	/csi/cbrdb/webarc/atlas-data/dataload/test/E-PFIZT/E-PFIZT-14/E-PFIZT-14_idf.txt	Successfully finished in 0:02:59
27/05/2011 14:45:51	Started	berubh	Load experiment [Restart]	/csi/cbrdb/webarc/atlas-data/dataload/test/E-PFIZT/E-PFIZT-14/E-PFIZT-14_idf.txt	
27/05/2011 14:45:51	Scheduled	berubh	Load experiment [Restart]	/csi/cbrdb/webarc/atlas-data/dataload/test/E-PFIZT/E-PFIZT-14/E-PFIZT-14_idf.txt	By web request from 10.47.197.135
27/05/2011 14:32:27	Finished	berubh	Load experiment	/csi/cbrdb/webarc/atlas-data/dataload/test/E-PFIZT/E-PFIZT-13_idf.txt	Successfully finished in 0:05:37
27/05/2011 14:26:50	Started	berubh	Load experiment [Restart]	/csi/cbrdb/webarc/atlas-data/dataload/test/E-PFIZT/E-PFIZT-13_idf.txt	

Figure 9.5 Atlas administration interface

Published by Woodhead Publishing Limited, 2012

Map 2.0 data set (E-MTAB-513). These experiments call for different statistics and visualization approaches entirely, and the Atlas team have started work on making this possible.

9.4 Deploying Atlas in the enterprise

9.4.1 Rationale and requirements for an 'in-house' Atlas

The EBI public Gene Expression Atlas web site has proven to be a valuable resource for mining public expression data. Having an 'in-house' installation of the system provides the added advantage of combining internal, proprietary data with the public content in one easy to use system. Furthermore, an internal system can be customized and extended and may also be integrated with other internal systems.

An open source system will typically have a shorter release cycle than most commercial systems and is likely to require some additional testing. Therefore, it is highly desirable to have a research computing environment where databases, data files and applications can be deployed without a formal change control processes. Pfizer Research Business Technology supports global and local computational biology groups with an application hosting environment that uses a hybrid support model where system management and configuration are under the control of the data center platform and operations teams using standard processes. Applications are deployed using non-privileged service accounts used by a deployment team. Databases are managed in a similar fashion, using an Oracle instance backed-up and managed by the corporate database operations team. A local representative in the application support group has system privileges enabling them to create, extend, and tune the Atlas database. These environments are used to support both test and quasi-production environments, enabling new releases to be evaluated offline and quickly released for general use after testing and evaluation.

9.4.2 Deployment and customization of the Pfizer Gene Expression Atlas

Deployment of the standalone Atlas follows procedures provided by the EBI Atlas team. The database file is downloaded and deployed on a

Oracle server (Oracle version 11.2) supporting research class applications. Typically, two environments are supported as described in the previous section. The netCDF data files and indexes are managed on general NAS filers that are available to the application hosts, this is where the bulk of the data resides (the database size is small being a few GB). The Tomcat web application is deployed on an Enterprise Red Hat Linux system, (HP Proliant DL 580 with 4 CPUs and 64 GB memory) running under a locally installed Java Runtime Environment.

A corporate system that includes proprietary data should have access control. In the Pfizer deployment, login and authentication was implemented by developing a standard J2EE-based request filter that integrates with a single sign-on system used throughout the company. The filter class and supporting libraries are simply installed to Tomcat's standard directories for shared libraries or classes. The filter is referenced in the web application deployment file, so no custom extensions or modifications are required.

A couple of minor extensions have been developed to date to facilitate the Pfizer deployment. First, it was desired to add a corporate logo to the in-house installation to distinguish the site from the public Atlas. This was done initially through modification of an application JSP page. The EBI Atlas team then developed a template model that allows the site to be 'branded' for local installations without modification of source code. A minor code change has also been implemented to facilitate capturing of application usage metrics, so code for the system is currently obtained from the source-code control system supporting the open source project and used to compile a custom build containing the local customizations.

9.4.3 Loading custom and private data sets into the in-house Atlas

Although the EBI public Atlas contains thousands of curated public experiments, the wealth of public data submitted historically and currently will require local curation to include additional studies of interest. In addition, public and private research institutions will have access to internally generated and proprietary data to add to their local Atlas installation. Therefore, an institution interested in setting up an in-house Atlas instance will want to consider workflow and resource for curation of internal data. The Pfizer installation of Gene Expression Atlas has been supplemented with public studies of interest to scientists and

Published by Woodhead Publishing Limited, 2012

clinicians supporting oncology research. Around 50 studies of interest were selected based on key cancer types, treatments or systems of interest to the research unit. Although a number of these studies were found to be in the public Atlas, the remaining studies where expression data and meta-data were available from public sources were curated in-house.

Atlas includes an administrative utility page that allows experiments to be loaded in MAGE-TAB format, which is the preferred format for data submission to the EBI Array Express database. The MAGE-TAB Sample and Data Relationship Format (SDRF) file allows specification of characteristics describing sample properties. These typically include general properties of samples that may hold for all samples in the experiment (such as tissue type) as well as factors under study, such as treatment or clinical outcome. In addition, the Experimental Factor identifies the primary variables of interest in the study, and these variables are used to drive the analytics in Expression Atlas.

Many of the public studies submitted to Array Express specify the key sample and experiment meta-data through Comment or Characteristics fields in the SDRF file. So for these types of studies, study curation can be achieved by formatting or transforming the key variables of interest as Experimental Factors in the SDRF file. Once these files are parsed and validated by the Atlas loader, the analytics workflow is automatically triggered to identify genes having significant association with the variables identified. A number of public studies include only some of the meta-data for the experiment, or even none at all! In these cases, original publications or other online resources have to be consulted, and data curation becomes more time-consuming or problematic.

One approach to loading internal data would be to export local expression studies of interest in MAGE-TAB format [7] for loading by the sample loader. This has proven to be feasible, although some manual curation is still required to identify the key experimental factors in a given study and to map meta-data terms used on local systems to the Experimental Factor Ontology used in the Expression Atlas. If all data for a given public or internal study are available in electronic format, curation of a typical experiment having 20–50 samples can generally be accomplished in an hour or two. Studies requiring re-entry of data from publications or other non-electronic sources will require considerably longer to properly curate, if, in fact, the meta-data can be obtained. Of course, a number of other more general considerations of quality have to be taken into account when entering internal and external studies, including experimental design and quality of samples and data files.

Installation, operation, and support of a local Atlas installation has proven to be very feasible and is certainly a cost-effective option to consider. Supplementing internal data has also proven to be feasible and worthwhile. As will be discussed in the next section, efforts to reduce the amount of manual curation in data entry are under active development.

9.4.4 Next steps and future considerations for an in-house Atlas

Manual curation of experimental meta-data for custom studies has proven to be feasible, but this is certainly not a high-value activity for in-house resource and is prone to human error. It would be highly desirable to map standard terms used for annotation of in-house data with the EBI EFO. Better yet, the EFO could be used in systems to capture experiment and sample meta-data so that data are complete and captured in a standardized format from the start! As the EFO is inherently dynamic, update and exchange mechanisms would be highly desirable to bring in new terms of general interest such as standard strain, cell line and treatment names.

As described above, public studies of interest to Pfizer have been curated in-house and loaded into the local Atlas. Although this activity adds value, it is likely to be of pre-competitive value to Pfizer and other commercial entities. Therefore, it would be of mutual benefit to public and private institutions having local installations to be able to publish locally curated public studies back to the public Atlas, subject to review of curators of the public Atlas. As a considerable portion of the available public expression data is currently uncurated, it would be mutually beneficial for the Atlas community to consider a one-time effort to review and curate all public studies meeting a defined quality standard. Such a project would benefit commercial and non-commercial partners alike, and it would be fitting to have an open source software system developed to share public expression data contribute to an open science model for gene expression and other 'omics data.

9.5 Conclusion and learnings

The open source EBI Atlas platform has proven to be a great project to bring curated expression studies into an easy to use portal system for scientists. Initial analysis of a gap in vendor solutions in this area pointed to either developing an in-house portal or adapting open source code. In collaborating with the Atlas team, a great deal of expertise in curation and data management of a large collection of gene expression data has been gained, while at the same time developing a usable prototype that serves as a proof-of-concept for future development. There were several learnings we can take away from this experience.

The first is the importance of data curation to the development of a useful portal. The EBI Atlas uses an ontology to control terms used in data curation of the various gene expression studies contained therein. This ontology make the data searchable in a semantically meaningful way, and also opens the door for using other algorithms which can understand semantics to be deployed later onto this large collection of data. One example is that we could go through the entire data collection and generate signatures of gene changes between normal and various disease states. Once the signatures have been computed, a gene set enrichment analysis [15] could be performed on each disease state to reveal pathways involved in particular diseases. Being able to make a query, for example, like 'Find all genes upregulated consistently in breast cancer' and then combining it with other data such as drug target databases would also be a very powerful use of this kind of semantically aware data. Of course this is a great experience when all the data are curated by subject matter experts, but what about in-house legacy data sets? It is difficult to get consistent annotations and curations of these data sets where in many cases the original scientists may have left the company a long time ago. It is also helpful to have good user-friendly tools to enable easy curation of experimental meta-data, and these are sorely lacking as yet. In this project, a combination of in-house bioinformatics expertise and computational biologists who are familiar with the subject matter (e.g. cancer) were a prerequisite to getting high-quality annotations. As stated above, we believe that for public data sets this could be a pre-competitive activity where one could outsource the data curation to experts and the work could be funded through a variety of organizations.

One of the other learnings in terms of collaborating with an open source project is that of the provision of support. Typically when dealing with a vendor, certain service levels of support and documentation to help us install, configure, and deploy a system such as the Atlas are expected. Even then, it is likely that there will be certain teething problems in going from a relatively open environment to a large controlled corporate data center. One of the essential components of a successful collaboration is good communications, and this was certainly the case

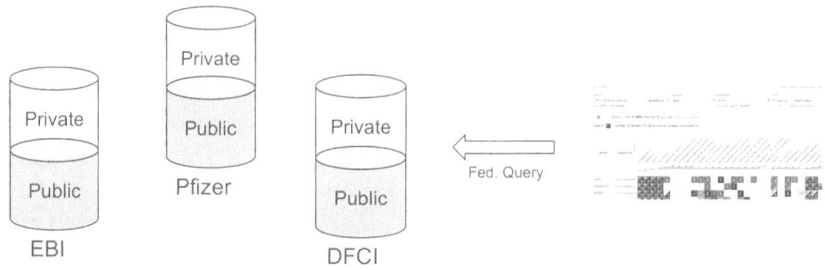

Figure 9.6 Federated query model for Atlas installations

here. The Atlas team was responsive and accommodating to our frequent requests for information and support. Had this not been the case, the project could have become very frustrating and difficult. It is also worth pointing out that in our initial agreement, we stipulated that the outcome of our collaboration would be free and open source, and that set the tone for an open collaborative relationship. It is important and worth stating that for corporations to make more use of open source software, they should also be willing to fund and contribute code to these efforts.

Finally, there is the learning in terms of ongoing maintenance of this standalone Atlas. We realized in setting up our own in-house version of this repository how similar our installation is to the public instance, and how with some tweaking we could move to more interesting models. One model is to have a federated solution whereby each institution could have their own Atlas repository, and that each repository would have a 'public' and 'private' portion (see Figure 9.6). The public portion of each repository could then be federated into the public instance, so that queries against that instance would also search all public repositories in the federation. This, however, does not solve the problem of maintenance, as each institute would still need to maintain their own individual installations and update them with patch releases as newer features are released.

Another model is the 'cloud' model where each institute hosts their separate installation in the cloud, but the infrastructure is maintained by one entity. Public/private firewalls would still exist but all data would be in the cloud, and the private data would be in a well-secured sandbox area within the larger cloud. This hosted solution has the advantage that maintenance of the infrastructure is no longer the problem of the individual institutes or companies, but that of the hosting provider. Furthermore, the institutes get instant updates and new features without having to implement patch releases and maintain in-house code and hardware infrastructure. If the system scales and has good security it is hard to argue that this would not be a boon for both companies and academic institutions. One common gripe of hosted solutions is the lack of direct control; for example what if you wanted to run your own algorithms on the data sets? A solution to this complaint is to build an easy to use yet comprehensive enough API to allow bioinformaticians and biostatisticians access to the data sets and annotations so that they can run these algorithms on the data sets in the cloud. We think that such a hosted solution is not only possible but is coming in the not-too-distant future.

9.6 Acknowledgments

The authors would like to thank the members of the Atlas Development team, namely: Tomasz Adamusiak, Tony Burdett, Aedin Culhane, Anna Farne, Alexey Filippov, Ele Holloway, Andrey Klebanov, Nataliya Kryvych, Natalja Kurbatova, Pavel Kurnosov, James Malone, Olga Melnichuk, Robert Petryszak, Nikolay Pultsin, Gabriella Rustici, Andrew Tikhonov, Ravensara S. Travillian, Eleanor Williams, Andrey Zorin, Helen Parkinson and Alvis Brazma. The authors would also like to thank Ami Engineer Khandeshi, Rajesh Molakaseema and Hugo Berube for their invaluable support of the Pfizer implementation. Thanks also to Lee Harland for critical review of the manuscript and helpful suggestions.

9.7 References

[1] Bates. The role of gene expression profiling in drug discovery. *Current Opinion in Pharmacology* 2011.
[2] Yang, et al. Target discovery from data mining approaches. *Drug Discovery Today* 2009;14(3–4):147–54.
[3] Lamb, et al. The Connectivity Map: using gene-expression signatures to connect small molecules, genes, and disease. *Science* 2006;313(5795): 1929–35.
[4] Classen, et al. Use of genome-wide high-throughput technologies in biomarker development. *Biomarkers in Medicine* 2008;2(5):509–24.
[5] *http://www.mged.org/Workgroups/MIAME/miame.html*
[6] *http://www.mged.org/minseqe/*
[7] Rayner, et al. A simple spreadsheet-based, MIAME-supportive format for microarray data: MAGE-TAB. *BMC Bioinformatics* 2006;7:489.

[8]   http://www.obofoundry.org

[9]   http://bioportal.bioontology.org

[10] http://www.ebi.ac.uk/efo

[11] Malone, et al. Modeling Sample Variables with an Experimental Factor Ontology. *Bioinformatics* 2010;26(8):1112–8.

[12] Kapushesky, et al. Gene Expression Atlas at the European Bioinformatics Institute. *Nucleic Acids Research* 2010;38(Database issue):D690–8.

[13] http://www.unidata.ucar.edu/software/netcdf/

[14] Goncalves, et al. A pipeline for RNA-seq data processing and quality assessment. *Bioinformatics* 2011;27(6):867–9.

[15] Subramanian, et al. Gene set enrichment analysis: a knowledge-based approach for interpreting genome-wide expression profiles. Proceedings of the National Academy of Sciences USA 2005;102(43):15545–50.

Published by Woodhead Publishing Limited, 2012

## 10

Setting up an 'omics platform in a small biotech

Jolyon Holdstock

Abstract: Our current 'omics platform is the product of several years of evolution responding to new technologies as they have emerged and also increased internal demand for tools that the platform provides. The approaches taken to select software are outlined as well as any problems encountered and the solutions that have worked for us. In conjunction lessons learnt along the way are detailed. Although sometimes ignored, the range of skills required to use and maintain the platform are described.

Key words: microarrays; disaster recovery; bioinformatic tools; skill sets.

10.1 Introduction

Founded in 1995 by the pioneer of Southern blotting and microarray technologies, Professor Sir Edwin Southern, Oxford Gene Technology (OGT) is focused on providing innovative clinical genetics and diagnostic solutions to advance molecular medicine. OGT is based in Yarnton, Oxfordshire and is a rapidly growing company with state-of-the-art facilities and over 60 employees. The key focus areas of OGT are:

- tailored microarray [1] and sequencing services enabling high-quality, high-throughput genomic studies;
- cytogenetics products and services delivering the complete oligonucleotide array solution for cytogenetics;

- delivering tailored biomarker discovery solutions to identify and validate the best biomarkers for all diagnostic applications;
- licensing fundamental microarrays and other patents;
- developing new diagnostic techniques with potential for in-house commercialisation or partnering with diagnostic and pharmaceutical organisations;
- delivering a comprehensive sequencing service using next-generation sequencing technologies.

Here, I will describe the approach that OGT has taken in the development of our in-house, 'omics platform. This covers the initial stages of setting up a commercial platform to service customer microarrays, through to developing our own microarray products, running high-throughput projects to recently setting up a pipeline to assemble and annotate data from next-generation sequencers. Each of these has required a range of software applications. Whenever OGT requires new software, a standard approach of evaluating all possible options is followed, be they FLOSS, commercial or even commissioned; weighing up the advantages and disadvantages of each solution. Obviously an important consideration is cost, and where there are comparable products then this becomes an important part of any discussion to decide which products to use. However, cost is just one of a series of important factors such as ease of use, quality of output and the presence/absence of specific skills within the group. Furthermore, any likely impact on infrastructure must also be borne in mind before a decision is made. In reality, more often than not a FLOSS alternative will be selected, indicating just how much high-quality FLOSS software is now available.

10.2 General changes over time

The computational biology group at OGT has evolved hugely over this time. In 2003, OGT was a small biotechnology company with a staff numbering in the low 20s, servicing mostly small academic projects for customers within the UK. Since then, the company has grown to more than 70 full-time employees and has customers from both industry and academia based around the globe. Concomitantly, the typical value of projects has greatly increased from the low thousands of pounds involving a handful of arrays, to hundreds of thousands based on thousands of microarrays. To cope with this growth, our internal procedures and processes have had to develop and evolve.

Indeed, the consequences of the improvements in microarray technology have been the increased demands placed on the IT infrastructure that supports it. For example, in 2003 the densest microarray that OGT produced contained 22 500 oligonucleotide probes. Following hybridisation and processing, this generated a data file of about 12 MB. Since then, DNA spotting has improved such that more features can be printed, and the resolution of scanners has similarly increased so these features can be smaller. This frees up real estate on the slide allowing for increases in feature number. At the time of writing, the latest arrays from our microarray provider, Agilent, have 1 000 000 features and generate a data file around 1.4 GB. This dramatic increase impacts across the IT system; files take longer to move around the network, backups take longer to run and files are now too big to deliver by email or via traditional methods such as FTP. Furthermore, we have recently expanded our offerings to include next-generation sequencing (NGS). This has necessitated a large investment in hardware to provide the processing power to deal with the terabytes of data generated. The chapters by Thornber (Chapter 22) and MacLean and Burrell (Chapter 11) provide good background on NGS data for those unfamiliar with this topic.

10.3 The hardware solution

Within the company there is a requirement for our hardware system to integrate the needs of the computational biology team, who use both Windows and Linux, with those outside the group, who work purely within a Windows environment. For example, the PCs in the labs used to process the microarrays must be able to write to a location that can be accessed by the Linux-based applications used to process the data. The approach taken at the start of OGT, and still in place today, was to keep all data on disk arrays controlled by a Linux server. These directly attached storage (DAS) arrays mounted on the Linux platform are then made available to all users over the company network. This is accomplished using the Linux utility Samba [2]. Samba comes with a great deal of supporting information to allow users to understand this powerful system. Essentially, there are folders defined within Samba to which Windows users can map a drive. The appearance of these drives is of a normal Windows folder, but under the bonnet the information is being made available across the network from a Linux server. This integration of Windows and Linux is reinforced by the fact that all authentications are handled by Windows Active Directory. It is the Active Directory that controls all aspects of user permissions, irrespective of the platform being used, making it much easier to manage the system. The only exceptions to this are root access to the servers, which is independent and will authenticate as usual via the standard/etc/passwd mechanism.

Originally, the first hardware purchased for the OGT 'omics platform was a complete system comprising a Beowulf compute cluster of a master server and four nodes. The master server handled all processes and farmed out jobs to the nodes on the cluster. All data was stored on a 1 TB DAS disk array and there was tape library to handle backups. This is shown in Figure 10.1.

The advantage of buying a pre-configured system was that it was ready to go on arrival. It reduced the time required to get up and running and make the 'omics platform operational. Also, although some of the group had advanced Linux skills, these were short of the level needed to

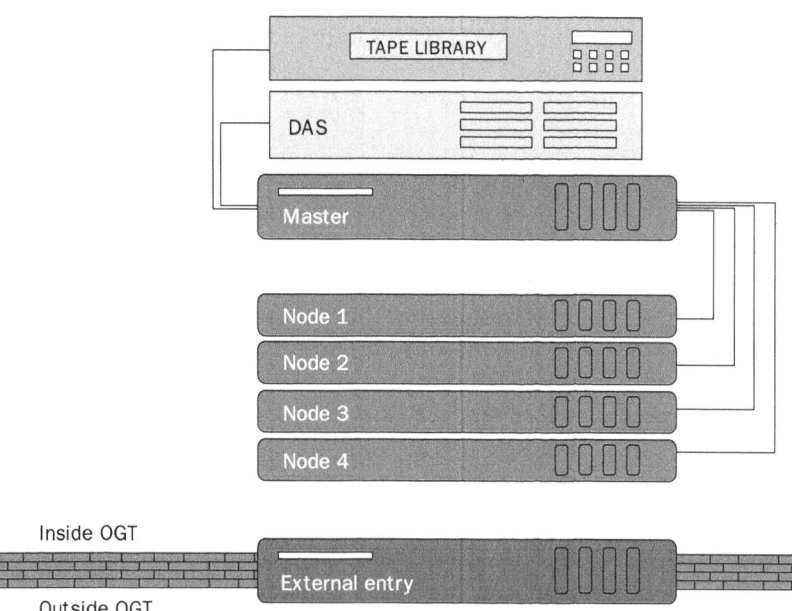

Figure 10.1 Overview of the IT system showing the Beowulf compute cluster comprising a master server that passes out jobs to the four nodes. The master server also provides access to the DAS disk array and performs data backups using a tape library

act as a true system administrator and setup a system from scratch. Following installation by our supplier, all day-to-day maintenance could be handled by a nominated member of the computational team who had access to an external consultant if required. The disadvantage of buying a pre-configured system became apparent when there was a problem after the warranty had expired. The master server developed a problem in passing jobs to the compute nodes. The supplier would no longer support the system and our Linux consultants could not resolve the issue because they did not understand how the cluster was meant to be working.

In the end, the situation was resolved by the purchase of a new server. With this server, and all subsequent servers, we have taken the approach of buying them without any operating system (OS) installed. We are therefore able to load the OS of our choice and configure the server to our own individual requirements. This can delay the deployment but it does mean when there are problems, and there have been some, we are much better able to resolve the issue.

Lesson: Understand your own hardware or know somebody who does.

This highlighted two other important points. First, the design of our system as it was with a single server meant there was no built-in redundancy. When the server went down, the disk arrays went down. Consequently, none of the Windows folders mapped to the disk array were available; this had an impact across the whole of the company. Second, it showed that IT hardware needs replacing in a timely manner even if it is working without fault. This can be a difficult sell to management as IT hardware is often viewed with the mantra 'if it ain't broke don't fix it'; however, at OGT we take the approach that three years is the normal IT lifespan and, although some hardware is not replaced at this milestone, we ensure that all the hardware for our business critical systems is. This coupled with a warranty that provides next business day support, means that OGT has a suitable level of resource available in the event of there being any IT hardware issues.

Lesson: Take a modular approach and split functions between systems.

Subsequently, we have developed a more component-based approach of dividing operations between different servers so that even when one server goes down unexpectedly, other services are not affected. For example, there is a single server providing file access, another for running a BLAT server, another running our MySQL database, etc. This is illustrated in Figure 10.2.

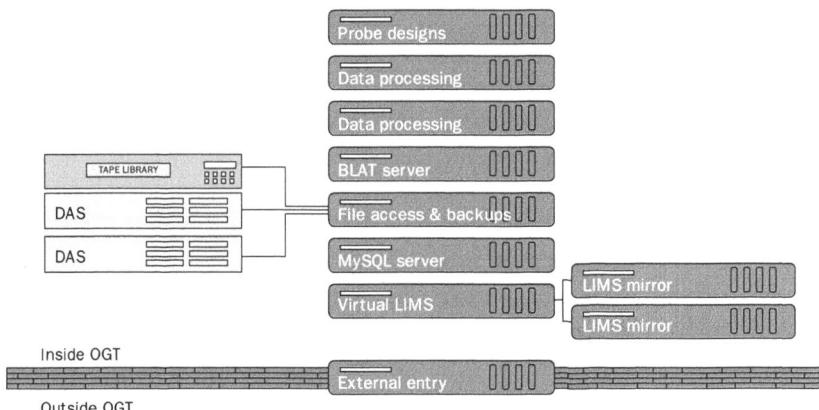

Probe designs

Data processing

Data processing

TAPE LIBRARY

BLAT server

DAS

File access & backups

DAS

MySQL server

LIMS mirror

Virtual LIMS

LIMS mirror

Inside OGT

External entry

Outside OGT

Figure 10.2 The current IT system following a modular approach with dedicated servers. As OGT has expanded we have been able to divide functionality between hardware

10.4 Maintenance of the system

Responsibility for the day-to-day administration of the 'omics platform resides within the computational biology group. There is a nominated system administrator, but we also have a support contract with a specialist Linux consultancy who are able to provide in-depth knowledge when required. For example, when we buy a new server, they install the operating system and configure it for our needs. Also, they provide a monitoring service so that all our systems are constantly being assessed, and in the event of a problem being detected are alerted and able to resolve it. Further to this, they have been able to provide sensible and practical advice as we have expanded. From the OGT administrator's point of view, this reduces time spent as a system administrator, freeing up time for other important functions in the group.

One early issue was that the Linux and Windows systems were maintained by separate people, with little communication flow in either direction. Given the overlap between these systems, this was not ideal. A recent development has been the formation of an IT steering committee comprising the Linux and Windows system administrators and some other IT literati from within OGT. With this has come regular fortnightly meetings at which any problems can be discussed, as well as any plans for the future discussed. Usually these meetings do not last very long, but they have significantly oiled the IT machinery within OGT.

Published by Woodhead Publishing Limited, 2012

Finally, with regard to being a system administrator, the following are some general rules of thumb I have found that hold true in our experience.

- Never do something you can't undo.

- Always check the backups. Never assume they are working and make sure you can restore from them.

- Write down what you did, even if you know you will never forget it, you will. In our experience, DokuWiki [3] is an excellent documentation tool.

- If you do something more than once, write a script to do it.

- Get to know your users before there is a problem. Then, when there is, they will know who you are and maybe have a little understanding.

- Remember you are performing a service for your users, you don't own the system, you just get to play with it.

- Check your backups.

- Before running rm –rf *, always check your current directory; especially if you are logged in as a superuser.

- Never stop learning; there is always something you should know to make your job easier and your system more stable and secure.

- Check your backups, again.

10.5 Backups

Given our commercial projects as well as our internal development projects, there is an absolute requirement to ensure that all necessary data are secured against the possibility of data loss. We settled on tapes as the medium for this, and our chosen backup strategy is to run a full backup of all data over the weekend and incremental backups Monday to Thursday. This covers all OGT's current projects; once a project is completed it is archived to a separate tape volume and retained, but not included further in the weekly backups as it is no longer changing.

Initially, our pre-configured IT system was delivered with a commercial product installed, this was NetVault from BakBone [4]. As this was already installed and configured on the system, worked well and was easy to use, there was no need to change to an alternative. However, as the amount of our IT hardware expanded so did the licence requirements for running NetVault. As a result, we considered open source alternatives. The main candidates were Bacula [5] and Amanda [6] (the Advanced

Maryland Automatic Disk Archiver). In the end, we chose to use Bacula primarily because it was the package with which our support was most familiar. Tools such as Bacula illustrate a common factor when considering FLOSS against commercial solutions, namely the level of GUI usability. However, in this case this was not such a factor as the console-based command-line controller offers a rich set of functionality. Although the learning curve is clearly steeper, this option means that administrators have no choice but to develop a better understanding of the tools capabilities as well as providing much finer control of the processes involved. Finally, the data format used by NetVault is proprietary and requires the NetVault software to access any previous archives. In contrast with this, Bacula provides a suite of tools that allow access to any archives without the requirement that Bacula is installed.

10.6 Keeping up-to-date

Taking this modular approach has helped significantly in managing the setup, as well as making the system much more robust and resistant to failure. Although the result is a more stable hardware platform, it has still been important to stay up-to-date with new technologies. For example, there is a great deal written about cloud computing and its impact on anything and everything (see Chapter 22 by Thornber, for example). Indeed, at a recent conference on 'end-to-end' infrastructure, most suppliers were offering products using virtualisation and cloud capabilities. This is something often discussed internally and something we feel does not offer any benefits significant enough to make it worth implementing at OGT at the moment. Of course, as this technology matures its capabilities will only increase and it will almost certainly have a future role to play at OGT. However, given the amount of data processed on a daily basis, the use of the cloud as a data repository is neither practical nor cost-effective at the moment. Where it may play a role for OGT is in providing a platform as a service; for example, if there was a short-term requirement for additional processing power that could not be met by our current systems. The ability of the cloud to provide such additional resources quickly and easily could prove very useful.

On a simpler level, there are alternatives to the DAS disk arrays currently in use. One of these is a network attached storage (NAS) device, as shown in Figure 10.3. As described previously, DAS disk arrays are used to provide the necessary data storage facility within OGT. However, some aspects of the system are being updated to include NAS devices as

Published by Woodhead Publishing Limited, 2012

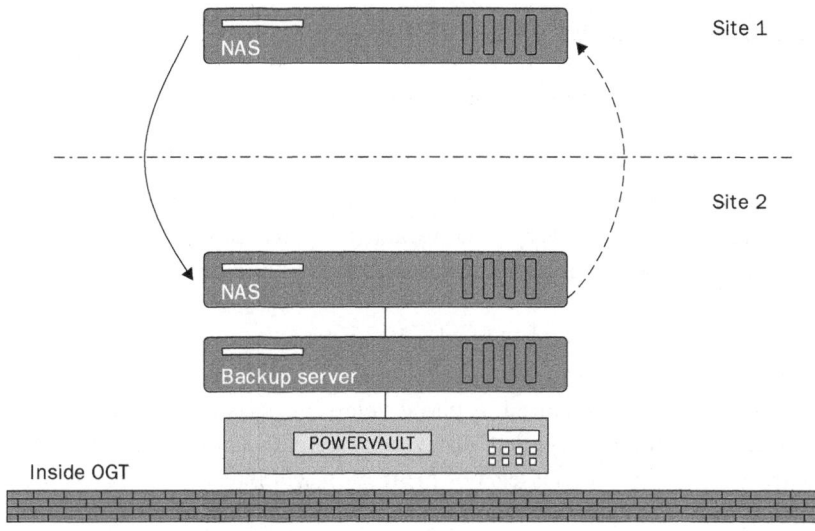

Inside OGT

Outside OGT

Figure 10.3 NAS box implementation showing the primary NAS at site 1 mirrored to the secondary NAS at a different site. Backups to tape are also made at the second site

well. As the name suggests, a NAS device is a computer connected to the network providing storage, and, as such, performs a job similar to a DAS; but there are some significant differences. The key difference is that a NAS is a self-contained solution for serving files over a network, whereas a DAS is a set of disks made available by an existing server. The NAS units currently used already come with an embedded Linux operating system. Consequently, from the initial setup on the network through to its day-to-day management the NAS will require less administration than a DAS. Furthermore, all administration will be handled through GUI-driven management software. As such, it is close to a plug-and-play solution with all the necessary software pre-installed, for instance, for it to act as a file server, web server, FTP server or printer server. There is also the option of replication between NAS devices, which will further strengthen the platform. We employ one NAS box as the primary file server and use the replication software on the NAS to maintain a duplicate instance at another site. There are some disadvantages of using a NAS. As a pre-configured system it is generally not as customisable in terms of hardware software as a general-purpose server supplied with DAS. But, as all we require is a means to make files available over the network, the NAS is currently the best option.

Published by Woodhead Publishing Limited, 2012

10.7 Disaster recovery

One of the best definitions of disaster recovery is from Wikipedia [7] and is as follows:

> Disaster recovery is the process, policies and procedures related to preparing for recovery or continuation of technology infrastructure critical to an organisation after a natural or human induced disaster.

'What happens if the building is hit by a plane?' is an often-quoted but fortunately rare scenario, but it remains the benchmark against which all recovery plans are measured. A perfect plan for disaster contingency will deliver zero data loss coupled with zero system downtime. Although this is always possible, such a level of protection will likely come at too high a price to be viable from a business perspective. At OGT, we have taken a flexible view of our needs and ensured that business critical systems have mechanisms in place to ensure they can be maintained. Other less important systems are given a lower level of support corresponding to their importance to the organisation.

There are different types of measure that can be employed to defend against disaster:

- preventive measures – these controls are aimed at preventing an event from occurring;
- detective measures – these controls are aimed at detecting or discovering unwanted events;
- corrective measures – these controls are aimed at correcting or restoring the system after a disaster or event.

These controls should be always documented and tested regularly. Documentation is particularly important; if the recovery plan does not accurately describe the process to restoring the system, the value of the entire plan is significantly lessened. Modularity again plays a key role here, allowing the plan to be tailored for different elements of the system, such as those described below.

10.7.1 Servers

Disk images of all servers are maintained such that they can be rebuilt on new hardware with a minimum of effort. The biggest delay that would be experienced would be in obtaining any new hardware.

10.7.2 Disk arrays

All disk arrays operate at RAID 6 level. In technical terms, this means this provides 'block level striping with double distributed parity'; in practical terms, an array can lose up to two disks without any data loss. In the event of a disk failure (which has happened more than once), the faulty disk can be replaced with a new disk. The disk array automatically re-builds and the hardware is repaired.

Lesson: Buy spare disks when you buy the hardware to ensure compatibility or replacements.

A further enhancement is the use of hot spares. This is a disk physically installed in the array but in an inactive state. If a disk fails, the system automatically replaces the failed drive with a hot spare and re-builds the array. The advantage of this is that it reduces the recovery time. However, it can still take many hours for a disk to re-build; in our disk arrays 24 hours would normally be allowed for a disk to complete re-building. If there is a fault with the unit other than with a disk, then downtime will be longer. One of the advantages gained from having duplicate NAS boxes is that, in the event of a such a fault, we can switch over from the primary NAS to the secondary unit in order to maintain a functional system. With the new duplicate NAS boxes should there be a fault with the unit as opposed to a disk, then we can switch to the backup NAS box while the primary unit can be repaired.

10.7.3 Data

Besides ensuring that the hardware is stable, the data on the disks can also be secured using backups. It is worth considering what data absolutely must be saved and what could be lost. For example, prioritising the TIFF image files generated by the scanners from the processed microarray slides. All the data used in the analysis of the microarray are extracted from the image file and so can easily be regenerated as long as the original TIFF image file is available.

An additional measure we take is in backing up relational databases. The MySQL administration command mysqldump provides a useful mechanism to create an archive of all the information in the database. The resultant file contains everything required to recreate the database and restore any lost data. Similarly, all software development is managed with Subversion [8], a software versioning and revision control system.

This not only provides a suitable backup, but also keeps track of all changes in a set of files, maintaining a record of transactions.

10.7.4 'omics tools

Running an 'omics platform requires the use of a range of tools. A good place to start exploring the range currently available are the resources provided by the European Bioinformatics Institute (EBI). There are online versions of most tools, but local installations provide greater throughput for those run most frequently, also ensuring high-priority analyses are not subject to delays sometimes experienced on shared resources. Typically at OGT, there are local installations of major systems such as the Ensembl [9] *Homo Sapiens* genome database, providing immediate query responses as part of our processing pipeline. Pointing the same scripts to the same databases or web services held externally at the EBI would result in queries running much more slowly. When there are only a few queries this does not matter so much, but as the number of queries increases to hundreds and thousands then the impact is greater. A local installation is also critical from a security point of view, particularly where the data are confidential. Customers are always happier being told that all systems are internal, and there have been situations where this has been stipulated as an absolute necessity. Of course, there is the overhead of maintaining these installations, but in the long run this is a cost worth meeting. The following is a list of general FLOSS tools that are routinely used in our pipelines.

10.7.5 Bioinformatics tools

- BLAST [10] – almost certainly the most used 'omics application. The Basic Local Alignment Search Tool, or BLAST, allows comparison of biological sequence information, such as the amino acid sequences of different proteins or the nucleotide sequence of DNA. A BLAST search enables a researcher to compare a query sequence with a library or database of sequences, and identify library sequences that resemble the query sequence;

- MSPcrunch [11] – this is a less well known but powerful tool that makes processing of large-scale BLAST analyses straightforward. Data can be easily parsed on a range of parameters. Also, there is a useful graphical view of the BLAST alignments that can be used in report generation applications;

- EMBOSS [12] – this is a freely available suite of bioinformatics applications and libraries and is an acronym for European Molecular Biology Open Software Suite. The EMBOSS package contains around 150 programs covering a variety of applications such as sequence alignment, protein domain analysis and nucleotide sequence pattern analysis, for example, to locate repeats or CpG islands;

- Databases – a relational database provides the backend to many applications, storing data from many sources. There are options available when choosing a suitable relation database system. These can be free and open source such as MySQL [13], PostgreSQL [14] and Firebird [15], or licence-driven products such as Oracle [16] and SQL Server [17]. We have chosen to use MySQL, primarily on the basis that it is the system we have most knowledge of so database administration is more straightforward. A strong secondary consideration was that it is free of charge. Also, the requirements for our database were relatively simple so pretty much any product would have been suitable.

10.7.6 Next-generation sequencing

At OGT we have compiled a pipeline capable of processing data from next-generation sequencers. The pipeline takes in the raw data reads and runs a series of quality control checks before assembling the reads against a reference human sequence. This is followed by a local re-alignment to correct alignment errors due to insertions/deletions (indels) and re-calibration of quality scores. The data are now ready for annotation of SNPs and indels as well as interpretation of their effects. This is a comprehensive and thorough pipeline, which makes use of a series of FLOSS utilities.

- FastQC [18] – FastQC aims to provide a simple way to do some quality control checks on raw sequence data coming from high-throughput sequencing pipelines. It provides a modular set of analyses, which you can use to give a quick impression of whether your data has any problems before doing any further analysis;

- Burrows-Wheeler Aligner (BWA)[19] – BWA is an efficient program that aligns the relatively short nucleotide sequences against the long reference sequence of the human genome. It implements two algorithms, bwa-short and BWA-SW depending on the length of the query sequence. The former works for lengths shorter than 200 bp and the latter for longer sequences up to around 100 kbp. Both algorithms do gapped alignment. They are usually more accurate and faster on queries with low error rates;

- SAMtools [20] – SAM (Sequence Alignment/Map) tools are a set of utilities that can manipulate alignments from files in the BAM format, which is the format of the raw data output file of the sequencers. SAMtools exports to the SAM (Sequence Alignment/Map) format, performs sorting, merging and indexing of the sequences and allows rapid retrieval of reads from any region;

- Picard [21] – Picard (*http://picard.sourceforge.net/*) provides Java-based command-line utilities that allow manipulation of SAM and BAM format files. Furthermore there is a Java API (SAM-JDK) for developing new applications able to read in as well as write SAM files

- VCFTools [22] – the Variant Call Format (VCF, *http://vcftools.sourceforge.net/*) is a specification for storing gene sequence variations and VCFtools is a package designed for working with these VCF format files, for example those generated by the 1000 Genomes Project. VCFtools provides a means for validating, merging, comparing and calculating some basic population genetic statistics;

- BGZip [23] – BGZip is a data compression utility that uses the Burrows-Wheeler transform and other techniques to compress archives, sounds and videos with high compression rates;

- Tabix [24] – Tabix is a tool that indexes position sorted files in tab-delimited formats such as SAM. It allows fast retrieval of features overlapping specified regions. It is particularly useful for manually examining local genomic features on the command-line and enables genome viewers to support huge data files and remote custom tracks over networks;

- Variant Effect Predictor [25] – this is a utility from Ensembl that provides the facility to predict the functional consequences of variants. Variants can be output as described by Ensembl, NCBI or the Sequence Ontology;

- Genome Analysis ToolKit (GATK) [26] – the GATK is a structured software library from the Broad Institute that makes writing efficient analysis tools using next-generation sequencing data very easy. It is also a suite of tools to facilitate working with human medical re-sequencing projects such as The 1000 Genomes and The Cancer Genome Atlas. These tools include tools for depth of coverage analysis, a quality score re-calibration, a SNP and indel calling and a local sequence re-alignment.

10.7.7 Development tools

An integrated development environment (IDE) is a software application that provides comprehensive facilities for programmers to develop software. An IDE normally consists of a source code editor, a compiler, a debugger and tools to build the application. At OGT several different IDEs are currently in use; this is partly due to personal preferences as well as variations in functionality.

- NetBeans [27] – the NetBeans IDE enables developers to rapidly create web, enterprise, desktop and mobile applications using the Java platform. Additional plug-ins extend NetBeans enabling the development in PHP, Javascript and Ajax, Groovy and Grails, and C/ C++;
- Eclipse [28] – Eclipse was originally created by IBM and, like NetBeans, provides an IDE that allows development in multiple languages. One advantage Eclipse has over other development platforms is its support of a wide variety of plug-ins that allow the same editing environment to be used with multiple languages including Perl and XML;
- R Studio [29] – R Studio is an IDE purely for R. (R is a statistics processing environment, see Chapter 4 by Earll, for example.)

10.7.8 Web services

Ubiquitous within any IT offering, data must be presented to consumers and web-based systems offer much in terms of flexibility and speed of development. Currently, OGT uses packages that will be familiar to many readers of this book, namely:

- Apache [30] – the Apache HTTP Server Project is an effort to develop and maintain an open source HTTP server for modern operating systems including UNIX and Windows NT. It ships with most Linux distributions and it is our default web server;
- Tomcat [31] – like Apache, Tomcat is an open source web server but it is also a servlet container, which enables it to deploy Java Servlet and

Published by Woodhead Publishing Limited, 2012

JavaServer Pages technologies and it is for this that it has been used at OGT;

■ PHP [32] – PHP is a widely used general-purpose scripting language that is especially suited for web development as it can be embedded into HTML. In conjunction with Apache and MySQL, we have used it extensively to develop our internal web services.

In addition to the free tools listed above, there are also those that require a commercial licence (but can be free to academic and non-profit groups). As this adds to the cost of implementation, each application is evaluated on a case-by-case basis. If a tool can provide additional 'must-have' functionality, the investment can often be justified. Below are some commercial non-FLOSS tools that we have employed:

■ BLAT [33] – the Blast Like Alignment Tool or BLAT is an alternative to BLAST and is available from Kent Informatics, a spin-out of the University of California at Santa Cruz (UCSC). BLAT is similar to BLAST in that is used to find related sequences of interest. However, it has been optimised for finding better matches rather than distant homologues. Thus, BLAT would be a better choice to use when looking for sequences that are stronger and longer homologues. Unlike BLAST, a BLAT query runs against a database that has been pre-loaded into memory. Coupled with BLAT's 'close-relatives' algorithm, the result is a search tool that runs very rapidly. In many applications where distant sequence relationships are not required, BLAT provides a highly efficient approach to sequence analysis. Unlike BLAST, however, BLAT is not free and commercial users require a licence (although it is free to non-profit organisations);

■ Feature Extraction – Agilent's Feature Extraction software reads and processes the raw microarray image files. The software automatically finds and places microarray grids and then outputs a text file containing all the feature information. There is currently no FLOSS alternative;

■ GenePix [34] – GenePix from Molecular Dynamics is another microarray application. It can drive the scanner as well as feature extract data from the resultant image file;

■ GeneSpring – Agilent's GeneSpring is an application to process feature extraction files from microarrays used in gene expression studies. It is easy to use and has useful workflows that make it easy analyse this type of data. A disadvantage is that the workflows hide many parameters that might otherwise be adjusted within the analysis. However, performing similar analyses with R is not straightforward and although some in the computational biology group have such a capability there is still, for the moment, at least, the need to licence GeneSpring;

- JBuilder [35] – JBuilder is a commercial IDE for software development with Java. This was used initially until other IDEs which were FLOSS became available;

- Pathway Studio – the Ariadne Pathway Studio provides data on pathway analyses allowing, for example, the opportunity to view a gene of interest in its biological context and perhaps draw further insights.

Lastly, there have been instances when a suitable FLOSS or paid-for option was not available, requiring new software to be commissioned. An example of this occurred when there was a requirement to add additional functionality to our Java-based CytoSure™ software [37]. A clear user need had been identified to provide a method that could take data from comparative genome hybridisations (CGH) and make a call as to the copy number of the regions being analysed. Thus providing a means to translate noisy microarray genomic data into regions of equal copy number. There was a pre-existing algorithm available that could do this but no Java implementation had been created. At times such as this, the bespoke approach is the best way to proceed as this provides the clear advantage of having something tailored to an exact specification.

Whereas the tools described above fall into the 'install and use' category, there are also many instances when more specialised tools need to be developed. In particular, BioPerl [38] modules for Perl, the equivalent Java library, BioJava [39] provide a firm foundation for development in the bioinformatics area. Similarly, the BioConductor [40] library is invaluable for rapid development of R-based workflows. Although OGT has custom software written in a variety of languages, Perl is perhaps the most common. There are several reasons for this; first, it is one of the easier languages to learn; new colleagues and interns with little prior programming experience can make substantial contributions in a relatively short period of time. Second, the BioPerl set of computational tools is the most mature, often providing much of the core requirements for new scripts. As Perl is predominantly a scripting language, when there is a requirement for a user interface then Java has proved more applicable. Figure 10.4 shows a screenshot from internally developed software designed to view data from ChIP-on-chip microarrays. Our ability to create such software is greatly increased by the availability of free libraries such as BioJava, which was used extensively in this example to process the underlying data.

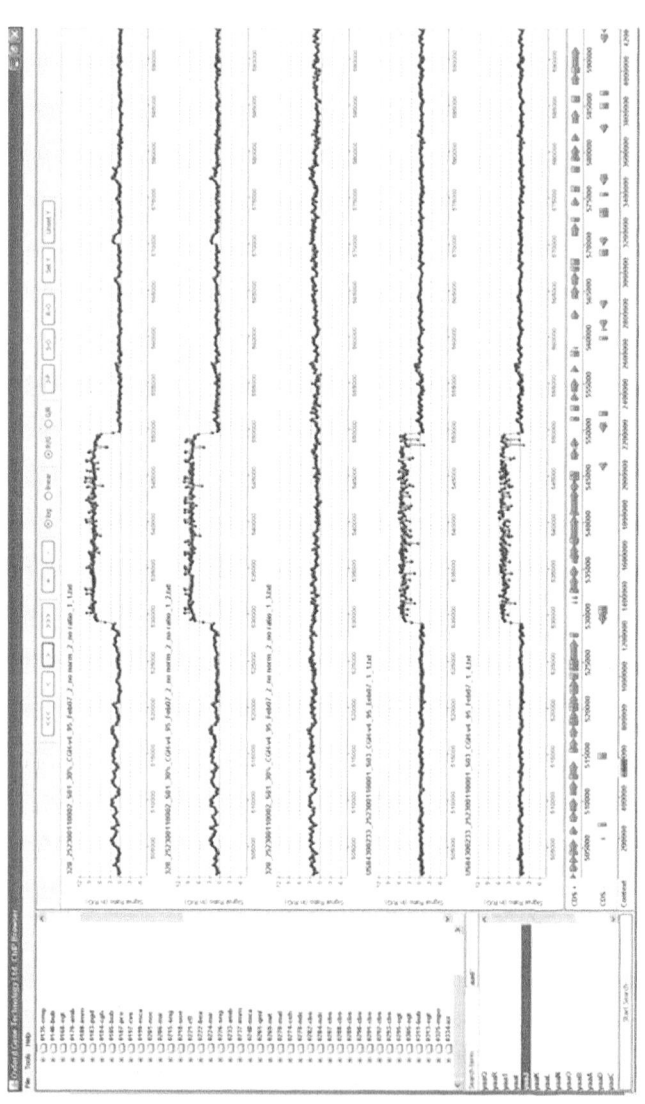

Figure 10.4 A screenshot of our ChIP-on-chip microarray data viewing application. The ChIP data is displayed aligned against a genomic reference sequence, which provides an easy interpretation in a genomic context of where proteins are bound

10.8 Personnel skill sets

The need for bioinformatics programming skills is becoming a necessity within biology and medicine [41]; there is now a requirement for scientists to have access to bioinformatics support and there is currently a lack of such expertise. As reported at the UK parliament by the Science and Technology Committee in their *Second Report on Genomic Medicine* [42], Professor Sir John Bell stated that 'the really crucial thing to train in the UK will be bioinformaticians—people who can handle data—the truth is we have now hit the wall in terms of data handling and management'.

This requirement is due to the exponential growth in data being generated, coupled with the increasingly detailed questions being asked of it. Although there are some degree and postgraduate courses in bioinformatics, there is also the opportunity for personal development through self-education and experience. What was once defined as 'would be nice to have', bioinformatics programming skills are now more of an 'essential' across many fields of biology. This is due to an increased awareness of what may be done and also the massive increase in the amount of data being generated. This all needs to be processed, stored and re-used as required. Similarly, biological information is distributed now thus adding web skills to an already bulging list of requirements in the bioinformatics tool bag.

There are a wide range of skills within the group, and these can be either technical or soft skills. On the technical side there is an absolute requirement to be computer-literate, being comfortable working on the command-line in a Linux environment in the first instance and then building up knowledge and experience with all the built-in tools. Utilities such as the stream editor Sed and the data extraction utility Awk are invaluable in manipulating basic data files but, as per the Perl motto (TIMTOWTDI; pronounced 'Tim Toady'), there is more than one way to do it. Following on from this is proficiency in at least one programming language; in my experience this is normally a scripting language such as Perl or Python, which as well as being easy to learn have vast repositories of (free) libraries available. The same can be said for other languages such as Java that offer object-orientated desktop and server-based applications. There is a tendency to perhaps 'forget' just how rich the landscape of libraries for software developers really is, and how much this has contributed to powerful scientific software.

Of course, as much biological and chemical information is stored in relational databases then knowledge of the Structured Query Language (SQL) is also a key skill for anyone routinely working with such information. Here again, the availability of free database systems (see above) provides the opportunity for anyone to explore and teach themselves without fear of breaking someone else's database. Free SQL clients such as SQuirrel [33], Oracle SQL Developer [34] and in the MySQL world, MySQL Workbench [35] and PhpMyAdmin [36], provide many capabilities for zero cost. Again, the availability of powerful database management software is incredibly enabling and is another factor aiding companies whose primary goal is science, rather than software development.

Finally, data analysis is also a key skill. More and more, a good understanding of the basics of statistics will help understand the validity and outcome of the many large-scale analyses now available within biology. As such, knowledge of tools, particularly the R project for statistical computing, as well as bioinformatics applications, can be crucial. If more advanced machine-learning approaches are required, the Weka [37] system offers an extensive range of algorithms and techniques, often more than enough to test an initial hypothesis.

10.9 Conclusion

FLOSS fulfils many critical roles within OGT that would otherwise require investment in alternative commercial products. In selecting FLOSS over these other solutions, there is, of course, a cost element to be considered, but that is not the whole story. If any FLOSS product did not meet the required functional criteria then it would not be used; similarly, if a better commercial product becomes available then this would be considered. However, besides reducing expenditure on licences, there are additional advantages from taking a FLOSS approach.

First, if there is a problem, then the very nature of open source software means that, assuming you have the technical ability, it is possible to see directly where the problem is. This was experienced directly at OGT when the format of some NCBI Genbank records changed. The downstream effect of this was that an application using a BioJava file parser threw an error. Because the code was available it was possible to find the problem and fix it locally so the parsing could complete. But beyond that the fix could be passed to the developers for incorporation into the distribution and made available to the whole community. In the

Published by Woodhead Publishing Limited, 2012

event of a problem that we are not able to resolve ourselves then there are usually many other users who can offer a solution or make suggestions on how to proceed. Encountering an error with a commercial product typically requires registration of the error and waiting for someone to do something about it. Even then you may have to wait for the next release before the fix is available to implement.

Similarly, by having access to the source code, there is the opportunity to adapt or extend the functionality to individual requirements as they may arise. In contrast, the functionality of a commercial product is decided by the manufacturers and any additions to this are decided by them alone. FLOSS also avoids the vagaries of a commercial manufacturer, for instance they may choose to amalgamate distinct products into a single offering. This then forces the users to use this new product if they want to keep using the most up-to-date version of the software. We have had experience of this with one piece of software that was incorporated into a more enterprising solution. Interestingly, after a couple of intermediate releases, the standalone product is now being made available. Similarly, commercial vendors control support for older versions of their products and decide when they will stop supporting older releases. Users are powerless to prevent this and helpless to its effects, which can mean they are forced to upgrade even when they would not choose to do so.

In conclusion, FLOSS has significant market share, and performs as well as commercial products with increased flexibility and adaptability. As such, I foresee it retaining a central role in OGTs 'omics platform.

10.10 Acknowledgements

I would like to thank Volker Brenner and Daniel Swan at OGT for reviewing and their helpful suggestions. Also, I would like to thank Lee Harland for his valuable editorial suggestions on the submitted manuscript.

10.11 References

[1] 'Microarrays: Chipping away at the mysteries of science and medicine.' [Online]. Available: *www.ncbi.nlm.nih.gov/About/primer/microarrays.html*.

[2] 'Samba – Opening Windows to a wider world.' [Online]. Available: *www. samba.org*.

[3] 'DokuWiki.' [Online]. Available: *www.dokuwiki.org*.

[4] 'NetVault – Simplified Data Protection for Physical, Virtual, and Application Environments.' [Online]. Available: *www.bakbone.com*.

[5] 'Bacula® – The Open Source Network Backup Solution.' [Online]. Available: *www.bacula.org*.

[6] 'Amanda Network Backup.' [Online]. Available: *www.amanda.org*.

[7] 'Wikipedia.' [Online]. Available: *www.wikipedia.org*.

[8] 'Apache™ Subversion®.' [Online]. Available: *http://subversion.apache.org/*.

[9] 'Ensembl.' [Online]. Available: *www.ensembl.org/index.html*.

[10] 'National Center for Biotechnology Information.' [Online]. Available: *www.ncbi.nlm.nih.gov*.

[11] 'MSPcrunch.' [Online]. Available: *www.sonnhammer.sbc.su.se/MSPcrunch.html*.

[12] 'EMBOSS – The European Molecular Biology Open Software Suite.' [Online]. Available: *www.emboss.sourceforge.net*.

[13] 'MySQL.' [Online]. Available: *www.mysql.com*.

[14] 'PostgreSQL.' [Online]. Available: *www.postgresql.org*.

[15] 'Firebird.' [Online]. Available: *www.firebirdsql.org*.

[16] 'Oracle.' [Online]. Available: *www.oracle.com*.

[17] 'SQL Server.' [Online]. Available: *www.microsoft.com/sqlserver*.

[18] 'FastQC – A quality control tool for high throughput sequence data.' [Online]. Available: *www.bioinformatics.bbsrc.ac.uk/projects/fastqc/*.

[19] 'Burrows-Wheeler Aligner.' [Online]. Available: *http://bio-bwa.sourceforge.net/*.

[20] 'SAMtools.' [Online]. Available: *http://samtools.sourceforge.net/*.

[21] 'Picard.' [Online]. Available: *http://picard.sourceforge.net/*.

[22] 'VCFtools.' [Online]. Available: *http://vcftools.sourceforge.net/*.

[23] 'BGzip-Block compression/decompression utility.'

[24] 'Tabix – Generic indexer for TAB-delimited genome position files.' [Online]. Available: *http://samtools.sourceforge.net/tabix.shtml*.

[25] 'VEP.' [Online]. Available: *http://www.ensembl.org/info/docs/variation/vep/index.html*.

[26] 'The Genome Analysis Toolkit.'

[27] 'NetBeans.' [Online]. Available: *www.netbeans.org*.

[28] 'Eclipse.' [Online]. Available: *www.eclipse.org*.

[29] 'RStudio.' [Online]. Available: *www.rstudio.org*.

[30] 'Apache – HTTP server project.' [Online]. Available: *httpd.apache.org/*.

[31] 'Apache Tomcat.' [Online]. Available: *www.tomcat.apache.org/*.

[32] 'PHP: Hypertext Preprocessor.' [Online]. Available: *www.php.net*.

[33] 'Kent Informatics.' [Online]. Available: *www.kentinformatics.com*.

[34] 'GenePix.' [Online]. Available: *www.moleculardevices.com/Products/Software/GenePix-Pro.html*.

[35] 'JBuilder.' [Online]. Available: *www.embarcadero.com*.

[36] 'Ingenuity.' [Online]. Available: *www.moleculardevices.com*.

[37] 'CytoSure.' [Online]. Available: *www.ogt.co.uk/products/246_cytosure_interpret_software*.

[38] 'The BioPerl Project.' [Online]. Available: *www.bioperl.org*.

[39] 'The BioJava Project.' [Online]. Available: *www.biojava.org*.

[40] 'BioConductor – Open source software for bioinformatics.' [Online]. Available: *www.bioconductor.org*.

[41] Dudley JT. 'A Quick Guide for Developing Effective Bioinformatics Programming Skills,' *PLoS Computer Biology* 2009;5(12).

[42] 'Science and Technology Committee – Second Report Genomic Medicine.' [Online]. Available: *www.publications.parliament.uk/pa/ld200809/ldselect/ldsctech/107/10702.htm*.

## 11

Squeezing big data into a small organisation

Michael A. Burrell and Daniel MacLean

Abstract: The technological complexity involved in generating and analysing high-throughput biomedical data sets means that we need new tools and practices to enable us to manage and analyse our data. In this chapter we provide a case study in setting up a bioinformatics support service using free and open source tools and software for a small research institute of approximately 80 scientists. As far as possible our support service tries to empower the scientists to do their own analyses and we describe the tools and systems we have found useful and the problems and pitfalls we have encountered while doing so.

Key words: genomics; NGS; institute; bioinformatics; infrastructure; pipeline; sequencing data.

11.1 Introduction

To paraphrase Douglas Adams, biological and medical data sets can be big, really big, I mean mind-bogglingly huge, and are only getting bigger. Recent years have seen an intimidating increase in the quantity of data collected in experiments performed by researchers across all fields of the medical and life sciences. As geneticists, genomicists, molecular biologists and biomedical scientists, we have seen the nature of our work transformed by huge advances in DNA sequencing, high-throughput drug discovery, high content screening and new technologies for fluorescence microscopy, but this transformation has come at a price. We have seen data costs plummet and yield grow at a rate of change in excess of that of the famous Moore's Law [1] (Figure 11.1), stretching the hardware and software we use to its limits. Dealing with these data is a truly mammoth task and contextualising the amount of data that must be handled for a new entrant to the field or a scientist with little or no informatics background is difficult; inevitably we end up talking about abstract quantities like terabytes (TB), which for most people, even technically minded people like scientists, has little connection to daily

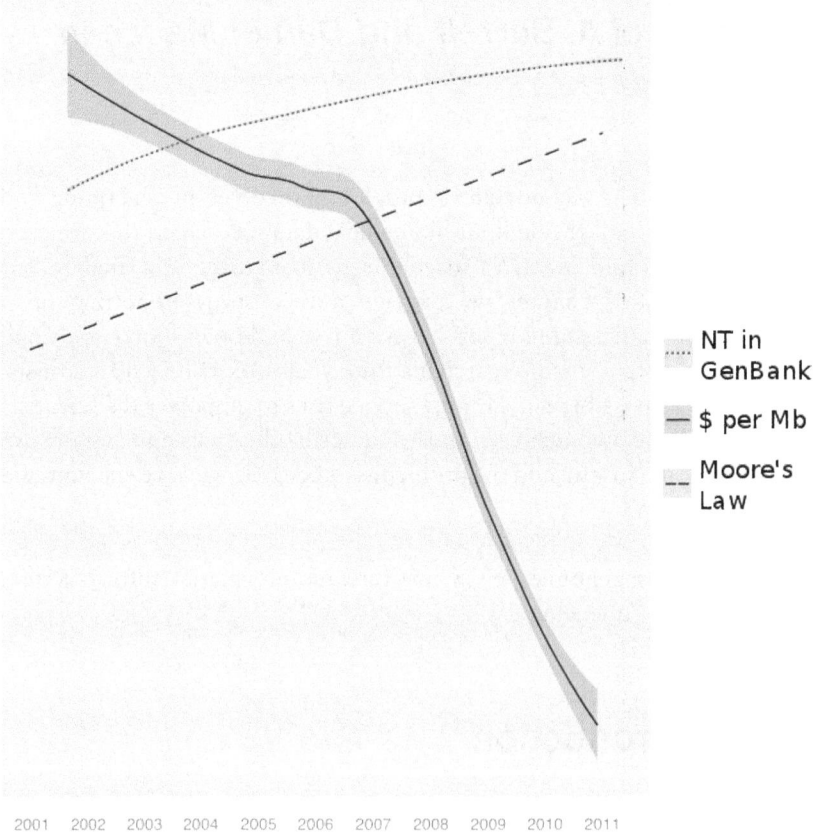

NT in GenBank

$ per Mb

Moore's Law

2001 2002 2003 2004 2005 2006 2007 2008 2009 2010 2011

Figure 11.1 Changes in bases of sequence stored in GenBank and the cost of sequencing over the last decade. Series have been altered to fit on a single logarithimic scale. Change of number of nucleotides in GenBank is according to GenBank Release notes *ftp://ftp.ncbi.nih.gov/genbank/gbrel.txt*. Cost of sequence per MB according to NHGRI at *http://www.genome.gov/sequencingcosts/*

experience. As a more real yardstick for the quantities being produced, it is interesting to note that the United States Library of Congress aims to keep a copy of each written work in all languages. It carries about 15 TB of data in this form. In comparison, a single DNA sequencer such as the Illumina HiSeq 2000 can easily generate 55 000 000 000 nucleotides (55 gigabases) [2] of sequence in one day and would take only ten months to exceed the total written output in the Library of Congress, essentially the writings of the whole of humanity since the dawn of time. Given that there are many thousands of DNA sequencers and other high-throughput machines in active use across the world, then the word deluge does not even begin to describe the current data situation.

The human and monetary resources that bioinformatics departments and core services can call on to deal with the data are being stretched ever thinner as the number of projects that make use of these technologies in their departments increases. To add to this difficulty, even if budgets could be stretched to employ more people there is currently a short-fall in the number of qualified bioinformaticians in the job market who can carry out the required analyses. There is not likely to be a change in this situation very soon so it is necessary that service providers and research groups find a way to prioritise the many challenges they face and apply the resources that they have to 'work smarter'. The wide range and flexibility of free and open source tools available now provide a great opportunity for us to create environments and pipelines with which we can tackle the data deluge. In this case study we shall describe how our small core bioinformatics service has implemented a service model that is more scalable and adaptable than previously extant schemes by making use of free and open source software.

11.2 Our service and its goals

The Sainsbury Laboratory (TSL) [3] is a part publicly, part privately funded research institution that concentrates on cutting-edge basic and translational research into plant and pathogen interactions. Conceived by Lord Sainsbury, the former UK Science Minister and funded by The Gatsby Charitable Foundation [4], TSL is a focussed laboratory of about 80 scientists in five research groups. The bioinformatics group was created in 2003 and had expanded to its current size of two full-time members by 2006 when we took delivery of our first Illumina Genome Analyzer (GA) Next Generation DNA sequencing (NGS) machine, since upgraded to GAII and we have expanded with multiple mass spectrometers and an Opera High Content Imaging System. All our high-throughput projects occur continuously, at time of writing we are working on 16 genomics projects concurrently, with 20 people on these projects. Projects can run over 1–3 years (or more) being the main focus of a post-doc's project perhaps because a new model organism needs to have a comprehensive set of genomic resources, or it can be a small component because we need to do a new RNAseq gene expression analysis with an existing model organism. Typically, our workflows would involve going from raw sequence reads to generating a rough draft assembly, which we would annotate *de novo* or with RNAseq methods or using NGS sequence to identify SNPs in both model or non-model species, which would be verified in the laboratory and developed into markers for disease resistance. We use our new models in genomics analyses including comparative genomics, the genes we predict will be classified and certain important domains and signatures identified to create candidates that may confer disease or resistance to disease for further lab testing.

The hardware we use is sufficient to cover the needs of a group of our size, given our wide interests. Our computer systems comprise 21 TB 'live' storage on a mirrored NetApp [5] device (aside from a substantial archive for older data) with 22 compute nodes with 2–16 processor cores each and from 8 GB up to 128 GB onboard RAM, these all run Debian 6.0. Our bioinformatics service has maintained its small size throughout this period of expansion and our raison d'etre is to provide the rest of TSL with the knowledge, expertise and facility to carry out the bold research that is their remit. Our philosophy on how we carry out this provision has shifted over this period, from being a primarily reactionary on-demand service to being a scalable, department-wide initiative to create a culture of informatics knowledge, that is to bridge the somewhat false gap between the bench scientist and the bioinformatician.

We consider that a successful bioinformatics team helps the department best by ensuring that whatever the biological question or source of the data we are able to continually understand, manage and analyse the data. We take responsibility for providing the bench scientist with the skills and tools to carry out analyses much more independently than in a reactionary service model. We have found that we can execute the bioinformatics work for at least as many projects as we have scientists (not just as many as can be handled by two core informaticians). In this context a large part of the job comes in facilitating that exercise. Partly this is through training – much in the manner that a post doc would look after a new graduate

Published by Woodhead Publishing Limited, 2012

student, rather than very formal tutoring – but mostly through structuring the data and tool organisation so that the biologist does not need to think too much about the computer administration side of the deal and can think about the data and the biological question alone. There are a great deal of free and open source tools that can be applied to this end.

11.3 Manage the data: relieving the burden of data-handling

Much of our data is generated internally on our high-throughput sequencer and image analysis machines, which send the raw data straight to the large central storage device. Actually getting at and using large data sets on a local machine so that they can be analysed in any sort of software is often much more of a trial than we would want it to be. Typically, something like FTP is used to copy the files over a network from some central filestore to the machine with the analysis software. FTP clients and command-line transfer can be powerful but cumbersome, result in unnecessary copies of the file and, often, the biologists are unaware of the concepts required to understand the transfer (it can often be surprising to an informatician how little a typical biologist knows of the nuts and bolts of dealing with computer systems). It is much more practical to integrate the data and transfer as far as possible with the client machine's operating system, and although Windows and Mac OS do provide some functionality for this, they are not much better than FTP programs. Instead we have made use of user space file system drivers, which allow a user to connect to an external filesystem without additional privileges to their machine or the remote machine/filesystem. In this way we have been able to make our central file store accessible to the desktop environment as if it were any other mounted device, like an external DVD drive or a USB stick (Figure 11.2).

11.4 Organising the data

Once the data can be seen by the computers they are working on, biologists still need to be able to find the files and folders that they need. We have used a simple but effective trick to create a directory structure

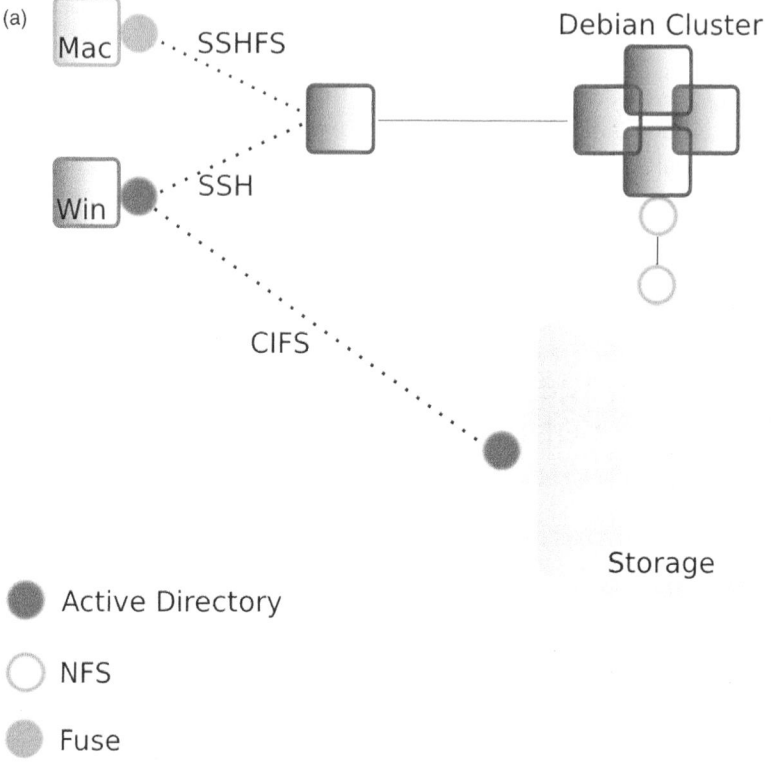

(a)

Mac ···· SSHFS

Win ···· SSH

CIFS ····

Debian Cluster

Storage

● Active Directory

○ NFS

● Fuse

Figure 11.2 Our filesharing setup. (a) Machines and network connections. We use a multiprotocol environment on our servers and file store, which is shared to Windows devices via Common Internet File System (CIFS) [35], to Mac OS X devices via Secure Shell File System (SSHFS) [36] via various routes. Macs are provided with a client program, MacFUSE [37], which allows connection to an intermediary Linux computer via SSHFS. The intermediary Linux machine has the file store directly mounted through Network File System (NFS) [38] and passes a connection to the client. In the Windows environment we provide CIFS configured on the storage device directly allowing the Windows machines to mount it. Authentication for Windows and Mac is done with the Active Directory service and via the Network Information Service for Linux. This configuration has been the most stable in our tests, and allows the user to be flexible about which machine they work with. Early on we attempted to deploy WinFUSE [37] in the Windows environment but found the drivers to be very

(b) **/home/data**
```
|-- alignments
|    `-- plants
|         `-- arabidopsis
|-- blastdb
|    |-- bacteria
|    |    `-- whole_genomes
|    |-- fungi
|    |-- other
|    |    |-- ncbi
|    |    |    `-- nr
|    |    `-- uniprot
|    `-- plants
|-- gene_features
|    `-- plants
|         `-- arabidopsis
|-- indexes
|    `-- plants
|         `-- arabidopsis
|              `-- landsberg
|                   |-- bwa
|                   `-- sam
`-- sequences
     |-- bacteria
     |-- fungi
     |    `-- aspergillus
     `-- plants
          `-- arabidopsis
               |-- landsberg
               |-- tair8
               |    |-- multi_fasta
               |    `-- single_fasta
               `-- tair9
                    |-- 2bit
                    `-- multi_fasta
```

Figure 11.2 (Continued) unstable in our setup. A further consideration in deciding which method to use was our need to access our data on portable devices via Virtual Private Networks (VPN) [39] from off our main site. These protocols provide facility for this behaviour and work as if on-site when a user tries to access remotely. (b) A fragment of our shared folder structure for files that will be required by multiple users, such as BLAST [6] databases or GFF files [7], that is easy to navigate and helps to keep itself populated properly. Inspired by the six-degrees of separation principle, we have broad top-level folder names that describe the general data type, for example 'features' or 'sequences', then a subdirectory of broad organism taxon, for example 'plants' or 'oomycetes', then a subdirectory of species before getting data-specific with, for example 'gff' or 'multi-fasta'. This structure can bring most files within four or five clicks of the root data directory (Figure 11.2). We specify the skeleton structure of the data directory but afterwards any user is able to add (and delete) what they like in any of these directories. The structure of the directory is kept sane by requiring users to follow simple instructions for adding a record to a text file in the root of the data directory. We have small scripts that watch and alert us to any discrepancies between the records in the file and the actual contents of the directory. This simple system is much less error-prone in practice than it may seem. In truth the system is mostly about letting the users know in a clear fashion where and how to deposit their data for easy sharing and retrieval, they are usually pretty good about doing this as long as it is clear. The threat of data deletion helps persuade users to stick to the scheme too (although in reality we never delete a file, we move it to a recycling bin and wait to see if anyone asks for it). The frequency with which new files are updated in our experience is sufficiently low to ensure that we are not constantly running around chasing files, we easily manage a couple of hundred shared files in this way. One important caveat is that this sort of directory structure really must be kept for shared files, and not an individual user's working files; assemblies from which many users may predict genes or calculate coverage of RNAseq reads would work well, but a spreadsheet of results derived from these for an individual result may be best left in the user's allotted working space. Also, we do not keep primary data from sequencers and mass specs in these folders, these are kept separate in a read-only folder on live storage for a useful amount of time (which will vary according to project), before being moved to archive. Our data retention policy for our Illumina sequence data is straightforward; we keep what we need to work with and what we need to submit to sequence repositories on publication. In practice we keep the FASTQ [8] sequence files and minimal technical meta-data on the run. Reads are kept on live storage until the allotted space is full and then operate a one in, one out policy, moving the older files to archive. Usually this means we can keep raw reads for around eight to twelve months.

11.5 Standardising to your requirements

We found it necessary to 'get tough' with the way that users format and describe their data. Sequence data have pretty good and straightforward file formats: FASTA and FASTQ are pretty clear, even if some things like quality encodings are sometimes a bit obscure, but most other data types are much more variable and harder to work with and important meta-data is always difficult. Bad descriptions and formats make it hard to work with internally, even to the level that we do not know what a file has in it and cannot therefore work with it. Every bioinformatician has had to frustratedly wrangle somebody else's poorly formatted file because it messed up their perfectly workable pipeline. Eventually, lots of the data we collect will have been analysed enough to warrant publication and we will need to submit to public repositories. Typically, the worker who generated the data will have moved on to new projects and getting the right meta-data will be difficult. To prevent these sorts of frustrations and eventual mistakes in submitted meta-data, we require workers to stick to certain standards when creating their data or we refuse to work with it.

We produce specifications for different file and data types too, meaning certain pieces of information must be included, for example we require all our GFF files to include a version number for the feature ontology used, each record to have a certain format for dbxref attribute if it is included. In doing this we found we had to provide a reliable and simple way of describing the data when they make it. At the outset we hoped that existing software would be fine, Laboratory Information Management Systems (LIMS) seem to be the right sort of tool but we have found these systems either far too cumbersome or just lacking in the right features for our needs. Inevitably in developing easy procedures we have had to create our own software, this is probably something that all bioinformatics groups will have to do at some point, even though generic LIMS and data management systems exist, the precise domains you need to capture may be harder to model in these systems than in a small custom tool of your own. We turned to Agile development environments, in particular the Ruby-on-Rails [9] Model-View-Controller paradigm web application framework for producing tools around popular Structured Query Language (SQL) databases (commonly called 'web apps' or 'apps'). Rails makes application development quick and easy by providing intelligent scaffolding and many built-in data interfaces. Most usefully, data-mart style REST [10] interfaces and XML [11] and JSON [12] responses are built-in by default, and others are very easy indeed to implement. Thanks to this it is fabulously easy to share data between tools in this framework. It is useful to note that the Rails project changes significantly between versions 2 and 3, so initiating a project in Rails 3, although giving you the most bleeding-edge code base has somewhat lower community support, for the time being at least. There is lower coverage of Rails 3 in the many blog articles about Rails and in the web forums that coders use to discuss and help each other with coding problems. In these forums most problems you encounter starting out will already have been asked, or will get answered in some fashion after a couple of days. Rails 2 applications are inherently incompatible with Rails 3 applications and upgrading is not straightforward.

Once the app is developed, serving it has its own issues; the in-the-box server Webrick or a vanilla Mongrel setup are not designed for a production environment. The traditional approach to hosting Rails applications is to have a server running Mongrel or even Webrick, the Ruby servers built into Rails, attached to local server ports with Apache [13] to proxy requests to the Rails application servers. This setup helps to avoid the limited concurrent connections and stability problems in Mongrel and Webrick. However, running multiple Rails application servers uses lots of machine resources and is time-consuming to configure and maintain. A far better approach is to allow Apache to run the Rails application code, which we do with the Phusion Passenger module (aka mod_rails) [14]. With this approach applications are only running when they are needed, can be deployed with minimal configuration and multiple Rails applications can be deployed under the same Apache server (each just requiring a separate RailsBaseURI declaration in Apache config). New apps can be deployed in minutes without time-consuming proxy configuration or ongoing maintenance efforts to ensure that individual application servers are running. We enjoy using Rails and do indeed find it is a very productive environment, but it is a little over-hyped, although increasingly easy to code with and prototype an application on your local machine, it can be a struggle to push to the production environment. Not many Sys-Admins are experienced enough with the system to host it easily and some work needs doing at deployment to successfully host an app with many users moving lots of data about, but once the right trick is learned for your production environment it is easy to scale.

For the high-throughput image data that we generate, initially we were using a flat-file archive system on our server, although the rate at which we generate images precludes this being viable for long. We are now moving over to a more versatile Omero server [15]. Omero is an image repository and API produced by the Open Microscopy Environment

Consortium with great mechanisms for storing microscope and other technical meta-data. Omero provides excellent capability for connectivity and there is a wealth of client applications that are able to interface with it. With these applications a user can select images for manipulation, annotation and analysis. Image analysis procedures such as segmentation, object identification and quantification can be performed within the client applications direct on the user's desktop machine pulling data automatically from the Omero database.

11.6 Analysing the data: helping users work with their own data

Once we have our data under control and in our apps/databases, we visualise with external tools. For the majority of our next-generation sequencing experiments, genome browsers are very popular with our users. We completely eschew the use of GBrowse-style browsers [16] that require a lot of setup on an external server and require the bioinformatician to upload (and prepare) data for a user before it can even be looked at. Instead we opt for the very useful and user-friendly SAVANT [17] and Integrated Genome Viewer (IGV) [18] desktop genome browsers. SAVANT works well in most cases, presenting a clean and straightforward interface for viewing the alignment of next-generation reads on reference or custom assemblies, which is mostly what we want. More can be done with IGV, it is particularly suited to filtering and comparing data. Both work well but have some minor network issues, getting either to start from behind a firewall or proxy is initially difficult. On first loading in such an environment SAVANT takes a while to initialise (long enough to make a user think the machine has crashed), patience and turning off auto-update once it is loaded will make subsequent loads much quicker. IGV relies more on files downloaded over the internet and needs details of the proxy to function. Early versions of SAVANT could crash with an out-of-memory error when trying to load in reference sequence, later versions have fixed this problem.

For some applications, such as hosting a next-generation data set to be viewed by collaborators over the web, we have used other web service-style browsers like JBrowse [19] (part of the GMOD project) and AnnoJ. Both of these are fast, flexible and somewhat attractive, but neither is anywhere near as easy to setup as the desktop browsers above. AnnoJ [20] is a web service that renders feature data. It makes requests of another data-providing web service via a URL and some GET parameters in its own non-standard format. The data-providing service must return AnnoJ-readable objects and meta-data in JSON for AnnoJ to render (Figure 11.3). Feature rendering is fast for gene models but somewhat slower for track types like the histograms that show the coverage of aligned reads over

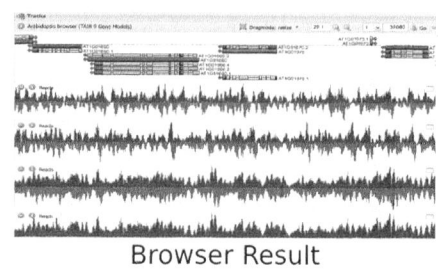

Browser Result

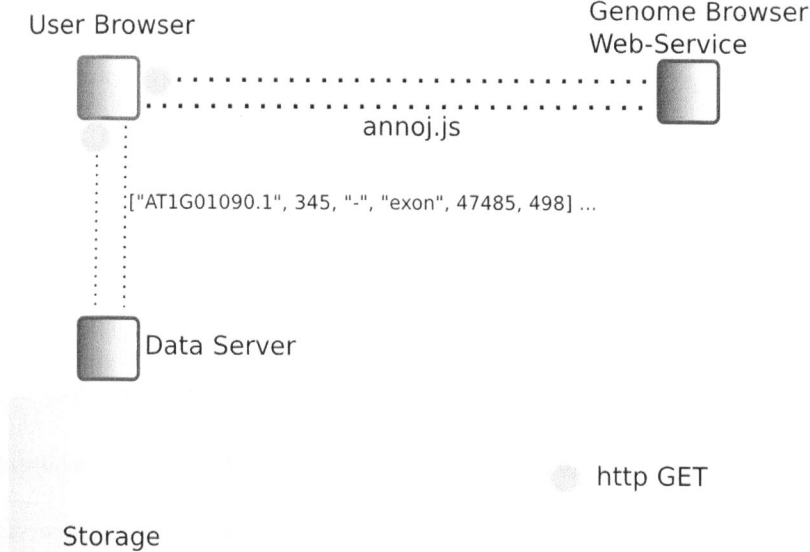

Figure 11.3 Connectivity between web browsers, web service genome browsers and web services hosting genomic data. A user client (web browser) makes a GET request from a URL with certain parameters to, for example, the AnnoJ web service (*www.annoj.org*), which returns a Javascript library. This then requests genomics data from the web service that hosts the genomics data (e.g. that served by a Gee Fu instance). AnnoJ can render and allow interactions with the genomics data in the web browser

Published by Woodhead Publishing Limited, 2012

contigs. AnnoJ is limited in the feature types it can display and is extremely inflexible with regard to styles, no real documentation on styling the features exists and AnnoJ is closed source, the actual JavaScript must also be retrieved over the internet as a web service. The core library for AnnoJ cannot be downloaded and is 'mini-fied', that is compressed in such a way that download times are reduced but the code in it cannot be read by a human. In truth, we actually began using AnnoJ with help from the author, who was a collaborator of a collaborator, so got a head start on setting it up. Sadly, nowadays, the author seems to have gone quiet on the project. Despite these drawbacks, we still use it. In particular, it has found utility as a quick browser option for our gene feature and assembly versioning database, Gee Fu [21]; as a pure web service it does have that flexibility we see in no other browser (as Gee Fu is a Rails application, it is automatically a web service and together the two provide a very quick way to provide feature data over a network for many users in an easily extended platform). JBrowse is superficially similar to AnnoJ, as it is also a Javascript project that renders gene feature data in a fast and attractive way. The main downside to JBrowse is that the data files (JSON files) have to be pre-generated and are not stored in a database. This includes images representing features, which is fine for multi-use elements like gene models but coverage information represented by histograms is not dynamically generated and PNG files for these must be pre-rendered too. Documentation provided with web applications and tools does not necessarily cover all installation problems that you may have in your particular setup. The FireBug plug-in for Firefox (and the developer tools in Chrome to a lesser extent) are extremely useful for de-bugging the requests made between browsers, CGI scripts/databases and web servers.

Usability is a constant concern when it comes to the tools that we present to the biologist, and as far as possible we stick to graphical tools. Free and open source software that has a graphical user interface never comes along with iPod-esque levels of ergonomic or aesthetic design finesse, but we have not come across any whose main interface is so terrible that no analysis can be done with it. Phylogenetic analysis software seems to be among the worst culprits for poor usability. The strength of free and open source software is actually in its limitations; quite often the software does not try to do too many things and this makes the workflow within the program straightforward, so most packages are at least usable for their main task. Sadly, programs with graphical interfaces are in the minority. We find that in practice, after a bit of training, the worst symptoms of 'command-line aversion syndrome' can be fought and biologists are not put off the command-line interface, especially for small tasks. To help with usability we have installed a local Galaxy [22] instance, this workflow engineering and tool integration environment has been described as the 'missing interface' for bioinformatics and it is well designed. Biologists can get up and running command-line tools through Galaxy in no time at all.

11.7 Helping biologists to stick to the rules

In our experience, bioinformaticians can be somewhat recalcitrant when it comes to telling the biologist that they must format data in a certain way or follow certain naming conventions. Perhaps this is because they feel that there will be resistance and that the biologists will not appreciate this intrusion into their work. Biologists work according to similar sorts of rules all the time and are quite flexible in their acceptance of these. For example, when carrying out their bench experiments, the molecular biologist will often need to create a piece of DNA that comes from some organism of interest and fuse it to DNA control sequences that perhaps come from a different organism, a chimaeric piece of DNA. To use this it may need to be inserted and activated in a third organism, often bacteria, creating a genetically modified organism with DNA from two donors. It is not too hard to imagine that this procedure requires lots of rules, standard procedures and paperwork so biologists are primed to jump through hoops when getting things done. When we introduced our new rules for data and meta-data standards, we did not believe that we would get much resistance from the biologists, although we were prepared for quite a lot. We armed ourselves with arguments as to what savings we would make in time and convenience and future protection against confusion and data loss and we took these arguments straight to the group leaders. Convincing the project leaders gave us some authority and official backing and made it much easier to get the rules adhered to. We get good uptake of our rules and standards, biologists quickly come to appreciate the reasons why and after a little while (usually after having to deal with someone else's files) are glad that they have been implemented.

11.8 Running programs

With the environment setup the new user can log on and start to do their analyses. We use a Debian-based [23] compute cluster to which users

Published by Woodhead Publishing Limited, 2012

may connect from their local machines, usually Macintosh computers and a small number of Windows machines. Simple shell scripts can be double-clicked mounting all devices that are needed including the NetApp storage device holding our data. Because of the age of the OS, this trick cannot be achieved on our Windows XP machines, on which WinSCP [24] is used instead, but can be done on Windows 7. All data can be interacted with as if on the local machine if needed, and the user has access to the high-performance compute cluster through the automatically connected terminal. Jobs requiring the cluster are submitted and managed with Platform's LSF system, a proprietary system. Within this framework we allow all users full freedom in a public folder, into which they may install and compile whatever software they need. This works well to allow them to try out any solutions they like. Of course, this sort of setup requires that the user has experience with the command-line interface for the majority of analysis software, and, although biologists can be brought quite quickly up to speed with this style of working, in most cases there are now workflow and tool integration pipelines that take away much of the pain a brand new user may feel.

The single most useful piece of software we have installed to this end is Galaxy. Galaxy is a workflow-engineering environment in which a user may easily combine data sets and command-line bioinformatics tools in a graphical user interface to create and save analyses. The environment is very powerful and intuitive to a biologist familiar with the point-and-click paradigm of computer interfaces and they can immediately get down to the work at hand. Other such environments exist, but in our experience Galaxy is the simplest for the user and requires less intervention from us. A further advantage is the extensive training videos that the Galaxy Team provide to teach users (not administrators) how to carry out common analyses. An awful lot of software comes bundled with Galaxy and we have yet to come across a command-line tool that cannot be integrated. However, the bias in the software provided and in the themes of the training videos is very much towards the analysis of next-generation sequencing data.

Administering a Galaxy installation can be quite straightforward, but we found there are a few points worthy of comment. The community is large and a lot of problems will have been encountered and solved before. In testing the Galaxy system in a local install as a newbie then typically it will get installed on a single machine, this will run fine and probably use SQLite [25] for its job's database. The most useful lesson we learned in moving to a production environment was to make sure we switched to a PostGreSQL [26] database. The SQLite was able to cope for a few weeks under test with a limited number of users but reached a point where it started to cause unexpected behaviours and errors in Galaxy (behaviours one would not expect from a slow database), switching to PostGreSQL fixed all previous odd behaviour. Be warned, there is no migration script from SQLite to PostGreSQL in Galaxy and upgrading in this way is not supported at all by the Galaxy team. We had an arduous week testing and re-testing a custom migration script to switch our precious two-man month's worth of work to a new database management system.

According to its development team, Galaxy is always at version 1. This reflects a commitment to backward compatibility that results in daily incremental updates to the codebase on their server, rather than significant new releases. Often these updates will be as little as a few lines. Sometimes many megabytes of code will change. Most importantly, the database schema can change. Galaxy provide database migration scripts between updates, but do not provide scripts for arbitrary jumps, say from schema 27 to schema 77. The practical implication of this is that it is wise to update often. The community at large seem to update on average once every 12 weeks, to provide a good balance between workload and ease of upgrading. Leaving upgrades too long can make Galaxy painful to upgrade as many merges and schema changes and update scripts need to be run and tested sequentially to ensure a smooth running path.

Running Galaxy, essentially a job server-based system, on a compute cluster requires a touch of planning but is made easier by the fact that most cluster systems are supported by reliable libraries. In our cluster Galaxy itself runs on a head-node, which is visible to the outside world and the machines that accept jobs from the queue. We used the free DRMAA libraries from Fedstage [27] compiled against LSF (a Sun Grid Engine version exists too) and merely had to configure job-runners to ensure that Galaxy jobs went into the cluster rather than being executed on the head-node machine.

When dealing with big files, as is inevitable in NGS analyses, it is best to ensure that upload and download do not run in the cluster as the client (web browser in this case) must be able to connect to the machine doing the job. Galaxy generates large output files, which end-users take as their results. Galaxy's facility for allowing users to download data occupies a lot of processing time within the main Python process, which can cause Galaxy to slow and fail when sending data to the web browser. The solution is to use the apache_xsendfile module, which provides a mechanism for serving large static files from web applications. When a user requests a large file for download, Galaxy authenticates the request and hands the work of sending the file to the user to Apache, without needing the Galaxy processes to be involved again. The apache_xsendfile module does not currently have an upload feature. Because we use a closed system we do not provide for upload of large files into Galaxy as this often results in data-duplication (instead users requiring large files to be put into the system must seek the bioinformatician and have the files put into a data library). The Galaxy team recommend the use of nginx [28] as a server for the transfer of large files up to Galaxy. An FTP server solution is also provided.

During analyses, Galaxy creates large data files that can only be disposed of when the user decides that they are finished with. Thus in a production environment Galaxy can use a lot of disk space. Our instance runs comfortably in 1.5 TB of allotted disk space provided that the cleanup scripts are run nightly. The timing of cleanup runs will depend on use but sooner is better than later as running out of disk space causes Galaxy to stop dead and lose all running jobs. When running scripts weekly we found that 3 TB of disk space was not enough to prevent a weekly halt.

11.9 Helping the user to understand the details

With all these new powerful tools at their disposal, it would be remiss of us not to teach the biologists how to understand the settings and how to interpret the output, and, most importantly, what are the technical caveats of each data type. It is quite possible to train biologists to do their own analyses and they can quickly get the hang of command-line computer operation and simple scripting tasks. Surprisingly though, a common faltering point is that biologists often come to see the methods as a 'black-box' that produces results but do not see how to criticise them. It is often counterproductive at early stages to drag a discussion down to highly technical aspects, instead introducing simple control experiments can work well to convince biologists to take a more experimental approach and encourage them to perform their own controls and optimisations. A great example comes from our next-generation sequence-based SNP finding pipelines. By adding known changes to the reference genomes that we use and running our pipelines again, we can demonstrate to the biologist how these methods can generate errors. This insight can be quite freeing and convinces the biologist to take the result they are getting and challenge it, employing controls wherever possible.

Training syllabuses or programmes will vary massively and be extremely site-dependent. The most important thing about the training programme we have implemented is that the people giving the training really see the benefit of working in this sort of service model and therefore really want to do the training. From an enthusiastic trainer all else should flow. We have found that making resources for training constantly available and operating an open-door question policy are vital in preventing any frustration a newbie may otherwise feel when learning a new topic. It goes without saying that web sites are great places to disseminate links to resources including course recordings and 'how-to' screencasts. We have used free video servers including YouTube for shorter videos and blip.tv for longer recordings. Syndication and RSS feeds can be disseminated via iTunes for free. Screen capture software such as iShowU-HD [29] is essential; this software has a free version (although it admittedly prints a watermark onto the video, we have never found this to be intrusive). Users are very fond of the longer videos, particularly when they have access to the accompanying course materials and can work through the course at their own pace later. The most popular training materials are those that are self-contained, that can be taken away and worked on when opportunity presents, such as our own PDFs, which deliberately follow a 'principle-example-exercise' structure and are quite focussed on discrete tasks. The least effective seem to be the longer reference books that aim too wide in scope, the effort required to get much out of these sorts of resources can be somewhat dispiriting to a beginner.

11.10 Summary

With all these free and open source tools and tricks, we have been able to implement a bioinformatics service that is powerful and flexible and, most importantly, clearly accessible to the user at the levels at which the user actually cares. By deploying tools and providing sensible and custom infrastructure in which the user can access what they need without needless layers of user involvement in between, we can lift the burden of data management from a scientist and allow them just to get on with the job of research. Thanks to the flexibility and interoperability of the free and open source tools that are now available, we can put together pipelines and environments for any sort of research our institute is likely to encounter. Using free and open source software in this way frees a bioinformatician from having to perform every last analysis themselves, and thanks to the software and hardware infrastructure the bioinformatician can have much more time to pursue their own research, such as developing new methods. As the approach stands as well for one user as it does for one thousand, then a sensible infrastructure is scalable to the coming demands of our data-flooded field.

It's not all easy when relying on free and open source software. As much of it is developed for other people's purposes, there can be significant shortcomings if your immediate purpose is slightly different from the creators. The first concern when sourcing software is 'does it do what I want?', and all too often the best answer after surveying all the options is 'nearly'. As plant scientists and microbiologists, our model organisms often do not fit some of the assumptions made by analysis software, for example SNP finding software that assumes a diploid population cannot work well in reads generated from an allotetraploid plant or the formats in which we receive data from genome databases are somewhat different from those the software expects. One feature we would love to have but never do is BioMart-style [30] automatic grabbing of data over the web, such software never supports our favourite databases. This reflects the fact that the main source of investment in bioinformatics is from those working in larger communities than ours, but, in general, lack of exactly the right feature is an issue everyone will come across at some point. Typically we find ourselves looking for a piece of software that can handle our main task and end up bridging the gaps with bits of scripts and middleware of our own, one of the major advantages of Galaxy is that it makes this easy.

It is surprising to us that there is lack of useable database software with simple pre-existing schemas for genomics data. There are of course the database schemas provided by the large bioinformatics institutes like EMBL or the SeqFeature/GFF databases in the Open Bioinformatics Foundation [31] projects, but these are either large and difficult to work with because they are tied into considerable other software projects like GBrowse [16] or ENSEMBL [32] browsers or just complicated. Often, the schema seems obfuscated making it difficult to work with on a day-to-day basis. Others, like CHADO [16], have been a nightmare to just start to understand and we have given up before we begin. In this case we felt we really needed to go back to the start and create our own solution, the Gee Fu tool [21] we described earlier. We cannot always take this approach, when we are stuck, we are stuck. It is not in the scope of our expertise to re-code or extend open source software. Our team has experience in Java and most scripting languages, but the time required to become familiar with the internals of a package is prohibitive, with busy workloads we must always pick our battles and sadly the extent of our contribution to Open Source projects is often little more than to report bugs to the project owner. A saving grace is that sometimes we are able to become part of the wider development community and have some influence on the way that software we use gets made or just get inside information on how to use it and how to code with it. Such contacts are worth their weight in gold.

Given such issues, why would we not use commercial software? If you pay, surely this results in better support? The answer is sometimes true, with commercial software providing the only practical solution. Usually this will be the case when buying a new piece of hardware, we are stuck with the vendor's software for running all of our sequencing machines, job management systems and storage devices. There is no open source alternative software that could be relied on for these absolutely critical parts of the pipeline. For the downstream analyses, though, the simple fact is that commercial software is rarely up to the task. In aiming for the largest markets, commercial tools often try to incorporate too many functions. This can often result in the tool being 'jack of all trades, master of none', with results being unsatisfactory or too much of a black box for scientists to be comfortable with. Tying users in to proprietary standards is another annoyance, legacy data in undocumented or proprietary formats that no other software reads makes it impossible to re-analyse in other packages. Science moves fast and new methods appear all the time, commercial software seems rarely to innovate.

Ultimately, this need for speed is why free and open source software will always be a vital component of the software pipelines of most scientific research institutes. The need of the community to innovate could never really be matched by commercial efforts that need to provide solidity and make a return on their investment. As painful as the time and meddling aspect of using open source is, it does help us to keep as current as we need to be. The best middle ground will come from open source projects like Galaxy and Taverna [33], and standard data format descriptions like SAM [34] or Omero-TIFF [15], that help us to tie disparate bits together and make both interoperability and usability of the rapidly appearing and evolving tools much greater where it really counts, with the scientist.

11.11 References

[1]   Moore's Law – *http://en.wikipedia.org/wiki/Moore's_law*
[2]   Illumina Hi Seq – *http://www.illumina.com/systems/hiseq_2000.ilmn*

[3]   The Sainsbury Laboratory – *http://www.tsl.ac.uk*

[4]   The Gatsby Charitable Foundation – *http://www.gatsby.org.uk*

[5]   NetApp – *http://www.netapp.com*

[6]   Altschul SF, Madden TL, Schäffer AA, Zhang J, Zhang Z, Miller W, Lipman DJ. Gapped BLAST and PSI-BLAST: a new generation of protein database search programs. *Nucleic Acids Research* 1997;17:3389–402.

[7]   GFF3 – *http://www.sequenceontology.org/gff3.shtml*

[8]   FASTQ – *http://en.wikipedia.org/wiki/FASTQ_format*

[9]   Ruby on Rails – *http://www.rubyonrails.org/*

[10] REST – *http://en.wikipedia.org/wiki/Representational_State_Transfer*

[11] XML – *http://en.wikipedia.org/wiki/XML*

[12] JSON – *http://en.wikipedia.org/wiki/JSON*

[13] Apache – *http://httpd.apache.org/*

[14] Phusion Passenger – *http://www.modrails.com/*

[15] Omero – *http://www.openmicroscopy.org/site/products/omero*

[16] GMOD – *http://gmod.org/*

[17] SAVANT – *http://genomesavant.com/*

[18] IGV – *http://www.broadinstitute.org/software/igv/home*

[19] JBrowse – *http://jbrowse.org/*

[20] AnnoJ – *http://www.annoj.org/*

[21] Ramirez-Gonzalez R, Caccamo M and MacLean D. Gee Fu: a sequence version and web-services database tool for genomic assembly, genome feature and NGS data. *Bioinformatics* 2011; in press.

[22] Galaxy – *http://usegalaxy.org*

[23] Debian – *http://www.debian.org/*

[24] WinSCP – *http://winscp.net/*

[25] SQLite – *http://www.sqlite.org/*

[26] PostGreSQL – *http://www.postgresql.org/*

[27] FedStage DRMAA – *http://sourceforge.net/projects/lsf-drmaa/*

[28] nginx – *http://nginx.net/*

[29] iShowU-HD – *http://www.shinywhitebox.com/ishowu-hd/*

[30] BioMart – *http://www.biomart.org*

[31] OBF – *http://www.open-bio.org/*

[32] ENSEMBL schemas – *http://www.ensembl.org/info/data/ftp/index.html*

[33] Taverna – *http://www.taverna.org.uk/*

[34] SAMTools – *http://samtools.sourceforge.net/*

[35] CIFS – *http://www.samba.org/cifs/*

[36] SSHFS – *http://en.wikipedia.org/wiki/SSHFS*

[37] FUSE – *http://fuse.sourceforge.net/*

[38] NFS – *http://en.wikipedia.org/wiki/Network_File_System_[protocol]*

[39] VPN – *http://en.wikipedia.org/wiki/Virtual_private_network*

## 12

Design Tracker: an easy to use and flexible hypothesis tracking system to aid project team working

Craig Bruce and Martin Harrison

Abstract: Design Tracker is a hypothesis tracking system used across all sites and research areas in AstraZeneca by the global chemistry community. It is built on the LAMP (Linux, Apache, MySQL, PHP/Python) software stack, which started as a single server and has now progressed to a six-server cluster running cutting-edge high availability software and hardware. This chapter describes how a local tool was developed into a global production system.

Key words: MediaWiki; Python; Django; MySQL; Apache; Nginx; high availability; failover.

12.1 Overview

Process management systems and software are now used across many industries to capture, integrate and share pertinent data for a given process. Given the increase in the quantity of data available to project teams in the pharmaceutical industry, as well as the drive to improve productivity and reduce timelines, new approaches and solutions are needed to help the project scientists manage, capture, view and share data across the design – make – test – analyse (DMTA) cycle.

Here, we describe Design Tracker and its software stack that specifically manages some of the various processes that occur in the DMTA cycle [1]. The software allows the tracking of design ideas (hypotheses) and associated compounds from within individual projects or across projects, known as a DesignSet. Candidate compounds associated with a design hypothesis can be prioritised and proposed for synthesis. Synthetic chemists are assigned and it is they who then progress the compounds (and associated design hypothesis) through the complete synthetic work on the design hypothesis. Figure 12.1 shows a sample DesignSet, which includes the hypothesis and proposed compounds; in this case four compounds have been requested and two are currently in synthesis. The compounds are automatically assigned as registered once they appear in the AstraZeneca compound collection and can then be sent for testing. Analysis of the data is then performed by scientists for the specific design hypothesis and the outcome is captured. Based on the outcome and learning, a further round of the DMTA cycle can be started.

Design Tracker stores projects, design hypotheses and compounds in a database together with associated data and creates automated mappings

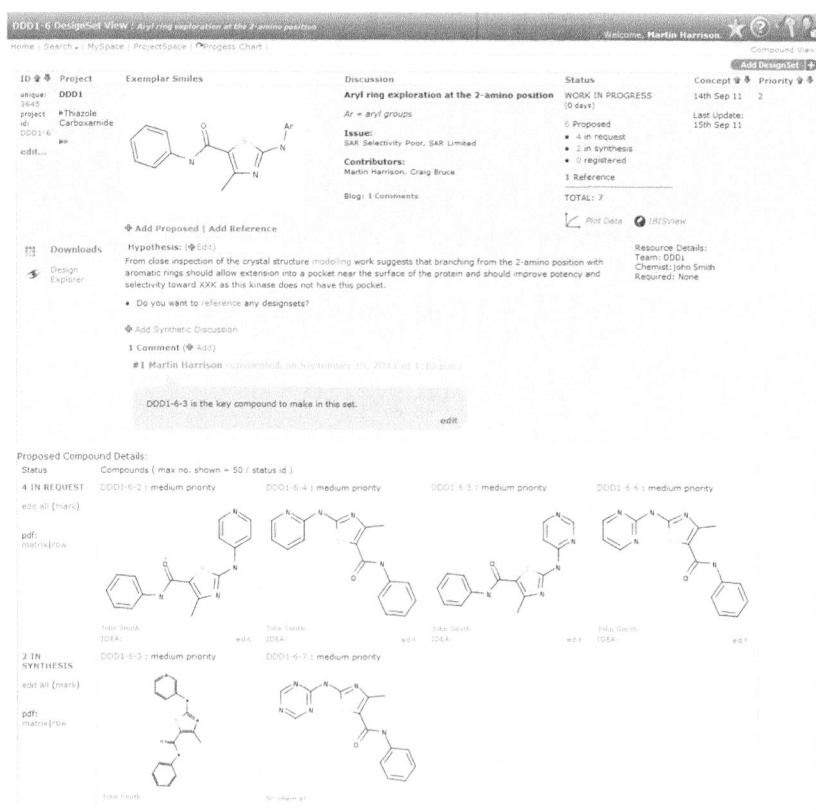

Figure 12.1 A sample DesignSet

to other AstraZeneca databases to pull in calculated and biological test data for the individual compounds, design hypotheses or projects. Design Tracker creates associations between chemists, compounds, design hypotheses, designers, projects, project issues, project series, etc., to allow different views of the information stored on the database. Views are tailored for individual group requirements and working practices, be they compound designers, synthetic chemists, synthetic team leaders, etc.

Design Tracker is the second generation of software for the capturing of design hypothesis and associated compounds. Design Tracker originated in 2006 when there was a desire for project chemists, medicinal chemists, computational chemists and physical chemists to work together more closely on projects. One of the barriers to this was a lack of clarity about what each discipline was doing. There was a clear need for an online document of the project team's activities that would allow the project team members to review and input into each other's work. The original technology chosen was a wiki as it had lots of built-in functionality for templates, revisions, etc. As the individual project wiki pages grew larger, we extracted data from the page and populated a database for easier searching and navigation. This eventually led to the creation of a progress chart for visual planning. This visual planning board concept is still in use today as part of Design Tracker, which is shown in Figure 12.2. The wiki implementation, version 1, was known as the Compound Design Database (CoDD) [2].

At the same time within AstraZeneca there was also a move towards 'hypothesis-based design'. Rather than attempt to enhance design by improving the odds through ever-increasing numbers of compounds, relevant questions were asked of the data and compounds were designed in order to answer these questions. This hypothesis-driven approach allows the prosecution of design systematically, potentially with a reduced number of compounds and design cycles. Clearly, this approach would benefit from the designers from each discipline working more closely together. The wiki-based approach described allowed better communication of design rationale within these teams and was, therefore, being developed and used at just the right time for the business.

The wiki-based approach to capture design hypothesis and associated compounds worked well up to a point but was limited by the fact that the project wiki pages tended to become difficult to manage and navigate. Once the number of hypotheses increased beyond a few tens of design ideas, then the wiki page became large and editing became tricky. It was clear that the concept proved invaluable to projects in capturing their design hypotheses. However, a more bespoke solution was then required

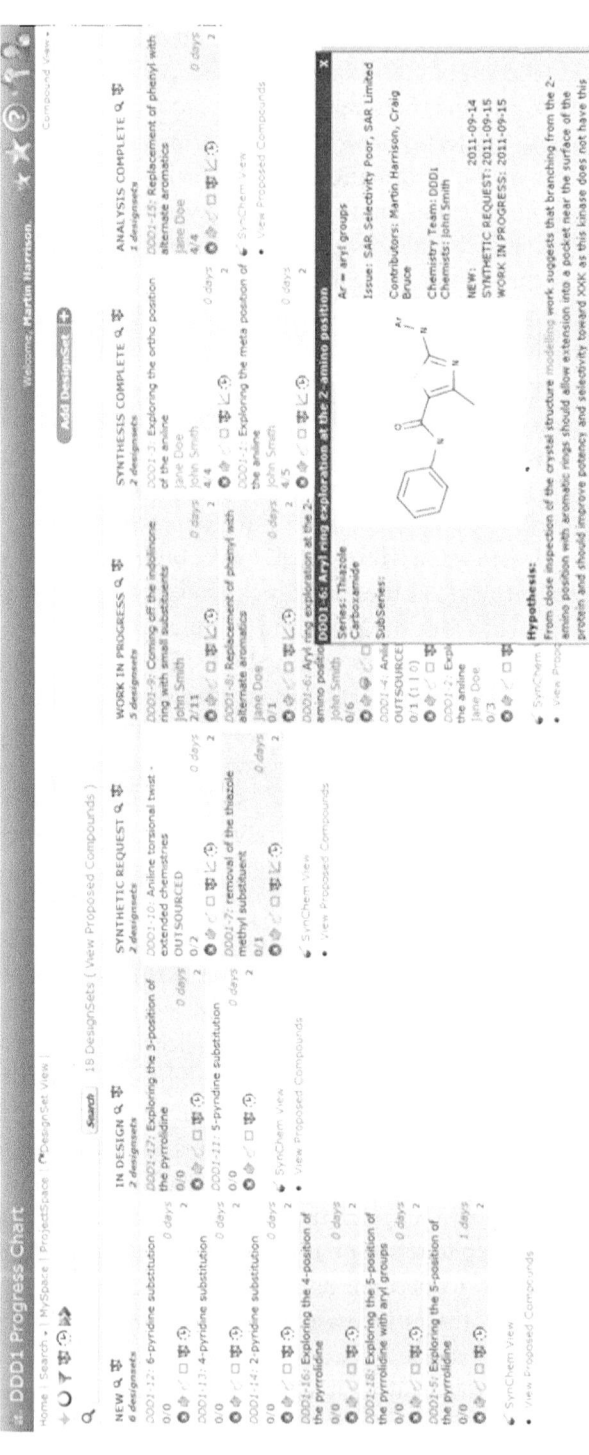

Figure 12.2 The progress chart for the DDD1 project. All DesignSets for the project are visible organised by status. A hover over provides a quick overview for each DesignSet

to build on this success and allow further information to be added and tracked through the DMTA cycle as well as methods for creating different views into the data.

12.2 Methods

Computational chemistry groups have long had access to Linux environments; it is core to the specialist software used in this field. Compared with the corporate Windows environment, Linux workstations and server systems may be fewer in number. Being dedicated to small R&D groups, there is generally far greater freedom to exploit their full functionality, such as running Apache web servers [3]. This is how we first started to explore web-based applications via Apache on our Linux network. As the technologies behind web sites have developed so have our web sites. Initial HTML hand-coded web sites have been replaced by wikis and content management systems. Although the end result is more functional and Web 2.0-like web sites, they are also far more complicated, utilising databases and multiple frameworks (Javascript, CSS and Ajax, etc.).

Traditionally, all of our applications are thick clients – this works well for our vendors. In-house code tends to primarily consist of scripts or command-line binaries, optimised to run on our high performance computing (HPC) clusters. We do have some GUI applications, but the effort to build a GUI is not trivial. Using web sites offers GUI-like functionality with relative ease. HTML tables are ugly but they can be populated from a database using Perl or Python [4]. New web technologies such as DataTables enable interactive tables [5]. Graphing is another example of improvement. We do not want to write and maintain graph libraries for each language we use. Originally we used Emprise Javascript Charts [6], but have switched from this commercial licence to the open source flot library [7]. The move was partly motivated by functionality we needed already being present in flot. An important aspect for pharmaceutical research is the visualisation of molecules. We are now in the fortunate position to have a multitude of chemistry toolkits to assist. Like the graphing libraries we have no desire to write our own chemistry toolkit. The combination of libraries and toolkits available mean that computational chemists can pick them up and make something to suit our specific needs. Our output is typically intended for our local users or for the computational chemistry community throughout the company, for example a relatively small number. Building web-based applications means every R&D scientist can use our software, without the need for

Linux. This has had the unexpected impact of creating a global system, which was originally targeted for a few local users.

12.3 Technical overview

Our initial implementation of CoDD was part of our group wiki, which is a MediaWiki installation [8]. MediaWiki is written in PHP [9] and runs using Apache and MySQL. Our server had everything needed to run this: Red Hat Enterprise Linux [10]. What makes our wiki stand out is the integrated chemistry via the MediaWiki plug-in system. We wrote a small PHP plug-in to enable to the use of <smiles></smiles> and <az></az> tags. Both would return an image of the SMILES specified or the AZ number. An AZ number is a reference to a compound in the AstraZeneca compound collection. Figure 12.3 shows the use of the <smiles> tag and how we can alter the size and title of the image.

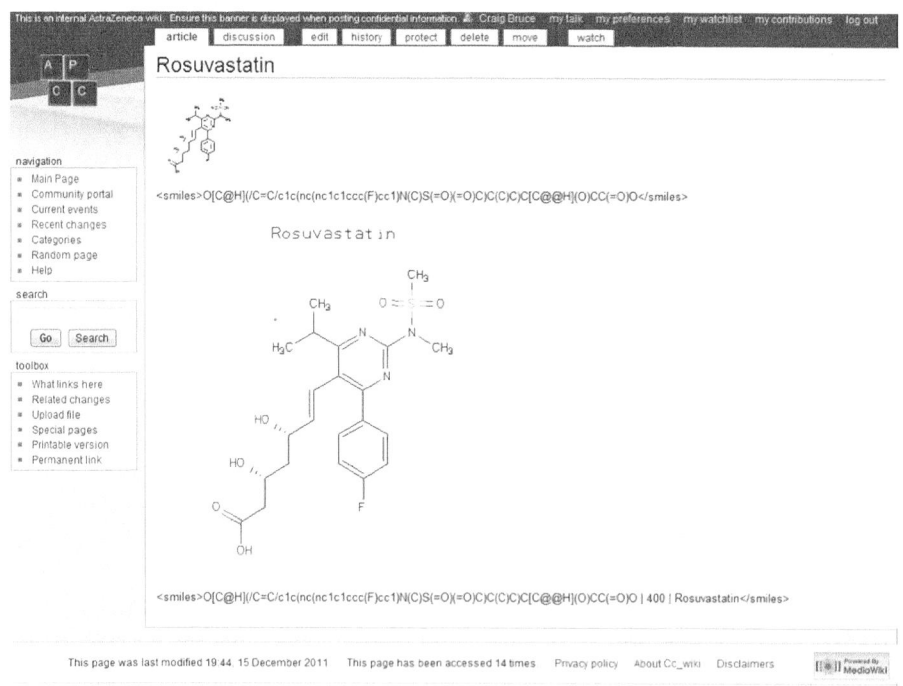

Figure 12.3 Using the smiles tag within our internal wiki. Options such as image dimensions and title can optionally be specified

Published by Woodhead Publishing Limited, 2012

In this case the PHP MediaWiki plug-in was not doing the chemistry; we actually utilised Python web services. We chose Python because our toolkit of choice, OpenEye, has C++, Java or Python bindings [11]. Python made the most sense in this scenario, as it was straightforward to add to Apache. There are, of course, other toolkits available with different bindings for your language of choice. The web service would either depict the provided SMILES string or perform a MySQL database lookup of the AstraZeneca compound collection to retrieve and depict the relevant SMILES. We wrote various web services – the wiki uses two, Smiles2Gif and AZ2Gif. We have since updated them to write high-resolution PNG images as opposed to the original low resolution GIF images. To achieve this we use Qt, as this gives us additional control over the images, for example the thickness of the bonds, via the pen controller [12].

Once the decision to write Design Tracker was taken we needed to pick the platform to develop on. Given our previous success with Python, we opted to use Django, one of many web frameworks available for Python [13]. Using Python natively meant we could also use the OpenEye toolkits from directly within the application. In terms of the deployment it is the same model as MediaWiki: Apache and MySQL. The only difference is using a Python-Apache link, opposed to PHP-Apache link. Again our single server was able to handle this, but as the popularity of Design Tracker spread across the company's global sites it became apparent that a more robust solution was required. The existing system had no failover or high availability component; redundant hardware would not protect against MySQL crashing. Our system became locked at fixed and dated versions of all software which was hindering both development and expandability. The decision was taken to invest in a new system to host our numerous Django applications, Design Tracker being the biggest. The platform needed to be robust, scalable, highly available and future-proofed.

At the time Django was a web framework solution that was in its infancy. When we began to start to write Design Tracker, Django version 0.96 was the latest version. What attracted us to Django was the fact that it was open source and was clearly being developed by a relatively large community. There was excellent up-to-date documentation and tutorials online, a roadmap to version 1.0 was clearly defined, and a number of Google groups were discussing Django and offering advice and tips. Most importantly of all, Django is written in Python.

Django also makes the developer create code in a very structured way. This was important as the code base would be written by computational chemists and not dedicated software programmers. Our computational chemists had used python and databases to create web pages before. Moving to Django (or any web framework) was the next step in this evolution to create more advanced web applications.

Django also provides a flexible admin interface that allowed us to quickly investigate different database schemas and relationships. After multiple iterations in this environment, we created the core relationships that still survive today. For pre-existing databases you can use the inspectdb option, as opposed to defining a new schema [14].

Using the admin interface we produced many of the forms and views that enabled us to speak to the customers and stakeholders who would be using Design Tracker. This gave us an opportunity to gather requirements and also to consider capturing additional data that would be useful in further versions. One example of this was the discussions we had with the synthetic chemistry community to try and understand their requirements to support greater usage of the Design Tracker software compared with the CoDD. It was from these efforts and the fact that we could easily capture information using Django's admin forms that we were able to map out their requirements and how the data would flow through the process. The Django admin views quickly allowed the synthetic chemists and teams to see the potential of developing the software further.

The vision for Design Tracker changed at this point due to this collaboration and we began to develop the software to not only capture design hypotheses and associated compounds but also the ability to assign chemists to compounds and design ideas. The chemists were a part of synthetic teams that were created in Design Tracker and consequently synthetic team views were developed to allow teams to see who was working on what at any given time as well as seeing what compounds and chemistries were coming through the design cycle. This allows synthetic teams and projects to effectively prioritise work based upon resource available and allows designers and synthetic chemists to improve their working relationships through making compound design a more transparent process.

We released version 0.9 (beta release) of Design Tracker at the beginning of September 2008 after four months of development effort, including time to learn Django, by one FTE computational chemist. We started working with a single drug project team to fix bugs for six weeks, releasing minor updates on a weekly basis. At the end of the six weeks a further project was added and we opened the development to enhancement requests from the projects. We released new versions regularly every two weeks to allow the projects to test the new features and give direct feedback. Using this approach the projects felt an integral part of the development and responded well to being the testers of the code. After ten weeks a further two projects were added to Design Tracker, resulting in four projects entering data and tracking their design hypotheses and compounds through the DMTA cycle. One of these projects was being run at another site, which allowed us to gain feedback and enhancement requests from a remote location before we released version 1.0 across multiple sites. In the two months before releasing to the wider community we again released on a two week schedule.

After essentially 16 weeks of beta release testing and enhancements we released version 1.0. This release included documentation and training materials – essentially tutorials for designers and synthetic chemists. Training and support are now carried out by the development team and the super user network [15]. The super user network is sponsored by the PCN (Predictive Chemistry Network) and is represented on each R&D site, unlike the development team. At this stage we did not formally announce the release but allowed Design Tracker to grow by word of mouth. We ran formal training sessions for on-site users in designated training rooms. This involved users going through a series of scenarios on a training version of Design Tracker so users could add, delete and update without fear of corrupting any production data.

Other R&D sites began to request access more regularly so a similar approach was adopted to start with a single project on that site, give demonstrations of the software to a limited group of people, encourage use by responding to their requests for training and minor enhancements promptly and develop relationships with a few key users on each site. The way the different chemistry disciplines were starting to interact with each other gave us the idea to develop the collaborative aspects of the software including commenting (akin to blogs) and voting systems. Design Tracker was starting to become a social network for collaborative drug design.

Throughout 2009 we released regular updates to the software and by the end of 2009 there were a large number of projects and registered users, as shown in Figure 12.4. At this stage at some research areas and sites all projects were using Design Tracker. It was at this point that a standard operating procedure was agreed by both the designers, synthetic chemists and project teams to give a clearer understanding of each other's roles and responsibilities. An example of this is that the designers have agreed not to add more compounds once they have moved a

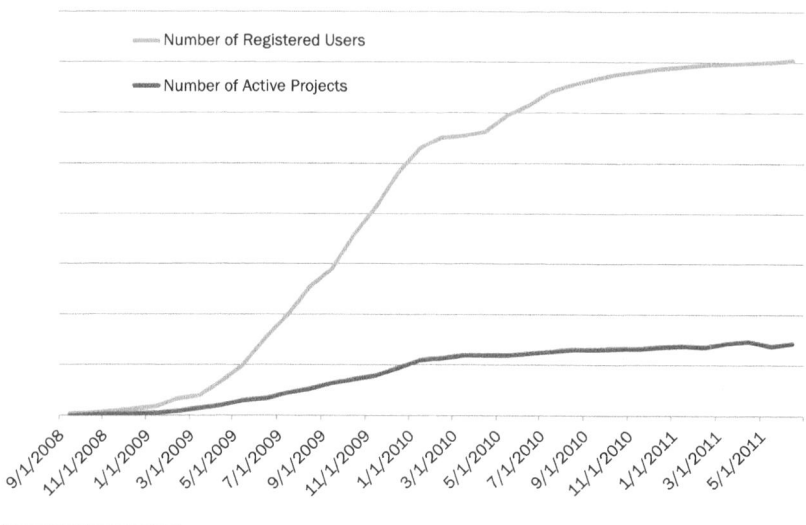

Figure 12.4 Adoption of Design Tracker by users and projects

DesignSet into Synthetic Request without speaking directly to the synthetic chemists first. This allows chemists to order reagents and develop the route without having to repeat the process when more compounds were added. The synthetic team leaders and chemists can, therefore, have an understanding of the relative worth of each designset and assign resource appropriately.

12.4 Infrastructure

Now we discuss the infrastructure required to deliver this. Our production server is a single server, which cannot provide a robust solution, it was clear we needed multiple servers. In addition other computing design principles became apparent. For example the database should run on different hardware to the application layer. Within the database layer read and write operations should be restricted to distinct MySQL instances to improve performance. In the application layer dynamic and static requests should be separated. Additionally hardware should be redundant. Software must manage high-availability, loads, failover and all seamlessly work together. This is not new: Wikipedia, Facebook and Google all have similar infrastructures to deliver the most accessed web sites in the world using a similar software stack to us.

Our initial server was a RAID1 system disk (RAID1 requires two disks so the complete data is mirrored at all times), dual power supplies, dual fibre cards, dual network ports, fibre access to the SAN for MySQL binlogs and backups. MySQL binlogs store every SQL statement run since the last backup and can be used to restore a corrupted database in conjunction with the last SQL backup.

Our new cluster consists of blades and storage in one chassis. Shared storage is provided via SAS (serial attached SCSI) connections to each blade. As well as RAID5 a parallel file system is used: GPFS, IBM's General Purpose file system [16]. The chassis includes dual power supplies, dual network switches and dual SAS switches. Ultimately, the hardware and system software choices are not important, provided that they enable system redundancy and remove any single point of failure. Our choices are based on previous experience and expertise held by our group. Although we want the same basic software stack – Linux, MySQL, Apache, Python – we need to expand it to handle the complexities of failover. One key design decision was to split the database and application operations. Now we have three blades running Apache to handle the application layer and three blades for the MySQL database.

There are multiple topologies available to design a MySQL cluster. We opted for master-slave, in our case a single master and two slaves. Additional software is required to handle this cluster. After evaluating several solutions we opted for Tungsten Enterprise from Continuent [17]. Tungsten Replicator sits on our MySQL blades and ensures the transactions carried out on the master are also repeated on the slaves. Tungsten Connector sits on the application blades and is where all the MySQL connections point to. Depending on the request (read or write) it then directs to the master for writing and a slave for reading. Separating the read and write commands offers substantial performance benefits. As the system grows, further slaves can be added to accommodate more traffic. If the master crashes one of the slaves is promoted automatically. The ability to move the master around means zero down-time for upgrades and maintenance.

For the application layer the setup is more complex than the database layer. We need highly available web servers that can scale. We have long used Apache and continue to do so. Using mod_wsgi we can link to Python [18]. Our application blades are setup with Apache but there is no failover. Nginx is an upcoming web server, which offers a proxy function [19]. We run a single instance of Nginx, which is the point for all URL requests. It then splits requests between the Apache servers. To improve the performance of the web server, dynamic and static requests should be split. Django makes many dynamic requests, and we let Apache handle these. However, Nginx is excellent at serving static files up so it now handles these requests. In this model we can add more Apache servers later as required, and again it allows easy maintenance.

Both Nginx and Tungsten Connector run as a single instance, there is no high availability protection. To ensure these services are always running we use Red Hat Cluster Suite [20]. This ensures that if the blade crashes, they are restarted on another blade. By abstracting Apache under Nginx, we are free to alter web servers later. The open source world changes far more quickly than enterprise software so having the flexibility and freedom to switch the building blocks of our system is paramount. We anticipate that several software components may change within the next few years. Importantly, unlike our previous system we can add more hardware as required and switch software as required. A large upgrade and migration cannot be imposed by external factors.

Although we have opted to use MySQL, OpenEye and Python, it is easy to mix databases, toolkits or languages of choice. Applications like Nginx, Apache and Red Hat Cluster Suite are not tied to a single software stack.

12.5 Review

The next steps for Design Tracker are introducing memory caching via memcached in the application layer for faster performance. MySQL 5.5 is beginning to utilise memcached as well. Currently we use MySQL 5.1 but Tungsten will let us run multiple versions while we upgrade the cluster at our own pace. We are keen to upgrade to the latest version, but with a global production system we test thoroughly on the development cluster first.

Looking forward we will investigate master–master replication across our geographical sites for disaster recovery purposes and to provide enhanced database performance where the limiting step is a slow network connection between sites. We are not tied to MySQL – we could switch to PostgreSQL – and Continuent is building support for other databases as well, including Oracle. From the application perspective would should add more Nginx and Tungsten Connector instances and use the LVS (Linux Virtual Server [21]) to load balance. This has the drawback of requiring more servers to setup LVS, therefore another solution which we are also investigating is using a hardware load balancer which would host

Published by Woodhead Publishing Limited, 2012

the virtual IP addresses then distribute requests to the multiple application servers, each running their own copy of Ngnix and Tungsten Connector.

Design Tracker as an application is widely used and is now a key component of our DMTA cycle. We work constantly to tighten integration of it with other in-house systems. Our users provide a constant stream of enhancement requests, which serve to make Design Tracker better. As we progress we now have a platform that can accommodate our current and future needs.

12.6 Acknowledgements

Design Tracker has been made possible by the efforts and funding of the computational chemistry groups at Alderley Park with notable contributions from Richard Hall, Huw Jones, Graeme Robb, David Buttar, Sandra McLaughlin, David Cosgrove, Andrew Grant and Oliver Böcking.

12.7 References

[1] Plowright AT, Johnstone C, Kihlberg J, Pettersson J, Robb G, Thompson RA. Hypothesis driven drug design: improving quality and effectiveness of the design-make-test-analyse cycle. *Drug Discovery Today* 2012; 1–2: 56–62.

[2] Robb G. *Hypothesis-driven Drug Design using Wiki-based Collaborative Tools.* Paper presented at: UK QSAR & Chemoinformatics Spring Meeting, 2009; Sandwich, Kent, UK.

[3] Apache Software Foundation. Apache HTTPD Server Project. November 29, 2011. Available at: *http://httpd.apache.org.* Accessed 29 November 2011.

[4] Python Software Foundation. Python. November 29, 2011. Available at: *http://www.python.org.* Accessed November 2011.

[5] Jardine A. DataTables. November 29, 2011. Available at: *http://datatables. net.* Accessed November 2011.

[6] Emprise Corporation. Emprise JavaScript Charts. Available at: *http://www. ejschart.com.* Accessed 29 November 2011.

[7] Laursen O. Google Code: flot. Available at: *http://code.google.com/p/flot.* Accessed 29 November 2011.

[8] Wikimedia Foundation. MediaWiki. Available at: *http://www.mediawiki. org/wiki/MediaWiki.* Accessed 29 November 2011.

[9] The PHP Group. PHP. Available at: *http://www.php.net.* Accessed 29 November 2011.

[10] Red Hat, Inc. Red Hat Enterprise Linux. Available at: *http://www.redhat. com/rhel.* Accessed 29 November 2011.

[11] OpenEye Scientific Software, Inc. OpenEye. Available at: *http://www. eyesopen.com*. Accessed 29 November 2011.

[12] Nokia Corporation. Qt. Available at: *http://qt.nokia.com*. Accessed 29 November 2011.

[13] Django Software Foundation. Django project. Available at: *https://www. djangoproject.com*. Accessed 29 November 2011.

[14] Django Software Foundation. Django Documentation. Available at: *https:// docs.djangoproject.com/en/1.3/ref/django-admin/#django-admin-inspectdb*. Accessed 29 November 2011.

[15] Cumming J, Winter J, Poirrette A. Better compounds faster: the development and exploitation of a desktop predictive chemistry toolkit. *Drug Discovery Today*; in press.

[16] International Business Machines Corporation. IBM General Parallel File System. Available at: *www.ibm.com/system/software/gpfs*. Accessed 29 November 2011.

[17] Continuent Inc. Continuent. Available at: *http://www.continuent.com*. Accessed 29 November 2011.

[18] Dumpleton G. Google Code: mod_wsgi. Available at: *http://code.google. com/p/modwsgi*. Accessed 29 November 2011.

[19] Sysoev I. *Nginx*. Available at: *http://nginx.org*. Accessed 29 November 2011.

[20] Red Hat, Inc. Red Hat Cluster Suite. Available at: *http://www.redhat.com/ rhel/add-ons/high_availability.html*. Accessed 29 November 2011.

[21] Zhang W. Linux Virtual Server project. Available at: *http://www. linuxvirtualserver.org*. Accessed 29 November 2011.

Published by Woodhead Publishing Limited, 2012

## 13

Free and open source software for web-based collaboration

Ben Gardner and Simon Revell

Abstract: The ability to collaborate and share knowledge is critical within the life sciences industry where business pressures demand reduced development times and virtualisation of project teams. Web-based collaboration tools such as wikis, blogs, social bookmarking, microblogging, etc. can provide solutions to these challenges. In this chapter we shall examine the use of FLOSS for web-based collaboration against the backdrop of a software assessment framework. This framework describes the different phases associated with an evolutionary model for the introduction of new IT capabilities to an enterprise. We illustrate each phase of this framework by presenting a use-case and the key learnings from the work.

Key words: MediaWiki; collaboration; Web 2.0; social networking; teamworking.

13.1 Introduction

Pfizer spends in excess of $7bn annually on research and development across multiple therapeutic areas in research centres across the globe. Within each therapeutic area are a number of separate projects, each working to identify new medicines to treat a specific condition or disease. The people working on each project come from different disciplines (e.g. chemistry, biology, clinical, safety), may be members of more than one project and may move between projects depending on their skills and the requirements of the project. This results in a complex, ever-changing matrix of individuals, who may not even be co-located at one site but who all need to share information to drive decisions. In response to these challenges Pfizer needed to introduce new tools to mitigate the inefficiencies associated with a geographically distributed team, specifically support for virtual working and collaboration.

During the first half of the last decade, 2000–2005, a new class of tools had emerged on the web that were enabling users to create, share and comment on web content without the need for technical skills such as HTML. These tools became referred to as Web 2.0 and evolved to power the social computing revolution we see now. As Web 2.0 culture developed, it rapidly became apparent that solutions such as wikis, blogs, social networking, social bookmarking, RSS, etc., which were supporting collaboration on the web, could equally be employed within business to solve the type of challenges described above. Andrew McAfee coined the term Enterprise 2.0 to describe this utilisation of Web 2.0 tools within the enterprise [1]. Web-based examples such as Wikipedia, Facebook, Twitter, Delicious, Blogger, Google Reader, etc. readily demonstrated that they could provide a solution to the collaboration and virtual working challenges businesses were facing. However, it also was very apparent that the user drivers and working practices associated with the success of these systems were very different to those seen within most places of work. Some examples of these differences are outlined in Table 13.1. As a consequence it was clear that if they were going to be successfully utilised to solve the virtual working and collaboration challenges Pfizer was facing, then it was necessary to experiment in order to understand how these differences affected deployment and integrated into colleagues' workflows.

Many of the best tools available around 2006/7, and still even now, had a narrow focus on performing a single task and do not suffer from

Table 13.1 Comparison of the differences between Web 2.0 and Enterprise 2.0 environmental drivers

	Web 2.0	Enterprise 2.0	Comment
User	Millions	Hundreds	On the web only a small percentage of the total user population needs to adopt a tool to achieve network effect. Within a business you may need the majority of users to become involved to see the same benefits

Mind set	Fun	Work	At home users do things for fun but at work users do things because they are paid
Organisational structure	Flat	Hierarchical	Flat organisational structures encourage collaboration, whereas hierarchical ones hinder [2]
Attitude	Sharing	Hoarding	At home users share information without expectation of recompense, whereas at work all too often people ask 'what is in it for me?'
Skill set	Digitally savvy	Digitally averse	On the web the users are all those who are web savvy by their nature. At work the user base covers the complete spectrum from web guru to technophobe
Visibility	Anonymity	Recognition	On the web you are one of the herd, the majority of users can assume that there is anonymity in a crowd. At work people seek recognition for their contribution as career progression can depend on it
Society	Public	Private	On the web users are able to control the information they share and are free to create alternative personas, masks, behind which they can obfuscate their identity. Within the business environment there is no anonymity, everything you say and do online can be traced back to you
Cultural	Innovative	Mundane	At home people are free to experiment and try new things. At work they have to use the tools they are given and are often told what to do

Published by Woodhead Publishing Limited, 2012

feature bloat often seen in more traditional enterprise software packages. As a consequence of this these tools provided a clean and simple user interface and the competition for users on the web placed an emphasis on quality of user experience. In our opinion these factors, user interface and user experience, were critical to the success of these tools and in some respects actually more important than their feature lists.

Finally at this time, the commercial market for these tools was quite limited and very immature. In contrast though there were numerous free and open source software (FLOSS) alternatives. Often the FLOSS alternatives were in fact powering some of the most successful social software on the web, or were clones of existing popular offerings. The fact that these tools could be downloaded and installed freely meant that they readily enabled the experimental approach we wanted to follow. The natural way to introduce these capabilities was through an evolutionary path akin to agile/scrum development methodology rather than a more traditional 'big bang' launch. This meant that the introduction of these capabilities would start with a phase of experimentation and progress to proof of concept followed by production deployment and finally transfer of support to a central shared service group. As a result we developed the framework illustrated in Figure 13.1 to introduce, evaluate, develop and deploy these new capabilities.

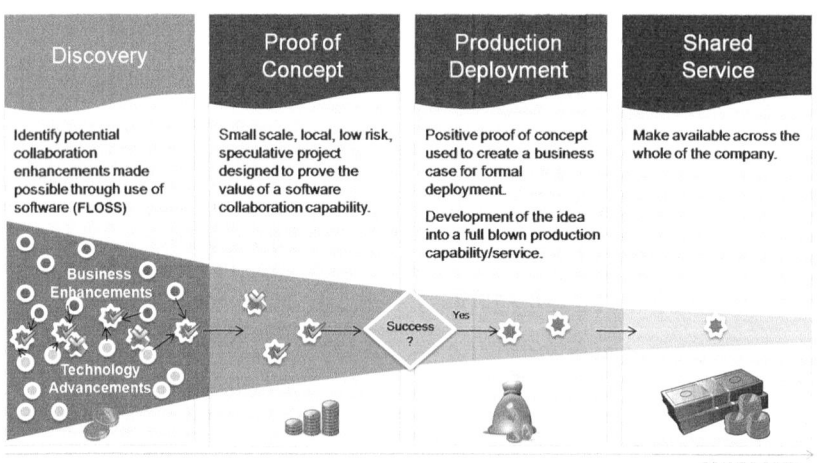

Figure 13.1 The FLOSS assessment framework. Describing the different phases associated with an evolutionary model for the introduction of new IT capabilities to an enterprise

Published by Woodhead Publishing Limited, 2012

FLOSS can be used at any or all of the stages but it certainly provides distinct cost and availability benefits over commercial software in the initial 'discovery' and 'proof of concept' stages. In some instances FLOSS collaboration systems may be shown to be superior to commercial alternatives in terms of functionality, integration opportunities and scalability, resulting in the solution being deployed into production. In this chapter we shall examine the use of FLOSS for web-based collaboration against the backdrop of our software assessment framework (Figure 13.1). We will illustrate each of these phases in turn, by presenting a use-case and the key learnings from the work.

13.2 Application of the FLOSS assessment framework

13.2.1 Discovery phase

The objective of this first phase is to identify a tool to explore the capability of interest. In addition to simply trying out various solutions, assessing how intuitive the user interface, user experience and features are, it is important to determine compatibility with the production environment and the stability of the solution. Finally, it is also critical to determine whether the solution can be hosted internally, as collaboration inherently requires the creation and sharing of content. Ultimately, to use these tools within real workflows precludes the use of third-party hosted solutions due to the risk of exposing intellectual property.

In assessing/exploring a Web 2.0 capability it is obviously highly advantageous to use FLOSS solutions, as by their nature their use does not impose any of the usual overhead associated with vendor negotiation. In addition, the open source communities have built various solutions that help with the distribution/implementation of FLOSS solutions. In our case we were looking to assess various Web 2.0 capabilities and the availability of virtualised versions of the LAMP stack (Linux OS, Apache server, MySQL and PHP) that could be installed on a Windows desktop i.e. Web On a Stick [3], Server2GO [4] and BitNarmi [5]. The existence of these 'plug and play' self-contained hosting packages made it possible for us to stand up and perform a rapid comparison and assessment of different capabilities, often without the requirement for significant technical support.

However, as tools such as Web On a Stick are self-contained virtualisation solutions, they really only support the basic assessment of different software offerings by an individual. Thus, once a specific solution had been identified as a potential candidate the next stage was to explore the multi-user experience/features it offered. To do this we were able to build a development environment utilising scavenged older hardware on which we either installed Fedora and Xampp [6], or if using a Windows box used a VMware image of Fedora downloaded from Thought Police VMWare Images [7] and then installed Xampp. These solutions proved to be very robust and in many cases these same environments were used to support the Proof of Concept phase.

One important further consideration was compatibility with Pfizer's standard browser, at the time Internet Explorer 6, and testing of any related extensions. In addition before any candidate solution was progressed to Proof of Concept, it was tested for stability, a high-level review of the code base was performed and the activity of the open source community was assessed. Our criteria are in line with those described by Thornber (Chapter 22) and consider issues such as current development status, responsiveness of the support network and general activity on the project.

13.2.2 Research phase use-case: Status.net – microblogging

With the explosion of Twitter [8], it was logical that some involved in research and development in Pfizer would start to wonder if and how microblogging might provide a simple and convenient method to facilitate a free-flow of information, updates and news across their organisational and scientific networks. With this in mind we looked to explore the opportunities of microblogging with the following technology options considered for our first microblogging experiment.

- Twitter [8] was quickly discarded because any data posted on that site would be open to all users on the web and hence could not be used to share any company proprietary information.

- Yammer [9], a Software-as-a-Service (SaaS) offering a Twitter-like application for which access could be contained to employees from a single company, was considered but rejected at this stage because of the time that would be needed to investigate the security and costing of this option.

- Workgroup Twitter for SharePoint 2007, a very slick proof-of-concept created by Daniel McPhearson of Zevenseas [10], was considered and although we liked the solution it was rejected for the simple reason that it was limited to providing microblogging capability to a workgroup/team and could not scale to support organisation-wide microblogging.

- An internally developed desktop client that allowed users to send short messages to and receive updates from a user specified SharePoint list was also available. This tool was used extensively for a number of different applications but for our initial purposes was rejected for the same reason of scope as the Zevenseas Workgroup Twitter application.

- Status.Net [11] a FLOSS microblogging platform that was being actively developed and supported was our ultimate choice. This solution provides all of the ease-of-use and similar functionality to Twitter but is based on the OStatus standard [12] for interoperability between installations.

Within not much more than a day, Status.net was downloaded and installed on a virtualised environment running on a spare Windows PC. A new icon was created to replace the default Status.net icon and launched to the alpha testing community as 'Pfollow' (Figure 13.2). Various other aspects of Status.net were quickly tweaked, most significantly the new user creation and login modules were customised to utilise our existing company-wide authentication infrastructure and a completely new URL shortening service based on YOURLS [13] was implemented and integrated into our Pfollow instance.

At this stage of exploring a microblogging capability, we had identified Status.net as our go forward solution and had performed preliminary testing with our alpha testing community. On the basis of this we had confidence that it was ready to move into Proof of Concept phase being both stable and scalable. In the next phase we would be opening up Pfollow to a wider audience and focusing on understanding how microblogging could deliver business value. Details of these learnings are presented below.

13.2.3 Proof of Concept phase

The key deliverable from this phase is a decision as to whether a capability delivers business value and if so what the requirements of the service are. In making this decision the following questions needed to be answered. Does the technology actually work in the workplace? Can it integrate into real world workflows? What use cases demonstrate business value?

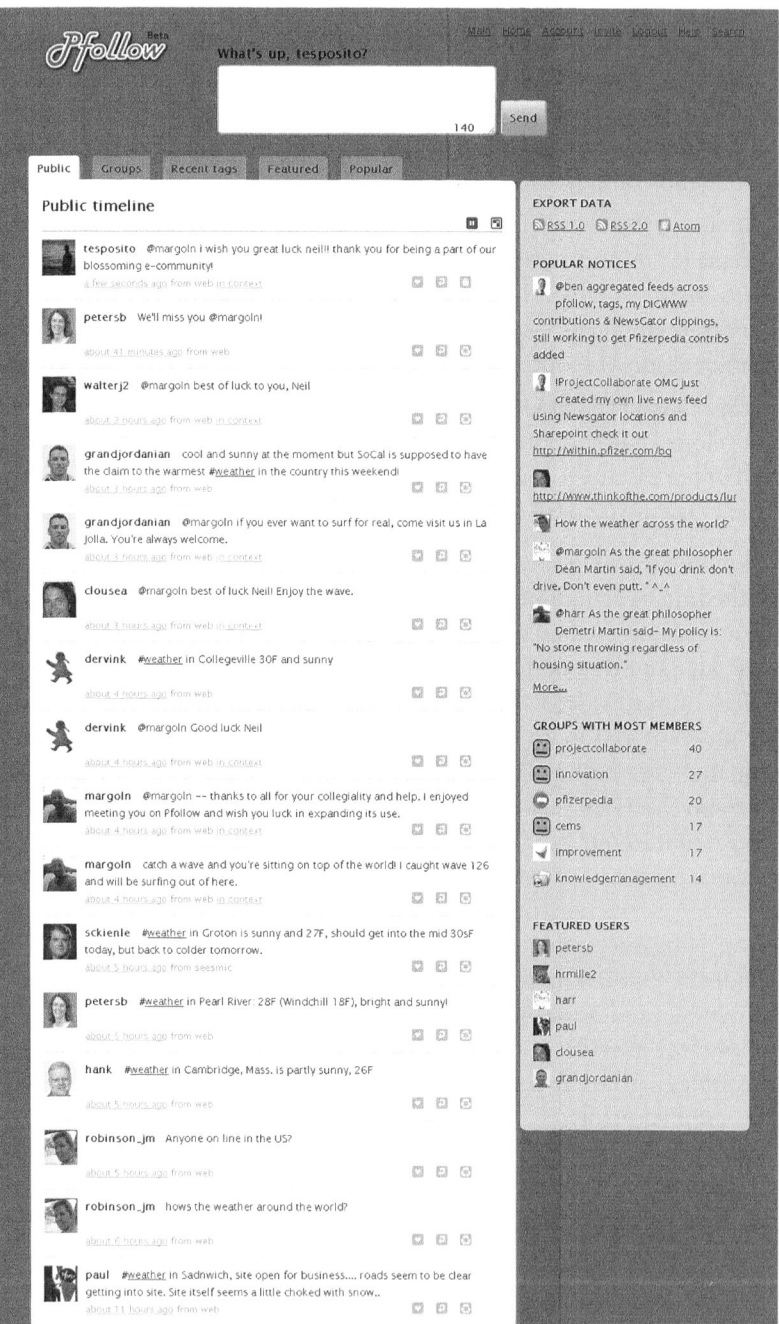

Figure 13.2 A screenshot of Pfollow showing the 'tweets' within the public timeline. This microblogging service is based on the FLOSS solution Status.net

What are the shortcomings of the solution/capability? What adoption patterns do we observe?, etc. In order to answer these questions we needed to perform safe-fail experiments, that is small-scale deployments that do not build critical business dependencies on the capability.

This was a more detailed exploration of the capability than the previous research phase but still did not involve significant development or modification of the chosen software. However, it was recognised very early on that LDAP integration, that is enabling users to use their corporate Windows network ID and password, was critical for getting adoption beyond members of the alpha testing community. In practice, this modification was generally very simple and had the added benefit that it significantly simplified the effort associated with migrating and merging user's content between test instances. Further, a consistent user ID across tools later allowed us to experiment with lightweight integration/cross surfacing of content between tools without the complexity of mapping user accounts, that is creating activity streams.

The data generated in this phase are used to build the business case for funding of a production capability and the requirements that service needs to meet. The purpose is not to promote the solution used during this phase but to enable an accurate assessment of the capability.

13.2.4 Proof of Concept phase use-cases: microblogging and social bookmarking

Pfollow – microblogging

Having identified Status.net as our solution for exploring microblogging (see previous section), the final step was to obtain the Pfollow internal domain name. At this point the new service was advertised to prospective users via existing social media and emails to relevant networks within the company. These users were invited to take part, but were made aware of the experimental nature of the software through a set of terms and conditions related to its use. The primarily underlining fact being that due to the nature of our deployment we needed to reserve the right to switch Pfollow off at any point once our experiment had reached a logical conclusion. Consequently, users needed to be made aware that they should not build business critical dependencies on this service.

The new Pfollow community was fast growing and over a matter of weeks we started to see both individual contributors making an impact and a large number of online groups coming into existence. Initial posts were largely non-work related – instead the equivalent of small talk around the water cooler really, allowing key participants to introduce themselves and get to know each other. For instance, a post each day relaying what the weather was looking like in their neck of the woods, an often common icebreaker at the start of teleconferences, when participants are dialling in from different parts of the world. This provided an easy entry for users to get to grips with the software and get comfortable with the idea of posting to such a public forum where the recipients are largely unseen and unknown.

Some specific work-related use-cases emerged from our experiment.

- Ask your extended network a question. Either directed at the general community, sometimes accompanied by a specific hashtag, or targeted to a specific group. The benefit seen here was that users realised they now could quickly and easily call on an extended network of professionals for help.

- Share a project/work/status update. This was seen as a good way of publicising progress or the completion of a piece of work, perhaps more widely than might be possible using traditional communication channels.

- Advertise something that has been posted online elsewhere. For example, a blog posting, paper or web page. Cross-posting was a popular use of Pfollow as, again, it opened up the existence of a resource or achievement to a much wider audience and was an effective way for an individual to communicate their speciality, knowledge gained or general expertise in a subject area.

Our experimentation with Status.net served to allow us to quickly identify the following key requirements if we were to proceed with a microblogging service.

- The ability to create groups proved to be a highly important feature – we saw special interest groups, project teams and communities of practice utilise Pfollow to quickly and easily form and share information. The fact the groups that had been created were available to view in public listing was also key as it allowed and encouraged people to connect based on their interests and specialities rather than arriving at a connection solely through their particular existing organisational and/or geographical associations.

- Hashtags were important to allow users to structure and interlink posts and allow areas of interest to emerge and be tracked by interested parties.

- Support for RSS proved important so that users could subscribe to distinct parts of the overall discussion, for example subscribe to specific groups, users or hashtags, depending on their particular areas of interest.

Overall our microblogging experiment was a success and provided us with the confidence to take the concept further. Status.net was critical in that it provided us with the full set of features required to see the benefits of using this type of technology, allowed us the ability to tweak it as our specific organisational needs dictated and, importantly, provided an ease of use that encouraged our users to embrace the software.

Tags.pfizer.com – social bookmarking

The goal of this project was to explore the use of social bookmarking within Pfizer. By allowing people to share their bookmarks online, social bookmarking services, such as Delicious [14], are turning the world of bookmarking (browser favourites) upside down. Instead of storing bookmarks on desktops, users store them online in a shared site. Not only can users access their bookmarks anywhere, but also so can anyone else. This allows colleagues to discover new sources of information by looking at what others working in similar areas are bookmarking. Furthermore, as users bookmark things that are important to their work or their area of expertise their bookmark collection becomes a tacit repository describing their interests. As a result social bookmarking services can also enable social networking. By looking at bookmarks saved by others, users can discover new colleagues who share their areas of expertise/interest.

To evaluate this capability Scuttle [15], an open source social bookmarking tool, was selected, judged by much the same criteria as described above for Status.net. Scuttle provides similar capabilities to the popular web service Delicious. In addition it supports the Delicious API, which allows bookmarks to be imported from that service and, in principle, allows any Delicious-compatible tools to work with it such as browser toolbar extensions. Further, Scuttle also provides RSS support with feeds automatically generated and filtered by user, tag and multitag/user combination. Finally, the company Mitre had used Scuttle as the codebase on which they developed Onomi, an inhouse social bookmarking solution [16] giving us confidence that Scuttle was robust, scalable and could work in a business environment.

Initially, a development instance was installed, which was used during the discovery phase for functionality evaluation and stability/load testing.

Having established that this tool was stable and offered the core functionality of a social bookmarking service the team created a second instance, tags.pfizer.com, for the internal community (Figure 13.3).

During the earlier discovery phase, the development instance of tags. pfizer.com had become heavily populated with technical bookmarks. As a result it was decided that during the Proof of Concept phase two active instances would be maintained. This would allow two communities to develop one IT technical and one with a drug discovery focus. Later these two instances were merged once the business facing community had been established.

To start, a 'private beta' model was used for user recruitment to tags. pfizer.com, targeting known Web 2.0/Enterprise 2.0 early adopters and colleagues familiar with social bookmark services. This was followed by a word of mouth viral campaign, targeting key influences, and finally formal presentations were made to the research community. This approach was used to help overcome the cold start issue, namely an empty bookmark database. To further ameliorate this problem, one of the first groups asked to contribute were information science colleagues. This group worked to populate tags.pfizer.com with bookmarks for key resources used by the colleagues they supported and provided rich and valuable content for other users.

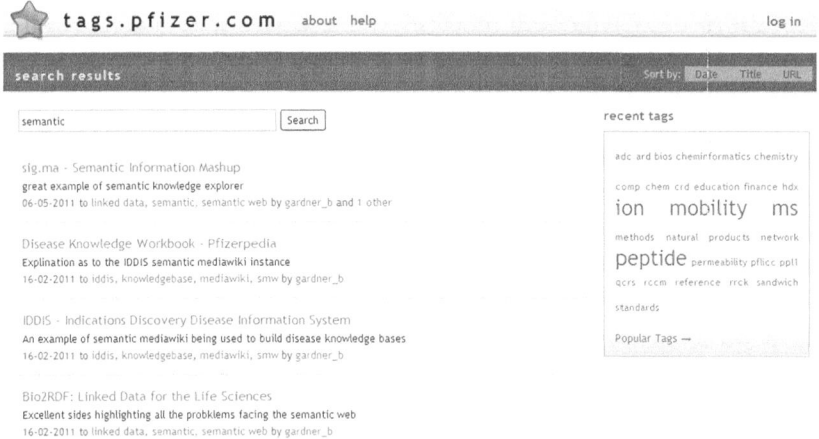

Figure 13.3 A screenshot of tags.pfizer.com. Bookmarks 'tagged' with the term 'semantic' are shown. This social bookmarking service is based on the FLOSS solution Scuttle

Published by Woodhead Publishing Limited, 2012

The use of this service was evaluated over a six month period. During this time the number of users, across the two instances, grew to 220, of which ~20 per cent were adding new bookmarks on a regular basis. After six months a total of ~5500 bookmarks had been added, although this did include bookmarks imported from users' Delicious accounts. Although these metrics indicate that users were obtaining value from this service, more detailed feedback was sought via an online questionnaire and user interviews. The key message that came through from these results was that a majority considered they were obtaining value from this service and would recommend tags.pfizer.com to their colleagues. However, it was also felt that tags.pfizer.com would benefit from increased scale; the more people who used it the more valuable it would become.

This issue of scale has been highlighted by Thomas Vander Wal [17] where he points out that as a social bookmarking service grows (both in the number of users and the number of bookmarks), the value and use of the system changes. This can be broken down into four phases. In the first phase the use is personal, a user saves, tags and re-finds bookmarks. As the community of users grows we move into the second phase, serendipity. Here the user can start to explore other users' bookmarks but the searching is hit or miss. In addition groups/project teams will start to coordinate and use common tags to highlight items of interest to the rest of the group/team. The next phase is social tagging maturity, in this phase users can regularly search and find new and related bookmarks. We would also expect to see the bookmarking service start to support social networking as users also begin to identify colleagues with similar interests based on their tag and/or bookmark fingerprints. In essense the service has evolved from purely a bookmarking service and has become a key tool in social networking that supports/enables complex social networks.

At this point tags.pfizer.com was in the first phase. As such most users were getting value from the service based on the improvement in ability to store and search their bookmarks over that offered via their 'favourites' folder in their browser. Although the benefits of this were evident, the greater value from being able to search/explore other people's bookmarks and identify kindred colleagues was generally not being realised. However, despite this a number of 'islands' of specialty began to build, each starting to exhibit the behaviours/functionality associated with the more mature phases of a social bookmarking service.

In terms of overall functionality, Scuttle provided the core capabilities and successfully enabled social bookmarking within the company. However, to achieve wider adoption some key issues still need to be resolved. First among these is the integration of user accounts with users' Windows network ID and passwords, as well as integration with the browser, that is toolbar extensions. However, at that time Pfizer's standard desktop was based around Internet Explorer 6 and such browser extensions would have required bespoke development. This highlights the conflicts we often see between the rapidly moving external world, where technologies are evolving at a rapid pace, and the internal world where technologies move slower [18]. Finally, this experiment highlighted that the utilisation of this service within the enterprise requires integration with internal systems in ways not required on the web. On the internet, web pages are bookmarked via a browser button or a scriptlet added to the browser links toolbar. This works well in this environment, but within the enterprise users often want to store links to other content, such as documents within content management systems. In this case, the URL for these documents is often not displayed in the browser toolbar but rather accessed via a right mouse click or complex Javascript (or even Flash) functionality. At best, capturing these bookmarks can be achieved through a manual cut and paste; additional complexity that is far from ideal for users. Given these limitations it was decided that if social bookmarking was to be developed as a production service it would require significant development of the Scuttle codebase, development of browser extensions and potentially changes to the user interface of a number of document management systems.

13.2.5 Production deployment phase

At this point in the FLOSS assessment framework, the value of a collaboration tool has been proven and the requirements are well understood. Here, in this phase, the production solution is selected and deployed. Successful usage patterns identified in the previous phases are used to promote the software into the business and drive adoption. New usage patterns will continue to emerge and these need to be captured and utilised to continue to promote the system. At the same time, the wider usage of the web site, service or tool may highlight areas were development/extension of the capability is required.

Obviously one of the critical decisions at this stage is to determine whether the software explored within earlier Discovery and Proof of Concept phases is suitable to go forward into a production environment. In taking this decision, each potential tool must be assessed with respect to a standard capability requirements document as well as any other factors that influence future support and development of the system. Up to this point, the flexibility provided by the open culture of the FLOSS solutions is generally seen as a positive. However, once deployment to a production environment is being considered, the associated business-critical workflows result in some of these same advantages becoming significant risks.

For example, a FLOSS solution might be considered robust but without an active developer community attached to it, the solution might become non-viable over time. This is particularly true if it is not being updated relative to the technology stack on which it is hosted. Alternatively, a large and active support community can fragment the codebase, resulting in multiple, possibly conflicting branches and variants. In some cases, these risks maybe mitigated by employing third-party vendor support models, a model which is becoming increasingly common. Of course, an organisation can also decide to fork the codebase itself and develop some features specific to its business requirements. Consequently, the burden of full support, along with all the inherent costs falls to that organisation alone. In light of these risks and costs, continuing with a FLOSS solution into production is not a trivial decision.

13.2.6 Production Deployment phase use-case: Pfizerpedia – Enterprise Wiki

MediaWiki is the FLOSS that was originally developed for Wikipedia, the free encyclopaedia web site used by many millions of users every day. As well as powering the family of Wikipedia web sites it is also used by hundreds if not thousands of web sites elsewhere on the web and within a number of large company intranets. The MediaWiki software [19] is available to download for free from *http://mediawiki.org* and contains a number of class-leading wiki features. Because of the high visibility of the Wikipedia [20] site, the MediaWiki software is very familiar to most users and the software has been explored by a number of life science organisations (see Chapter 16 for another example).

Although MediaWiki is FLOSS, it is clearly high quality and has many traits suitable for deployment in the enterprise. For instance, scalability of MediaWiki has been proven quite clearly by the implementation of Wikipedia on the web. For an enterprise organisation, this scalability is not an advantage solely in terms of performance. It also brings the functions necessary for large numbers of users to happily coexist on the same platform, and system administration tools required to manage the large number of pages that will (rapidly) be created within the wiki.

On the performance front, MediaWiki has been designed from the first to be quick and responsive and its authors have built in many routines, such as aggressive caching, to ensure sustained high performance.

Another key facet of the scalability of MediaWiki is that it has been designed to be easily extended and as a result there are a large number of extensions that can also be freely downloaded from the MediaWiki.org site to expand how it can be used even further. Indeed, a single instance of MediaWiki can take the place of an untold number of individually coded applications that would naturally exist within any organisation, for example see Table 13.2. Note that MediaWiki can be taken to another level of sophistication altogether with the introduction of the Semantic MediaWiki family of extensions and this is covered in detail elsewhere in this book (see Chapters 16 and 17).

Table 13.2 Classifying some of the most common uses of MediaWiki within the research organisation

Application of MediaWiki	Description
Sharing knowledge	One of the most high value uses of MediaWiki within a research organisation is as a way of sharing knowledge derived from various projects. A 'one-stop shop' MediaWiki page on a specific subject that multiple authors from across an organisation contribute to can be a very powerful tool
Help and how-to pages	Ease of authoring and cross-referencing content means capturing how-to and help guides are a popular use of MediaWiki. For instance, on the check out the Wired How-To Wiki [21] or the Open Wetware Wiki [22]
Lessons learnt/best practices	The low cost of MediaWiki as an application and its complete openness makes it a perfect place to store lessons learnt for future reference
Catalogues	A single MediaWiki instance within an organisation is an excellent place to investigate what someone is working on or the service a group can provide. The power comes from being able to access related content, e.g. the 'Innovation' page links to a generic topic of 'Social Problem Solving', which in turn leads to a tool available within the organisation that provides a solution for this need

Yellow and white pages	It is incredibly easy to create formal structured directories of people and groups within MediaWiki using the MediaWiki categories feature. If the names of organisational groups should change, MediaWiki handles this in an elegant fashion by allowing you to 'move' pages – automatically creating a link from the old name to the new for anybody who subsequently attempts to access it through an out-of-date cross-link or URL
Collaborative authoring	Wikis are unique at providing true collaborative authoring, many authors can improve the same page at the same time. Every MediaWiki page has a 'history' tab, providing a log of exactly who has changed what and a 'discussion' tab to discuss changes. With features like RSS and MediaWiki Email Alerts, users can be notified when specific pages are updated by other authors
Profiles	One of the great strengths of MediaWiki is the ability to see exactly who is contributing to what topics. To enhance this feature even further, users can create a page describing themselves, populated with links to areas within the wiki that reflect their work. MediaWiki will automatically aggregate and display their contributions; this can be very valuable when you can remember the name of somebody who contributed to a piece of work but not the name of the topic they contributed to

Inside Pfizer, a single general-purpose instance of MediaWiki was installed and named Pfizerpedia (Figure 13.4). At inception it was envisioned that this would develop into a scientific encyclopaedia, akin to Wikipedia, for the research and development community. However, it was rapidly recognised that wikis could enable collaboration in a much broader sense and the idea that Pfizerpedia would be restricted to being an encyclopaedia was relaxed. Pfizerpedia evolved to become a knowledge-sharing tool supporting a wide range of use-cases, including encyclopaedia pages, hints and tips pages, user/team/department/ organisation profiles (Figure 13.5), application/process support/help pages (Figure 13.6), acronym disambiguation, portfolio/service catalogues

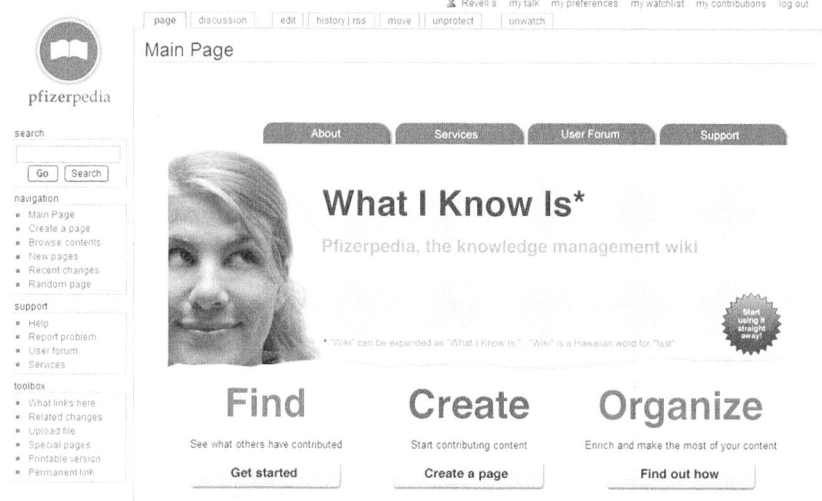

Figure 13.4 A screenshot showing Pfizerpedia's home page. This wiki is based on the FLOSS solution MediaWiki

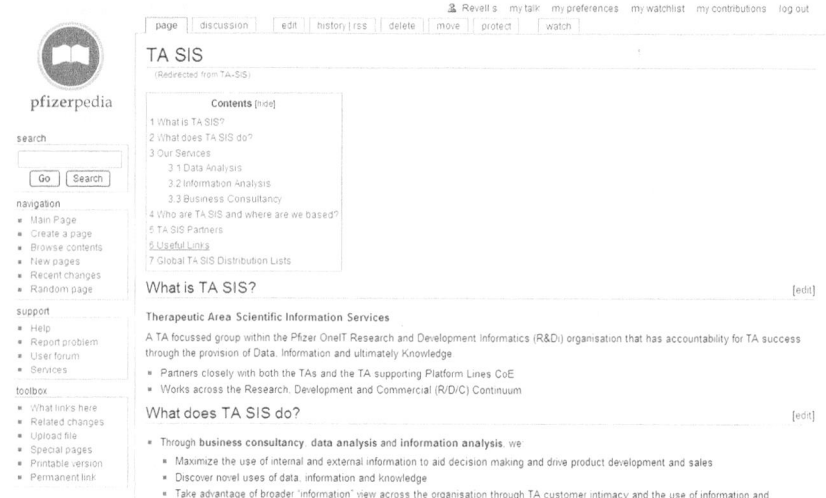

Figure 13.5 A screenshot showing an example profile page for the Therapeutic Area Scientific Information Services (TA SIS) group. Profile pages typically provide a high-level summary of the person/group/organisation along with links to resources and information often held in other systems

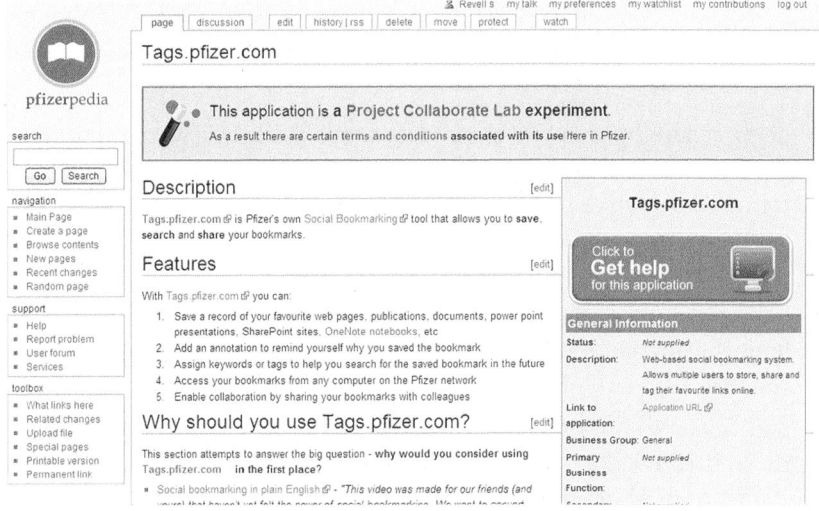

Figure 13.6 A screenshot of the tags.pfizer.com social bookmarking service page from the R&D Application catalogue. This catalogue contains a collection of pages describing the various applications available to the R&D community. Each page provides a high-level description of the application along with links to full help and support pages also within Pfizerpedia

and social clubs. Thus, Pfizerpedia is much more than a collection of factual pages; it is a tool that enables groups to collaborate, share and find information.

Pfizerpedia has grown from an experimental instance for the R&D community (originally hosted on a PC under a colleague's desk) into a solution that is being adopted across the whole company. As the number of users and volume of content has grown, we have started to examine how we can automate some of the more common tasks associated with managing a wiki, known as 'wiki gardening'. This involves prompting when content may need to be updated and applying appropriate categories, templates and banners across many pages. MediaWiki caters for this via an Application Programming Interface (API), allowing for the creation of 'bots' that can be scheduled to automatically retrieve and update pages. Bots can be programmed in a number of different computer languages including Java, .Net and Python.

In choosing to use MediaWiki as the solution for Pfizerpedia, we made the conscious decision to stay as close to the core MediaWiki distribution as possible to limit our exposure to maintenance and support costs/overheads. This has meant we have:

- never redeveloped any of the codebase for the MediaWiki application itself nor any third-party extensions used. This way the upgrade path for our installation of MediaWiki is pain-free and low effort as we can deploy new versions without the need to reapply company specific customisations;

- been conservative in our adoption of third-party extensions, again to limit the amount of effort required for our upgrade path and ongoing support;

- tried to limit any third-party extensions installed to being from only those used on the Wikipedia family of web sites, maintained by the same group of developers of the core system itself. This should mean that the development of these extensions is more likely to be in sync with the main MediaWiki software. This should negate the need for revision of extension code by internal developers whenever a new version of MediaWiki is issued.

As a result of adopting these principles, we have faced no software issues or outages, low to non-existent support and maintenance costs, and a set of happy and very satisfied users. Our experience confirms MediaWiki as a powerful piece of software that can be applied to many different knowledge-sharing and collaboration scenarios. It boasts features that facilitate connections to people, knowledge and information that do not exist in many other forms of collaboration and content management software. Couple this with the large and highly active support/development community and it can readily be seen that MediaWiki is an example of a FLOSS solution that can be readily deployed into production with minimal risk.

13.2.7 Shared Service phase

In many respects the challenges and objectives for this final phase are common to all application deployments within large organisations. Here we see a transition of responsibility from the team that developed/implemented a solution over to the sustain/support team. Typically, in large organisations, this involves two groups; those delivering back-end support and those providing direct end-user support (i.e. a 'helpdesk'). The former requires the technical details about the implementation; which

Published by Woodhead Publishing Limited, 2012

servers the solution is on, how to perform upgrades, what are the dependencies, etc. In contrast, the user support team need to be provided with training material, a knowledge base covering basic troubleshooting, any installation guides, etc. Yet, in the case of collaboration solutions we have found that users often require more than just a technical training session. Focusing only on the function of each feature, how to log in, how to save, etc., misses educating users on how the platform can be used in a social and collaborative context at work. Thus, the training aspect in adopting these tools often requires more emphasis than traditional IT solutions. Many of the basic reasons for this are associated with overcoming the differences in our expectations and behaviours between work and home life (Table 13.1). However, software collaboration tools also require users to develop the social skills and confidence associated with formal community-building and a willingness to explore new ways of working.

Successful collaborative working requires more than just building an online space. People willing and able to engage and manage the community or team and to encourage the members to adopt new working practices are critical. Even better, colleagues able to think beyond existing tools and envision how combinations of individual services can support or enable new working practices are invaluable. The openness and flexibility of web-based collaborative capabilities is a great strength but to take advantage of this capacity requires users to think differently. This really represents a core difference between the Web 2.0 type tools and the traditional solutions with which knowledge workers have been provided. Typically traditional tools designed for knowledge works included hard-coded workflow process, a one-size-fits-all assumption. In contrast, the Web 2.0 tools are open and flexible and allow users to superimpose their own way of working onto a blank canvas. However, it is our experience that users confronted with a blank canvas find this lack of signpost/ guidance quite intimidating. Often, this is enough to prevent them exploring the potential within these capabilities to significantly enhance/ simplify their working. In response to the challenges discussed above, we recognised the need to provide users not just with technical training on tools but also to support community building/developing new working practices. To do this we developed 'consultancy workshops' targeted to those colleagues involved in continuous improvement activities and end-user close IT colleagues. These workshops aimed to provide colleagues with a technical understanding of the new capabilities and skills required to build and maintain online communities. In particular, examples of how these collaboration capabilities have been successfully used with business groups were presented. The key learning here was that these tools can revolutionise the way people collaborate but if you only focus on 'building it', the users might not follow you. Ultimately, success requires enthusiasm and partnership between developers and users, allowing iterative evolution of the software system.

13.2.8 Principle for collaboration

Finally, a few words need to be said about our general learning concerning this type of software. During our experimentation with FLOSS collaboration capabilities, we identified a number of key principles that define the core architecture of collaboration. These form a framework, which applies equally to the technology and the culture. It is our experience that adoption of these principles is required if a culture of collaboration and openness is to develop. The four core principles we developed are:

- freedom – the easiest way to prevent collaboration from occurring is to impose overly burdensome control around how colleagues work. If collaboration is to flourish we need to trust colleagues and not impose rigid workflows, inappropriate approval processes (moderation), restriction on who can collaborate with whom (association) and have an open attitude towards sharing information;

- emergence – no two collaborations are the same, each team/group will have different requirements and will develop different working practices. Given this we need to allow patterns and structures to emerge as collaborations develop. This is not to say we should not stimulate behaviours we want or share experiences but rather we should accept this and recognise that we need to avoid a one-size-fits-all approach;

- clarity of purpose – in this case, colleagues are confused as they are presented with multiple tools, all of which seem to do the same or a similar task. In the case of Pfizer we had a plethora of different tools that enable various degrees of collaboration; Enterprise Document repositories such as Insight, Documentum, SharePoint, eRooms, Pfizerpedia and so on. The lack of consistent advice around how and when to use these tools inevitably leads to adoption of Outlook for information management, fragmented silos of project data and a lack of any real knowledge management strategy.

- ease of use – collaboration is about enabling conversations between people. It is not about technology. Therefore, it is critical that technology does not get in the way of collaboration. If a collaborative culture is to be enabled then it must be ensured that colleagues find the tools are intuitive, integrate into their workflows and require minimal training.

13.3 Conclusion

The availability of free and open source software was a critical enabler in our ability to introduce and build collaborative capabilities for research and development at Pfizer. In the early phases where experimentation and agility were critical, FLOSS allowed us to rapidly explore a capability by examining various different approaches. This was further enabled through the availability of packaged distributions that incorporated the complete technology stack. The ability to download and run a personal version of a tool, on your desktop or even via a USB drive, meant that participation in the Discovery phase was opened up to many more participants than possible with 'traditional' software. This allowed a diverse community of interest to develop and actively contribute from the very beginning. In the Proof of Concept stage, these strengths of FLOSS were also enabling but the fact that many are also of high quality allowed us to explore how collaboration capabilities really add value within the organisation. The ability to test the ideas and 'stories' that were circulating on the web at this time was invaluable and lead us to really appreciate the different challenges that implementing these tools produce. All of this done without one licensing discussion or vendor engagement and for zero investment costs.

As we move into the Production Deployment phase, the emphasis switches from enabling experimentation to deploying a robust solution on which business critical processes may depend. The experience gained from working with the FLOSS solutions was central to making the tools selection at this stage. In many cases, FLOSS solutions scored highly, and often outscored vendor offering in terms of features, user experience and user interface. However, at this point we also needed to take into account how a tool would be supported, developed and maintained. In this respect, many FLOSS solutions scored poorly, and as a consequence compromises may have to be made. Specifically, where a capability fits with a niche or specialist need, the FLOSS community supporting that the tool tends to be small (Scuttle or Status.net, for example). In these cases, the risks and burden in selecting these solutions for a production deployment can be too high. However, where a capability has wide application (e.g. MediaWiki, the Lucene search engine, etc.) then the FLOSS community tends to be large and active. Consequently, third-party support vendors will often have emerged to support these products, which means there is often little to choose from the sustain/maintenance side of the equations between the FLOSS and a commercial solution.

Overall, FLOSS has been critical in our developing collaboration capacity. It is certainly the case that without the availability of software that can be rapidly deployed with minimal initial overhead, we would not have been able to experiment in the way we did during the early phase. Being able to build up experience and understanding as to what collaboration with social media tools really means within a business environment was critical. The learning provided from our FLOSS projects has proved invaluable during the Production Deployment and Shared Service phases and allowed us to create solutions that fit with the new working practices in a Web 2.0-enabled world.

13.4 Acknowledgements

The authors would like to acknowledge the many colleagues who contributed to the various projects described here. In particular Paul Driscoll, Scott Gavin, Chris Bouton, Stephen Jordan, Andrew Berridge, Steve Herring, Jason Marshall, Daniel Siddle, Anthony Esposito and Nuzrul Haque.

13.5 References

[1] McAfee A. '*Enterprise 2.0: The Dawn of Emergent Collaboration*,' MIT Sloan Management Review 47(April 2006): 21–28.; and McAfee A. '*Enterprise 2.0: New Collaborative Tools for your Organisation's Toughest Challenges*,' (Boston: Harvard Business School Press, 2009).

[2] Patterson R. 'Social Media and the Organisation' FastForward Blog, *http://www.fastforwardblog.com/2007/08/29/social-media-and-the-organisation-part-1/*, updated August 2007.

[3] MoWeS Portable CH Software, *http://www.chsoftware.net/en/mowes/mowesportable/mowes.htm*

[4] Server2Go, *http://www.server2go-web.de/*

[5] BitNami, *http://bitnami.org/*

[6] Apache Friends, *http://www.apachefriends.org/en/xampp.html*

[7] Thought Police VMWare Images, *http://www.thoughtpolice.co.uk/*

Published by Woodhead Publishing Limited, 2012

[8]   Twitter, *http://twitter.com/*

[9]   Yammer, *https://www.yammer.com/*

[10] McPherson D. 'Twitter for SharePoint?' Zevenseas Point2Share, *http:// community.zevenseas.com/Blogs/Daniel/Lists/Posts/Post.aspx?ID=93*, updated 24 April 2009.

[11] StatusNet, *http://status.net/*

[12] OStatus, *http://ostatus.org/about*

[13] YOURLS, *http://yourls.org/*

[14] Delicious, *http://www.delicious.com/*

[15] Scuttle, *http://sourceforge.net/projects/scuttle/*

[16] Damianos L, Griffith J, Cuomo D. Onomi: Social Bookmarking on a Corporate Intranet. *http://www.mitre.org/work/tech_papers/tech_papers_06/06_0352/06_0352.pdf*

[17] Vander Wal T. Bottom Up Tagging. *http://www.slideshare.net/vanderwal/bottom-up-tagging*

[18] Wingfield, N. It's a Free Country. . . . So why can't I pick the technology I use in the office? *The Wall Street Journal* 15 November 2009, *http://online.wsj.com/article/SB10001424052748703567204574499032945309844.html*

[19] MediaWiki.org, *http://www.mediawiki.org/wiki/MediaWiki*

[20] Wikipedia.org, *http://www.wikipedia.org/*

[21] Wired How-To-Wiki, *http://howto.wired.com/wiki/About_How-To_Wiki*

[22] Open Wetware – Share Your Science, *http://openwetware.org/wiki/Main_Page*

## 14

Developing scientific business applications using open source search and visualisation technologies

Nick Brown and Ed Holbrook

Abstract: In addition to increased scientific informatics demands from high-throughput and more complex scientific applications, solutions are being developed in pharmaceutical companies that focus on business needs. In AstraZeneca, our New Opportunities unit focuses on disease indications of unmet patient need, outside of existing therapy areas. Over the past few years, systems have been developed using commercial and open source technologies, where many of these systems leverage a foundation layer – a publication index, based on SOLR. Systems such as automatic Key opinion leaders (KOL) identification, drug repositioning and an atlas-of-science can analyse and combine disparate data with other sources to aid decision making in our business.

Key words: SOLR; New Opportunities; search; tagging; visualisation.

14.1 A changing attitude

Over the course of the past decade, there has been an evolution in the landscape of IT systems in pharmaceutical companies. Our scientists would have state-of-the-art solutions at work, whereas at home, only for the technically gifted. Nowadays with information technology becoming integrated into everyone's daily lives, with sites like Google, Facebook, LinkedIn, Twitter and Ebay, and tools like iPad, Sky+, TiVo and Kindle, the situation has started to reverse. With intuitive, tailored systems in our personal lives, everyone expects a lot more in our working life.

In addition to increased expectations, there is significantly more demand for scientific informatics within the pharmaceutical environment. With the rise of genomics, developments in metabolomics and proteomics, the explosion of externally generated scientific knowledge, high-throughput functional and phenotypic screening and the advent of next-generation sequencing technologies, informatics is at the heart of most, if not all, projects. This increased burden, coupled with a continual squeezing of resources and budgets, has meant that open source and often free alternatives are starting to be considered alongside the mainstream commercial platforms.

14.2 The need to make sense of large amounts of data

In AstraZeneca, our New Opportunities Innovative Medicines (iMED) group focuses on disease indications of unmet patient need, outside of the existing therapy areas (oncology, infection, respiratory, inflammation, cardiovascular, gastrointestinal, pain). As a small, virtual team, this is achieved through in-licensing late stage opportunities, out-licensing assets with world experts in different disease areas and evaluating new hypotheses with our drugs with external contract research organisations (CRO), key opinion leaders (KOL) and specialist disease companies.

Working in novel disease indications, without the years of knowledge that is normally intrinsically captured, our informatics systems need to be able to aggregate, standardise, analyse and visualise the huge wealth of information. This is normally hidden in unstructured texts and requires teasing out, either as structured text analytic queries or by generating themes and trends.

Our approach initiated in 2008, was to provide a foundation layer that would aggregate common unstructured, external scientific content into a single, unified search index. The initial programme focused on external scientific abstracts, leveraging publications, patents, conference reports, posters, PhD theses, UK/US grants and external clinical trials. Our original licence agreements were re-negotiated to ensure the ability to text-mine and analyse these sources. With new agreements in place, each individual data type was parsed into a unified format (described below), loaded into a search index and then different business applications built over the content, typically using a variety of different visualisation approaches for simple, end-user interaction.

14.3 Open source search technologies

Working with small volumes of data is relatively simple and a wide variety of search solutions can handle this requirement. When dealing with an index of millions of scientific abstracts, there can be issues with time; both in terms of indexing content and also speed of retrieval. Many search platforms were investigated, but SOLR stood out as it provided robust results in milliseconds.

SOLR [1] was developed as part of the Apache Lucene project. It provides a search platform that is fast, easy to implement and potentially enterprise capable. It has been adopted by a widening number of mainstream organisations including AT&T, AOL, Apple, IBM, Ticketmaster, Ebay, NASA and Netflix. The best-known example of SOLR in action is the purchase of goods on Amazon, where on selection of an item, the user is provided with a series of alternatives that other people who had bought that item had also purchased. This is SOLR's optimised faceted search in action.

The searching capabilities that AstraZeneca have used at present include:

■ powerful full-text searching – including searching for exact text, text with slight spelling errors and logical functions (e.g. AND, OR and NOT);

■ key word searching, that can be linked with text searching (e.g. search for documents that contain the word 'cancer' and are written by a given author);

■ faceted searches – quickly provide a list of all the 'facets' associated with a search (e.g. provide a list of the authors for the documents returned in a given search, and the number of documents that they have authored – without having to run through all the documents individually).

SOLR also has a number of other features that allow future integration into our enterprise IT, including:

■ database integration. This is especially useful when the documents indexed need to be annotated after they have been indexed. If an update to the original document needs to be indexed again (e.g. an update is supplied for a document from the external suppliers), then all annotations on the document would be lost without it being stored and re-applied from the database;

- rich document (e.g. Word, PDF) handling. This is provided by integrating SOLR with the Apache Tika system [2], and is useful when indexing a diverse set of documents in a file system. The text extracted can also then be passed through other systems for further processing using a single text-based pipeline;

- geospatial search. Basic geospatial search has been appended to our index, but there are possibilities of marking documents with the author's location (e.g. country, postcode, longitude, latitude). This means that clustering of work can be investigated. In addition, it may be possible to utilise this capability to look for elements that are similar to each other.

SOLR is highly scalable, providing distributed search and index replication and at the time of writing our index currently provides sub-second searching on over 50 million documents. It is written in Java (utilising the Lucene Java search library) and runs as a server within a servlet container such as JeTTY or Tomcat. It has REST-like HTTP/XML and JSON APIs that has made it easy for us to integrate it with a number of applications. In addition, it has been extended by experimenting with Java plug-ins to allow us to implement specialist searches, should this be required.

The SOLR index has been found to be reliable, scalable, and quick to install and configure. Performance-wise, it has proved to be comparable to similar commercial systems (better in some cases). In addition, the faceting functionality provided has proved invaluable. The API has proved easy to integrate into a number of applications, ranging from bespoke applications to excel spreadsheets. However, the main disadvantages found have been the lack of tools to configure, manage and monitor the running system, and the lack of support.

14.4 Creating the foundation layer

The AstraZeneca SOLR publication management system was first developed to provide a data source to the KOL Miner system (more information given later). This was to be a system that automatically would identify KOLs, especially important for groups that work effectively by knowing which external partners to work with.

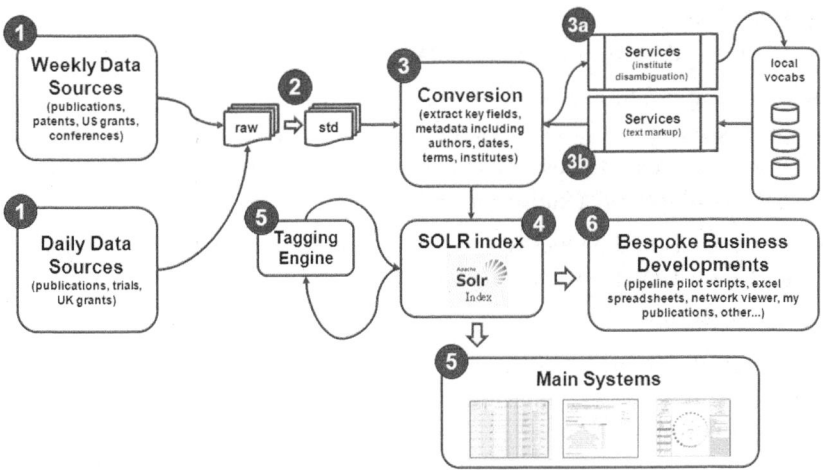

Figure 14.1 Schematic overview of the system from accessing document sources (1), standardising them into a single format (2), converting these and performing any disambiguation clean-up steps (3), indexing (4), re-tagging as appropriate (5) and finally development into business-driven systems (6)

The first version of KOL Miner utilised Lucene indexing technology, but this proved to be too restrictive in that it did not provide an *enterprise-class* index and rapid search. In addition, it proved to be more scalable, especially as the faceting feature of SOLR enabled the structured querying of a large number of documents without having to load individual document details.

A document pipeline process was developed to process the documents as available. This is summarised in the Figure 14.1.

Each process was quite specific and kept componentised for efficiency as steps could be replaced easily as better solutions were identified.

14.4.1 Raw documents

Data feeds are collected from a number of different sources on a regular basis (either daily or weekly). By default, all these feeds come in the form of XML files. However, they are collected using a number of different technologies, including:

- FTP files from a secure remote server. This was normally only necessary when the data feeds had been made available on a subscription basis;

- accessing publically available web pages to identify new XML files to be downloaded via a generic Python script. A further enhancement to this system enabled users to log into the web site where the feeds were only available on a subscription basis;

- in some cases, data files were sent by suppliers directly to the company, either on CDs/DVDs, or via email, but these were normally replaced by an automated service as described above.

Although the *standard* document feeds ended up being XML files, other smaller SOLR indices have also been created from other file types, as described in step 2.

14.4.2 Normalise documents

The initial step in preparing the document data is to convert the documents into a standard XML format, known internally as 'Unified XML'. Once this process has completed, all other processing can be performed using a single, standard pipeline. Initially, the conversion was performed using PERL scripts, but this has mostly been replaced by Python scripts utilising the 'xml.etree.ElementTree' library [3] to simplify the development. However, other one-off mechanisms have been used to convert non-XML files to the unified form, including standard Bash shell scripting and the Apache Tika text extraction utilities to extract text and meta-data out of different documents (such as PDF and Word documents).

Where possible, meta-data is extracted from the documents so that it can be easily searched in the indices. This includes:

- author data – including whether an author is the 'first author' (probably the person who did most of the work) or the 'last author' (probably the department head or senior research fellow). This also utilises our author disambiguation scripts, which try to clean these into a standardised format. There is no ideal solution and different geographies provide different challenges for this problem. A combination of institute, date and scientific tags can help in this challenge;

- author institute data;

- publication dates;

- title;

- any tags or key words that have already been attributed to the document. For instance, information delivered by the MedLine fed included 'MeSH' terms, which can be used to identify key concepts that the document is addressing.

14.4.3 Process documents

Once in a standardised form, the documents are converted into an XML form that can be loaded into the SOLR index. In addition to this, the documents are further processed to extract more information that may be of interest. For example:

Institute disambiguation

The authors of the documents are normally associated with a particular institute (e.g. university, hospital, company). However, the same institute is often presented in many ways. For example, an author from the Department of Cell Biology, Institute of Anatomy, University of Aarhus, Denmark may publish under a number of different institute names, including:

- Department of Cell Biology, Institute of Anatomy, University of Aarhus;

- Department of Cell Biology, University of Aarhus;

- Institute of Anatomy, University of Aarhus;

- Aarhus University Hospital;

- University of Aarhus;

- Aarhus Universitet.

Each of these institutes are linked to the 'University of Aarhus' – meaning that a single search will find all documents from this university, despite the multiple original institute terms. Significant investment was made initially to bring together a comprehensive set of 'stem' terms for each pharmaceutical company, top biotechnology companies and top 500 world universities. These were used to bring together a significant set of synonyms across multiple sources. All of this information was used to identify consistent synonyms for each company or academic institute.

In addition, some institutes are hard to disambiguate due to multiple institutes with the same or similar information. For instance, 'Birmingham University' could either be 'Birmingham University, England' or

'Birmingham University, Alabama'. Other clues need to be used to identify which institute is actually being referenced.

A Java program to perform this disambiguation was developed around a number of institute names and locations. These locations are held in another SOLR index, which provides fast lookup to allow rapid disambiguation. This service is also made available via a web interface to allow other programs to use it. It includes a number of steps, each designed to 'hone' the search further if no individual match is found in the previous steps:

- check the address to see if it contains any 'stem' words that uniquely identify key organisations (e.g. 'AstraZeneca', 'Pfizer'). If so, match against that organisation;

- split the provided address into parts (split on a comma) and identify the parts that reference any country, state and city information from the address (by matching against data in the index);

- at the same time check to see if any of the address parts contain an 'institute-like term' (e.g. 'University'), in which case only that part is used in the next steps of the search (otherwise all non-location address parts are used);

- look for an exact match for the institute name with the information held in the SOLR index. This can also be honed by state/country, if that information is available. In addition, common spellings of key institute-type names (e.g. 'Universitet') are automatically normalised by the SOLR searching process;

- if still no match is found, a 'bag-of-words' search is performed to check to see if all the words in the institute name match the words of an institute appearing in any order within the institute synonym.

If no match is found or multiple matches still exist, the application returns the address part that had an institute-like term in it as the institute (as well as any location information extracted from the full address). In future, more advances could be incorporated to handle mergers and acquisitions.

Text tagging

The main body of text from the documents is automatically scanned for key scientific entities of interest to AstraZeneca (e.g. genes, diseases and biomedical observations). This uses a text markup system called Peregrine, which was developed by the BioSemantics group at Erasmus University

Medical Centre [4]. This tool is used to identify and normalise the terms so that they can all be referenced as one. For instance, a gene may be known by various terms, including:

- name – e.g. 'Vascular Endothelial Growth Factor A';
- official symbol – e.g. 'VEGFA';
- EntrezGene ID – 'EntrezGene:7422';
- other – e.g. 'VEGF-A', 'vascular permeability factor', 'vegf wt allele'.

Each of these synonyms will be mapped to the same entity. In addition, attempts are made to make sure that a synonym does not accidentally match another word. For instance, the 'Catalase' gene is also known as 'CAT', but this also matches many other things (including cat the animal, cat the company that makes diggers, cat cabling, and cat – Cambridge Antibody Technology – company acquired by AstraZeneca in 2006).

14.4.4 Creating SOLR index

This normally takes place in an overnight process. Although, technically, the SOLR system can cope with dynamic updates, it was found that this process is best performed while the index is offline. In future, it is envisaged that a dedicated SOLR server will be used to index new data, and the indexes swapped over at night, to improve up-time, and the number of documents that can be processed per day.

14.4.5 Enhanced meta-data

In addition to in-pipeline annotations, further data enhancement processes have been developed to allow extra annotations to be performed on the data, using the power of the SOLR index. For instance, a set of SOLR queries were developed to identify if any of the documents mentioned any standard chemical reactions (e.g. 'Baeyer-Villiger Oxidation', 'Schotten-Baumann Reaction'), and the relevant documents tagged with that information. This tagging process takes approximately five minutes for about 100 different reactions. Unfortunately, as this updating is performed on the SOLR index directly, any updates from the original source results in these annotations being over-written – which is why a possible enhancement to the system being considered is to tie the index into a database to hold the extra annotations.

14.4.6 Searching to build applications

Once the data are held within the index, they can be searched simply using either the RESTful HTTP interface, or using standard APIs developed for JAVA and other programming languages.

There are a number of systems that use the data, ranging from systems developed alongside third parties (e.g. KOL Miner), to Pipeline Pilot scripts and Excel spreadsheets that have been developed to interface with the Publications Index for specific purposes for individual scientists. Typically, many of our approaches use visualisation technologies to present the data in a more meaningful way back to the business users.

14.5 Visualisation technologies

Several commercial alternatives for building interactive, information visualisation systems have been evaluated. A thorough comparison against a variety of platforms demonstrated that open source alternatives tended to have at least as rich a set of features for data modelling, visualisation and interaction as well as layout and encoding techniques. More importantly, the frequently active community continued to improve and grow the options, often resulting in many examples that could be relatively quickly deployed and customised for business applications.

There are a variety of open source visualisation technologies including Visualisation Toolkit (VTK), Prefuse, Flare, Simile, Java Universal Network Graph (JUNG), InfoVis, yFiles, HCIL and AXIIS; each with their differences and unique features.

The visualisations that are described today focus predominantly around the utility of Prefuse [5], although our group has build systems using Flex, Flare and Simile, as well as commercial offerings such as Flex.

14.6 Prefuse visualisation toolkit

Prefuse is a Java-based toolkit providing a set of libraries for rich visualisations, including network diagrams, heat maps and trees. It is licensed under the terms of a BSD license and can be used freely for commercial and non-commercial purposes.

The online Prefuse gallery provides a great overview of approaches that can be considered. Figure 14.2 highlights some approaches that have been piloted.

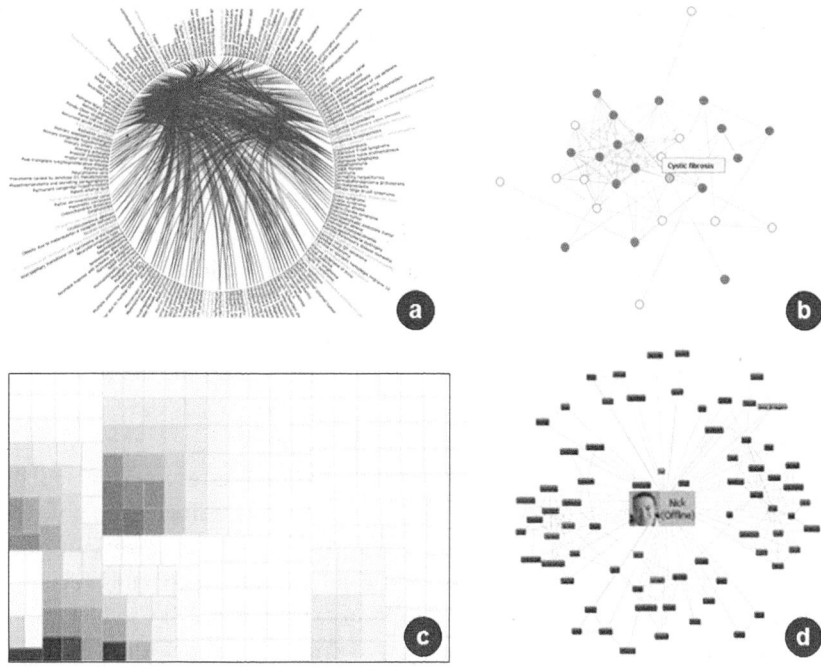

Figure 14.2 Node/edge networks for disease-mechanism linkage (a), radial networks for disease co-morbidities (b), treemaps for identifying trends (c) and photo networks (d) for social network interaction maps

As well as simple customisation, it is possible to combine the variety of components to build quite advanced systems. The rest of this chapter describes some example applications that have been built for our drug projects in AstraZeneca.

14.7 Business applications

Our approach has always been to use the right technology for fit-for-purpose business applications. This means that a single technology is not used for every problem and open source technologies are combined with commercial offerings. Some of the examples described here use such a combination.

14.7.1 KOL Miner – identifying influential key opinion leaders

When working in new disease areas, rapid and up-to-date identification of key experts is critical. Our approach was to leverage this enriched SOLR index to automatically identify the KOLs and academic institutes on-the-fly. As a virtual iMED, we work effectively by partnering with an external company – being able to identify KOLs rapidly is critical and helps us identify new people to work with. In fact, the system was developed in partnership with another company. OpenQ are a world leader in KOL management and together developed a system known internally as KOL Miner that was released externally as OpenIdentify™.

For any given search, over 30 faceted search queries are performed at the same time and aggregated together. Very few technologies are able to provide the extremely efficient and fast faceted search required. In this case, the SOLR index usually takes between 10 and 15 seconds for any search, even large queries retrieving millions of records. For each KOL or institute returned, the top meta-tags are identified, allowing an understanding of a KOLs main area of interest without reading a single article as shown in Figure 14.3.

As well as providing an interactive tabular view, very simple interaction network maps and vertical timeline plots are also generated. These let our scientists understand KOL spheres of influence and temporal trends, respectively, as seen in Figure 14.4. Using visualisation technology was critical to our success here. In this situation, the solution was developed with a commercial partner but many valuable lessons were learnt. Some visualisation problems are simple to solve using tables and summary figures, but understanding trends over time or interactions between often hundreds of individuals are things that cannot be easily comprehended by the human brain. Visual representation of the same data is therefore intuitive and often simpler for AstraZeneca scientists to interpret. In addition, a simple report was not sufficient. Users wanted the ability to view the high-level results over a given search, but then drill down into the underlying evidence once they had finished experimenting with search. For the first time, this technology provided a mechanism that could rapidly investigate many different areas and identify the key players within a matter of minutes.

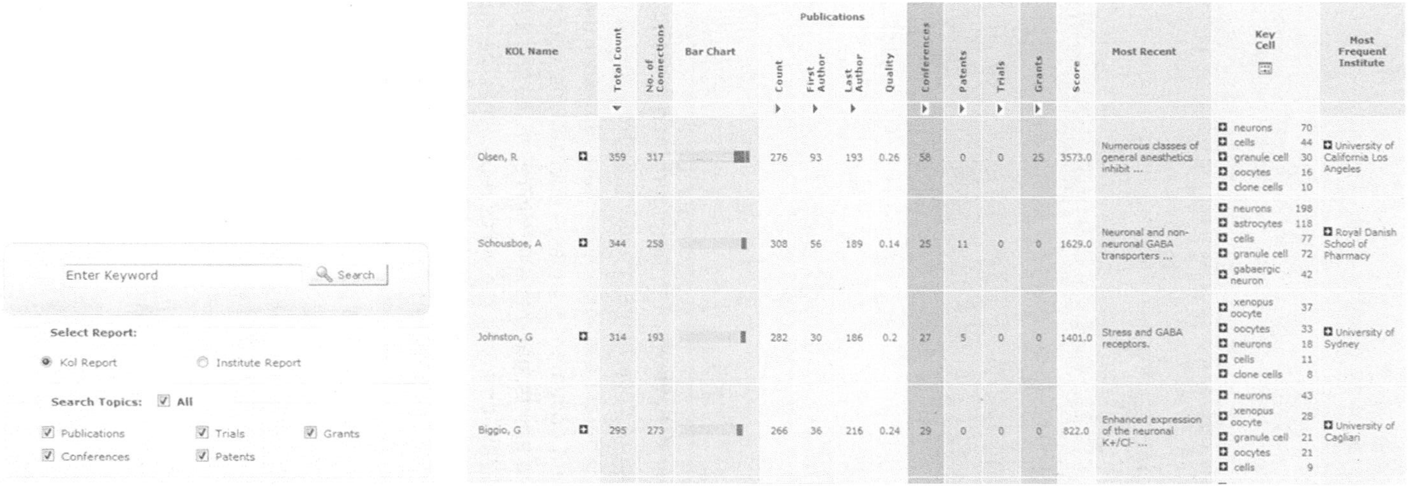

Figure 14.3

KOL Name	Total Count	No. of Connections	Bar Chart	Publications Count	First Author	Last Author	Quality	Conferences	Patents	Trials	Grants	Score	Most Recent	Key Cell	Most Frequent Institute
Olsen, R	359	317		276	93	193	0.26	58	0	0	25	3573.0	Numerous classes of general anesthetics inhibit ...	neurons 70, cells 44, granule cell 30, oocytes 16, clone cells 10	University of California Los Angeles
Schousboe, A	344	258		308	56	189	0.14	25	11	0	0	1629.0	Neuronal and non-neuronal GABA transporters	neurons 198, astrocytes 118, cells 77, granule cell 72, gabaergic neuron 42	Royal Danish School of Pharmacy
Johnston, G	314	193		282	30	186	0.2	27	5	0	0	1401.0	Stress and GABA receptors.	xenopus oocyte 37, oocytes 33, neurons 18, cells 11, clone cells 8	University of Sydney
Biggio, G	295	273		266	36	216	0.24	29	0	0	0	822.0	Enhanced expression of the neuronal K^+/Cl^-	xenopus oocyte 43, oocyte 28, granule cell 21, oocytes 21, cells 9	University of Cagliari

Enter Keyword — Search

Select Report:
◉ KoI Report   ○ Institute Report

Search Topics: ☑ All
☑ Publications   ☑ Trials   ☑ Grants
☑ Conferences   ☑ Patents

KOL Miner in action. Users have a very simple search interface (although advanced search options are available) and the results are visually appealing with different categories highlighted as both data and graphical charts. All results are clickable so users can drill down into the raw evidence at any point or add terms into the search to follow a train of thought

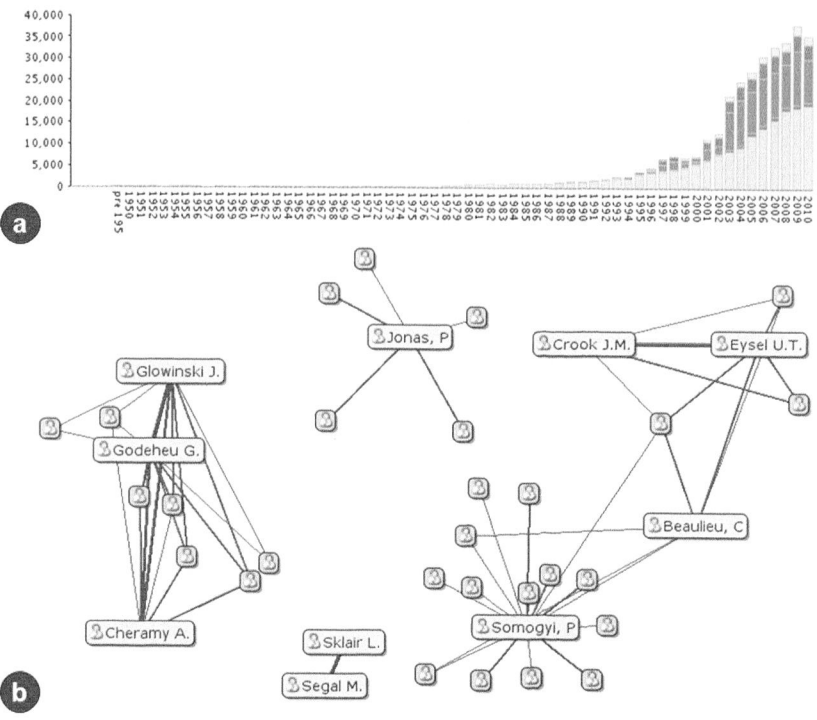

Figure 14.4 **Further KOL views.** The bar chart (a) represents the number of documents reflecting a given search over time. Different sections of the bars represent the different document sources – publications, conference reports grants, patents and clinical trials. The x-axis represents time and the y-axis represents the number of documents in total. The network visualisation (b) represents the top ten authors for a given search. Here the top ten authors actually interact by their co-authorships, and it is possible to gain an understanding about who is working with who, which groups are isolated and which individuals might have more influence – even if are not the top KOL

14.7.2 Repositioning matrix – finding new indications for our drugs

AstraZeneca has over 500 compounds that have been tested clinically, representing over 170 different target mechanisms. Historically these

Published by Woodhead Publishing Limited, 2012

projects have focused on less than 100 disease indications, yet the standard disease ontologies have thousands to consider (for example, Medical Subject Headings (MeSH) [6] has over 2850 diseases). To consider all of the possibilities would be unrealistic without informatics.

In New Opportunities, the problem was simplified and initially less than 150 additional indications were considered, which had clear unmet patient needs. A system was therefore developed that leverages both the unstructured data within SOLR (>50 M scientific documents) and the structured content within an aggregated collection of competitive intelligence databases. This brings together an understanding of biological and competitive rationale by searching for sentence level co-occurrence between disease term(s) of interest and the drug's mechanism term(s) as outlined in Figure 14.5.

This information is then stacked on top of each other in an interactive heatmap visualisation. Figure 14.6 illustrates this approach and combines this mining result with additional internal ideas. By providing a single visualisation of over 30 000 serious opportunities, our scientists can prioritise and evaluate new indications for our compounds, or identify disease area themes to consider. Figure 14.7 shows an example of the sentence level extraction that scientists are able to view when they drill down into the results. In this way, compounds were identified that had potential across a range of eye and skin disorders. In 2009, a deal was signed with Alcon, experts in ophthalmology, and in 2011 with Galderma, experts in dermatology.

This system was built entirely using Pipeline Pilot [7], although there are alternative workflow tools such as KNIME (see Chapter 6 by Meinl and colleagues) and Taverna that could be considered. In this situation, Pipeline Pilot provides extremely good flexible web services that can be

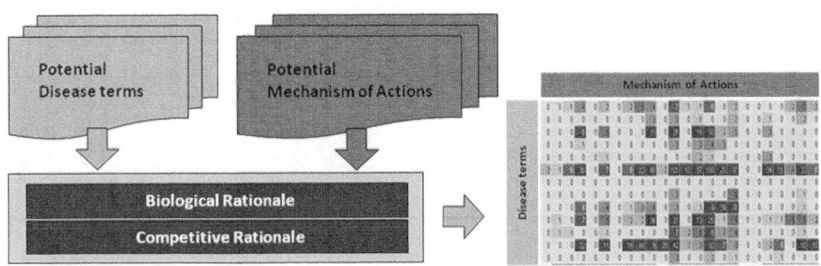

Figure 14.5 Drug repurposing matrix. Schematic representing the approach to search scientific literature and combine with competitive intelligence for over 60 000 potential combinations identifying hot spots to investigate further

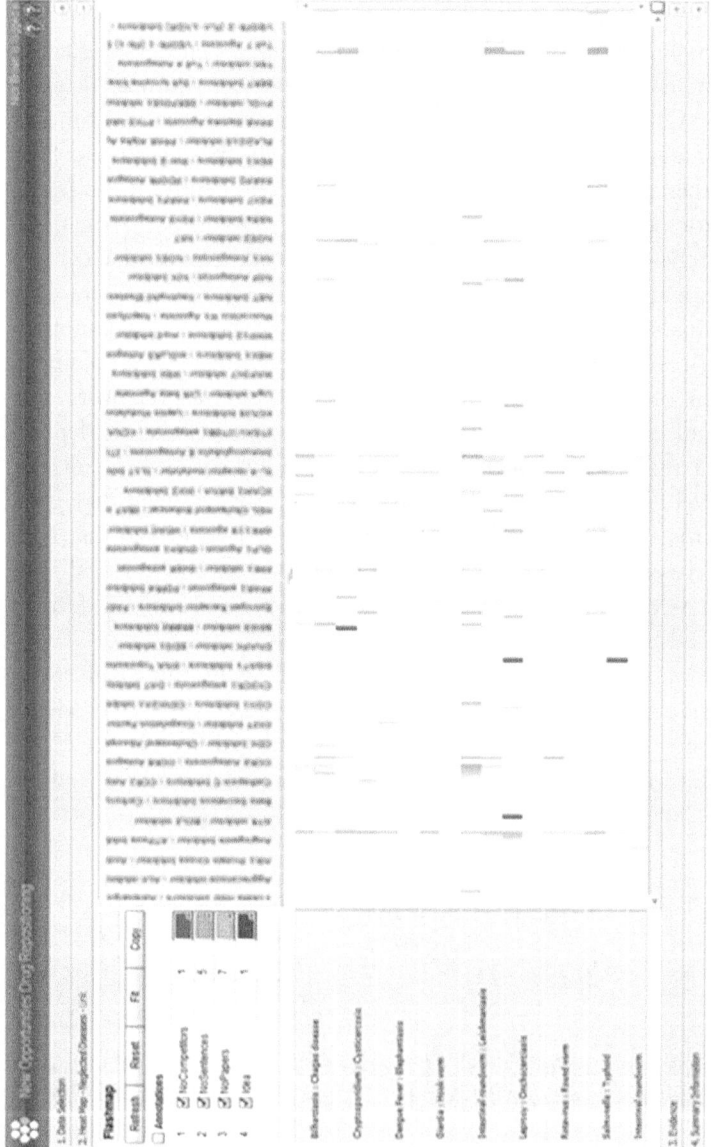

Figure 14.6 Early snapshot of our drug-repositioning system. Disease indications (y-axis) and drug mechanisms (x-axis) are plotted against each other. In the application, different colours reflect the information identified – such as competitive intelligence and scientific rationale at sentence or abstract level

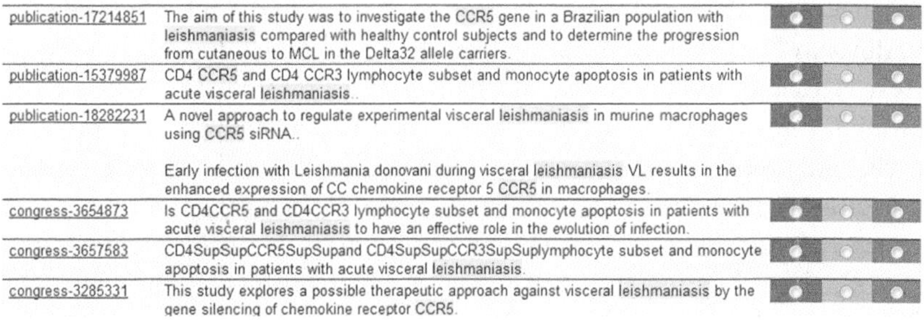

publication-17214851	The aim of this study was to investigate the CCR5 gene in a Brazilian population with leishmaniasis compared with healthy control subjects and to determine the progression from cutaneous to MCL in the Delta32 allele carriers.	
publication-15379987	CD4 CCR5 and CD4 CCR3 lymphocyte subset and monocyte apoptosis in patients with acute visceral leishmaniasis.	
publication-18282231	A novel approach to regulate experimental visceral leishmaniasis in murine macrophages using CCR5 siRNA..	
	Early infection with Leishmania donovani during visceral leishmaniasis VL results in the enhanced expression of CC chemokine receptor 5 CCR5 in macrophages.	
congress-3654873	Is CD4CCR5 and CD4CCR3 lymphocyte subset and monocyte apoptosis in patients with acute visceral leishmaniasis to have an effective role in the evolution of infection.	
congress-3657583	CD4SupSupCCR5SupSupand CD4SupSupCCR3SupSuplymphocyte subset and monocyte apoptosis in patients with acute visceral leishmaniasis.	
congress-3285331	This study explores a possible therapeutic approach against visceral leishmaniasis by the gene silencing of chemokine receptor CCR5.	

Figure 14.7 Article-level information. Here the disease and mechanism terms are highlighted as well as any synonyms on record

called for each of the different interactions that could be performed. The main visualisation was built using a commercial platform – Adobe Flex. This technology was already in use in AstraZeneca and had been proven to handle over 1 million data points yet still be extremely responsive.

In addition, many more dimensions can be incorporated into this view, including target efficacy, safety liabilities, patent position, internal knowledge and tissue expression. SOLR is flexible enough to handle many different types of data and could be leveraged to accommodate this, whereas the Flex visualisation has already been tested and can handle tens of dimensions with millions of rows of data.

Before this approach, our re-positioning opportunities would be opportunistic and not systematic. This technology enables an impressive summary of a huge volume of complex data and the generation of mechanism-disease landscapes. Without SOLR and Flex, this would be possible, but SOLR enables a rapid up-to-date system to be repeated weekly and Flex provides sufficient speed even when faced with large data numbers or statistical measures.

14.7.3 Arise – biological process maps of how AstraZeneca drugs work

Another approach taken to professionalise our drug repositioning efforts was to develop internal knowledge of understanding around how the company's drugs work at the cell and tissue level.

Successful repositioning in the past has been through one of two routes: (1) serendipity or (2) biological understanding. Using the classic and well-known repositioning example of sildenafil, it was originally designed for the treatment of angina. Unexpected and serendipitous findings in Phase I clinical trials meant that it was repositioned in 1998 to treat erectile dysfunction and sildenafil sold as Viagra had peak sales of around $2B by 2008. This drug works as a PDE5 inhibitor and protects cyclic guanosine monophosphate (cGMP), leading to smooth muscle relaxation, causing vasodilation that increases the inflow of blood into the spongy, penile tissue [8]. By understanding how this drug works at the biological level of the cells and tissue, it was then repositioned again in 2005 into pulmonary arterial hypertension (PAH). This rare lung disorder affects patients whose arteries that carry blood from the heart to the lungs become narrowed, making it difficult for blood to flow. This was launched as Revatio (sildenafil) as a lower dose, which by relaxing the smooth muscle leads to dilation of the narrowed pulmonary arteries [9].

When new indications are considered for our AstraZeneca compounds, having a detailed understanding about how each of our drugs works is critical. As a large pharmaceutical company, with over 12 000 R&D scientists that contribute to the collective knowledge about our drugs in all of our existing and historical projects, having this in a single repository would be incredibly valuable. Although corporate, global data repositories do hold millions of documents that describe every finding about all of our compounds, this is held in unstructured texts, project minutes, drug filings, product labels and even marketing material. Unfortunately the rich, invaluable and often tacit knowledge about how these drugs work is not stored systematically in a structured format, but in our brains! Luckily, our project scientists are always keen to discuss their compounds and how the biological function can be used in new potential indications.

To capture this knowledge once and for all, a project was initiated to describe how our compounds and drugs work at the biological level. Historically our focus has been at the signalling pathway level, with gene targets, trafficking and receptor interactions. This project tried to understand what the drug did to the cells, the tissues and the pharmacology at the pathophysiological level. Although run as an informatics project to store this information in new ontologies and visualise the results in new biological process pathway maps, the information was gathered by coordinating hundreds of interviews across the company. Similar issues within other major companies are discussed in this book by Harland and colleagues (Chapter 17) and by Alquier (Chapter 16).

The technology used to build these visual network maps was Prefuse, which easily lends itself to network objects and attaching relationships. Rather than a static network diagram, these biological process maps (Figure 14.8) have been designed to be built within the

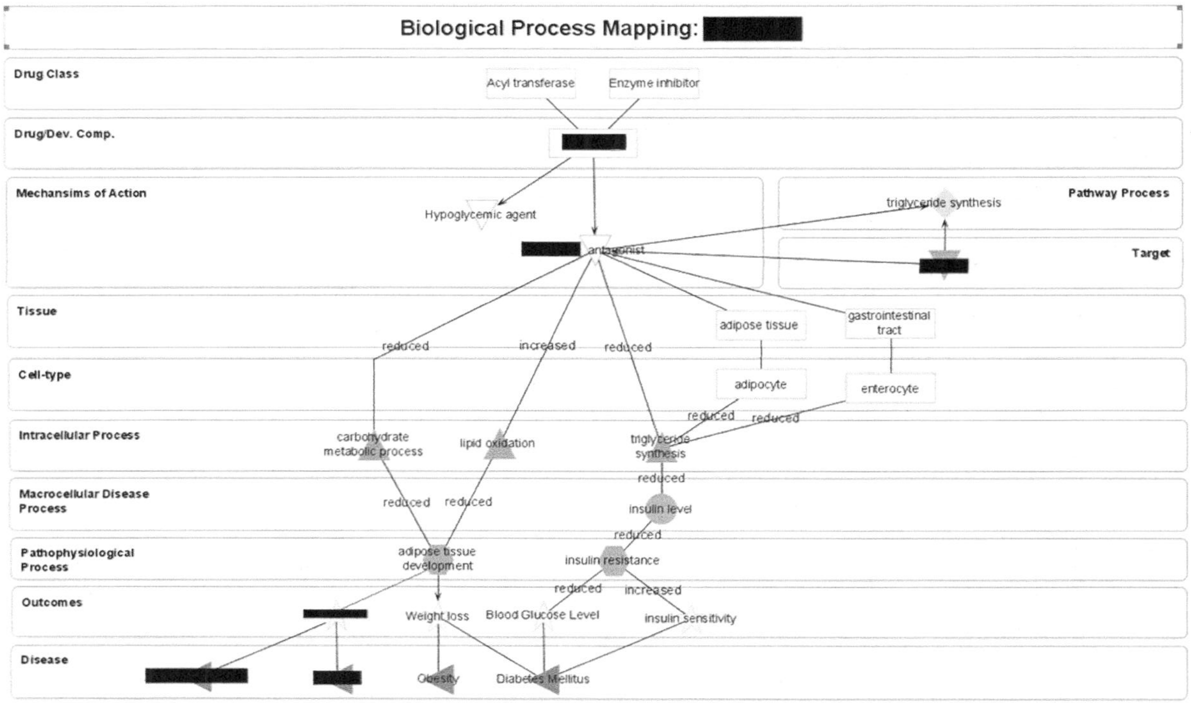

Figure 14.8 An example visual biological process map describing how our drugs work at the level of the cell and tissue. This approach can be used to record not only the scientific rationale for projects and drugs, but also potential indications to consider through the understanding of disease and drug

application and our annotators scribe the information directly into the system now.

In addition to capturing the knowledge about a drug, any new ideas scientists have can be captured and overlaid by processes that are connected to other diseases or drugs. This work is usually done by an informatician working with the scientists and building the maps in the system together.

Approaches such as this have not been initiated before because to attempt this approach without robust visualisation technology would be extremely difficult, and it would be virtually impossible to develop the interactivity required in alternative technologies such as mind mapping or presentation software. In truth, the underlying Prefuse technology has now been applied to many different areas across AstraZeneca because it is simple to adopt, easy to customise and intuitive for end-users to use. For example, this has been further developed into a business-led, multifaceted system for computational biology networks with vast sets of these networks together.

14.7.4 Atlas Of Science – searching for the missing links

Atlas Of Science is a software tool written in Java and based on the Prefuse toolkit. It provides a simple and interactive mechanism for searching and dynamically visualising information contained within large sets of documents. The original version of Atlas Of Science was intended for the one master set of publications held in the SOLR Index (described elsewhere in this chapter), but it can also be used to look at any set or subsets of documents from other projects, using either a Lucene index or a SOLR index.

The user enters into a document search either a direct SOLR query (e.g. all documents containing the work 'cataract'), those matching a list of key words (e.g. those that have previously marked as talking about 'pancreatic cancer'; documents written by a particular author), or a combination of a number of these criteria. The numbers of documents that match the query are displayed, along with the top key words found in those documents as seen in Figure 14.9.

The system allows the user to visualise these key words in a number of ways:

- key word lists. The key word list panel shows the key words that have been found across your result set. They can be ordered by total number

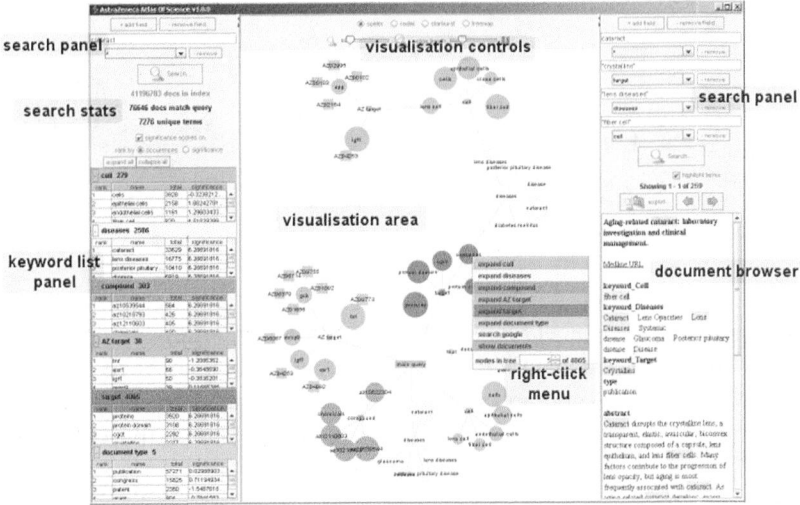

Figure 14.9 A screenshot of the Atlas Of Science system. Key elements are the ability to search (top left), identify individual facets (bottom left), main interaction window (centre), query terms (top right) and converted XML results (bottom right). This approach allows users to surf over the information rapidly to find new trends and themes

found or by significance. The significance calculation gives extra weighting to terms that appear more often with the main search query than without;

- hierarchical visualisations. There are three different hierarchical views, force, radial and starburst as shown in Figure 14.10. These all show a tree that represents the key words in the current result set. Each one uses a different layout algorithm for the tree. The centre of the tree represents the main search that you have typed into the search panel. A sub group is added to the tree for each category of key words, and the top key words shown for each category are shown. The size of each node is weighted to represent the total or significance score relative to the rest of the nodes in its category.

Atlas Of Science is also highly interactive. At any time it is possible to view the documents associated with the data in a search just by right-clicking on the relevant part of the key word list or visualisation.

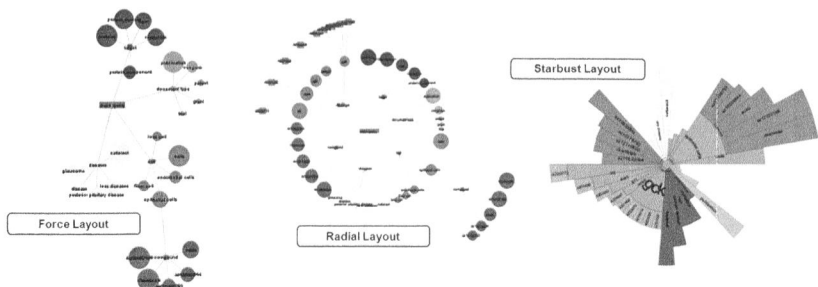

Figure 14.10 Typical representation of three layout approaches that are built into Prefuse. Force layout uses force repulsion to spread the nodes effectively, radial layout pushes the nodes into a series of circles and starburst layout builds a space-filled radial tree

Alternatively, it is possible refine the search to include a key word by clicking on it, or to do subsearches within the visualisation by 'spidering out' searches.

For scientists, this allows them to surf across the enormous complexity of data held in documents and find new facts or trends without reading individual articles. It also means that instead of searching for positive hits in a standard Pubmed search, this approach lets users find verifiable new connections to investigate, something that is incredibly difficult to do when you have a lot of information to sift through.

14.8 Other applications

A number of other applications have been developed to make use of the SOLR Publication Management Index. These include:

- 1Alert. This system allows scientists to quickly set up a regular email or RSS feed to show new developments in a given area. For example, the system can provide a weekly email with a list of all the new publications, conference proceedings and clinical trials pertaining to a particular disease or disease area. In addition, more complicated search techniques can be applied across the whole of the index to enable the scientists to be informed of significant events in the publications. 1Alert makes use of the advanced tagging utilised within the SOLR Publication Management Index to improve the document feedback produced for this alerting;

- My Publications. In an organisation the size of AstraZeneca, it is difficult to track what experience the employees have, especially when that experience was gained before the employee joined the company. For scientists, one way of tackling this problem is to have a view of what publications these people have authored during their career.

However, even viewing this is difficult as, very often, the author's name is not unique (e.g. try searching for 'N. Brown'). One of the first searches most scientists perform when they come across the SOLR Publications Management Index, is for their own publications. A small system was therefore developed known as My Publications that allows users to quickly perform this search and then to tag the documents with their unique AZ identifier. It is then possible for other people within AZ to search for publications that are authored by AZ employees and to discover their areas of expertise based on the areas in which they have published.

This concept is to be extended to include internal documents that allow users to quickly identify other people within the company who have been (or are) working in a particular area in which they are interested. It is hoped that this will encourage a culture of interaction and collaboration that helps stimulate innovation.

14.9 Challenges and future developments

The SOLR Publications Management system has proved to be a robust and reliable system. However, if it is to be made available to the wider AstraZeneca community, the following enhancements need to be developed.

- The system should be made more robust. It may be possible to use the new SOLR Cloud development to provide this [10]. This will allow data replication across multiple servers, and failover facilities. In addition, it is desirable to implement separate index and search servers, to improve the amount of data that can be indexed, and the frequency at which the index can be refreshed. This is especially true if internal content is integrated with up to 200+ million documents.

- A system management facility should be implemented. Our current systems check to see if the servers are still running, and re-start them if necessary. However, this is still a mainly manual process, and the search for third-party tools to support this (e.g. LucidWorks [11]) is desirable.

- A search security model should be implemented. At the moment every user can search every document. However, in the future, the intent is to implement a search strategy that restricts the results returned depending on whether the user is allowed to access the documents, This will be required especially if the system is to be used to index internal documents, which may, for instance, include HR or restricted project documents.

In addition, the following enhancements to the existing systems are being considered.

- Allowing users to provide feedback/annotations on individual documents. This would require the SOLR Index to be integrated with a database to hold the annotations, so that subsequent updates to the documents from the original source did not result in the loss of the annotations.
- The standardising of the pipeline, utilising other tools such as Nutch [12] (to crawl file and web infrastructures) and Tika (to extract text and meta-data from a wide range of document types), to extend the reach of the documents indexed.

In addition, more uses of a single, enhanced data source are being discovered on an almost daily basis, meaning that new simple business applications are being developed by various informaticians and developers within AZ.

14.10 Reflections

Overall, our experience of open source technologies as applied to solve business problems has been extremely positive. Even after recent comparisons to commercial software, SOLR still appears highly competitive and truly optimised for faceted search. Prefuse continues to grow and, with the advent of Flare, has made steps to wider adoption with improved visualisation performance.

If we were to do this again, we would invest in both SOLR and Prefuse/ Flare, with a capability to be able to build business applications over the scientific content. Looking back, it appears that we have several different technologies in the mix and in hindsight, much of what we have developed could probably be designed through using just open source approaches.

One area that we would change would be to have a closer connection to the open source community. Currently the relationship is very much one-directional and although we consume open source technology and focus on the must-win problems that immediately affect our business, we would hope to work more closely with the open source community and of course make any improvements to any code publically available.

14.11 Thanks and Acknowledgements

The work described here has been completed over the course of three years. There are many people that have contributed to this and we can't hope to thank everyone. We would especially like to thank the following people.

Mark Robertson, without his scientific acumen we might still be building IT systems for the sake of IT, Chris Duckett and Graham Cox (Tessella) for their patience and great coding skills, Caroline Hellawell for her continued support and biological understanding, Jamie Kincaid (Accelrys) for being able to rationalise and interpret our ideas, Ian Dix for sponsoring our activities through Knowledge Engineering activities in AstraZeneca, John Fish for believing in us and providing us the original funding that got this off the ground in the first place, and last but certainly not least, Ruben Jimenez (OpenQ) and Otavio Freire (OpenQ) who originally introduced us to SOLR three years ago. Little did we realise how important that technology choice would prove to be in making any of this possible.

14.12 References

[1]   SOLR. Lucene.apache.org at *http://lucene.apache.org/solr/*
[2]   Apache Tika. Tika.apache.org at *http://tika.apache.org/*
[3]   'xml.etree.ElementTree' library at *http://docs.python.org/library/xml.etree.elementtree.html*
[4]   Schuemie MJ, et al. *Peregrine: lightweight gene name normalization by dictionary lookup*. Proceedings of the Biocreative 2 workshop. Madrid 2007.
[5]   Prefuse at *http://www.prefuse.org*.
[6]   Medical Subject Headings (MeSH). nlm.nih.gov at *http://www.nlm.nih.gov/mesh/*
[7]   Pipeline Pilot. Accelrys at *http://accelrys.com/*
[8]   Langtry HD, Markham A. Sildenafil: A review of its use in erectile dysfunction. *Drugs* 1999;57(6):967–89.

[9] Murray F, Maclean MR, Insel PA. Role of phosphodiesterases in adult-onset pulmonary arterial hypertension. *Handbook Experimental Pharmacology* 2011;204:279–305.

[10] SolrCloud: wiki.apache.org at *http://wiki.apache.org/solr/SolrCloud*

[11] Lucidworks: Lucid Imagination at *http://www.lucidimagination.com/products/lucidworks-search-platform/enterprise*

[12] Nutch: at *http://nutch.apache.org/*

## 15

Utopia Documents: transforming how industrial scientists interact with the scientific literature

Steve Pettifer, Terri Attwood, James Marsh and Dave Thorne

Abstract: The number of articles published in the scientific literature continues to grow at an alarming rate, making it increasingly difficult for researchers to find key facts. Although electronic publications and libraries make it relatively easy to access articles, working out which ones are important, how their content relates to the growing number of online databases and resources, and making this information easily available to colleagues remains challenging. We present Utopia Documents, a visualization tool that provides a fresh perspective on the scientific literature.

Key words: semantic publishing; visualization; PDF; collaborative systems.

It is a curious fact that, throughout history, man's capacity to create knowledge has far outstripped his ability to disseminate, assimilate, and exploit that knowledge effectively. Until recently, the bottleneck was very much a physical one, with hand-written scrolls, and then printed manuscripts and books, being deposited in libraries to which access was limited, either by politics, social factors, or simply the need to physically travel to a particular location in order to access the content. Digital technologies have not solved the problem – and, it could be argued, have in some ways made matters worse.

Although it is no longer necessary to visit a library in person to read a scholarly work, so much material now exists that making sense of what is relevant is becoming increasingly difficult. In the life sciences alone, a new article is published every two minutes [1], and the number of articles has been growing exponentially at around 4% per year for the past 20 years [2]. In 2011, for example, over 14 000 papers were published relating in some way to Human Immunodeficiency Virus (admittedly, this is a much broader topic than any one researcher would care to explore in detail; nevertheless, it gives some indication of the scale of the problem) [3]. If we add to the peer-reviewed literature all the information that is deposited in databases, or written informally on personal blogs, matters get more complex still. The problem has been characterized, at a broad level, as being one of 'filter failure' [4], that is it's not that there is too much information per se – as, surely, lots of information has to be a good thing, and there has always been more available than we are able to consume – but, rather, that we are lacking mechanisms to convert information into meaningful knowledge, to find what is relevant to our particular needs, and to make connections between the disjoint data.

The problem is particularly acute in the life sciences. As pharmaceutical companies race to find new drugs, it has become clear that mastery of the literature is at least as important as mastery of the test tube or the sequencing machine. It has been documented that some failures of pharmaceutical company projects – which are costly on a mind-boggling scale, and typically result in failure – could have been avoided if scientists within the company had simply been able to find the relevant scientific article beforehand [5]. This inability to find the 'right article' is not for the want of trying: pharmaceutical companies employ large teams of scientists to scour the literature for just such information; it is simply that, even with the best tools and the most proficient scientists, there is more information 'out there' than can currently be processed in a timely way.

The danger of characterizing this as simple 'filter failure' is that it makes the solution sound trivial: get a better 'filter', and the problem goes away. The difficulty, of course, is that no filter yet exists that is sophisticated enough to cope with the complexities of life science data and the way in which it is described in 'human-readable' form. The simple key word searches or faceted browsing techniques that work so well for buying consumer goods or for booking a holiday online, break down quickly when faced with complex, context-sensitive, inter-related, incomplete and often ambiguous scientific data and articles. In many cases information gleaned from articles is already known in some part of an organization, but its relevance to some other new project is not discovered until after the fact. For organizations such as pharmaceutical companies, as the volume of information grows, the problem of 'knowing what we already know' is set to become worse [6].

At the heart of this problem lies a simple truth. We are now completely reliant on machines to help process the knowledge we generate; but humans and machines speak radically different languages. To perform efficiently, machines require formal, unambiguous and rigorous descriptions of concepts, captured in machine-readable form. Humans, on the other hand, need the freedom to write 'natural language', with all its associated ambiguities, nuances and subtleties, in order to set down and disseminate ideas. The trouble is, other than trained 'ontologists', mathematicians, and logicians, very few scientists can turn complex thoughts into the kind of formal, mathematical representations that can be manipulated computationally; and very few ontologists, mathematicians, and logicians understand biology and chemistry in enough detail to capture leading-edge thinking in the pharmaceutical or life sciences. For the time being, at least, we are faced with the challenge of bridging this divide.

Text- and data-mining techniques have made some progress here, attempting to extract meaningful assertions from written prose, and to capture these in machine-readable form (e.g. [7]). But this is far from being a solved problem – the vagaries of natural-language processing, coupled with the complex relationships between the terminology used in the life sciences and records in biomedical databases make this a non-trivial task.

In recent years, in what we consider to be a frustrated and frustrating distraction from the real issues, many have pointed the finger of blame at the file format used by publishers to distribute scientific articles: Adobe's PDF. It has variously been described as 'an insult to science', 'antithetical to the spirit of the web', and like 'building a telephone and then using it to transmit Morse Code', as though somehow the format itself is responsible for preventing machines from accessing its content [8]. Although it is true that extracting text from PDFs is a slightly more unwieldy process than it is from some other formats, such as XHTML, to accuse the PDF as the culprit is to entirely miss the point: the real problem arises from the gulf between human- and machine-readable language, not merely from the file-format in which this natural language is being stored. It is worth setting the record straight on two PDF-related facts.

1. Although originating in a commercial environment, PDF is now – and has been for some considerable time – an open format. The specifications of the PDF have been publically available for many years, and have been formally 'public domain' since 2008, when Adobe released the format as ISO 32000-1 and officially waived any patents relating to creating or consuming PDF content [9]. It is, however, undoubtedly a more verbose, and arguably less-elegant format than more recent offerings.

2. The PDF has the potential to be as 'semantically rich' as any of the other alternatives (e.g. eBook, XHTML), and, since version 1.4 in 2001, has had the facility to include arbitrarily sophisticated meta-data. The fact that virtually no publishers use this feature is not a fault of the PDF, and such semantic richness is just as conspicuously absent from the other formats used to publish scientific material. Speculation as to which link in the chain between author, publisher, and consumer causes this omission, we leave entirely to the reader.

Regardless of whether the PDF is a suitable vehicle for disseminating scientific knowledge, there are two irrefutable facts: first, it is by far the most popular medium by which scientific articles are consumed (accounting for over 80% of downloaded articles); and second, even if publishers move to a more 'web friendly' format at some point in the future, vast numbers of legacy articles will remain only as PDFs (many of which are digital scans of back-catalogues that pre-date even this ancient format, some of which have had text extracted via optical character recognition, while many exist only as images) – at the very least, these would require converting into something more 'semantic'. In spite of significant effort by publishers to enhance their online/HTML content with additional links, meta-data, data and multimedia extensions, the PDF stubbornly remains the format of choice for most readers. The recent growth in modern tools for managing collections of PDFs (e.g. Mendeley [10] and Papers [11]) gives additional backing to the view that the PDF is not in any danger of disappearing yet.

In an attempt to bridge the divide between what scientists actually read, and what computers need to support the process of taming the scientific literature, we embarked on a project to build a new tool for linking the content of articles with research data. The resulting software – Utopia Documents [12, 13] – combines all the advantages of the PDF with the interactivity and interconnectedness of a blog or web page. Acting as an alternative desktop application to Adobe Acrobat and other

PDF viewers, Utopia Documents forms two-way links between the traditional human-readable PDF and any available semantic models produced by publishers or the community, and public or in-house data sources. The act of reading an article triggers heuristics that attempt to automatically identify features within the PDF, which can then be semantically annotated, and these annotations published to augment the original article.

The software 'reads' a PDF much like a human does, recognizing document content and features (title, authors, key words, sections, figures, etc.) by their typographical layout, and ignoring less-important artifacts and non-document content (such as watermarks and additional information in the header or margins). Having gleaned this information from the PDF, Utopia Documents is then able to automatically link citations and references to online repositories.

As well as creating inbound and outbound links, Utopia Documents incorporates a range of interactive features and visualizations that effectively bring static PDF files to life. For example, the software can turn tables of data into live spreadsheets (Figure 15.1); it can generate charts – already linked to the source data – on the fly; it can turn objects and images (e.g. protein sequences and 3D structures) into live, interactive views; and it can dynamically include data- and text-mining results through customization with appropriate 'plug-ins'. It also allows readers to add comments to articles, which can then be shared (and commented on) with other users instantaneously, without the need to redistribute the PDF file itself. The system currently links to several major life science and bioinformatics databases, as well as to more general online resources, such as Wikipedia. By means of its plug-in architecture, Utopia Documents can be integrated with any web service-accessible resource, making it easily customizable and extensible. Although it is possible for journal publishers to enhance PDFs with content that Utopia can recognize (noted below), it is important to state that the majority of Utopia's features work on all modern scientific PDFs, and are not limited to only a few specialized documents.

15.1 Utopia Documents in industry

In a commercial context, Utopia Documents aims to provide a new way to enable 'joined-up thinking' within institutions, reducing the cost of relating knowledge from within scientific articles to existing

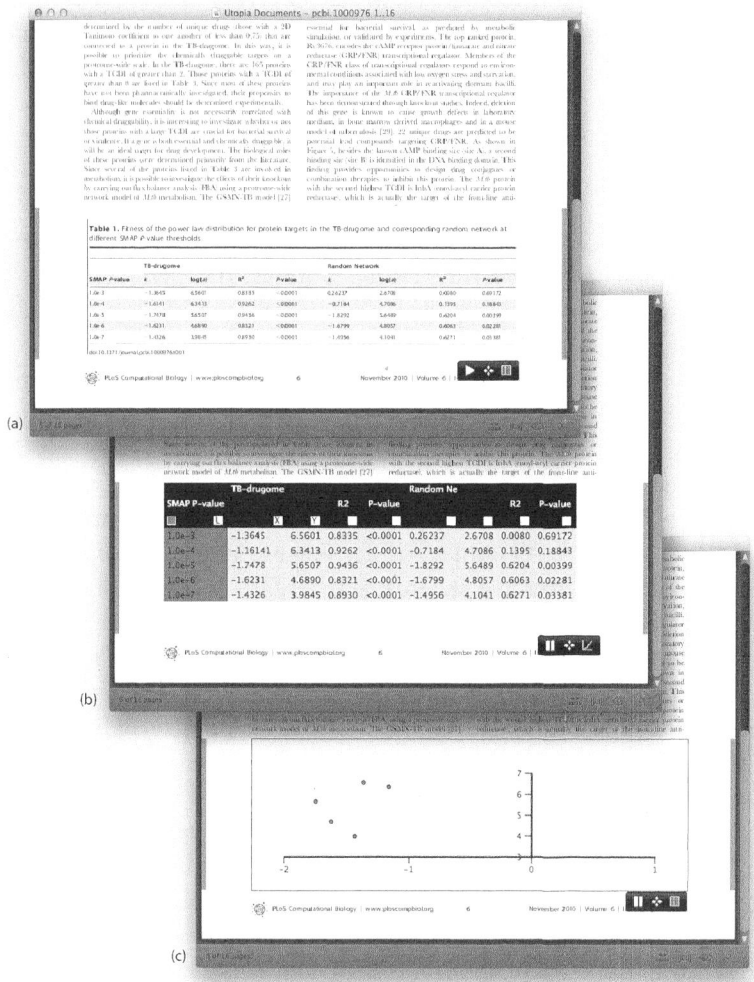

Figure 15.1 A static PDF table (a) is converted first into an editable 'spreadsheet' (b), and then into a dynamic scatterplot (c)

in-house and public data, and at the same time supporting auditable interaction with the literature. This results in simplifying the 'due diligence' processes essential in modern drug discovery. Other 'knowledge management' tools typically provide web-based interfaces that allow scientists to explore their content, but treat the scientific article itself as a distinct and separate entity, somewhat divorced from the discovery process. The consequence of this approach is that scientists are required to repeatedly switch from one application to another while pursuing a line of thinking that requires access both to the literature and to data. Often the situation is worse, and they are forced to print documents out, to scribble notes in the margin and thereby remove the potential for interconnected data completely. Not only does the act of consciously switching require a certain amount of mental inertia (leading inevitably to, 'Oh, I'll look that up next time when I'm using Application X'), it interrupts the workflow and thought processes associated with reading the literature. In turn, this makes documenting associations between concepts found in articles and data recorded in other systems difficult and fragmented.

Consider a typical industry use-case: a scientist is reading a paper describing a number of proteins and/or small molecules relevant to a disease or phenotype. Some of these might be new to the reader, who may wish to know a little more about them in order to better understand the findings in the paper. Traditionally, manual copy/pasting of protein and drug names into numerous search engines is the only solution. For compounds, this can also require tedious manual sketching of the molecule into a structure editor. The difficulty in performing what should be very simple tasks often results in either lost time or simply abandoning the enquiry.

Now, consider these same scenarios using Utopia Documents. Simply selecting the protein name using the cursor brings up a rich information panel about that entity (Figure 15.2). Alternatively, text-mining services can be invoked directly from within the Utopia environment to automatically identify interesting entities and relationships, and to link these to in-house data. The result is a highly visual and organized index of the genes, compounds, and diseases mentioned in the text. For compounds, Utopia goes even further, as Figure 15.3 shows. Here, the software has identified a number of small molecules mentioned only by name in the text of the paper. Using (in this example) the ChemSpider [14] API, the actual structures have been retrieved, along with other information, providing a completely new chemical index browser to the paper. Functionality developed as part of the EU Open PHACTS project [15] also allows the retrieval of activity data from public and corporate structure–activity relationship databases.

Finally, because of Utopia's unique software architecture, many other types of plug-in can be developed, including the ability to interact with tables and to embed applications, such as 3D protein structure viewers, directly in the document. Utopia provides a two-way API, allowing web services to retrieve structured content from the PDF ('give me all tables

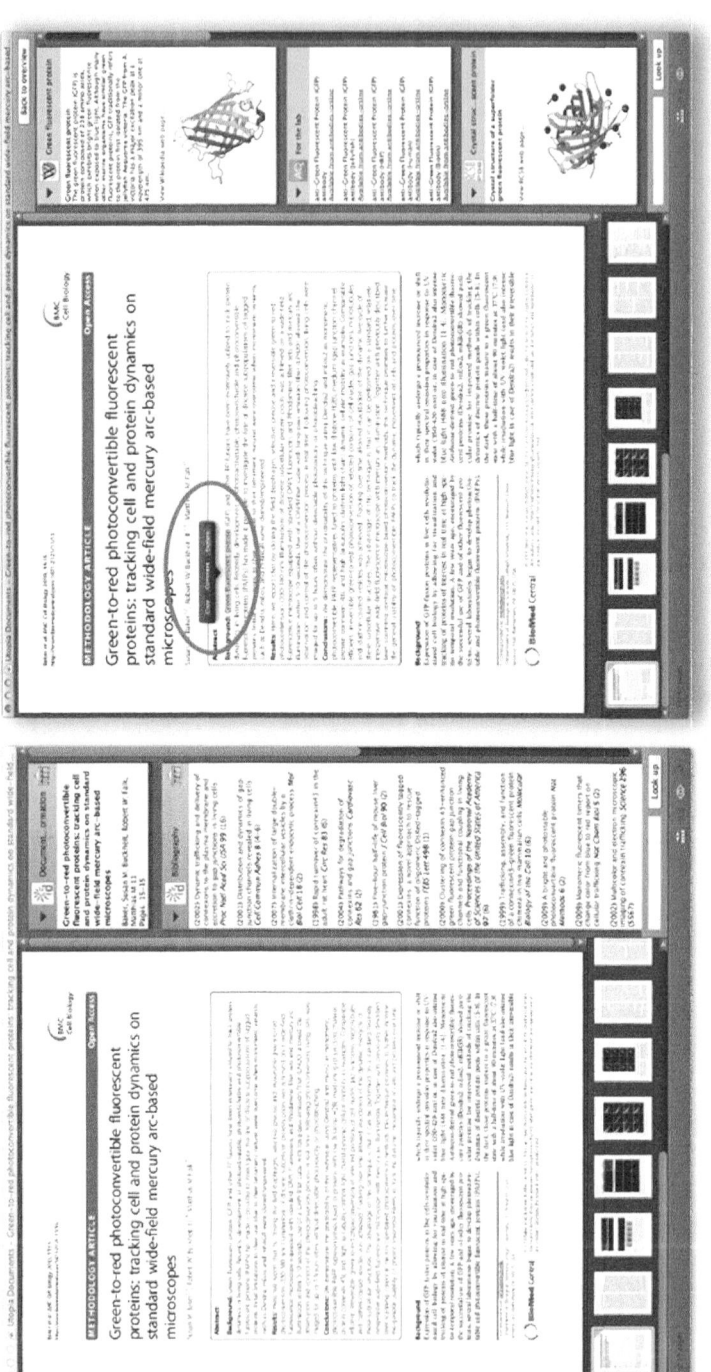

Figure 15.2 In (a), Utopia Documents shows meta-data relating to the article currently being read; in (b), details of a specific term are displayed, harvested in real time from online data sources

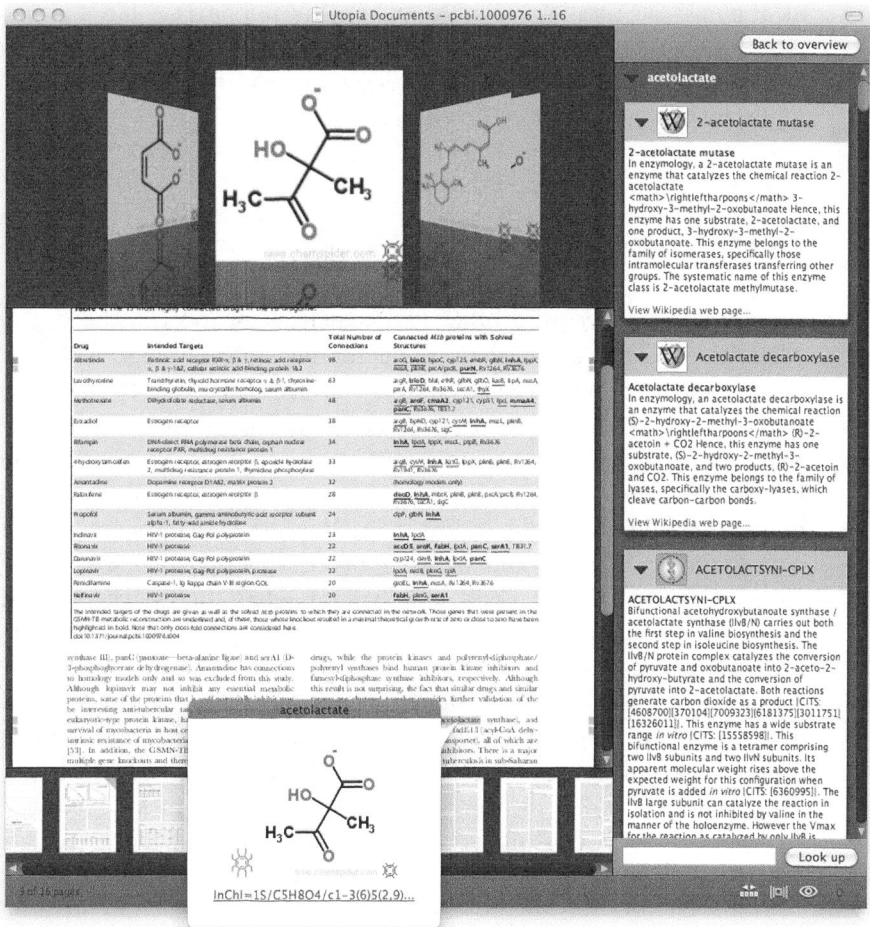

Figure 15.3 A text-mining algorithm has identified chemical entities in the article being read, details of which are displayed in the sidebar and top 'flow browser'

from the document') and inject content into the PDF ('show these data when the user highlights a compound'). This provides many opportunities to interconnect with internal systems within a company. For instance, highlighting a protein might call for a look-up of the internal project-management system to list which groups are actively researching this target for relevant indications. Similarly, highlighting a compound might invoke a structure-search of the corporate small-molecule database, identifying structurally similar compounds in the internal file. By these means, Utopia connects the content of the PDF into a company's systems, finally joining one of the most important pieces of the scientific process to the existing information infrastructure.

15.2 Enabling collaboration

Utopia stands apart from other PDF-management tools in the way in which it promotes collaboration and sharing of knowledge within an institution. Most life science industry projects are run by dynamic, ever-changing teams of scientists, with new members being introduced as specialist expertise is required, and scientists typically contributing to many projects at any given time. A researcher faces a constant battle to stay on top of the literature in each project, determining which papers should be read and in what order, and exchanging thoughts with colleagues on the good and bad of what they have digested. Often the solution to both these needs is ad hoc email, reading papers suggested by colleagues and replying with relevant critique. Of course, this relies on the researcher being copied into the email thread and, assuming that this has happened, on finding the email in an ever-growing inbox (in all probability, emailing a published article to a colleague, even within the same organization, is likely to be a technical violation of the license under which the original reader obtained the article).

Utopia changes this by providing a team-based approach to literature management. Unlike other document-management tools, this enables collaboration not just at the level of a paper, but rather, at the level of a paper's content. Individuals can manage their own paper collections in a familiar and intuitive interface; however, by simply creating team folders and sharing these with colleagues, interesting papers and their related data and annotations can be circulated within relevant groups, without the need to remember to send that email and get everyone's name right. Simple icons inform team members of new papers in the collection (Figures 15.4 and 15.5). As well as comments, annotations on articles can include data-files (e.g. where the group has re-interpreted a large data set), or links to data held in knowledge-management systems. Thus, anyone reading the paper will see that there is additional information from inside the company that might make a real difference to their interpretation.

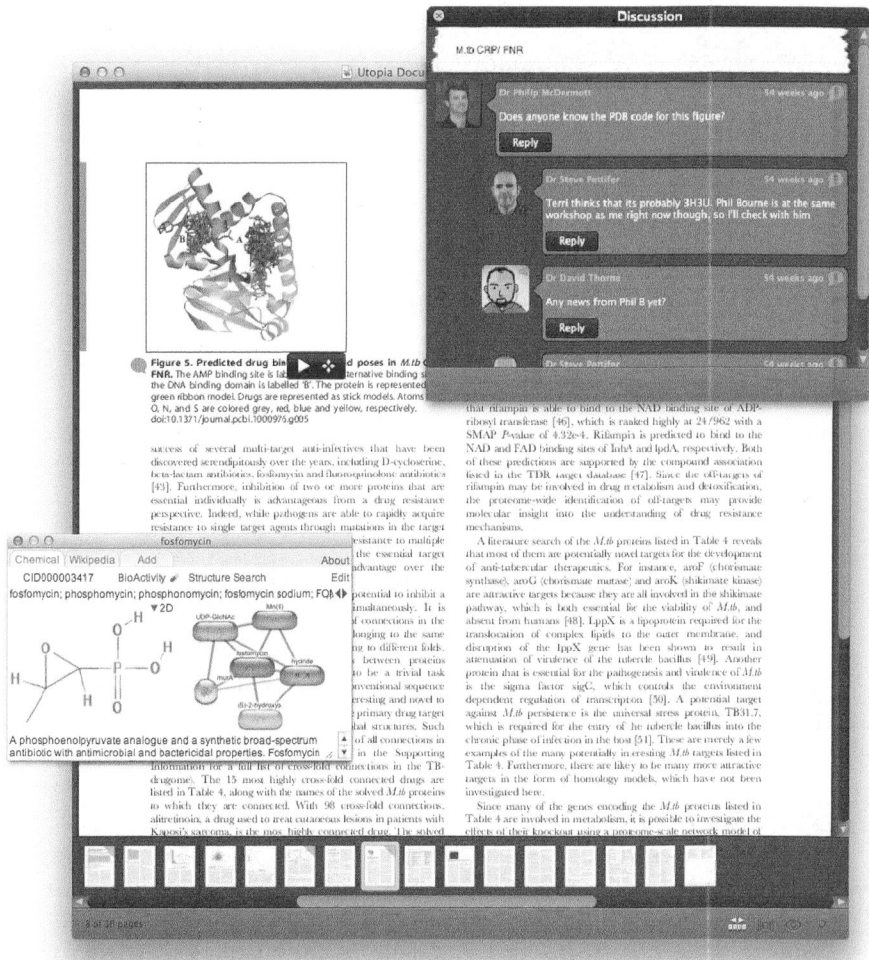

Figure 15.4 Comments added to an article can be shared with other users, without the need to share a specific copy of the PDF

15.3 Sharing, while playing by the rules

Unlike other tools, Utopia does not rely on users exchanging specific, annotated copies of PDF articles, or on the storage of PDFs in bespoke institutional document repositories (all of which raise legal questions around the storage and distribution of copyrighted PDFs and the breach

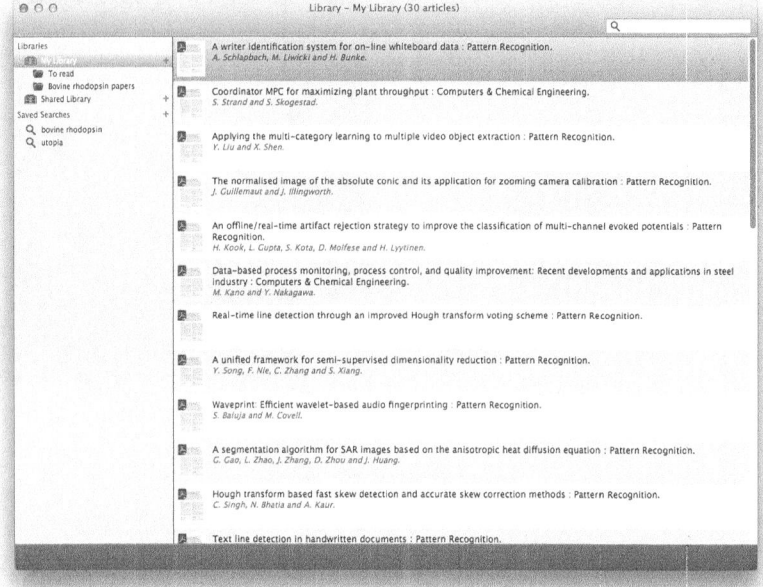

Figure 15.5 Utopia Library provides a mechanism for managing collections of articles

of publishers' licenses). When a user opens a PDF file in Utopia, a small semantic fingerprint is created that uniquely identifies that paper to the system. Any annotations or attachments are associated with this fingerprint and not with the PDF file itself. This has significant consequences for sharing and copyright within corporate environments.

- If a scientist on the other side of the world goes to the journal web site and downloads a completely new PDF of a specific article, it has the same fingerprint as any other copy of that article (even if the PDF is watermarked by the digital delivery system to include provenance information). When opened in Utopia, this fingerprint is recognized, and all enhancements and company annotations are retrieved and overlaid on the article in real time. This means that colleagues who may not even know that this paper was of interest to someone in the company are able to exploit the tacit knowledge of the organization that would otherwise be lost.

- A corporate PDF repository is not required. When user A posts a paper to the shared folder, user B sees the title, citation and so on, but does not get a physical copy of the paper until they decide to read that article. Then, a new copy of the PDF is automatically downloaded via the company's digital delivery service, allowing copyright to be respected and usage/access metrics to be accurately maintained.

15.4 History and future of Utopia Documents

Utopia Documents was originally developed in collaboration with Portland Press Ltd, who have been using the software since December 2009 to enhance the content of their flagship publication, the *Biochemical Journal* [16]. A publishers' version of the software allows the journal's editorial team to 'endorse' annotations, confirming the validity of specific associations between terms in articles and particular database entries. This has allowed thorough but rapid markup (typically, 10–15 minutes per article, depending on its suitability for annotation) as part of the normal publishing process. To date, around 1000 articles have been annotated in this way as part of the *Semantic Biochemical Journal* (*http:// www.biochemj.org*). More recently, the publications of the Royal Society of Chemistry [17] have been augmented by automatically generated annotations on the chemical compounds mentioned in articles, and a plug-in to highlight these in Utopia Documents is included in the latest releases (covering around 50 000 articles, at the time of writing). Other extensions include linking articles automatically to their Altmetric [18] statistics (giving an indication of the popularity of an article based on its citation in blogs, tweets, and other social media), and to online data repositories such as Dryad [19].

Recently, we have engaged in projects with pharmaceutical companies to install the system within an enterprise environment. Utopia consists of two main components, the PDF viewer installed on any user's PC, and a server, which holds the annotations, fingerprints, and other essential meta-data. Installation of this server behind the corporate firewall provides the security and privacy often required by commercial organizations. In addition, the system will then have access to secure in-house text-mining and data services. Ultimately, this allows commercial entities to use the software without fear of their activities being tracked by competitors, or their sensitive in-house knowledge 'leaking' into the public domain. We have explored two major approaches to this configuration: the traditional single installation on a company server, and a cloud-based approach. For the latter, a service provider hosts a secure, tailored instance of the server to which only users from a specific company have access. All of the advantages of an internal installation are preserved, but with much less overhead in maintaining the system. Our experience of the use of Utopia in industry has suggested that the team-based sharing is of greatest initial interest. Industry scientists urgently need tools to help them manage information more effectively, and to manage it as a team.

Deciding on an appropriate license for Utopia Documents has been an interesting and complex endeavor. With the vast majority of its funding so far having come from commercial bodies (primarily publishers and the pharmaceutical industry), rather than from research grants or the public purse, we do not feel the same ethical imperative to make the software open source as we might do had the development been funded through grants derived from public funds. Yet, as academics, we nevertheless find openness extremely attractive. The difficulty arises when the ideals of openness meet the reality of finding sustainable income streams that will allow us to continue the research and development necessary to maintain Utopia Documents as a useful and powerful tool for scientists, whether in commercial or academic environments. What has become clear is that very few of our end-users care about the detailed licensing of the software, as long as the tool itself is robust, reliable, freely available, and has some plausible longevity.

Although it is notionally possible to sustain software development and research on the back of 'consultancy' associated with open-source software, there appear to be relatively few concrete examples of where this has worked as a long-term solution for keeping a research team together. Our current model, therefore, is one in which the core software is – and will remain – free to use without restriction (in both academic and commercial settings), while we retain the rights to customize the tool for commercial environments, or to provide specific services to publishers. As academics, finding a model that both sustains our research and provides a freely available resource for the community and 'the good of science' has been challenging on many levels; we are nevertheless optimistic that our current approach provides a reasonable balance between long-term sustainability and free availability, and are excited about the potential of Utopia Documents as a means of making more of the scientific literature.

15.5 References

[1]   Attwood TK, Kell DB, Mcdermott P, Marsh J, Pettifer SR and Thorne D. Calling international rescue: knowledge lost in literature and data landslide! *Biochemical Journal*, Dec 2009.

[2] Lu Z. '*PubMed and beyond: a survey of web tools for searching bio-medical literature.*' Database 2011: baq036.

[3] Statistics generated Jan 2012 by searching *http://www.ncbi.nlm.nih.gov/pubmed/* for articles mentioning HIV, published in 2011.

[4] *http://blip.tv/web2expo/web-2-0-expo-ny-clay-shirky-shirky-com-it-s-not-information-overload-it-s-filter-failure-1283699*

[5] Korstanje C. Integrated assessment of preclinical data: shifting high attrition rates to earlier phase drug development. *Current Opinion in Investigational Drugs* 2003;4(5):519–21.

[6] Superti-Furga G, Wieland F and Cesareni G. 'Finally: The digital, democratic age of scientific abstracts.' *FEBS Letters* 2008;582(8):1169.

[7] Ananiadou S and McNaught J (Eds) *Text Mining for Biology and Biomedicine.* Artech House Books. ISBN 978-1-58053-984-5, 2006

[8] Pettifer S, McDermott P, Marsh J, Thorne D, Villéger A and Attwood TK. Ceci n'est pas un hamburger: modelling and representing the scholarly article. *Learned Publishing*, Jul 2011.

[9] International Organization for Standardization *Document management — Portable document format — Part 1*: PDF 1.7, ISO3200-1:2008.

[10] *www.mendeley.com*

[11] *http://www.mekentosj.com/papers*

[12] *www.utopiadocs.com*

[13] Attwood TK, Kell DB, Mcdermott P, Marsh J, Pettifer SR and Thorne D. Utopia Documents: linking scholarly literature with research data. *Bioinformatics* 2010;26:i540–6.

[14] *www.chemspider.com*

[15] *www.openphacts.org*

[16] *www.biochemj.org*

[17] *www.rsc.org*

[18] *www.altmetric.com*

[19] *www.datadryad.org*

Semantic MediaWiki in applied life science and industry: building an Enterprise Encyclopaedia

Laurent Alquier

Abstract: The popularity of Wikipedia makes it a very enticing model for capturing shared knowledge inside large distributed organisations. However, building an Enterprise Encyclopaedia comes with unique challenges of flexibility, data integration and usability. We propose an approach to overcome these challenges using Semantic MediaWiki, augmenting the mature knowledge sharing capabilities of a wiki with the structure of semantic web principles. We provide simple guidelines to control the scope of the wiki, simplify the users' experiences with capturing their knowledge, and integrate the solution within an enterprise environment. Finally, we illustrate the benefits of this solution with enterprise search and life sciences applications.

Key words: wikis; semantic web; user experience; usability; intranets; internal communication; collaboration; organisational memory; repositories; knowledge management; knowledge sharing.

16.1 Introduction

Silos are still very present in large decentralised organisations such as global pharmaceutical companies. The reasons they persist after decades of awareness are multiple: legacy systems too costly to change, cultural differences, political games, or simply because of disconnects between local and global needs. Far from improving knowledge sharing, omnipresent collaboration tools often turn into little more than fancy documents management systems. Knowledge remains hidden under piles of reports and spreadsheets.

As a result, users tend to spend a large amount of their time on trivial tasks such as finding a link to some team portal or looking up the meaning of an acronym. A 2009 survey found it took an average of 38 minutes to locate a document [1]. Overzealous search engines provide little relief: how useful are 2000 hits on documents including a team's name when all you are looking for is the home page of that team?

The majority of these tasks come down to questions such as 'where is what', 'who knows what', and the ever popular 'if we only knew what we knew'. In this context, the idea of an Enterprise Encyclopaedia becomes particularly attractive.

16.2 Wiki-based Enterprise Encyclopaedia

The goal of an Enterprise Encyclopaedia is to capture the shared knowledge from employees about a business: which entities are involved (systems, hardware, software, people, or organisations) and how they relate to one another (context).

Traditional knowledge management techniques have tried (and mostly failed) to provide solutions to this problem for decades, particularly in the collection, organisation and search of text and spreadsheet documents. In recent years, wikis have emerged as a practical way to capture knowledge collaboratively. However, they suffer from a reputation of turning quickly into unstructured and little used repositories. They are difficult to organise and even harder to maintain [2, 3].

The first instinct for users is to collect whatever documents they have into a digital library and rely on search to sort it out. To break that perception, it is necessary to teach them to make the wiki their document. This requires a lot of hand holding and training. Part of that perception comes from the lack of quick and easy ways to get people over the stage of importing what they know in the wiki. Wikis work particularly well once they are in maintenance mode, when users just add new pages or edit them when needed. They are more complicated when users are faced with the task of first uploading what they know into blank pages. There is also a tendency to use Wikipedia as a benchmark. 'If Wikipedia can be successful, setting up our own wiki will make it successful.' This approach overlooks Wikipedia's infrastructure and network of collaborators (humans and bots), developed over years of growing demands.

In an enterprise context, wikis need to provide more than a collaborative equivalent of a word processor. The content they hold has to be accessible both to individuals and other systems, through queries in addition to search. What we need is practical knowledge sharing; make it easy for people to share what they know, integrate with other sources and automate when possible. In our experience, Semantic MediaWiki provides a good balance of tools to that end.

16.3 Semantic MediaWiki

The Johnson & Johnson Pharmaceutical Research & Development, L.L.C. is exploring the use of Semantic MediaWiki (SMW) as a data integration tool with applications to guiding queries in a linked data framework for translational research, simplifying enterprise search and improving knowledge sharing in a decentralised organisation. Semantic MediaWiki was initially released in 2005 as an extension to MediaWiki designed to add semantic annotations to wiki text, in effect turning wikis into collaborative databases. MediaWiki itself enjoys some popularity in life sciences. It is used in many biowikis and more notably, in Wikipedia [4, 5, 14]. The Semantic MediaWiki extension is now at the core of a flurry of other extensions, developed by an active community of developers and used by hundreds of sites [6].

16.3.1 Flexibility

At the confluence of R&D and IT, change is unavoidable. Capturing and understanding concepts involved in a business is a moving target. Flexibility is assured by MediaWiki core functionalities:

- complete edit history on any page;
- redirects (renaming pages without breaking links);
- a differentiation between actual and wanted pages (and their reports in special pages);
- a vast library of free extensions;
- an advanced system of templates and parser functions;
- a background jobs queue for automated, mass updates.

These features are born out of real needs in collaborative knowledge management and make MediaWiki a solid base for an Enterprise Encyclopaedia. The addition of Semantic MediaWiki's simple system of annotations augments this flexibility, as it is generic enough to adapt to most applications. Semantic relationships are much more than the system of tags and simple taxonomies often found in content management systems. They can be seen as tags with meaning, turning each wiki page as the subject of these relationships. For example, the page about the acronym 'BU' may be annotated with an embedded semantic property in the format: *[[Has meaning::Business Unit]]*. This annotation creates a triple between the subject (the page titled 'BU'), the object ('Business Unit') and a predicate (the relationship 'Has meaning'). Figure 16.1 shows an example of semantic annotations using the 'Browse Properties' view of a wiki page. This default view allows for a visual inspection of annotations captured in a wiki page as well as links to view linked pages and query for pages with the same annotations. This view is generated automatically by Semantic MediaWiki and is independent from the actual wiki page.

Has description	ADNI uses MRI and PET imaging, as well as ... markers that can predict clinical outcome.
Has experimental design	Longitudinal disease progression
Has funding source	This project is funded for $60 million, wi ... t/About_Funding.shtml About ADNI funding].
Has home country	United States of America + ⊕
Has keyword	Alzheimer + ⊕
Has license URL	http://www.loni.ucla.edu/Common/TermsOfUse.pdf + ⚲
Has license status	Agreement + ⚲
Has literature reference	A full list of publication is available on the ADNI website: * [http://www.adni-info.org/Scientists/ADNIScientistsHome/ADNIPublications.aspx ADNI publications]
Has number of subjects	3,000 + ⚲
Has page author name	Laurent Alquier + ⊕, Tim Schultz + ⊕
Has page contributor name	Laurent Alquier + ⊕
Has page name	ADNI data + ⚲
Has pubmed ID	20042704 + ⚲, 19847051 + ⚲
Has relation to jnj	External + ⚲.

Figure 16.1 An example of semantic annotations. The 'Browse Properties' view of a wiki page provides a summary of semantic properties available for query inside of a wiki page

Published by Woodhead Publishing Limited, 2012

ADNI data

Summary	Origin	Content	Access	Clinical trial summary	Clinical data

Description:

```
ADNI uses MRI and PET imaging, as well as laboratory and cognitive testing
of 821 normal, mildly cognitively impaired and Alzheimer's disease patients.
Among the goals of the study is to provide better tools for carrying out
effective clinical trials and identifying biomarkers that can predict
clinical outcome.
```

Acronym: ADNI

Topics:

Scope:

Keywords: Alzheimer

Business owner: Mark Schmidt

Technical owner: Mike Farnum, Tim Schultz

Support contact:

Free text:

H1▾ | B I | ☰ ☰ ☰ | ▦ ∞▾ | ✎ | ▢ | ✒ ✏▾ | " ▤▾ | ▣▾ | ⟳

```
The Alzheimer's Disease Neuroimaging Initiative ([[ADNI]]) is a large five year (we began on Oct 1
2004) research project to study the rate of change of cognition, function, brain structure and
function, and biomarkers in 200 elderly controls, 400 subjects with mild cognitive impairment, and
```

Figure 16.2 Semantic Form in action. The form elements define a standard list of annotations for a page category while the free text area allows for capture of manual text and annotations

Semantic MediaWiki also provides a mechanism to query these properties, turning a wiki into a free form, collaborative database. Creating complete applications within the wiki comes down to selecting the right annotations. For example, adding a URL property to pages instantly turns a wiki into a social bookmarking tool. Add a date to create blogs or new feeds or add a geographic location to create an event planner. With the help of some extensions such as Semantic Forms and Semantic Results Formats, Semantic MediaWiki allows for a wide range of applications, such as question and answer tools, glossaries or asset management systems. Finally, a significant part of MediaWiki's flexibility comes from several hundred extensions developed by a very active community to extend its capabilities.

16.3.2 Import and export of content

Useful content for an Enterprise Encyclopaedia rarely exists in a vacuum. Fortunately, many MediaWiki extensions exist to simplify import of content.

Semantic Forms make manual entries as easy as possible. Using forms to encapsulate semantic annotations allows standardised capture and display of structured content and avoids presenting users with a blank page. Figure 16.2 shows an example of form. Form fields support several customisation parameters to mark them as required, control their display type (pull down list, text field, text area) and enable auto completion from categories, properties, and even external sources to the wiki. Using the optional HeaderTabs extension, it is also possible to simplify complex forms by grouping fields into tabs.

The Data Transfer extension allows the creation of pages in bulk from XML or CSV files. Batch creation of pages from files is critical to help users when content is already available in a structured form. The External Data extension allows embedding of live content from remote sources directly inside wiki pages. The extension supports XML or JSON feeds, as well as MySQL and LDAP connections.

Several specialised extensions are also available. For instance the Pubmed extension shown in Figure 16.3 allows the retrieval of full

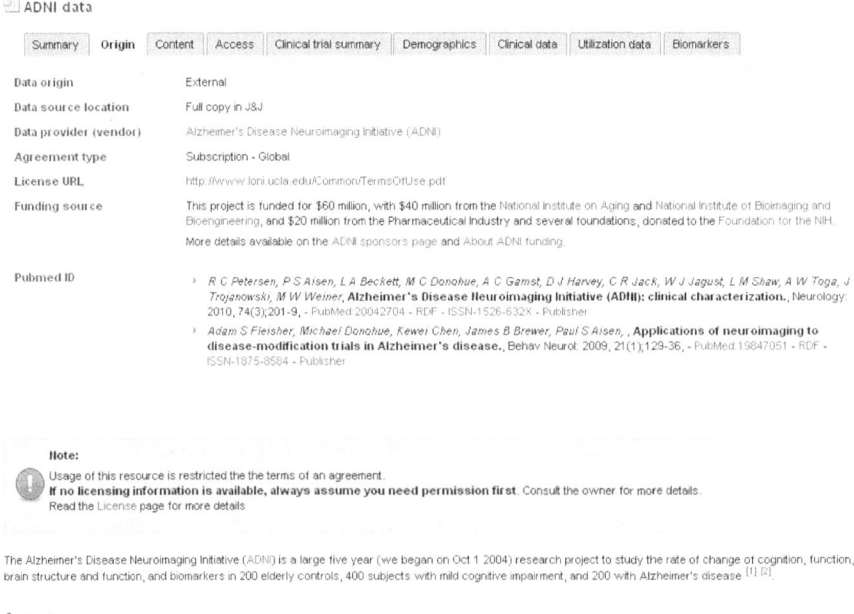

Figure 16.3 Page template corresponding to the form in Figure 16.2. The abstracts listed in the Pubmed ID field were automatically extracted from a query to the Pubmed database using only two identifiers

citations from Pubmed using a list of identifiers. Similar extensions exist for RSS feeds or social networks.

Finally, the Application Programming Interface (API) provided by MediaWiki allows more elaborate automation. Programs (or Bots) can perform maintenance updates in an unattended fashion and handle issues such as detecting broken links or correcting spelling and capitalisation [7]. Encyclopaedic knowledge is meant to be consumed in many ways, presented differently for different audiences. Similarly, several extensions allow export of content in various formats.

A key feature of Semantic MediaWiki is the capability to query semantic annotations like a database in addition to the traditional search. The default query mechanism allows the formatting of query results inside wiki pages as lists and tables, but also as links to comma separated values (CSV) files, RSS feeds and JSON files. The optional Semantic Result Formats extension adds many more formats such as tag clouds, graphs, charts, timelines, calendars, or maps.

16.3.3 Ease of use

The last category of extensions making Semantic MediaWiki such a good fit to build an Enterprise Encyclopaedia covers interface customisations and improvements to user experience. Extensions such as ContributionCredits improve trust and credibility by giving visible credits to the authors of wiki pages and providing readers with an individual to contact for more questions on the topic of a page. Other improvements include authentication frameworks, to remove the need to login and lower the barrier of entry for new users, navigation improvements such as tabbed content or social profiles complete with activity history and a system of points to encourage participation.

A particularly noteworthy effort to improve usability of Semantic MediaWiki is the Halo extension [9], which provides a complete makeover of advanced functionalities, including a customised rich text editor, inline forms and menus to edit properties and manage an ontology, an advanced access control mechanism and improved maintenance and data integration options. Halo and related extensions are both available as individual modules and as a packaged version of Semantic MediaWiki called SMW+.

In the end, the array of extensions available to MediaWiki makes it possible to customise the wiki to the needs of the users and provides the necessary tools to fit it easily in a growing ecosystem of social networks within the enterprise.

16.3.4 Design choices

Our internal wiki, named KnowIt, started initially as a small knowledge base to help with documentation of systems within J&J Pharmaceutical Research & Development, with categories such as applications, databases or servers. The home page for the wiki is shown in Figure 16.4, and Figures 16.5 and 16.6 show an example of an individual page.

KnowIt was launched in 2008, with content from previous attempts at a knowledge base about informatics systems dating back to 2002. That initial content provided a useful base to seed the wiki with a few thousand pages worth of useful information [8]. The transition to Semantic MediaWiki was prompted by the need to capture more contextual information about these systems, such as people and organisations involved, geographic locations or vocabularies. It was also necessary to keep the knowledge base flexible enough to adapt to inevitable changes in R&D. Gradually, the system has been outgrowing its initial scope and is turning slowly into an encyclopaedia of 'shared' knowledge within Johnson & Johnson.

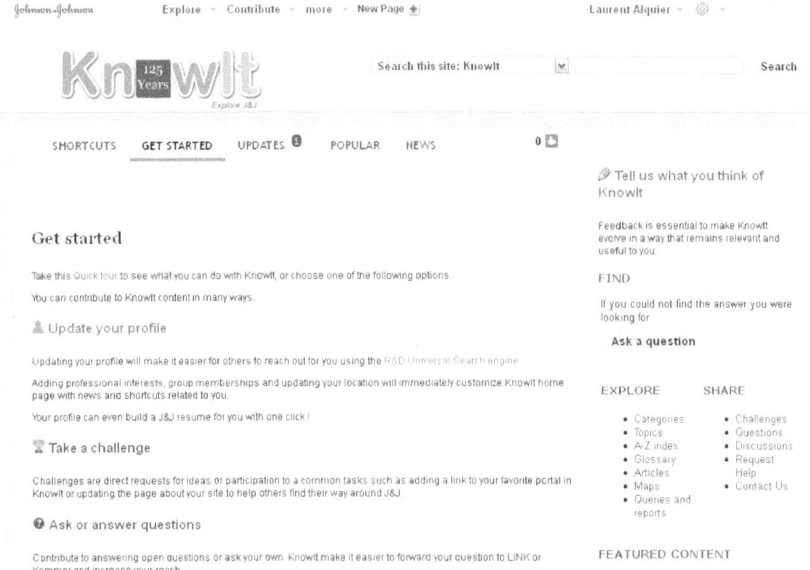

Figure 16.4 The KnowIt landing page. The home page of the wiki provides customised links based on semantic annotations captured in user profiles about their location, their group memberships and professional interests

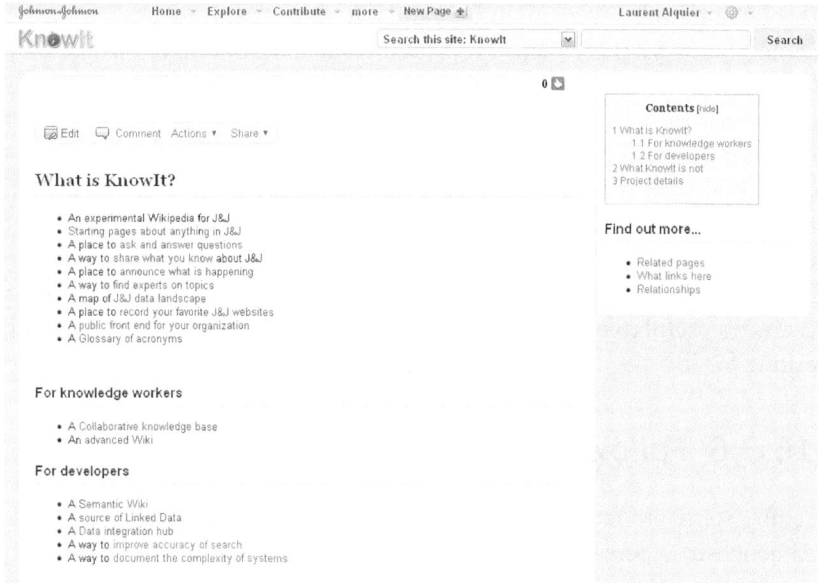

Figure 16.5 The layout of KnowIt pages is focused on content. Search and navigation options are always available at the top of pages

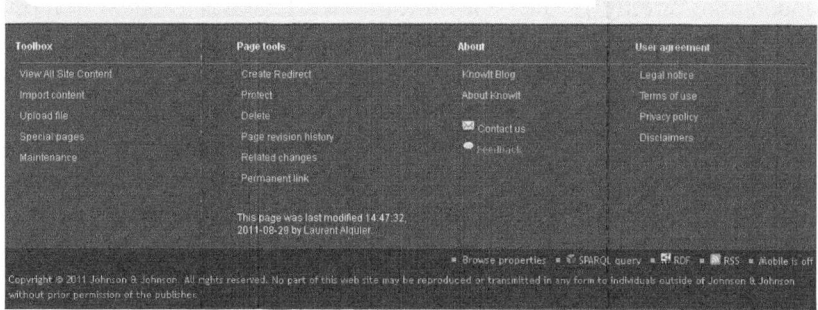

Figure 16.6 Advanced functions are moved to the bottom of pages to avoid clutter

16.3.5 Control the scope

From a strategic point of view, the choice of focusing on 'shared' information made sense as it reduced the scope of the wiki to content generic enough to be shared openly in the enterprise without going through multiple layers of access rights. It allows users to focus on what should be easily shared and keep sensitive material in places where they are already secured. From a more practical point of view, KnowIt acts as a directory of 'where is what' instead of trying to replicate every piece of information possible. In this context, 'shared' information includes categories such as links to internal portals, dictionaries of acronyms, names of teams and external partners. Given the potential breadth of topics, another key decision was to start small and let user needs drive the type of content made available. This ensures that the content of the wiki is always useful to users and that no expansion is made without someone asking for it.

16.3.6 Follow simple guidelines

Setting some ground rules was also helpful with the design of the wiki. Although it is possible to allow semantic annotations anywhere on wiki pages, doing so would require every contributor to be familiar with the names of properties. It is both impractical and unrealistic to expect users will be experts in the structure of the wiki. Using the Semantic Forms extension added a much-needed level of abstraction over semantic annotations and ensured that users can focus on content instead of having to struggle with annotations directly. At the same time, the Free Text area available in most forms allows more advanced users to capture additional annotations not covered by the forms. This is an important feature as it allows the wiki to adapt to exceptions to the structure provided by the forms.

Semantic Forms work best when wiki pages are modelled after the idea of classes through the use of MediaWiki's built-in system of categories. By restricting categories to represent only the nature of a page (what a page is about), and setting the limit to only one category per page, it is possible to take full advantage of Semantic MediaWiki's basic inference model. For example, if Acronyms can be defined as a subcategory of Definitions, a query for any page in the Definition category will also return pages in the Acronym category.

This design choice has a few consequences on the behaviour of the wiki.

- With one category per page, it makes sense to have only template per category and at least one form. Semantic Forms includes a convenient wizard to create forms and templates from a list of semantic properties. The association of a category, form, template and properties is what defines a 'class' in Semantic MediaWiki.

- Creating a single 'top' category for all categories associated with forms makes it possible to separate semantic content from other pages such as portals or index pages. This is useful when trying to get an accurate count of wiki pages.

Identifying a core set of semantic properties common to all categories is useful to simplify reporting and navigation between pages.

- Description: a short summary of what a page is about.
- Topics: tags to define a flexible classification of content inside pages (examples of topics are 'chemistry', 'biology', 'sequencing').
- Scope: tags to indicate for whom a page is relevant (examples of scopes are individuals, teams, functions, companies).

In addition to the core set, a few standard properties can help overcome some limitations of Semantic MediaWiki.

- Page title (text): pattern matching in semantic queries is allowed only for properties of type text. Creating a text version of the page title as a property allows queries such as 'all pages starting with the letter A', which can be useful for creating a table of contents or alphabetical indices.
- Class name (text): creating a redundant property for the category of pages allows running of queries on a single category, without inheritance.
- Related to: it is useful to define a 'catch all' property on certain pages to capture relationships that have not yet been formally defined. Regular reviews of values provided by users for this property are useful to identify properties that should be made more explicit inside forms.

16.3.7 Simplify the user experience

Choosing the right mix of extensions is not enough to ensure participation. In order to be successful, an Enterprise Encyclopaedia has to find a place in the flow of a casual user, which means it has to fit within the ecosystem of tools they use daily to find and share information. Customisation of the visual theme of the wiki (its skin) is necessary to allow content to be embedded in remote sites, viewed from mobile devices or printed if necessary. For example, the interface needs to provide options to share content using email, forums and whatever micro-blogging platform is used internally. By providing options to share content from the wiki to other tools in use in the enterprise, the number of necessary steps to share content is reduced and adoption is improved. Similarly, shortcuts need to be put in place to simplify the creation of pages such as bookmarklets or links from popular portals. Finally, content of the wiki needs to be available from enterprise search engines to allow users to navigate seamlessly to the wiki from their search.

Reducing the barrier to entry also means adjusting the vocabulary of the interface itself to be more readily understandable. Fortunately, MediaWiki provides simple ways to change just about every label and message used in its interface. Changing 'watchlists' to 'alerts' or 'talk pages' to 'comments' can make a difference between features being ignored or used. In some cases, it was also necessary to move away from the default pages provided by MediaWiki to display alerts or recent changes and provide our own, simplified version instead. Many of these advanced pages appear too technical and cumbersome for casual users. Some advanced maintenance tasks such as creating redirections to and from a page or managing disambiguation can be made more accessible with the help of a few forms and extensions. For example, the CreateRedirect extension provides users with a simple form to create aliases to pages instead of expecting them to manually create a blank page and use the right syntax to establish a re-direction to another page.

Finally, there is an on-going debate about the value of a rich text editor in a wiki. Rich text editors can simplify the user experience by providing a familiar editing environment. On the other hand, they encourage copy and paste from word processors or web pages, at the cost of formatting issues, which can quickly become frustrating. MediaWiki supports both options. Although CKeditor, as provided by the Halo extension [9], is currently one of the most advanced rich text editors supported by Semantic MediaWiki, we decided to use a markup editor such as MarkItUp for the convenience of features such as menus and templates to simplify common tasks related to wiki markup.

16.3.8 Implementation – what to expect

Semantic MediaWiki is a good illustration of the type of integration issues that can arise when working with open source projects. It can be difficult to find an extension that works exactly the way you need it to. The good news is that there are hundreds of extensions to choose from and it is relatively easy to find one that behaves very close to what you need, and customise it if needed. On the other hand, the lack of maturity of certain extensions means that making them fit seamlessly in a wiki can be a frustrating experience. It is necessary to decide early if features provided by an extension are absolutely necessary and to keep customisations to minimum. If customisations are required, keep track of changes to external code and work with the original authors of the extension to push for integration if possible. Furthermore, the lack of detailed documentation is an on-going issue with any open source project. Fortunately, the community around Semantic MediaWiki is active and responsive to questions and requests.

16.3.9 Knowledge mapping

One of the major obstacles to using Semantic MediaWiki as an encyclopaedia is that it requires the relatively hard work of breaking information into pages and linking them together. Semantic Forms does make this task easier, but it still requires a lot of effort to get people used to the idea. This is especially visible when someone is trying to transfer their knowledge into a relatively empty area of a wiki. Under ideal conditions, users would have that content already available in a spreadsheet ready to load with the Data Transfer extension, assuming there is not a lot of overlap with the current content of the wiki. In practice, users have reports, presentation slides or emails, which make it hard to break into a structured collection of pages. For example, a system configuration report would have to be mapped into individual pages about servers, applications, databases, service accounts and so on. Once that hurdle is passed, the payoff is worth the effort. Being able to query wiki content like a database is vastly superior to the idea of search, but, in order to show this to users in a convincing way, someone has to go through the painful job of manual updates, data curation and individual interviews to import that content for them.

Mapping relationships between people, organisations and other topics is a crucial step towards building a rich map of expertise in an organisation. As the graph of relationships is coming from both automated and manual inputs, the links between people and content tend to be more relevant than automated categorisations performed by search engines. The quality of these relationships was confirmed empirically by observing improved search results when the content of the wiki was added to an enterprise search engine. Another side effect of using Semantic MediaWiki is the increased pressure for disambiguation, shared vocabularies, and lightweight ontology building. As page titles are required to be unique, ambiguities in content quickly become visible and need to be resolved. This becomes visible, for example, when the same name is used by different users to reference a web application, a database or a system as a whole. The payoff from disambiguation efforts is that it allows content to be shared using simple, human readable URLs. The wiki provides stable URLs for concepts and can act as both a social bookmarking tool and a URL shortener in plain English. For example, the URL *http://wikiname.company.com/John_Smith* is easier to re-use in other applications and is easier to remember than *http://directory.company.com/profiles/userid=123456*.

16.3.10 Enterprise IT considerations

Even though KnowIt started as an experiment, it was deployed with proper security and disclaimers in place. We were fortunate to benefit from an internal policy of 'innovation time' and develop the knowledge base as an approved project. This allowed us to get buy-in from a growing number of stakeholders from the information technology department, management and power users. By deciding to let use cases and practical user needs drive the development of expansions to the wiki, it was possible to keep demonstrating its usefulness while at the same time, increasing exposure through internal communication channels and external publications. It also helped being prepared and anticipating the usual objections to new technologies, such as security, workflow and compatibility with environments and applications already in use.

Microsoft SharePoint

Microsoft SharePoint is difficult to avoid in an enterprise environment, in particular in life sciences. Most of the big pharmaceutical companies are using SharePoint as a platform to collaborate and share data [10]. SharePoint is often compared to a Swiss Army knife in its approach to collaboration. In addition to its core document management capabilities and Microsoft Office compatibility, it provides a set of tools such as discussion areas, blogs and wikis. These tools are simple by design and require consultants and third-party modules in order to complete their features. By Microsoft's own admission, 'About $8 should be spent on consultants and third party tools for every 1$ spent on licenses' [11, 12].

The document-centric design of SharePoint governance and customisations keep it from turning into another disorganised file storage. By contrast, the data-centric design of Semantic MediaWiki requires an effort of knowledge mapping from users, and in turn, a significant effort to assist them in extracting knowledge into individual wiki pages instead of uploading documents and relying on search alone to sort things out.

Semantic MediaWiki can coexist with SharePoint as a complement. Integration is possible in both directions. The wiki provides XML representations of queries to its content using simple URLs. SharePoint provides web parts capable of consuming that XML content from generic format (using XLST to render the results) or using specialised formats such as RSS feeds. Extracting content from SharePoint into the wiki is possible using web services. However, at this time, importing content from SharePoint requires some development, as there is no MediaWiki extension available to perform this task out of the box (although extensions such as Data Import provide basic mechanisms to connect to REST or SOAP services and could be used as the start of a SharePoint specific extension).

Security

MediaWiki has the reputation of being a weak security model, partly because of the many warnings visible on extensions designed to augment security. The security model used by MediaWiki is in fact robust when applied as intended, using groups and roles, although it lacks the level of granularity of other collaboration tools. However, more granular security models are available, such as the Access Control List from the Halo extension [9].

To reduce the likelihood of vandalism, already rare within a corporate environment, anonymous access was replaced by identified user names based on an Enterprise Directory. Using network accounts removed the need for keeping track of separate passwords.

Workflow

Based on earlier experience with collaborative knowledge bases, the editorial workflow model was deliberately left open, making full use of alerts and education instead of various levels of approval. This policy reduces bottlenecks, fosters collaboration and, overall, increases quality of the content and accountability of authors, as changes are made available immediately, with the name of authors visible to everyone.

Performance

On the administrative side, running MediaWiki as a server requires maintenance tasks such as keeping up with updates of extensions, managing custom code when necessary, and most importantly, tuning performance through various caching techniques at the database, PHP and web server levels [13]. Beyond simple performance tuning, scalability becomes a concern as the wiki grows. At this stage of development, KnowIt is far from reaching its limit. Wikis such as Proteopedia [14] are ten times larger with no noticeable performance issue. A wide range of options are available to evolve the architecture of the system with increasing demands, from investigating cloud-based deployments to distributing content across a network of synchronised databases or making heavier use of triple stores as more efficient storage and query solutions.

16.3.11 Linked data integration

With its many options to import and export content, Semantic MediaWiki is a good platform for data integration. In an encyclopaedia model, wiki pages become starting places around a topic, hubs to and from multiple resources. The wiki provides the means to bring content from remote sources together and augment it with annotations. Information about these sources still has to be collected manually, starting with the locations of the data sources, the formats of data they contain and proper credential or access rights. Once sources of content are identified, synchronisation of remote content with local annotations may become necessary. By providing a uniform Application Programming Interface (API) and support for multiple vocabularies, Linked Data principles can help simplify such synchronisations. Linked data is a term used to describe assertions encoded in a semantic format that is compatible with both humans and computers (and is discussed in Chapter 19 by Bildtsen et al. and Chapter 18 by Wild). The greatest source of flexibility in Semantic MediaWiki, and an under-utilised feature, is its capacity to produce and consume Linked Data.

Authoring Linked Data with Semantic Mediawiki

As of Semantic MediaWiki 1.6, semantic annotations can be stored directly in any SPARQL 1.1 compatible triple store. This will enable an easier integration with third-party triple stores. Earlier versions relied on a triple store connector, which provided live synchronisation of wiki content with a triple store running parallel to the wiki [15]. In both cases, the result is a near seamless production of Linked Data formats: RDF export of semantic annotations and queries, and OWL export of categories and properties defined in the wiki. This lets users focus on content through a familiar wiki interface and relegates Linked Data considerations to automated export formats. Most importantly, storing semantic annotations inside a triple store opens up more advanced capacities, such as the use of rules and inference engines. It also enables the use of the SPARQL query language to go beyond the limitation of the built-in query language in Semantic MediaWiki and enable more complex queries.

It is also worth noting that more lightweight approaches to making semantic annotations available externally are also emerging. Using RDFa or microdata formats, they will allow seamless augmentation of existing wiki pages in formats recognised by web browsers and search engines [16].

Consuming Linked Data content

Various options also exist to query and embed Linked Data content into wiki pages. From emerging extensions such as RDFIO or LinkedWiki, to more stable solutions such as the Halo extension, they do not stop at simply embedding the result of SPARQL queries inside wiki pages. Instead, they allow import of remote annotation to augment wiki pages [9, 15].

These Linked Data extensions are only steps toward using wikis to build mashups between multiple sources [17] or a fully dynamic encyclopaedia defined as collections of assertions with their provenance, such as demonstrated by Shortipedia [18].

16.3.12 Applications of Linked Data integration

KnowIt is making use of the default RAP triple store provided with Semantic MediaWiki prior to version 1.6. It is also using the SPARQL Query extension to form queries and provide simple links to results in several formats (CSV, XML, JSON) directly from the wiki [15]. Figure 16.7 gives an overview of the system. The Semantic MediaWiki side of the system provides an interface to capture manual annotations from users. Synchronised automatically when a user makes a change in the wiki, the triple store side of the system provides an interface to other applications using SPARQL queries.

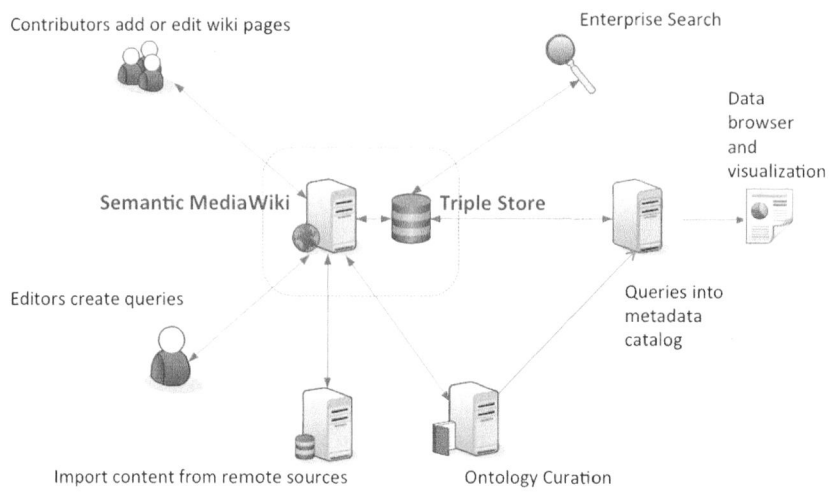

Contributors add or edit wiki pages

Enterprise Search

Data
browser
and
visualization

Semantic MediaWiki

Triple Store

Editors create queries

Queries into
metadata
catalog

Import content from remote sources

Ontology Curation

Figure 16.7 Semantic MediaWiki and Linked Data Triple Store working in parallel

16.3.13 Incremental indexing of enterprise search

As a first application, we created an XML connector from KnowIt to our enterprise search engine. SPARQL queries to the wiki can now be used to control which content should be indexed. Integration with existing search engines is important as it allows content of the wiki to be seamlessly available to users unfamiliar with the wiki. For instance, this allows scientists to search for data sources that contain terms such as 'gene sequencing' and 'proteins' and see a resulting list of all relevant data sources internal and external to the enterprise.

Although this application is possible using the basic query mechanism from Semantic MediaWiki, SPARQL queries make it possible to formulate elaborate filters on pages to update, and index them based on modification date or other criteria. Eventually, this mechanism will allow us to filter out pages that should not be indexed, and will provide the search engine complete semantic annotations when RDF content is available. This allows a reduction of the impact of crawling on the server, better use of network bandwidth and better control of what to index [19].

Published by Woodhead Publishing Limited, 2012

16.3.14 Linked Data queries for translational research

Finding sources of data with content of interest inside a large research organisation is often a challenging task. A typical experience for a scientist involved finding out about the data source and associated applications by word of mouth, looking for the owner of the data in order to gain access to it, which may require an in-depth discussion of the terms and conditions of the license agreement, and finally installation of new software tools on their desktop. If a scientist was interested in an external data source they would need access to IT resources to bring a copy of the data in-house and potentially provide an interface, as well as discussing the licensing terms and conditions with the legal department. These challenges will only become more prevalent as pharmaceutical companies increasingly embrace external innovation and are confronted with the need to reconcile many internal and external data sources.

KnowIt plays a central role in an internally developed Linked Data framework by exposing meta-data related to data sources of interest and by providing a catalogue of Linked Data services to other applications [19, 20]. To help build these services, we developed a C# library that provides other developers in the organisation with tools to discover and query the biomedical semantic web in a unified manner and re-use the discovered data in a multitude of research applications. Details provided by scientists and system owners inside the wiki augment catalogue this information such as the contact information for a data source, ontological concepts that comprise the data source's universe of discourse, licensing requirements and connection details, preferably via SPARQL endpoints. The combination of Semantic MediaWiki and the C# library create a framework of detailed provenance information from which Linked Data triples are derived. Eventually, paired with triple stores, and tools for mapping relational databases to RDF, this framework will allow us to access the mass of information available in our organisation seamlessly, regardless of location or format.

Two applications were designed to demonstrate this framework. First, we extracted a high-level, static visual map of relationships between data sources. Next, we developed a plug-in for Johnson & Johnson's Third Dimension Explorer, a research tool for visualising and mining large sets of complex biomedical data [21]. The plug-in utilises metadata in KnowIt to render an interactive visualisation of the data source landscape. Unlike the static map, the plug-in performs a clustering of data source attributes, such as ontological concepts and content, expressed by each data source in order to group similar data sources together.

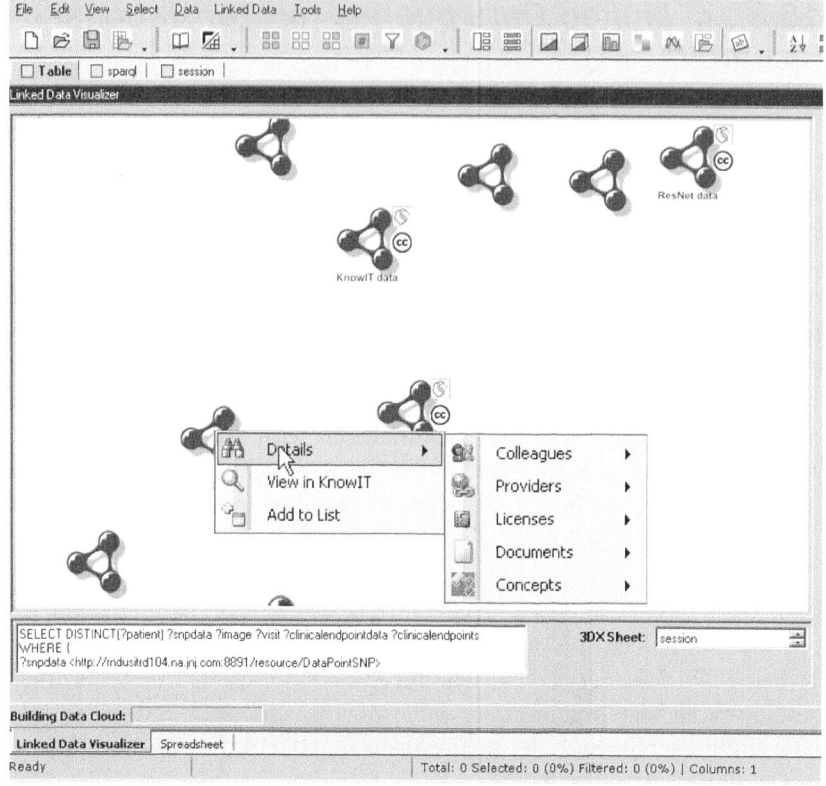

Figure 16.8 Wiki-based contextual menus. Information about data sources inside the query application is augmented with details managed inside KnowIt

From the Third Dimension Explorer interface, researchers can explore the visually rendered map of data sources, issue queries to each data source's SPARQL endpoint, and aggregate query results. Figure 16.8 shows an example of a contextual menu with information about the origin, contact information, licensing details and content extracted from KnowIt. Adding a new data source to the query tool becomes as simple as adding a new page in the wiki. The query application also provides access to query wizards to other internal biological, chemical and clinical data warehouses. This permits users to refine queries in a single data source as well as retrieve known information for a single entity across multiple domains.

This application illustrates the full benefits of building a semantic Enterprise Encyclopaedia. Thanks to Semantic MediaWiki, the encyclopaedia serves as a backbone for openly exposing meta-data about

Published by Woodhead Publishing Limited, 2012

biomedical data sources in a way that can be programmatically exploited across a multitude of applications within J&J.

16.4 Conclusion and future directions

Three years into using Semantic MediaWiki as a base for an Enterprise Encyclopaedia, usage is still ramping up, already expanding beyond the original informatics user base into topics relevant to the whole organisation. However, many areas of improvements remain.

Queries in many formats do not replace the power of an interactive, visual exploration of relationships graph captured in the wiki. Although many attempts have been made to visualise relationships in a wiki, very few make good use of semantic annotations provided by Semantic MediaWiki. Wiki-based visualisation tools at this point are mainly the product of academic research, using methods and tools developed for the occasion of the research. There is still a need for a robust browser to explore semantic relationships in an interactive way, with the layers, filters and links that would be required of that type of visualisation.

Similarly, although many bots exist to help automate maintenance tasks in a MediaWiki instance, we were not able to find examples of bots designed to work specifically for the maintenance of Semantic MediaWiki content. With the exception of the Gardening tools provided with the Halo Extension, creating a bot to perform batch updates of annotations or synchronisation of content with remote sources still requires the customisation of bots designed to work with text operations (search and replace strings) instead of semantic operations (remove or add values to semantic properties).

Finally, making queries more user-friendly would help a lot with adoption. Building queries still requires certain knowledge of available properties and how they relate to categories in the wiki. Additionally, there is no easy way at the moment to save the result of a query as a new page, which would be very helpful for users to share and comment on the results of their queries. An interesting example of this type of saved queries can be found in the Woogle extension, where users can persist with interesting search results, promote or remove individual results and comment the result page itself as a wiki page [22].

Regarding KnowIt as an Enterprise Encyclopaedia, several challenges remain. As is the case with many wiki projects, fostering the right levels of participation requires a constant effort of education, community outreach and usability improvements. Overall, Semantic MediaWiki provides a flexible yet structured platform to experiment with many aspects of enterprise data integration and knowledge sharing, in a way that can be immediately useful. The open source aspect of the software makes it ideal to experiment with ideas quickly, customise extensions to meet local requirements and provide unique features at a very low cost.

16.5 Acknowledgements

The system presented here is the result of contributions from many people over a long period of time. We would like to thank current and past contributors for their patience, ideas and support. In particular, Keith McCormick and Joërg Wegner for their ideas about user experience and tireless advocacy of KnowIt, Susie Stephens, Tim Schultz and Rudi Veerbeck for extending the use of a meta-data catalogue into a Linked Data framework, Joe Ciervo for developing the XML connector to the enterprise search engine and Dimitris Agrafiotis, Bob Gruninger, Ed Jaeger, Brian Johnson, Mario Dolbec, and Emmanouil Skoufos for their approval and support.

16.6 References

[1]   'Employees in Companies with 10,000+ Workers Spend 38 Minutes Per Document Search'. Recommind Inc. survey. 11 May 2009. 11 August 2011. *http://www.recommind.com/node/667*

[2]   Shehan Poole E and Grudin J. '*A taxonomy of Wiki genres in enterprise settings.*' In Proceedings of the 6th International Symposium on Wikis and Open Collaboration (WikiSym '10). ACM, New York, NY, USA, 2010;Article 14, 4 pages. DOI=10.1145/1832772.1832792.

[3]   Caya P and Nielsen J. '*Enterprise 2.0: Social Software on Intranets – A Report From the Front Lines of Enterprise Community, Collaboration, and Social Networking Projects*'.

[4]   Boulos, M. 'Semantic Wikis: A Comprehensible Introduction with Examples from the Health Sciences'. *Journal of Emerging Technologies in Web Intelligence* 2009;1(1).

[5]   Huss, 3rd JW, Lindenbaum P, Martone M, Roberts D, Pizarro A, Valafar F, Hogenesch JB and Su AI. 'The Gene Wiki: community intelligence applied to human gene annotation'. *Nucleic Acids Research* 2009.

[6]   Krötzsch M, Vrandecic D and Völkel M. '*Semantic MediaWiki*'. In Isabel Cruz, Stefan Decker, Dean Allemang, Chris Preist, Daniel Schwabe, Peter Mika, Mike Uschold, Lora Aroyo, eds: Proceedings of the 5th International Semantic Web Conference (ISWC-06). Springer 2006.

[7]   'Wikipedia: Creating a bot'. Wikipedia manual. 22 July 2011. 11 August 2011 *http://en.wikipedia.org/wiki/Wikipedia:Creating_a_bot*

[8] Alquier L, McCormick K and Jaeger E. '*knowIT, a semantic informatics knowledge management system*'. In Proceedings of the 5th International Symposium on Wikis and Open Collaboration (Orlando, Florida, 25–27 October 2009). WikiSym '09. ACM, New York, NY, 1–5. 2009. DOI= *http://doi.acm.org/10.1145/1641309.1641340*

[9] 'Halo extension'. Semanticweb.org. 29 October 2010. 8 July 2011. *http:// semanticweb.org/wiki/Halo_Extension*

[10] Basset H. 'The Swiss Army knife for information professionals'. Research Information review. August–September 2010.

[11] Sampson M. '*The cost of SharePoint*' Michael Sampson on Collaboration. 7 March 2011. 4 August 2011.

[12] Jackson J. 'SharePoint has its limits' *ComputerWorld* 28 April 2010. 04 August 2011.

[13] Hansch D. 'Semantic MediaWiki performance tuning' SMW+ forum. Ontoprise.com. 11 August 2011. *http://smwforum.ontoprise.com/ smwforum/index.php/Performance_Tuning*

[14] Bolser D. 'BioWiki'. BioInformatics.org. 1 April 2011. 10 August 2011. *http://www.bioinformatics.org/wiki/BioWiki#tab=BioWiki_Table*

[15] Krötzsch M. 'SPARQL and RDF stores for SMW'. SemanticMediaWiki. org. 2 August 2011. *http://semantic-mediawiki.org/wiki/SPARQL_and_ RDF_stores_for_SMW*

[16] Pabitha P, Vignesh Nandha Kumar KR, Pandurangan N, Vijayakumar R and Rajaram M, 'Semantic Search in Wiki using HTML5 Microdata for Semantic Annotation', *IJCSI International Journal of Computer Science* 2011;8(3):1.

[17] Lord P. '*Linking genes to diseases with a SNPedia-Gene Wiki mashup*', 5 August 2011 By Uduak Grace Thomas genomeweb.

[18] Vrandecic D, Ratnakar V, Krötzsch M, Gil Y. 'Shortipedia: Aggregating and Curating Semantic Web Data'. In *Semantic Web Challenge* 2010.

[19] Alquier L, Schultz T and Stephens S. '*Exploration of a Data Landscape using a Collaborative Linked Data Framework*'. In Proceedings of The Future of the Web for Collaborative Science, WWW2010 (Raleigh, North Carolina, 26 April 2010).

[20] Alquier L. 'Driving a Linked Data Framework with Semantic Wikis', Bio-IT World 2011, (Boston, Massachusetts, 12 April 2011).

[21] Agrafiotis DK, et al. 'Advanced biological and chemical discovery (ABCD): centralizing discovery knowledge in an inherently decentralized world'. *Journal of Chemical Information and Modeling* 2007;47(6):1999–2014. Epub 2007 Nov 1.

[22] Happel, H-J. 'Social search and need-driven knowledge sharing in Wikis with Woogle.' *WikiSym 09 Proceedings of the 5th International Symposium on Wikis and Open Collaboration* 2009.

## 17

Building disease and target knowledge with Semantic MediaWiki

*Lee Harland, Catherine Marshall,
Ben Gardner, Meiping Chang, Rich Head
and Philip Verdemato*

Abstract: The efficient flow of both formal and tacit knowledge is critical in the new era of information-powered pharmaceutical discovery. Yet, one of the major inhibitors of this is the constant flux within the industry, driven by rapidly evolving business models, mergers and collaborations. A continued stream of new employees and external partners brings a need to not only manage the new information they generate, but to find and exploit existing company results and reports. The ability to synthesise this vast information 'substrate' into actionable intelligence is crucial to industry productivity. In parallel, the new 'digital biology' era provides yet more and more data to find, analyse and exploit. In this chapter we look at the contribution that Semantic MediaWiki (SMW) technology has made to meeting the information challenges faced by Pfizer. We describe two use-cases that highlight the flexibility of this software and the ultimate benefit to the user.

Key words: Semantic MediaWiki; drug target; knowledge management; collaboration; disease maps.

17.1 The Targetpedia

Although many debate the relative merits of target-driven drug discovery [1], drug targets themselves remain a crucial pillar of pharmaceutical research. Thus, one of the most basic needs in any company is for a drug target reference, an encyclopaedia providing key facts concerning these important entities. In the majority of cases 'target' can of course, be either a single protein or some multimeric form such as a complex, protein–protein interaction or pharmacological grouping such as 'calcium channels' [2]. Consequently, the encyclopedia should include all proteins (whether or not they were established targets) and known multiprotein targets from humans and other key organisms. Although there are a number of sites on the internet that provide some form of summary information in this space (e.g. Wikipedia [3], Ensembl [4], GeneCards [5], EntrezGene [6]), none are particularly tuned to presenting the information most relevant for drug discovery users. Furthermore, such systems do not incorporate historical company-internal data, clearly something of high value when assessing discovery projects. Indeed, in large companies working on 50+ new targets per year, the ability to track the target portfolio and associated assets (milestones, endpoints, reagents, etc.) is vital. Targets are not unique to one disease area and access to compounds, clones, cell-lines and reagents from existing projects can rapidly accelerate new studies. More importantly, access to existing data on safety, chemical matter quality, pathways/mechanisms and biomarkers can 'make or break' a new idea. Although this is often difficult for any employee, it is particularly daunting for new colleagues lacking the IT system knowledge and people networks often required to find this material. Thus, the justification for a universal protein/target portal at Pfizer was substantial, forming an important component in an overall research information strategy.

17.1.1 Design choices

In 2010, Pfizer informaticians sought to design a replacement for legacy protein/target information systems within the company. Right from the start it was agreed that the next system should not merely be a cosmetic update, but address the future needs of the evolving research organisation. We therefore engaged in a series of interviews across the company, consulting colleagues with a diverse set of job functions, experience and grades. Rather than focusing on software tools, the interviews centred around user needs and asked questions such as:

- What data and information do you personally use in your target research?

- Where do you obtain that data and information?

- Where, if any place, do you store annotations or comments about the work you have performed on targets?

- Approximately how many targets per year does your Research Unit (RU) have in active research? How many are new?

- What are your pain points around target selection and validation?

Based on these discussions, we developed a solid understanding of what the user community required from the new system and the benefits that these could bring. Highlights included:

- Use of internet tools: most colleagues described regular use of the internet as a primary mechanism to gain information on a potential target. Google and Wikipedia searches were most common, their speed and simplicity fitting well with a scientist's busy schedule. The users knew there were other resources out there, but were unclear as to which were the best for a particular type of data, or how to stay on top of the ever-increasing number of relevant web sites.

- Data diversity: many interviewees could describe the types of data that they would like to see in the system. Although there were some differences between scientists from different research areas (discussed later), many common areas emerged. Human genetics and genomic variation, disease association, model organism biology, expression patterns, availability of small molecules and the competitor landscape were all high on the list. Given that we identified an excess of 40 priority internal and external systems in these areas, there was clearly a need to help organise these data for Pfizer scientists.

- Accessing the internal target portfolio: many in the survey admitted frustration in accessing information on targets the company had previously studied. Such queries were possible, particularly via the corporate portfolio management platforms. However, these were originally designed for project management and other business concerns, and could not support more 'molecular' questions such as 'Show me all the company targets in pathway X or gene family Y'. As expected, this issue was particularly acute for new colleagues who had not yet developed 'work arounds' to try to compile this type of information.

- Capturing proteins 'of interest': one of the most interesting findings was the desire to track proteins of interest to Pfizer scientists but that were not (yet) part of the official research portfolio. This included emerging targets that project teams were monitoring closely as well as proteins that were well established within a disease, but were not themselves deemed 'druggable'. For instance, a transcription factor might regulate pathways crucial to two distinct therapeutic areas, yet as the teams are geographically distributed they may not realise a common interest. Indeed, many respondents cited this intersection between biological pathways and the 'idea portfolio' as an area in need of much more information support.

- Embracing of 'wikis': we also learnt that a number of research units (RUs) had already experimented with their own approaches to managing day-to-day information on their targets and projects. Many had independently arrived at the same solution and implemented a wiki system. Users liked the simplicity, ease of collaboration and familiarity (given their use of Wikipedia and Pfizer's own internal wiki, see Chapter 13 by Gardner and Revell). This highlighted that many colleagues were comfortable with this type of approach and encouraged us to develop a solution that was in tune with their thinking.

- Annotation: responses concerning annotation were mixed; there were a range of opinions as to whether the time spent adding comments and links to information pages was a valuable use of a scientists' time. We also found that users viewed existing annotation software as too cumbersome, often requiring many steps and logging in and out of multiple systems. However, it was clear from the RU-wikis that there was some appetite for community editing when placed in the right context. It became very clear that if we wanted to encourage this activity we needed to provide good tools that were a fit with the scientists' workflow and provided a tangible value to them.

17.1.2 Building the system

The Targetpedia was intended to be a 'first stop' for target-orientated information within the company. As such, it needed to present an integrated view across a range of internal and external data sources. The business analysis provided a clear steer towards a wiki-based system and we reviewed many of the different platforms before deciding on the Semantic MediaWiki (SMW) framework [7]. SMW (which is an extension of the MediaWiki software behind Wikipedia) was chosen for the following reasons:

- Agility: although our interviews suggested users would approve of the wiki approach, we were keen to produce a prototype to test this hypothesis. SMW offered many of the capabilities we needed 'out of the box' – certainly enough to produce a working prototype. In addition, the knowledge that this same software underlies both Wikipedia and our internal corporate wiki suggested that (should we be successful), developing a production system should be possible.

- Familiarity: as the majority of scientists within the company were familiar with MediaWiki-based sites, and many of our specific target customers had set up their own instances, we should not face too high a barrier for adopting a new system.

- Extensibility: although SMW had enough functionality to meet early stage requirements, we anticipated that eventually we would need to extend the system. The open codebase and modular design were highly attractive here, allowing our developers to build new components as required and enabling us to respond to our customers quickly.

- Semantic capabilities: a key element of functionality was the ability to provide summarisation and taxonomy-based views across the proteins (described in detail below). This is actually one of the most powerful core capabilities of SMW and something not supported by many of the alternatives. The feature is enabled by the 'ASK' query language [8], which functions somewhat like SQL and can be embedded within wiki pages to create dynamic and interactive result sets.

Data sourcing

Using a combination of user guidance and access statistics from legacy systems, we identified the major content elements required for the wiki. For version one of Targetpedia, the entities chosen were: proteins and protein targets, species, indications, pathways, biological function annotations, Pfizer people, departments, projects and research units.

For each entity we then identified the types and sources of data the system needed to hold. Table 17.1 provides an excerpt of this analysis for the protein/target entity type. In particular, we made use of our existing infrastructure for text-mining of the biomedical literature, Pharmamatrix (PMx, [9]). PMx works by automated, massive-scale analysis of Medline and other text sources to identify associations between thousands of biomedical entities. The results of this mining provide a rich data source to augment many of the areas of scientific interest.

Table 17.1	Protein information sources for Targetpedia	
Name	Description	Example sources
General functional information	General textual summary of the target	Wikipedia [3] EntrezGene [6] Gene Ontology [10]
Chromosomal information	Key facts on parent gene within genome	Ensembl [4]
Disease links	Known disease associations	Genetic Association Database [11] OMIM [12] Literature associations (PMx) GWAS data (e.g. [13])
Genetics	Polymorphisms and mouse phenotypes	Polyphen 2 [14] Mouse Genome Informatics [15] Internal data Literature mining (PMx)
Drug discovery	Competitor landscape for this target	Multiple commercial competitor databases [2]
Safety	Safety issues	Comparative Toxicogenomics Database [16] Pfizer safety review repository Literature mining (PMx)
Pfizer project information	Projects have we worked on this target for	Discovery and Development portfolio databases Pfizer screening registration database
References	Key references	Gene RIFs [6] Literature mining (PMx)
Pathways	Pathways is this protein a member of	Reactome [17] BioCarta [18] Commercial pathway systems

Loading data

Having identified the necessary data, a methodology for loading this into the Targetpedia was designed and implemented, schematically represented in Figure 17.1 and discussed below.

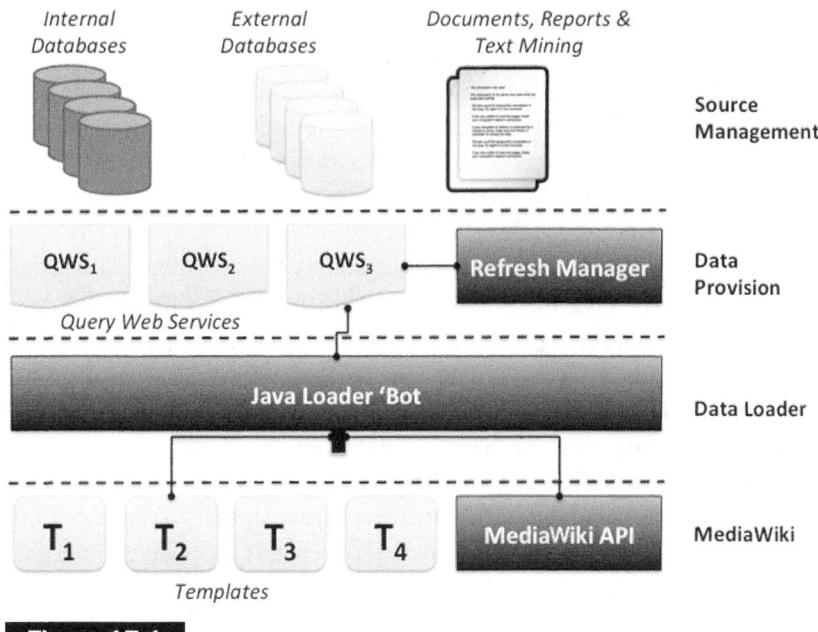

Figure 17.1 Data loading architecture

As Chapter 16 by Alquier describes the core SMW system in detail, here we will highlight the important Targetpedia-specific elements of the architecture, such as:

- Source management: for external sources (public and licensed commercial) a variety of data replication and scheduling tools were used (including FTP, AutoSys [19] and Oracle materialised views) to manage regular updates from source into our data warehouse. Most data sets are then indexed by and made queryable by loading into Oracle, Lucene [20] or SRS [21]. Data sources use a vast array of different identifiers for biomedical concepts such as genes, proteins and diseases. We used an internal system (similar to systems such as BridgeDb [22]) to provide mappings between different identifiers for the same entity. Multiprotein targets were sourced from our previously described internal drug target database [2]. Diseases were mapped to our internal disease dictionary, which is an augmented form of the disease and condition branches of MeSH [23].

- Data provision: for each source, queries required to obtain information for the wiki were identified. In many instances this took the form of summaries and aggregations rather than simply extracting data 'as-is' (described below). Each query was turned into a REST-ful web service via an in-house framework known as BioServices. This acts a wrapper around queries and takes care of many common functions such as load balancing, authentication, output formatting and metadata descriptions. Consequently, each query service ('QWS' in the figure) is a standardised, parameterised endpoint that provides results in a variety of formats including XML and JSON. A further advantage of this approach is that these services (such as 'general information for a protein', 'ontology mappings for a protein') are also available outside of Targetpedia.

- Data loading: a MediaWiki 'bot [24] was developed to carry out the loading of data into Targetpedia. A 'bot is simply a piece of software that can interact with the MediaWiki system to load large amounts of content directly into the backend database. Each query service was registered with the 'bot, along with a corresponding wiki template that would hold the actual data. Templates [25] are similar to classes within object-orientated programming in that they define the fields (properties) of a complex piece of data. Critically, templates also contain a definition of how the data should be rendered in HTML (within the wiki). This approach separates the data from the presentation, allowing a flexible interface design and multiple views across the same object. With these two critical pieces in place, the 'bot was then able to access the data and load it into the system via the MediaWiki API [26].

- Update scheduling: as different sources update content at different rates, the system was designed such that administrators could refresh different elements of the wiki under different schedules. Furthermore, as one might often identify errors or omissions affecting only a subset of entries, the loader was designed such that any update could be limited to a list of specific proteins if desired. Thus, administrators have complete control over which sources and entries can be updated at any particular time.

The result of this process was a wiki populated with data covering the major entity types in a semantically enriched format. Figure 17.2 shows an excerpt of the types of properties and values stored for the protein, phosphodiesterase 5.

17.1.3 The product and user experience

The starting point for most users is the protein information pages. Figure 17.3 shows one such example.

Published by Woodhead Publishing Limited, 2012

special page

Browse wiki

Phosphodiesterase 5A cGMP-specific (PDE5A)

Clinical project count	● + 🔍
Clinical project count active	● + 🔍
DNA Sequence	NM_033437 + 🔍
Description	phosphodiesterase 5A cGMP-specific (PDE5A) + 🔍
Drugs at highest phase	12 + 🔍
Entrez ID	8654 + 🔍
First letter	P + 🔍
GO:Biological Process	BP:positive regulation of oocyte development + ⊕, BP:relaxation of cardiac muscle + ⊕, BP:cGMP catabolic process + ⊕, BP:positive regulation of vasoconstriction + ⊕, BP:positive regulation of MAP kinase activity + ⊕, BP:positive regulation of apoptosis + ⊕, BP:vasodilation + ⊕, BP:negative regulation of T cell proliferation + ⊕, BP:response to testosterone stimulus + ⊕, BP:response to lipopolysaccharide + ⊕, BP:platelet activation + ⊕, BP:positive regulation of cardiac muscle hypertrophy + ⊕, BP:cyclic nucleotide metabolic process + ⊕, BP:short-term memory + ⊕, BP:blood coagulation + ⊕, BP:nervous system development + ⊕, BP:signal transduction + ⊕, BP:positive regulation of chronic inflammatory response + ⊕, BP:regulation of the force of heart contraction + ⊕, BP:response to hypoxia + ⊕
GO:Cellular Component	CC:cytosol + ⊕, CC:cytoplasm + ⊕
GO:Molecular Function	MF:3' 5'-cyclic-GMP phosphodiesterase activity + ⊕, MF:metal ion binding + ⊕, MF:cGMP binding + ⊕, MF:hydrolase activity + ⊕, MF:zinc ion binding + ⊕, MF:3' 5'-cyclic-nucleotide phosphodiesterase activity + ⊕, MF:catalytic activity + ⊕, MF:magnesium ion binding + ⊕, MF:nucleotide binding + ⊕
Gene Family	PDE5 + ⊕, Phosphodiesterase + ⊕, Enzymes + ⊕
Has GAD data	false + 🔍
Has MGI data	false + 🔍
Has RIF data	true + 🔍
Has Toxicomatrix data	True + ⊕
Has company	Boehringer Ingelheim + 🔍, Pfizer + 🔍, Lilly + 🔍, Dong-A + 🔍, Ipsen + 🔍
Has disease inum	Pressure and volume under development + ⊕, Diabetic foot ulcers painful diabetic neuropathy benign prostatic hyperplasia (16523) + ⊕,

targetpedia

navigation
- My Targets
- Help Searching
- General Help
- Recent changes
- Random page

search

[ ] [Go] [Search]

support
- Contact Us
- Leave Feedback

toolbox
- Upload file
- Special pages

Figure 17.2 Properties of PDE5 stored semantically in the wiki

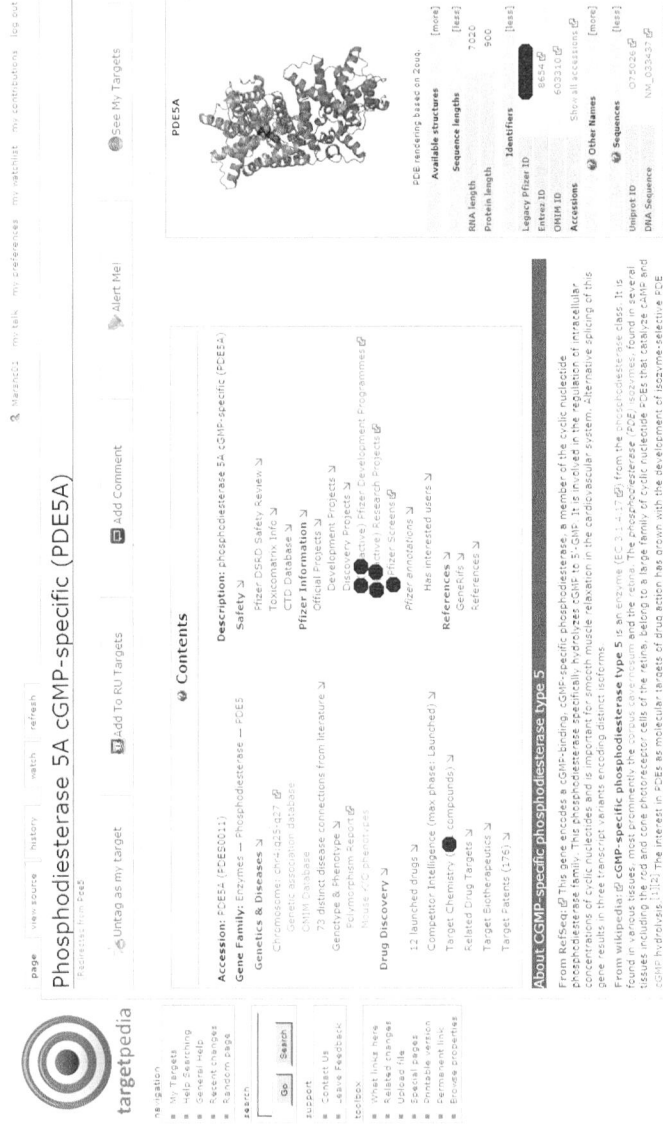

Figure 17.3 A protein page in Targetpedia (confidential information masked)

The protein pages have a number of important design features.

- Social tagging: at the top of every page, buttons that allow users to 'tag' and annotate targets of interest (discussed in detail below).

- The InfoBox: as commonly used within Wikipedia pages, customised to display key facts, including database identifiers, synonyms, pathways and Gene Ontology processes in which the protein participates.

- Informative table of contents: the default MediaWiki table of contents was replaced with a new component that combines an immediate, high-level summary of critical information with hyperlinks to relevant sections.

- Textual overview: short, easily digestible summaries around the role and function of this entity are provided. Some of this content is obtained directly from Wikipedia, an excellent source thanks to the Gene Wiki initiative [27]. For this, the system only presents the first one or two paragraphs initially; the section can be expanded to the full article with a single click. This is complemented with textual summaries from the National Center for Biotechnology Information's RefSeq [28] database.

The rest of the page is made up of more detailed sections covering the areas described in Table 17.1. For example, Figure 17.4 shows the information we display in the competitor intelligence section and demonstrates how a level of summarisation has been applied. Rather than presenting all of the competitor data, we present some key facts such as number of known small molecules, maximum clinical phase and most common indications. Should the user be sufficiently interested to know more, many of the elements in the section are hyperlinked to a more extensive target intelligence system [2], which allows them to analyse all of the underlying competitor data.

One of the most powerful features of SMW is the ability to 'slice and dice' content based on semantic properties, allowing developers (and even technically savvy users) to create additional views of the information. For example, each protein in the system has a semantic property that represents its position within a global protein family taxonomy (based on a medicinal chemistry view of protein function, see [29]). Figure 17.5 shows how this property can be exploited though an ASK query to produce a view of proteins according to their family membership, in this case the phosphodiesterase 4 family. The confidential information has been blocked out, Pfizer users see a 'dashboard' for the entire set of proteins and to what extent they have been investigated by the

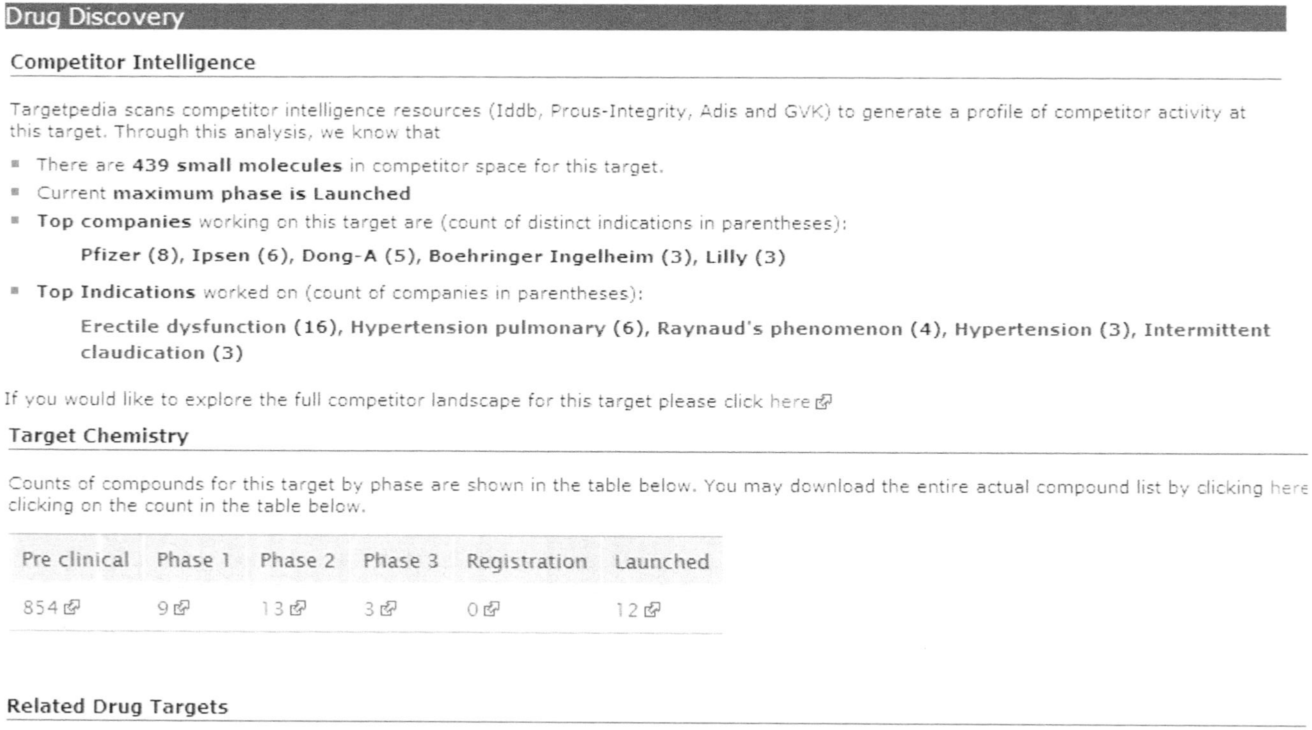

Drug Discovery

Competitor Intelligence

Targetpedia scans competitor intelligence resources (Iddb, Prous-Integrity, Adis and GVK) to generate a profile of competitor activity at this target. Through this analysis, we know that

- There are **439 small molecules** in competitor space for this target.
- Current **maximum phase is Launched**
- **Top companies** working on this target are (count of distinct indications in parentheses):

  Pfizer (8), Ipsen (6), Dong-A (5), Boehringer Ingelheim (3), Lilly (3)

- **Top Indications** worked on (count of companies in parentheses):

  Erectile dysfunction (16), Hypertension pulmonary (6), Raynaud's phenomenon (4), Hypertension (3), Intermittent claudication (3)

If you would like to explore the full competitor landscape for this target please click here 🗗

Target Chemistry

Counts of compounds for this target by phase are shown in the table below. You may download the entire actual compound list by clicking here clicking on the count in the table below.

Pre clinical	Phase 1	Phase 2	Phase 3	Registration	Launched
854 🗗	9 🗗	13 🗗	3 🗗	0 🗗	12 🗗

Related Drug Targets

Figure 17.4 The competitor intelligence section

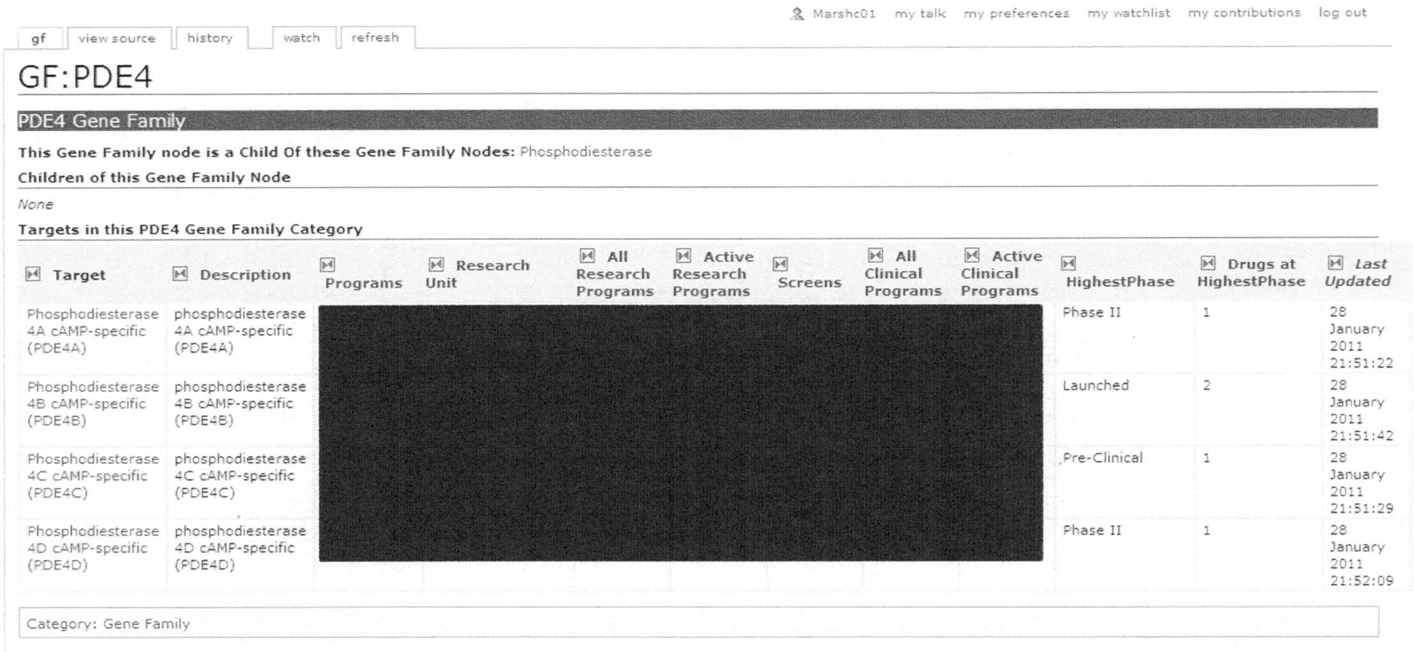

GF:PDE4

PDE4 Gene Family

This Gene Family node is a Child Of these Gene Family Nodes: Phosphodiesterase

Children of this Gene Family Node

None

Targets in this PDE4 Gene Family Category

Target	Description	Programs	Research Unit	All Research Programs	Active Research Programs	Screens	All Clinical Programs	Active Clinical Programs	HighestPhase	Drugs at HighestPhase	Last Updated
Phosphodiesterase 4A cAMP-specific (PDE4A)	phosphodiesterase 4A cAMP-specific (PDE4A)								Phase II	1	28 January 2011 21:51:22
Phosphodiesterase 4B cAMP-specific (PDE4B)	phosphodiesterase 4B cAMP-specific (PDE4B)								Launched	2	28 January 2011 21:51:42
Phosphodiesterase 4C cAMP-specific (PDE4C)	phosphodiesterase 4C cAMP-specific (PDE4C)								Pre-Clinical	1	28 January 2011 21:51:29
Phosphodiesterase 4D cAMP-specific (PDE4D)	phosphodiesterase 4D cAMP-specific (PDE4D)								Phase II	1	28 January 2011 21:52:09

Category: Gene Family

Figure 17.5 A protein family view. Internal users would see the table filled with company data regarding progression of each protein within drug discovery programmes

company. Critically, SMW understands hierarchies, so if the user were to view the same page for 'phosphodiesterases' they would see all such proteins, including the four shown here. Similar views have been set up for pathway, disease and Gene Ontology functions, providing a powerful mechanism for looking across the data – and all (almost), for free!

Collaboration in Targetpedia

One of the major differences between Targetpedia and our legacy protein/ target information systems are features that empower internal collaboration. For instance, Pfizer drug discovery project tracking codes are found on all pages that represent internal targets, providing an easy link to business data. Each project code has its own page within Targetpedia, listing the current status, milestones achieved and, importantly, the people associated with the research. The connection of projects to scientists was made possible thanks to the corporate timesheets that all Pfizer scientists complete each week, allocating their time against specific project codes. By integrating this into Targetpedia, this administrative activity moves from simply a management tool to something that tangibly enables collaboration. Further integration with departmental information systems provides lists of colleagues involved in the work, organised by work area. This helps users of Targetpedia find not just the people involved, but those from say, the pharmacodynamics or high-throughput screening groups. We believe this provides a major advance in helping Pfizer colleagues find the right person to speak to regarding a project on a target in which they have become interested.

A second collaboration mechanism revolves around the concept of the 'idea portfolio'. We wanted to make it very easy for users to assert an interest in a particular protein. Similar to the Facebook 'like' function, the first button in the tagging bar (Figure 17.3) allows users to 'Tag (this protein) as my target' with a single click. This makes it trivial for scientists to create a portfolio of proteins of interest to them or their research unit. An immediate benefit is access to a range of alerting tools, providing email or RSS updates to new updates, database entries or literature regarding their chosen proteins. However, the action of tagging targets creates a very rich data set that can be exploited to identify connections between disparate individuals in a global organisation. Interest in the same target is obvious, but as mentioned above, algorithms that scan the connections to identify colleagues with interests in different proteins but

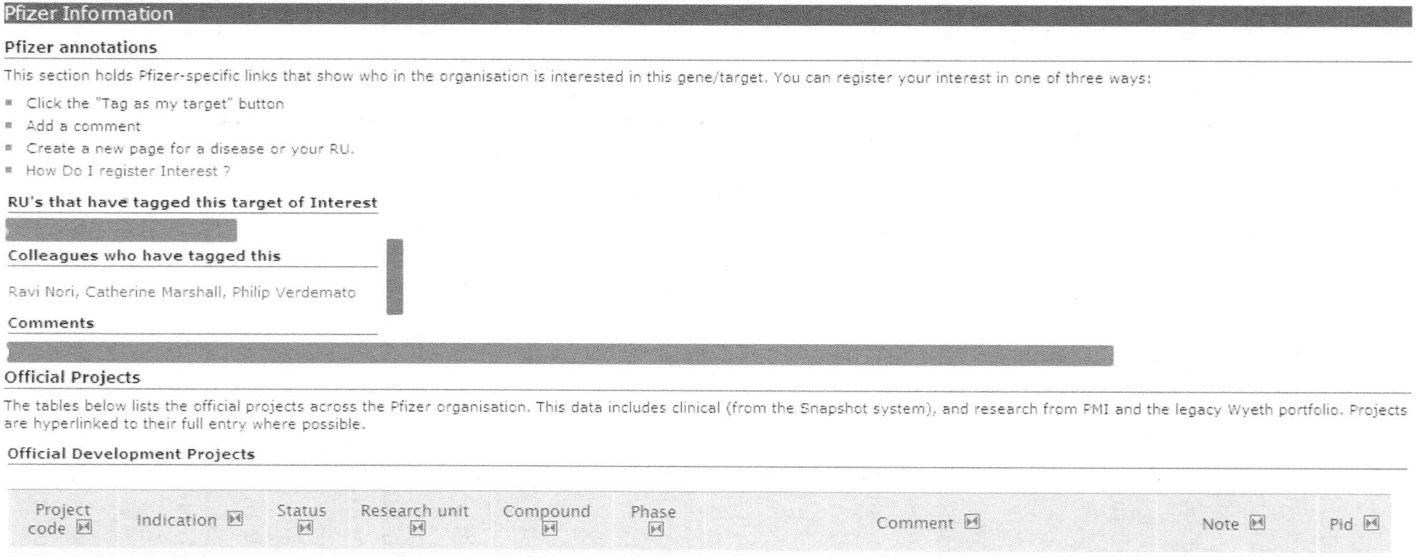

Pfizer Information

Pfizer annotations

This section holds Pfizer-specific links that show who in the organisation is interested in this gene/target. You can register your interest in one of three ways:

- Click the "Tag as my target" button
- Add a comment
- Create a new page for a disease or your RU.
- How Do I register Interest ?

RU's that have tagged this target of Interest

Colleagues who have tagged this

Ravi Nori, Catherine Marshall, Philip Verdemato

Comments

Official Projects

The tables below lists the official projects across the Pfizer organisation. This data includes clinical (from the Snapshot system), and research from PMI and the legacy Wyeth portfolio. Projects are hyperlinked to their full entry where possible.

Official Development Projects

Project code ⊠	Indication ⊠	Status ⊠	Research unit ⊠	Compound ⊠	Phase ⊠	Comment ⊠	Note ⊠	Pid ⊠

Figure 17.6 Social networking around targets and projects in Targetpedia

the same pathway present very powerful demonstrations of the value of this social tagging.

Finally, Targetpedia is of course, a wiki, and whereas much of the above has concerned the import, organisation and presentation of existing information, we wanted to also enable the capture of pertinent content from our users. This is, in fact, more complex than it might initially seem. As described in the business analysis section, many research units were running their own wikis, annotating protein pages with specific project information. However, this is somewhat at odds with a company-wide system, where different units working on the same target would not want their information to be merged. Furthermore, pages containing large sections of content that are specifically written by or for a very limited group of people erodes the 'encyclopaedia' vision for Targetpedia. To address this, it was decided that the main protein pages would not be editable, allowing us to retain the uniform structure, layout and content across all entries. This also meant that updating would be significantly simpler, not having to distinguish between user-generated and automatically loaded content.

To provide wiki functionality, users can (with a single click) create new 'child' pages that are automatically and very clearly linked to the main page for that protein. These can be assigned with different scopes, such as a user-page, for example 'Lee Harland's PDE5 page', a project page, for example 'The Kinase Group Page For MAPK2' and a research unit page, for example 'Oncology p53 programme page'. In this way, both the encyclopaedia and the RU-specific information capture aspects of Targetpedia are possible in the same system. Search results for a protein will always take a user to a main page, but all child pages are clearly visible and accessible from this entry point. At present, the template for the subpages is quite open, allowing teams to build those pages as they see fit, in keeping with their work in their own wiki systems. Finally, there may be instances where people wish to add very short annotations – a key paper or a URL that points to some useful information. For this, the child page mechanism may be overkill. Therefore, we added a fourth collaborative function that allows users to enter a short (255 character) message by clicking the 'add comment' button on the social tagging toolbar. Crucially, the messages are not written into the page itself, but stored within the SMW system and dynamically included via an embedded ASK query. This retains the simple update pattern for the main page and allows for modification to the presentation of the comments in future versions.

17.1.4 Lessons learned

Targetpedia has been in use for over six months, with very positive feedback. During this time there has been much organisational change for Pfizer, re-affirming the need for a central repository of target information. Yet, in comparison with our legacy systems, Targetpedia has changed direction in two major areas. Specifically, it shifts from a 'give me everything in one go' philosophy to 'give me a summary and pointers where to go next'. Additionally, it addresses the increasing need for social networking, particularly through shared scientific interest in the molecular targets themselves.

Developing with SMW was generally a positive experience, so much so we went on to re-use components in a second project described below. Templating in particular is very powerful, as are the semantic capabilities that make this system unique within its domain. Performance (in terms of rendering the pages) was never an issue, although we did take great care to optimise the semantic ASK queries by limiting the number of joins across different objects in the wiki. However, the speed by which content could be imported into the system was something that was suboptimal. There is a considerable amount of data for over 20 000 proteins, as well as people, diseases, projects and other entities. Loading all of this via the MediaWiki API took around a day to perform a complete refresh. As the API performs a number of operations in addition to loading the MySQL database, we could not simply bypass it and insert content directly into the database itself. Therefore, alternatives to the current loading system may have to be found for the longer term, something that will involve detailed analysis of the current API. Yet, even with this issue, we were able to provide updates along a very good timeline that was acceptable to the user community. A second area of difficulty was correctly configuring the system for text searches. The MediaWiki search engine is quite particular in how it searches the system and displays the results, which forced us to alter the names of many pages so search results returned titles meaningful to the user.

Vocabulary issues represented another major hurdle, and although not the 'fault' of SMW, they did hinder the project. For instance, there are a variety of sources that map targets to indications (OMIM, internal project data, competitor intelligence), yet each uses a different disease dictionary. Thus, users wish to click on 'asthma' to see all associated proteins and targets, but this is quite difficult to achieve without much laborious mapping of disease identifiers and terms. Conversely, there are no publically available standards around multi-protein drug targets (e.g.

'protein kinase C', 'GABA receptor', 'gastric proton pump'). This means that public data regarding these entities are poorly organised and difficult to identify and integrate. As a consequence, we are able to present only a small amount of information for these entities, mostly driven from internal systems (we do, of course, provide links to the pages for the individual protein components). We believe target information provision would be much enhanced if there were open standards and identifiers available for these entities. Indeed, addressing vocabulary and identity challenges may be key to progressing information integration and appear to be good candidates for pre-competitive collaboration [30].

For the future, we would like to create research unit-specific sections within the pages, sourced from (e.g.) disease-specific databases and configured to appear to relevant users only. We are also investigating the integration of 'live' data by consuming web services directly into the wiki using available extensions [31]. We conclude that SMW provides a powerful platform from which to deliver this and other enhancements for our users.

17.2 The Disease Knowledge Workbench (DKWB)

The Pfizer Indications Discovery Research Unit (IDU) was established to identify and explore alternative indications for compounds that have reached late stage clinical development. The IDU is a highly collaborative group; project leads, academic collaborators and computational scientists work together to develop holistic views of the cellular processes and molecular pathways within a disease of interest. Such deep understanding allows the unit to address key areas such as patient stratification, confidence in target rationale and identification of the best therapeutic mechanism and outcome biomarkers. Thus, fulfilling the IDUs' mission requires the effective management and exploitation of relevant information across a very diverse range of diseases.

The information assessed by the IDU comes from a range of internal and external sources, covering both raw experimental output and higher-level information such as the biomedical literature. Interestingly, the major informatics gap was not the management and mining of these data per se, as these functions were already well provided for with internal and public databases and tools. Rather, the group needed a mechanism to bring together and disseminate the knowledge that they had assessed and synthesised, essentially detailing their hypotheses and the data that led them there. As one would expect, the scientists used lab note books to record experiments, data mining tools to analyse data and read many scientific papers. However, like many organisations, the only place in which each 'story' was brought together was in a slide deck, driven by the need to present coherent arguments to the group members and management. Of course, this is not optimal with these files (including revisions and variants) quickly becoming scattered through hard drives and document management systems. Furthermore, they often lack clear links to data supporting conclusions and have no capability to link shared biology across different projects. This latter point can be especially critical in a group running a number of concurrent investigations; staying abreast of the major elements of each project and their interdependencies can be difficult. Therefore, the IDU and the computational sciences group began an experiment to develop tools to move disease knowledge management away from slides and into a more fit for purpose environment.

17.2.1 Design choices

Any piece of software that allows users to enter content could fall into the category of 'knowledge management'. However, there are a number of tools that allow users to represent coherent stories in an electronic representation. These range from notebook-style applications [32] to general mind-mapping software (e.g. [33]) and more specialised variants such as the Compendium platform for idea management [34]. A particularly relevant example is the I2 Analysts Notebook [35], an application used throughout law enforcement, intelligence and insurance agencies to represent complex stories in semi-graphical form. Although we had previously explored this for knowledge representation [2] and were impressed with its usability and relationship management, its lack of 'tuning' to the biomedical domain was a major limiting factor.

A further type of system considered was one very tuned to disease modelling, such as the Biological Expression Language (BEL) Framework produced by Selventa [36]. This approach describes individual molecular entities in a causal network, allowing computer modelling and simulation to be performed to understand the effects of any form of perturbation. Although this is an important technology, it works at a very different level to the needs of the IDU. Indeed, our aim was not to try to create a mathematical or causal model of the disease, but to provide somewhere for the scientists interpretation of these and other evidence to reside. This interpretation takes the form of a 'report' comprising free text, figures/images, tables, text and hyperlinks to files of underlying evidence. This is, of course, very wiki-like, and the system which most matched these needs was AlzSWAN [37], a community-driven knowledge base of Alzheimer disease hypotheses. AlzSWAN is a wiki-based platform that allows researchers to propose, annotate, support and refute particular pathological mechanisms. Users attach 'statements' to a hypothesis (based on literature, experimental data or their own ideas), marking whether these are consistent with the hypothesis. This mechanism to present, organise and discuss the pathophysiological elements of the disease was very aligned with the needs of our project, capturing at the level of scientific discourse, rather than individual molecular networks.

17.2.2 Building the system

Although AlzSWAN came closest to our desired system, it has a top-down approach to knowledge management; one provides a hypothesis and then provides the evidence for and against it. In the case of the IDU, the opposite was required, namely the ability to manage individual findings and ideas with the ultimate aim of bringing these together within a final therapeutic strategy. Thus, we concluded that a knowledge structure different from AlzSWAN was needed, but still took great inspiration from that tool. Given that the concept of the disease knowledge workbench was very much exploratory, a way to develop a prototype at minimal cost was essential. It was here that experience with SMW within Targetpedia enabled the group to move rapidly and create an AlzSWAN-like system, tailored to the IDU's needs, within a matter of weeks. Without the ready availability of a technology that could provide the basis for the system, it is unlikely that the idea of building the workbench would have turned into a reality.

Modelling information in DKWB

Informaticians and IDU scientists collaborated in developing the information model required to correctly manage disease information in the workbench. The principle need was to take a condition such as sepsis and divide it into individual physiological components, as shown in Figure 17.7.

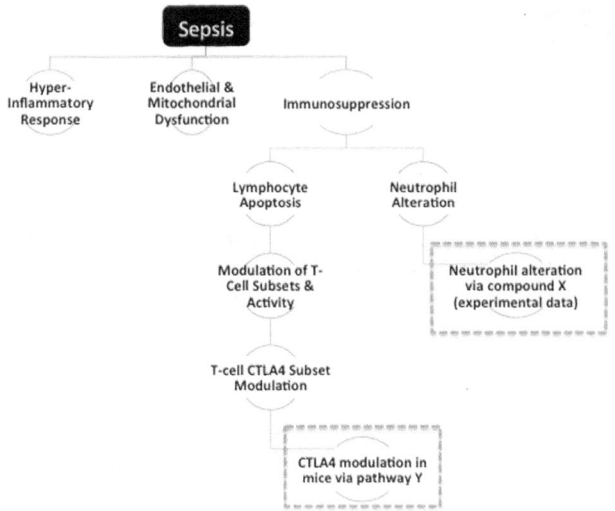

Figure 17.7 Dividing sepsis into physiological subcomponents. Dashed grey boxes surround components that are hypothesised and not yet validated by the team

Each component in Figure 17.7 represents an 'assertion', something that describes a specific piece of biology within the overall context of the disease. As can be seen, these are not simple binary statements, but represent more complex scientific phenomena, an example being 'increases in levels of circulating T-Lymphocytes are primarily the result of the inflammatory cascade'. There are multiple levels of resolution and assertions can be parents or children of other assertions. For instance, assertions can be observations (biology with supporting experimental or published data) or hypotheses (predicted but not yet experimentally supported). It should be noted that just because observations were the product of published experiments that does not necessarily mean that they are treated as fact. Indeed, one of the primary uses for the system was to critique published experiments and store the teams view of whether that data could be trusted. Regardless, each can be associated with a set of properties covering its nature and role within the system, as described in Table 17.2. Properties such as the evidence level are designed to allow filtering between clinical and known disease biology and more unreliable data from pre-clinical studies.

Table 17.2 The composition of an assertion

Name	Description	Example values	Semantically indexed
Type	Describes the nature of the assertion	Hypothesis or observation	Y
Role	The type of disease biology the assertion concerns	Disease mechanism; biomarker; therapeutic intervention; patient stratification	Y
Outcome	The ultimate effect of the assertion on disease pathology (where relevant)	Known negative† predicted negative; known positive; predicted positive; unknown	Y
Evidence level	The clinical relevance of the assertion	Established disease biology; clinical finding; pre-clinical finding	Y
Parent	The parent assertion		Y
Status	Current status of the curation around this assertion	Not started; in progress; under review; complete	Y
Informative text	Free text entry of synopsis, figures, and other pertinent information	Free text and image entry	N
Semantic tags	Tag entry with key entities of interest	Genes, cells, diseases etc.	Y
References	Literature references		Y

† in this context, 'negative' means making the disease or symptoms worse

Although we did not want to force scientists to encode extensive complex interpretations into a structured form, there was a need to identify and capture the important components of any assertion. Thus, a second class of concepts within the workbench were created, known as 'objects'. An object represents a single biomedical entity such as a gene, protein, target, drug, disease, cellular process and so on. Software developed on the Targetpedia project allowed us to load sets of known entities from source databases and vocabularies into the system, in bulk.

By themselves, these objects are 'inert', but they come alive when semantically tied to assertions. For instance, for the hypothesis 'Altered monocyte cytokine release contributes to immunosuppression in sepsis', one can attach the entities involved such as monocyte cells and specific cytokine proteins. This 'semantic tagging' builds up comprehensive relationships between pathophysiology and the cellular and molecular components that underlie these effects. They also enable the identification of shared biology. For instance, the wiki can store all known molecular pathways as observations; assigning all of the protein members to it via semantic tagging. Then, as authors of other assertions tag their pages with the principle proteins involved, an ASK query can automatically identify relevant pathways and present them on the page without any need for user intervention. This can be extended to show all assertions in the system that are semantically tagged with proteins in any shared pathway, regardless of what disease areas they describe. This aids the identification of common pathways and potential synergies across projects.

17.2.3 *The product and user experience*

To allow Pfizer scientists to fully describe and present their hypotheses and summaries, the assertions within the system are a mix of structured data (Table 17.2, semantic tagging) and free-form text and images. The semantic forms extension [38] provided a powerful and elegant mechanism to structure this data entry. Figure 17.8 shows how an assertion is created or edited.

The top of the edit form contains a number of drop-down menus (dynamically built from vocabularies in the system) to enable users to rapidly set the main properties for an assertion. Below this is the main wiki-text editing area, using the FCKEditor plug-in that enables users to work in a word-processor like mode without needing to learn the wiki markup language. Next is the semantic tagging area, where assertions can be associated with the objects of interest. Here, the excellent auto-complete mechanism that suggests as one types is employed, allowing the user to more easily enter the correct terms from supported vocabularies. The user must enter the nature of the association ('Link' on the figure), chosen from a drop-down menu. One of the most common links is 'Has Actor', meaning that the assertion is associated with a molecular entity that plays an active role within it. Finally, additional figures and references can be added using the controls at the bottom of the page.

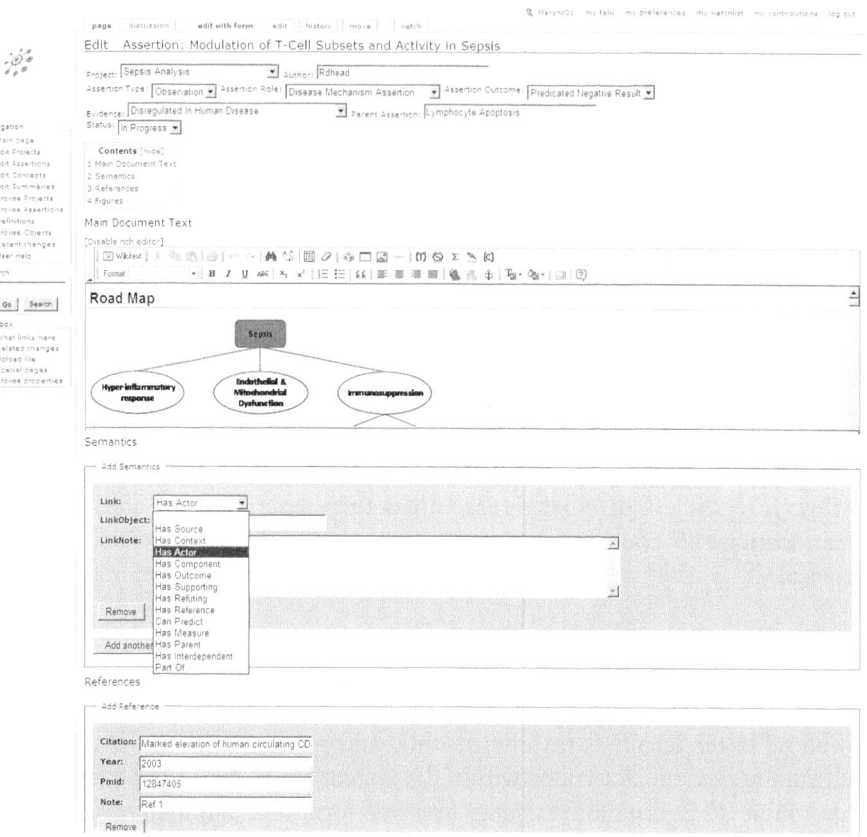

Figure 17.8 The Semantic Form for creating a new assertion. Note, the free-text entry area (under 'Main Document Text') has been truncated for the figure

Once the user has finished editing, the page is rendered into a user-friendly viewing format via a wiki template, as shown in Figure 17.9(a). As can be seen in Figure 17.9(b), the system renders semantic tags (in this case showing that CD4+ T-cells have been associated with the assertion). In addition, via an ASK query, connections to other assertions in the system are identified and dynamically inserted at the bottom of the page, in this case identifying two child observations that already exist in the system.

All assertions are created as part of a 'project', normally focused on a disease of interest. In some cases disease knowledge was retrospectively entered into the system from legacy documents and notes and subsequently further expanded. New projects were initiated directly in the wiki, following discussions on the major elements of the disease by the scientist.

Figure 17.9 (a) An assertion page as seen after editing. (b) A semantic tag and automatic identification of related assertions

From the main project page (Figure 17.10), one can get an overview of the disease, its pathogenesis and progression, diagnosis and prevalence, standard care and market potential, as well as key disease mechanisms. The assertions listed at the bottom of this page (Figure 17.10(b)) are divided into three sections: (1) 'top level disease mechanism' (i.e. those directly linked to the page); (2) 'all other disease mechanisms below the top level' (i.e. those which are children of parent associations in (1)); and (3) 'non-disease mechanism assertions such as target/pathway or therapeutic intervention'. One can drill into one of the assertion pages

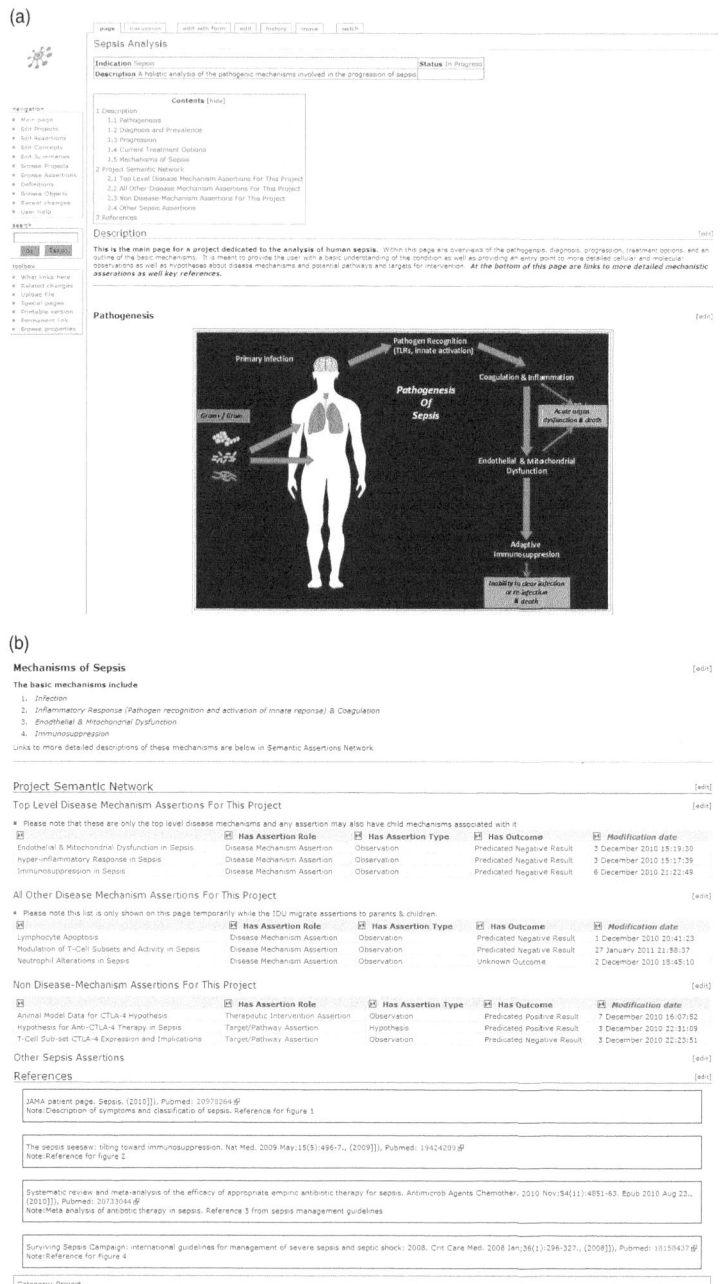

Figure 17.10 The sepsis project page. The upper portion (a) is generated by the user. The lower portion (b) is created automatically by the system

and from there navigate back to the project main page or any parent or child assertion pages using the semantic network hyperlinks. In this way the workbench is not only useful as a way to capture disease knowledge but also provides a platform to present learning on diseases and pathways to project teams across the company. As individual targets and diseases are semantically indexed on the pages, one can explore the knowledge base in many ways and move freely from one project to another.

In summary, we have developed a knowledge system that allows scientists to collaboratively build and share disease knowledge and interpretation with a low barrier to adoption. Much of the data capture process is deliberately manual – aiming to capture what the scientist thinks, rather than what a database knows. However, future iterations of the tool could benefit from more intelligent and efficient forms of data entry, annotation and discovery. Better connectivity to internal databases would be a key next step – as assertions are entered, they could be scanned and connected to important results in corporate systems. Automatic semantic tagging would be another area for improvement, the system recognising entities as the text is typed and presenting the list of 'discovered' concepts to the user who simply checks it is correct. Finally, we envisage a connection with tools such as Utopia (see Chapter 15 by Pettifer and colleagues), that would allow scientists to augment their assertions by sending quotes and comments on scientific papers directly from the PDF reader to the wiki.

17.2.4 Lessons learned

The prototype disease knowledge workbench brings a new capability to our biologists, enabling the capture of their understanding and interpretation of disease mechanisms. However, there are definitely areas for improvement. In particular, although the solution was based around the writing of report-style content, the MediaWiki software lacks a robust, intuitive interface for general editing. The FCKEditor component provides some basic functionality, but it still feels as if it is not fully integrated and lacks many common features. Incorporation of images and other files needs to evolve to a 'drag and drop' method, rather than the convoluted, multistep process used currently. Although we are aware of the Semantic MediaWiki Plus extensions in these areas [39], in our hands they were not intuitive to use and added too much complexity to the interface. Thus, the development of components to make editing content much more user-friendly should be a high priority and one that, thanks to the open source nature, could be undertaken by ourselves and donated back to the community.

As for the general concept, the disease knowledge capture has many benefits and aids organisations in capture, retaining and (re-)finding the information surrounding key project decisions. However, a key question is, will scientists invest the necessary time to enter their knowledge into the system? Our belief is that scientists are willing to do this, when they see tangible value from such approaches. This requires systems to be built that combine highly user-orientated interfaces with intelligent semantics. Although we are not there yet, software such as SMW provides a tantalising glimpse into what could be in the future. Yet, as knowledge management technologies develop there is still a need to look at the cultural aspects of organisations and better mechanisms to promote and reward these valuable activities. Otherwise, organisations like Pfizer will be doomed to a life of representing knowledge in slides and documents, lacking the provenance required to trace critical statements and decisions.

17.3 Conclusion

We have deployed Semantic MediaWiki for two contrasting use-cases. Targetpedia is a system that combines large-scale data integration with social networking while the Disease Knowledge Workbench seeks to capture discourse and interpretation into a looser form. In both cases, SMW allowed us to accelerate the delivery of prototypes to test these new mechanisms for information management and converting tacit knowledge into explicit, searchable facts. This is undeniably important – as research moves into a new era where 'data is king', companies must identify new and better ways of managing those data. Yet, developers may not always know the requirements up front, and initial systems always require tuning once they are released into real-world use. Software that enables informatics to work with the science and not against it is crucial, and something that SMW arguably demonstrates perfectly. Furthermore, as described by Alquier (Chapter 16), SMW is geared to the semantic web, undoubtedly a major component of future information and knowledge management. SMW is not perfect, particularly in the area of GUI interaction, and like any open source technology some of the extensions were not well documented or had incompatibilities. However, it stands out as a very impressive piece of software and one that has significantly enhanced our information environment.

17.4 Acknowledgements

We thank members of the IDU, eBiology, Computational Sciences CoE and Research informatics for valuable contributions and feedback. In particular, Christoph brockel, Enoch Huang, Cory Brouwer, Ravi Nori, Bryn-Williams-Jones, Eric Fauman, Phoebe Roberts, Robert Hernandez, Markella Skempri. We would like to specifically acknowledge Michael Berry and Milena Skwierawska who developed much of the original Targetpedia codebase. We thank David Burrows & Nigel Wilkinson for data management and assistance in preparing the manuscript.

17.5 References

[1]   Swinney D. How were new medicines discovered? *Nature Reviews Drug Discovery* 2011.

[2]   Harland L, Gaulton A. Drug target central. *Expert Opinion on Drug Discovery* 2009;4:857–72.

[3]   Wikipedia, *http://en.wikipedia.org*.

[4]   Ensembl, *http://www.ensembl.org*.

[5]   GeneCards, *http://www.genecards.org/*.

[6]   EntrezGene, *http://www.ncbi.nlm.nih.gov/gene*.

[7]   Semantic MediaWiki, *http://www.semantic-mediawiki.org*.

[8]   Semantic MediaWiki Inline-Queries, *http://semantic-mediawiki.org/wiki/Help:Inline_queries*.

[9]   Hopkins et al. *System And Method For The Computer-Assisted Identification Of Drugs And Indications*. US 2005/0060305.

[10] Harris MA, et al. Gene Ontology Consortium: The Gene Ontology (GO) database and informatics resource. *Nucleic Acids Research* 2004;32:D258–61.

[11] Genetic Association Database, *http://geneticassociationdb.nih.gov/*.

[12] Online Mendelian Inheritance In Man (OMIM), *http://www.ncbi.nlm.nih.gov/omim*.

[13] Collins F. Reengineering Translational Science: The Time Is Right. *Science Translational Medicine* 2011.

[14] PolyPhen-2, *http://genetics.bwh.harvard.edu/pph2/*.

[15] Mouse Genome Informatics, *http://www.informatics.jax.org/*.

[16] Comparative Toxicogenomics Database, *http://ctd.mdibl.org/*.

[17] Reactome, *http://www.reactome.org/*.

[18] Biocarta, *http://www.biocarta.com/genes/index.asp*.

[19] AutoSys, *http://www.ca.com/Files/ProductBriefs/ca-autosys-workld-autom-r11_p-b_fr_200711.pdf*.

[20] Apache Lucene, *http://lucene.apache.org/java/docs/index.html*.

[21] Biowisdom SRS, *http://www.biowisdom.com/2009/12/srs/*.

[22] van Iersel MP, et al. The BridgeDb framework: standardized access to gene, protein and metabolite identifier mapping services. *BMC Bioinformatics* 2010;11:5.

[23] Medical Subject Headings (MeSH), *http://www.nlm.nih.gov/mesh/*.

[24] MediaWiki 'Bot, *http://www.mediawiki.org/wiki/Help:Bots*.

[25] MediaWiki Templates, *http://www.mediawiki.org/wiki/Help:Templates*.

[26] MediaWiki API, *http://www.mediawiki.org/wiki/API:Main_page*.

[27] Huss JW, et al. The Gene Wiki: community intelligence applied to human gene annotation. *Nucleic Acids Research* 2010;38:D633–9.

[28] RefSeq, *http://www.ncbi.nlm.nih.gov/RefSeq/*.

[29] Schuffenhauer A, et al. An ontology for pharmaceutical ligands and its application for in silico screening and library design. *Journal of Chemical Information and Computer Sciences* 2002;42:947–55.

[30] Harland L, et al. Empowering Industrial Research with Shared Biomedical Vocabularies. *Drug Discovery Today* 2011 *doi:10.1016/j.drudis.2011.09.013*

[31] MediaWiki Data Import Extension, *http://www.mediawiki.org/wiki/Extension:Data_Import_Extension*.

[32] Barber CG, et al. 'OnePoint--' combining OneNote and SharePoint to facilitate knowledge transfer. *Drug Discovery Today* 2009;14:845–50.

[33] FreeMind, *http://freemind.sourceforge.net/wiki/index.php/Main_Page*.

[34] Compendium, *http://compendium.open.ac.uk/institute/download/download.htm*.

[35] i2 Analysts Notebook, *http://www.i2group.com/us/products--services/analysis-product-line/analysts-notebook*.

[36] Selventa BEL Framework, *http://www.selventa.com/technology/bel-framework*.

[37] AlzSwan, *http://www.alzforum.org/res/adh/swan/default.asp*.

[38] Semantic Forms, *http://www.mediawiki.org/wiki/Extension:Semantic_Forms*.

[39] Semantic MediaWiki +, *http://smwforum.ontoprise.com/smwforum*.

## 18

Chem2Bio2RDF: a semantic resource for systems chemical biology and drug discovery

David Wild

Abstract: This chapter describes an integrated semantic resource of drug discovery information called Chem2Bio2RDF. Heterogeneous public data sets pertaining to compounds, targets, genes, diseases, pathways and side effects were integrated into a single resource, which is freely available at chem2bio2rdf.org. A number of tools that use the data are described, along with the implementation challenges that derived from the project including details of where data is stored, how to organize it, and how to address data quality and equivalence.

Key words: semantic web; data integration; drug discovery; RDF.

18.1 The need for integrated, semantic resources in drug discovery

'How do we find the needles in the haystacks?' is a question that has been in the minds of pharmaceutical researchers since the early 1990s when high-throughput methods (screening, microarray analyses, and so on) began producing huge volumes of data about compounds, targets, genes, and pathways, and the interactions between them. The question is predicated on the assumption that somewhere in these vast haystacks of data can be found 'needles' – key pieces of knowledge or insight that could help find new drugs or new understandings of disease processes.

Unfortunately, producing the haystacks of data has turned out more straightforward than knowing how to sift through them, or even knowing if the needles are there or how to identify them when they are found. The 2000s saw this question move beyond pharmaceutical companies into the public arena, as huge volumes of data, particularly about chemical compounds and their biological activities, became available in the public domain, including such resources as PubChem, ChemSpider, and ChEMBL.

The needle-in-a-haystack analogy hides a subtler issue: that although most public data sets are centered on chemical or biological entities (compounds, genes, and so on), the most useful insights often lie in the relationships between these entities. Although some sets, such as ChEMBL, do represent a constrained set of relationships (e.g. between compounds and targets), these are not networked to other kinds of relationships, and so wider patterns cannot be seen. Yet it is these patterns that must be key to understanding the systematic effects of drugs on the body. A more appropriate analogy than haystacks is the Ishihara color blindness test, in which to find the hidden patterns one has to look at the whole picture with the right set of lenses.

It is with this in mind, that a research project at Indiana was developed to prototype new ways of representing publicly available entities and relationships as large-scale integrated sets, and new ways of data-mining them (new lenses in the analogy) to reveal the hidden patterns. Since this research began in 2005, many new relevant technologies have come to the fore (particularly in the area of the Semantic Web), but the problems have remained the same: finding ways to integrate public data sets intelligently; providing a common access and computation interface; developing tools that can find patterns across data sets; and developing new methodologies that make these tools applicable in real drug discovery problems.

Our initial work involved the development of ChemBioGrid [1], an open infrastructure of web services and computational tools for drug discovery, operating at the interface of cheminformatics and bioinformatics. This infrastructure allowed the quick development of new kinds of tools that integrated both cheminformatics and bioinformatics applications. However, this did not address the problem of data integration, which is addressed by the topic of this chapter, Chem2Bio2RDF. Data integration is a difficult problem [2, 3], as by definition it involves heterogeneous data sets that have often been developed from different disciplines and groups, each with their own terminology and ways of representing data. Traditionally, integration has been achieved using relational databases, a tortuous manual process that involves complex, formalized relational schema. The new technology that has made this process much more feasible is RDF (Resource Description Framework), a simple language that allows the representation of pairs of entities and the relationships between them (RDF triples). These take the form of Subject–Predicate–Object. For example, we could simply represent a fact about the author using the triple <David Wild><Is_a><Author>. Each RDF triple can be considered as two nodes of a network connected by an edge. In aggregate, the RDF triples describe a network of entities and relationships between them.

18.2 The Semantic Web in drug discovery

The power of RDF lies in its simplicity, but also in the advent of three other Semantic Web technologies which unleash its power: triple stores for efficient storage, access and searching of RDF; SPARQL, a language for querying RDF stores, and ontologies for the standardization of the content of RDF triples. More details on the RDF and semantic web approach are discussed in Chapter 19, where Bildtsen and co-authors describe the 'Triple map' application, and how it is used for semantic searching and knowledge collaboration in biomedical research. These technologies have only recently reached a point of maturity where they are practically useful, and in some ways the Semantic Web was over-hyped before it was mature enough, leaving an impression in some quarters that it is not technically up to the job of data integration. It is the opinion of the author that the technologies have reached a point where they are not only technically capable, but also are the only current technologies capable of tackling this problem. Indeed, as the chapters by Bildtsen and colleagues (Chapter 19), Harland and colleagues (Chapter 17) and Alquier (Chapter 16) show, this technology is now gaining traction within industry.

Recently, there has been an escalation in activity in the application of Semantic Web methods in drug discovery as a means of integration, including the well-funded EU OpenPhacts project [4], the W3 Semantic Web in Health Care and Life Sciences Interest Group [5], CSHALS [6] – a conference on Semantics in Health Care and Life Sciences, and Linked Open Drug Data (LODD) – a central repository for RDF related to drugs [7]. More widely in the biological sciences, the Semantic Web is having an impact through efforts such as Bio2RDF [8]. Bio2RDF covers a large number of genetic, pathway, protein, and enzyme sets, but does not cover chemical sets, except the KEGG ligand database. It thus overlaps with Chem2Bio2RDF, but the key differentiator of the latter is that chemical sets are mapped fully into genetic, protein, and pathway sets.

The Semantic Web provides a technical infrastructure for the large-scale integrated data-mining that is necessary, but advances have also been made in the data-mining techniques themselves, particularly in the fields of chemogenomics (the study of the relationships between chemical compounds – or drugs – and genes) and systems chemical biology [9] (a new term relating to the integrated application of cheminformatics and bioinformatics techniques to uncover new understanding of the systematic effects of chemical compound on the body). Recent research in chemogenomics has included the development of generalizable algorithms for predicting new compound–target interactions [10], creation of predictive networks in domain areas such as Kinases [11], and combination approaches that can be used to predict off-target interactions and new therapeutic uses for drugs [12]. However, application of these methods beyond the original research has been limited by lack of integrated access to public data. Little research has been done in Systems Chemical Biology outside of chemogenomics, for instance relating compounds to side effects, pathways, or disease states, primarily due to the lack of available tools and resources for relating compounds to entities other than targets or genes [9].

Chem2Bio2RDF was designed to address these gaps in data accessibility by providing access to a wide range of data sets covering compounds, drugs, targets, genes, assays, diseases, side effects, and pathways in a single, integrated format.

18.3 Implementation challenges

The main issues of implementing Chem2Bio2RDF were those that plague any implementation that uses emerging technologies: which particular technologies to employ and which to reject (at least initially). For Chem2Bio2RDF, this boiled down to four questions: (1) how should the data be stored and accessed; (2) where should the data be stored; (3) how to organize the data, and whether an OWL ontology should be used to semantically annotate the data; and (4) how to address data quality and equivalence across sets.

18.3.1 Data storage and access

Prior to our Semantic Web implementation, we stored a variety of public data sets in a PostgreSQL relational database enhanced with cheminformatics search functionality using the gNova CHORD cartridge

[13], with access directly through JDBC or ODBC database connections, or via web service interfaces (in our ChemBioGrid architecture). All of the data and services resided on the same machine, making access easier. Our initial version of Chem2Bio2RDF preserved this architecture, keeping the data in a relational database and using the D2R tool to provide an RDF SPARQL interface to the relational data set. However, this proved to have limitations that severely restricted the utility of Chem2Bio2RDF: namely (1) cross-data set searching is difficult to implement; and (2) it is hard to embed an ontology. So after initial testing we moved to use Virtuoso Triple Store as our basic representation, which is a true RDF-triple store, thus allowing all the data to be treated as a graph, enabling easy cross-data set searching, and permitting the later development of a Chem2Bio2RDF ontology. For migration, we used D2R to generate RDF for all the relations in PostgreSQL, and then exported them into the triple store. We also found that Virtuoso was much more efficient than the PostgreSQL/D2R implementation: as D2R eventually searches the data by SQL, speed of searching is highly dependent on the structure of the tables and associated indices. Additionally, Virtuoso provides a web interface for data management, and provides a REST web service allowing the data to be searched from other endpoints.

Using either D2R or Triple Store provides access to searching using SPARQL, but this has domain limitations, most notably it does not provide cheminformatics- or bioinformatics-specific searching capabilities such as similarity searching, substructure searching, or protein similarity searching. We thus had to extend SPARQL to allow such queries. This was done using the open source Jena ARQ [14] with cheminformatics functionality from the Chemistry Development Kit (CDK), ChemBioGrid, and bioinformatics functionality from BioJava.

18.3.2 Where data is stored

We opted to use a single medium-performance server (a Dell R510 with four quad-core Xenon processors, 1 TB storage) to store all of the data in our triple store and also to provide searching capabilities. Thus far, this has provided good real-time access to the data assuming no more than two searches are being performed at the same time. Keeping all of the data in one location (versus federating searches out) has proved useful in permitting query security (i.e. queries are not broadcast outside our servers) but a concern is that if we expand Chem2Bio2RDF to new, much larger data sources such as from Genome Wide Association Searches or patient data records, or if demand on our Chem2Bio2RDF server increases dramatically, we may not be able to scale searching to meet the needs. There is thus a likely future need to permit searches to be intelligently distributed in parallel, perhaps using cloud technologies.

18.3.3 How to organize the data

We decided to organize our data sets into six categories based on the kinds of biological and chemical concepts they contain. These categories are: chemical & drug (drug is a subclass of chemical), protein & gene, chemogenomics (i.e. relating compounds to genes, through interaction with proteins or changes in expression levels), systems (i.e. PPI and pathway), phenotype (i.e. disease and side effect), and literature. However, we did not initially develop an OWL ontology, instead depending on 'same-as' relationships between data sets (e.g. PubChem Compound X is the same as Drugbank Drug Y). This decision was made due to the difficulty in defining a scope for an ontology before we had a good idea how Chem2Bio2RDF was to be used. We subsequently developed a set of use-cases that allowed us to describe a constrained, implementable ontology for Chem2Bio2RDF. As we already had access to some other ontologies (such as Gene Ontology), this really boiled down to a chemogenomic ontology for describing the relationship between compounds and biological entities. This was aligned with other related ontologies, submitted to NCBO BioPortal, and will be described in an upcoming publication.

18.3.4 Data quality and equivalence

Addressing quality is fraught with numerous complexities in details – for example is a PubChem BioAssay IC_{50} result comparable with one in CheMBL or from an internal assay? Is an experimental result always more significant than a predicted result or an association extracted from a journal article? What happens when we get so many links between things that we cannot separate the signal from the noise? We are clearly constrained by the inherent quality of the data sources available. For Chem2Bio2RDF we decided on two principles: (1) we would not constrain users from making their own quality decisions (e.g. by excluding or including data sets or data types); and (2) we would not make judgments about equivalence beyond the very basic (two compounds equivalent in two data sets, etc.). Thus we pushed addressing primarily to the tool level, allowing users to select which data sets they are comfortable using, and understanding the caveats in doing so.

18.4 Chem2Bio2RDF architecture

Chem2Bio2RDF is available as a triple store with a SPARQL endpoint, can be accessed indirectly through a variety of tools, and all of the data can be downloaded in RDF format from the Chem2Bio2RDF website [15]. Our Chem2Bio2OWL ontology is also freely available for download.

As previously described, our data sets are organized into six categories based on the kinds of biological and chemical concepts they contain. Some data sources are listed in multiple categories. Some of the data used were previously employed in relational database format in our prior work and in this case they were simply converted into RDF/XML via the D2R server. For the rest of the data sets, we acquired the raw data set (by downloading from web sites), and converted the data into our relational database using customized scripts. These are then published as RDF in the Virtuoso Triple Store. The data can be queried via a SPARQL endpoint.

A list of data sets included in Chem2Bio2RDF is shown in Table 18.1, along with the number of RDF triples for each set. We have developed a streamlined process for the addition of new data sets. We adopted PubChem Compound ID (CID) as the identifier for compounds, and UniProt ID for protein targets. The compounds represented by other data formats (e.g. SMILES, InChi and SDF) were mapped onto the compound ID via InChi keys. All the triples are stored together and the whole set is called the Chem2Bio2RDF data set. Initially, we developed a schema to classify the concepts and the RDF resources in Chem2Bio2RDF. The RDF data can be explored and queried on our web site (*www.chem2bio2rdf. org*). Chem2Bio2RDF and its related tools rely heavily on open source software: a list of open source software used in Chem2Bio2RDF along with links for where the software can be downloaded is given in Table 18.2.

18.5 Tools and methodologies that use Chem2Bio2RDF

We have developed a variety of tools and algorithms that employ the Chem2Bio2RDF architecture. How some of these relate to

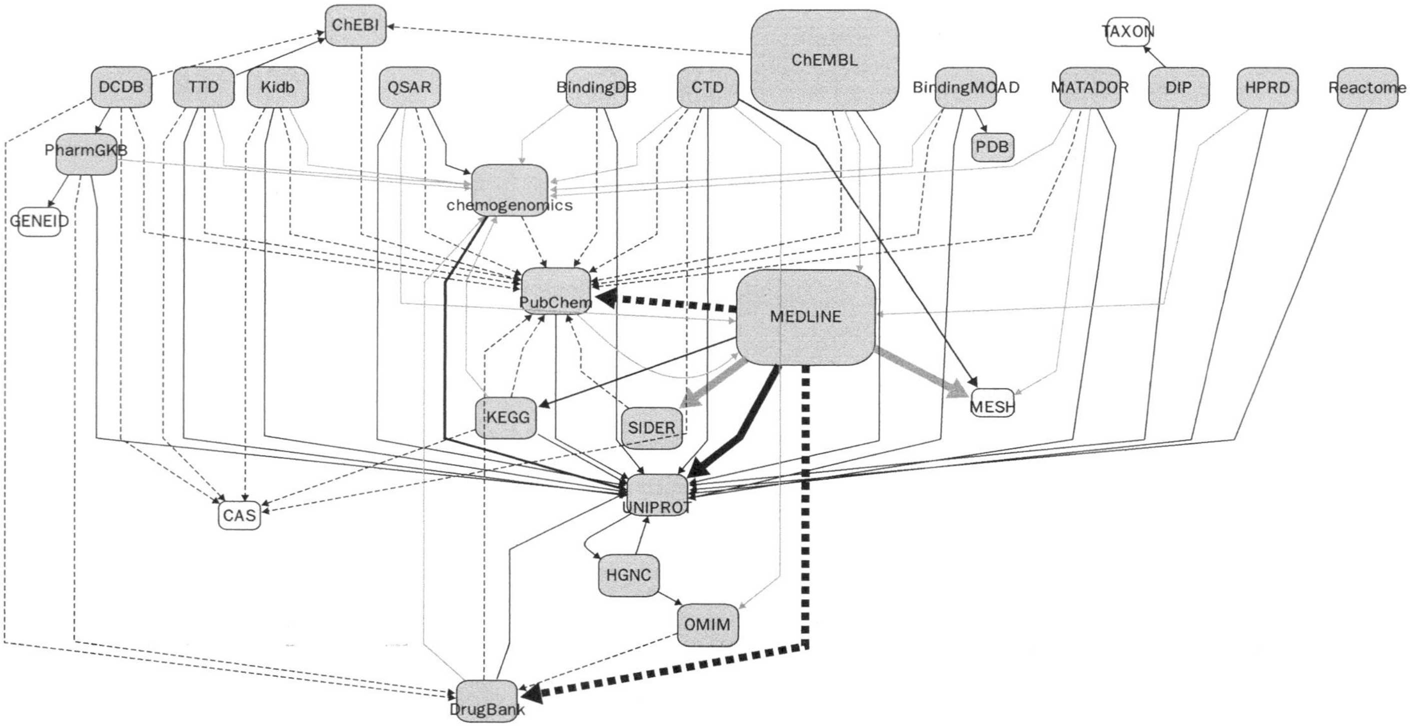

Figure 18.1 Chem2Bio2RDF organization, showing data sets and the links between them. Compound-related links are shown as dashed lines, protein/gene-related links are shown in dark gray, and other links are shown in light gray

Table 18.1	Data sets included in Chem2Bio2RDF ordered by number of RDF triples

Data set	Triples
ChEMB	57 795 793
PubChem Bioassay	5 908 479
Comparative Toxicogenomics Data set (CTD)	4 933 484
Miscellaneous Chemogenomics Sets	4 526 267
ChEBI	2 906 076
Database of Interacting Proteins (DIP)	1 113 871
BindingDB	1 027 034
HUGO HGNC (genes)	860 350
KiDB (CWRU)	745 026
UniProt	596 274
PharmGKB	512 361
Human Protein Reference Database (HPRD)	477 697
KEGG (Pathways)	477 697
MATADOR (Chemogenomics)	269 656
BindingMOAD	255 257
DrugBank	189 957
Sider Side Effects Database	127 755
Toxicogenomics Tracking Database (TTD)	116 767
Miscellaneous QSAR sets	32 206
Drug Combination Database (DCDB)	20 891
OMIM	17 251
Reactome (Pathways)	15 849

Chem2Bio2RDF is shown in Figure 18.2. All of the tools are freely available for public use, and where possible the code has been submitted into open source repositories.

Graph theory is well established for the analysis and mining of networked data, and lends itself naturally to application to RDF networks. We implemented an algorithm for computing semantic associations previously applied to social networks [16] for finding multiple shortest or otherwise meaningful paths between any two entities in a network, to enable all of the network paths within a given path

Table 18.2	Open source software used in Chem2Bio2RDF	
Software	**Purpose**	**Where to find it**
Virtuoso Open Source Edition	RDF Triple Store	*http://www.openlinksw.com/wiki/main/Main*
D2R	RDF interface to relational database	*http://www4.wiwiss.fu-berlin.de/bizer/d2r-server/*
Chemistry Development Kit (CDK)	Cheminformatics functionality in SPARQL search	*http://sourceforge.net/projects/cdk/*
BioJava	Bioinformatics functionality in SPARQL search	*http://biojava.org/wiki/Main_Page*
Jena ARQ	Adding functionality to SPARQL	*http://jena.sourceforge.net/ARQ/*
Cytoscape	Visualization of query results	*http://www.cytoscape.org/*
Protégé	Ontology Editing	*http://protege.stanford.edu/*
SIMILE	Data visualization in web pages	*http://www.simile-widgets.org/exhibit/*

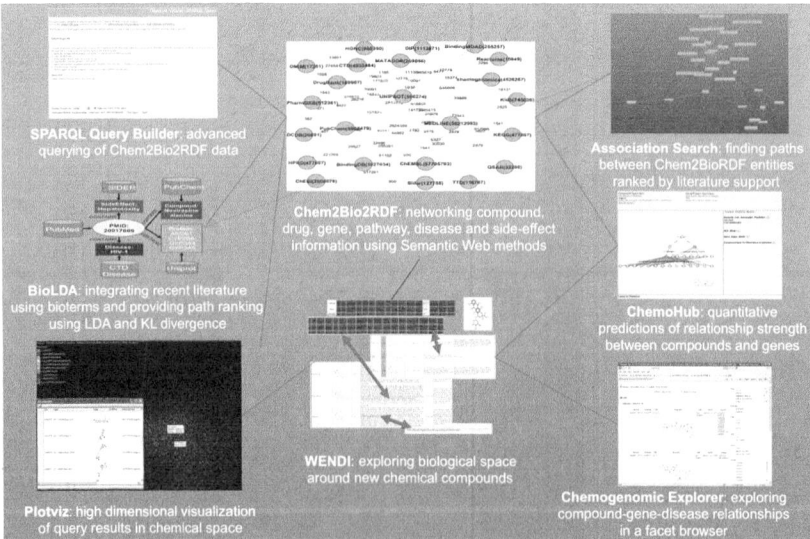

Figure 18.2 Tools and algorithms that employ Chem2Bio2RDF. More details and access to the tools can be obtained at *http://djwild.info*

length range between any pair of Chem2Bio2RDF entities to be identified. We have recently combined this with the BioLDA algorithm [17] described below into an association search tool that shows, for any pair of entities, the network paths between them that have the highest level of literature support. This has proven useful particularly for suggesting gene associations that can account for a drug's side effects or interactions with a disease, which has led us to develop a modification of the algorithm that allows the search to be restricted to only those paths that contain a particular type of entity (such as a gene).

We adapted a second algorithm from the social networking community to bring scholarly publications into our networks. A database of recent PubMed abstracts (for the last four years) was analyzed to identify Bioterms, that is terms that can be associated with entities in our existing network – for example, the name of a side effect, gene, compound, or drug. These Bioterms constitute an association between a PubMed article and an entry in one of our databases, producing an RDF association that can be mined. These Bioterms were further applied to a modified Latent Dirichlet Allocation algorithm (a method of identifying latent topics – or clusters – in a set of documents based on their word frequency) to identify latent topics in the PubMed literature. Publications and entities can then be probabilistically associated with these topics, and the product of multiple associations over a path used to create a measure of distance between entities (via topics) known as KL-divergence. The automatically identified topics along with their associated Bioterms show a surprising correlation with real areas of study, such as psychiatric disorders. Our resultant BioLDA algorithm [17] can be used for a variety of purposes, including identifying previously unknown Bioterm connections between research areas, constraining other searches to topic areas, and ranking of association paths by literature support as implemented in our association search tool.

We are currently taking the association search and literature-based methods a step further to provide quantitative measures of the association between any two items. We have developed a Semantic Link Association Prediction (SLAP) algorithm to provide such a quantitative measure based on the semantics and topology of the network. We developed an associated tool to provide both quantitative assessments of association strength along with graphical descriptions of the association paths. Initial studies, which will be published shortly, indicate a high rate of success in missed-link predictions (essentially leave-one-out studies using the network). Thus methods such as SLAP appear to be useful in predicting associations that might not already be known in the scientific literature or databases.

A large part of drug discovery involves the creation of new chemical compounds for possible therapeutic use, and new compounds can be integrated into the networks by chemical similarity (i.e. associating them with known compounds with similar chemical structures). This is the basis of our tool called WENDI [18] that embeds new compounds into our networks and permits examination of the potential therapeutic applications of the compounds. A recent development of WENDI, called Chemogenomic Explorer, extends the methods to create 'evidence paths' between compounds and diseases that constitute different paths through the network (via genes, pathways, and so on). These are generated using RDF inference tools, and in aggregate represent a cluster of independent or semi-independent evidence linking a compound to a disease. We think this evidence clustering is important as a way of mitigating the risks of errors in data, as well as the known propensity for individual pieces of published medical research to be later proved incorrect [19].

18.6 Conclusions

Our experience developing Chem2Bio2RDF has shown that Semantic Web technologies are becoming capable of meeting, at least to a reasonable degree, the challenges of data integration, and that the increasing availability of data management tools (including open source) make this process much easier. We consider Chem2Bio2RDF as a working prototype, but hope that it demonstrates that such complex, integrative networks can be built and exploited effectively for drug discovery applications. For robust, commercial applications, issues such as scalability and reliability will come into play, and it remains to be seen how well the technologies address these concerns.

18.7 References

[1] Dong X, Gilbert KE, Guha R, et al. Web service infrastructure for chemoinformatics. *Journal of Chemical Information and Modeling*, 2007;47(4):1303–7.
[2] Wild DJ. Mining large heterogeneous datasets in drug discovery. *Expert Opinion on Drug Discovery* 2009;4(10):995–1004.
[3] Slater T, et al. Beyond Data Integration. *Drug Discovery Today* 2008;13:584–9
[4] *http://www.openphacts.org*

[5]   *http://www.w3.org/2001/sw/hcls/*

[6]   *http://en.wikipedia.org/wiki/CSHALS*

[7]   Samwald M, et al. Linked open drug data for pharmaceutical research and development. *Journal of Cheminformatics* 2011;3:19.

[8]   Belleau F, et al. Bio2RDF: towards a mashup to build bioinformatics knowledge systems. *Journal of Biomedical Informatics* 2008;41(5):706–16.

[9]   Oprea TI, et al. Systems Chemical Biology. *Nature Chemical Biology* 2007;3:447–50.

[10] Keiser MJ, et al. Predicting new molecular targets for known drugs. *Nature* 2009;462:175–81.

[11] Metz JT et al. Navigating the Kinome. *Nature Chemical Biology* 2011;Web publication date 20 February 2011.

[12] Xie L, et al. Structure-based systems biology for analyzing off-target binding. *Current Opinion in Structural Biology* 2011;Web publication date 1 February 2011.

[13] *http://www.gnova.com*

[14] *http://jena.sourceforge.net/ARQ*

[15] *http://www.chem2bio2rdf.org*

[16] Tang J, Zhang J, Yao L, Li J, Zhang L and Su Z. *ArnetMiner: Extraction and Mining of Academic Social Networks*. In Proceedings of the Fourteenth ACM SIGKDD International Conference on Knowledge Discovery and Data Mining (SIGKDD'2008), pp. 990–8.

[17] Wang H, et al. Finding complex biological relationships in recent PubMed articles using Bio-LDA. *PLoS One*, 2011;6(3):e17243

[18] Zhu Q, et al. WENDI: A tool for finding non-obvious relationships between compounds and biological properties, genes, diseases and scholarly publications. *Journal of Cheminformatics* 2010;2:6.

[19] Ioannidis J. Why most published research findings are false. *PLoS Med* 2005;2(8):e124.

TripleMap: a web-based semantic knowledge discovery and collaboration application for biomedical research

Ola Bildtsen, Mike Hugo, Frans Lawaetz, Erik Bakke, James Hardwick, Nguyen Nguyen, Ted Naleid and Christopher Bouton

Abstract: TripleMap [1] is a web-based semantic search and knowledge collaboration application for the biomedical research community. Users can register for free and login from anywhere in the world using a standard web browser. TripleMap allows users to search through and analyze the connections in a massive interconnected network of publicly available data being continuously compiled from Linking Open Drug Data [2] sources, the entire corpus of PubMed abstracts and full text articles, more than 10 000 biomedical research relevant RSS feeds, and continuously updating patent literature. TripleMap represents everything in its data network as 'master entities' that integrate all information for any given entity. By searching for, and saving sets of entities, users build, share, and analyze 'dynamic knowledge maps' of entities and their associations.

Key words: semantic technologies; Big Data; RDF; analytics; visualization; semantic search; triples; knowledge collaboration.

19.1 The challenge of Big Data

In recent decades computationally based technologies including the internet, mobile devices, inexpensive high-performance workstations, and digital imaging have catalyzed entirely new ways for humans to generate, access, share, and visualize information. As a result, humanity now produces quantities of digital data on an annual basis that far exceed the amount of non-digital data previously generated over the entire span of human history. For example, it is estimated that in 2011 the amount of digital information generated will be 1 800 000 petabytes (PB) or 1.8 zettabytes (ZB) [3]. In comparison:

- 5 MB would contain the complete works of Shakespeare [4];
- 10 TB represents the entire printed collection of the US Library of Congress [4];
- 200 PB contains all printed material ever generated by humanity [4].

In other words, in 2011 alone, humanity will generate 9000 times more digital information than all of the information ever printed in physical books, magazines, and newspapers. Furthermore, humanity's ability to generate and handle digital data through advances in the means of its generation, transmission, and storage is increasing at an exponential rate [5].

This mind-boggling scale of data is commonly referred to as 'Big Data'. The phrase 'Big Data' is not meant to denote a specific data scale such as 'petabytes' or 'exabytes' (10^{15} and 10^{18} bytes, respectively), but instead to be indicative of the ever-increasing sea of data available to humans and the challenges associated with attempting to make sense of it all [6]. The Big Data explosion in the biomedical research sector has been further catalyzed by the completion of the human genome sequencing project [7], the sequencing of genomes from a number of species, and the emergence of a range of correlated, omics and next-generation technologies (e.g. microarrays, next-generation sequencing approaches) as discussed in a number of other chapters in this book.

The availability of Big Data presents humans with tremendous capabilities to 'see the big picture' and gain a 'bird's eye' view of patterns, trends, and interrelationships under consideration and it stands to reason that individuals, teams, and organizations that are able to harness Big Data for searching, analyzing, collaborating, and identifying patterns will be able to make faster and more effective decisions. In fact a new class of workers, often referred to as 'data scientists' or 'knowledge workers', is emerging to conduct exactly this type of Big Data analytical activity for both public and private organizations. However, Big Data also presents significant challenges related to the management, integration, updating, and visualization of information at this scale. Due to these challenges, computational tools that enable users to manipulate, visualize, and share in the analysis of data are currently, and will become ever-more, important for enabling the derivation of knowledge and insight from Big Data. In particular, software tools that allow users to share and collaborate in the derivation of high-level knowledge and insight from the vast sea of data available are critical. The need for such tools is particularly pressing in the biomedical research and development space where the integration and understanding of relationships between information from multiple disparate data sources is crucial for accelerating hypothesis generation and testing in the pursuit for disease treatments.

19.2 Semantic technologies

A range of software tools have been developed over the past couple of decades to address the growth of Big Data across many sectors. Highly prevalent and successful text-based search tools such as Google [8], Endeca [9], FAST [10], and Lucene/Solr [11] (and discussed in Chapter 14 by Brown and Holbrook) enable users to conduct key word searches against unstructured text repositories with the results of these searches being lists of matching text documents. Although this sort of search paradigm and the tools that enable it are highly valuable, these tools have a number of attributes that are not optimally suited to Big Data scenarios. These attributes include:

- As the amount of data grows, the results derived from text-based search systems can themselves be so numerous that the user is left needing to manually sift through results to find pertinent information.

- Text-based search systems do not provide a means with which to understand the context and interconnections between resulting text. This is because results are generated through key word matching, not a formalized representation of the things (i.e. entities) that a user may search for, their properties and the associations between them.

- Certain disciplines, such as biomedical research and development, are confronted with the challenge that any given entity of interest may have many names, synonyms, and symbols associated with it. Because text-based search systems are based on key word matching, users can miss important search results if they do not search for each of the various names, synonyms, and symbols used to represent a given entity of interest.

As an example of the difficulties associated with text-based search systems for biomedical research, let us imagine the scenario that a user wants to search for all of the information available for the gene ABL1. First, ABL1 has many names and symbols such as ABL1, c-abl, and V-abl Abelson murine leukemia viral oncogene homolog 1. Second, ABL1 can be both a gene and a protein (the translation of the gene). Third, ABL1 is a gene found in many species including human, mouse, rat, and primate. So, our hypothetical user would start in a text-based search system by typing ABL1 and they would receive a large set of document links as a result. That search would only be run across text documents (i.e. unstructured data sources) and thus the user would not be provided with any information about the gene from structured databases or other data sources. Additionally, how does the user know that they have found all documents relevant to ABL1 and all of its synonyms, symbols, and identifiers instead of only those that mention ABL1? Further, how do they refine their search to results matching the gene versus protein version of ABL1 and also the right species? Confronted with this set of challenges, it is often the case that the user simply does not address any of these issues and instead just accepts the results provided and attempts to sift through large quantities of results in order to identify those documents most relevant to their interests. In addition to the difficulties associated with finding all pertinent information for a given entity, it is often the case that the user is not only interested in the single entity but also for other entities with which it might be associated. For example, for a given gene product, like ABL1 protein, there may be compounds that are selective for it, diseases in which it is indicated, and clinical trials that might target it. All of the associations for a user's searched entity (e.g. ABL1) simply are not provided by text-based search systems, and instead the user once again must read through all resulting documents in order to try to glean associations for their search term. By forcing a user to manually read through large sets of documentation, critical properties of entities and knowledge about associations between entities that might be highly valuable to their research can be easily missed. To sum up, text-based search systems, although valuable tools, lack a range of attributes that would be useful in scientific and related disciplines in which a user is interested in searching for and analyzing structured data representations of entities and their associations.

Published by Woodhead Publishing Limited, 2012

Novel technologies have been developed over the course of the past two decades that make it possible to address many of the downsides of text-based search systems outlined above. These novel technologies adopt a 'semantic' approach to the representation of entity properties and their associations. With these technologies it is now possible to design and build next-generation systems for scientific research, which can handle large-volumes of data (Big Data), provide comprehensive search results to a user query no matter what names might be used to query for a given entity, and also automatically and dynamically show the user all of the other entities known to be associated with their queried entity.

19.3 Semantic technologies overview

The foundational set of 'semantic' technologies were originally proposed in 2001, by Tim Berners-Lee and colleagues in a *Scientific American* article entitled, 'The Semantic Web' [12]. The idea of the Semantic Web was to make information about entities and their associations with other entities accessible for computation. A range of flexible, open standard semantic technologies have been developed over the last decade under the leadership of the World Wide Web Consortium (W3C). These technologies enable novel forms of facile integration of large quantities of disparate data, representation of entities and entity associations in a computable form. Combined with comprehensive querying and inferencing capabilities (the ability to compute novel relationships 'on-the-fly'), semantic technologies promise collaborative knowledge building through joint creation of computable data. Furthermore, the foundational components of semantic technologies adhere to open, free standards that make them an excellent framework on which to build next-generation 'Big Data' search and analysis systems.

Semantic technologies utilize a data formatting standard called resource description framework or 'RDF' [13]. RDF defines a very simple but highly flexible data format called 'triples.' A triple is made up of a subject, a predicate and an object. Here is an example:

```
subject → predicate → object
converted into simpler terms produces:
something → does something to → something else
or
something → has some → property
```

Most human languages have evolved to require even the most basic sentence to include a subject and a predicate (grammatically speaking, a predicate is inclusive of the verb and object). To compose a sentence is to assemble a snippet of meaning in a manner that can be readily interpreted by the recipient. An RDF triple is the codification of this requirement in a format suitable for computation. The triple is, to a degree, a basic sentence that can be intelligibly parsed by the computer. In short, RDF provides a standard for assigning meaning to data and representing how it interconnects, thus using semantics to define meaning in computable form. This simple but powerful design allows for computable forms of not only entity information but also the associations between entities.

In addition to the advantages provided by the flexibility of the RDF format, technologies for use with RDF also offer some compelling advantages. For example, RDF storage systems, commonly referred to as 'triple stores,' require little design. In comparison, the data models in relational databases can comprise dozens of tables with complex relationships between both the columns of a single table and between the tables. Diagrams of these tables, their columns, and interconnections are called a relational schema. The complexity of relational schemas can mean that administrators intent on updating a relational data model may need days or weeks to fully comprehend what will happen when values are modified in any given record. Furthermore, attempting to integrate multiple schemas from multiple databases can take months of work. An RDF triple store on the other hand comprises nothing more than a series of triples. This means that the type and properties for all of the things represented in the triple store are codified in one format with a standardized structure. As a result, integration across RDF triple stores, or the inclusion of new triples in a triple store, is as easy as combining sets of triples. Simply put, relational database schemas are fundamentally a non-standardized data format because each schema is different. In comparison, RDF is fundamentally a standardized data format thereby enabling greater data integration flexibility and interoperability.

RDF triple stores utilize a specialized query language called SPARQL [14] that is similar to the query language for relational databases, SQL. Despite the similarity between SPARQL and SQL, triple stores are easier to query because their contents are not partitioned into tables as is usually the case with relational data models. Furthermore, triple stores that are version-compliant can be swapped in and out relatively easily. The same cannot be said of relational database technologies where schemas usually require careful modification when being migrated or interconnected with a different database.

RDF triple stores are also comparatively fast thanks to the simple and fixed triples data format that allows for optimized indexes. This coupled with the inherent scalability available in many triple stores means programs can be written that provide views derived in real time from data present in billions of triples. Many RDF triple stores offer linear performance gains by scaling across many computers without the complex setup that accompanies most traditional relational database clusters. Triple stores can also readily scale across machines, as they do not have to support the complex data relationships associated with SQL-type databases. This characteristic relieves the triple store of the heavy overhead associated with locking and transactions.

All of the properties of semantic technologies including both the flexible, extensible formatting of data as RDF triples and the handling of data with the types of technologies outlined above, make these technologies an excellent foundation for next-generation search and analysis systems for scientific research and related Big Data applications. Of course, in addition to a data format and technologies to handle data, a biomedical search and analysis system would need actual data to be a valuable tool in a researcher's toolbox. Thankfully, there exists an excellent set of publically available biomedical data referred to as the Linking Open Drug Data set for use as a foundation.

19.3.1 Linking Open Drug Data

In the biomedical research and development (R&D) sector, flexible data integration is essential for the potential identification of connections across entity domains (e.g. compound to targets, targets to indications, pathways to indications). However, the vast majority of data currently utilized in biomedical R&D settings is not integrated in ways that make it possible for researchers to intuitively navigate, analyze, and visualize these types of interconnections. Collection, curation, and interlinking public biomedical data is an essential step toward making the sea of Big Data more readily accessible to the biomedical research community. To this end, a task force within the W3C Health Care and Life Sciences Interest Group (HCLSIG) [15] has developed the Linking Open Drug Data (LODD) set [16]. Standards for the representation of data within the LODD set have been defined and all of the data have been made available in the RDF format. The LODD is an excellent resource for the biomedical research and development community because it provides the basis for the interconnection of valuable biological, chemical and other relevant content in a standard format. This standardization of formatting and nomenclature is the critical first step toward an integrated, computable network of biomedically relevant data. It also provides the foundation for the formatting and integration of further data sets over time.

19.4 The design and features of TripleMap

Based on experience working with Big Data and in the biomedical research and development sector, we set out over three years ago to design and develop a next-generation semantic search and analysis system, which we named TripleMap. Users are welcome to register and login to the free web-based version of TripleMap at *www.triplemap.com* in order to test out its various features. The public instance of TripleMap is completely functional, lacking only the administrator features found in Enterprise deployments within an organization. Below, we outline TripleMap's features and then provide additional details regarding three aspects of the system, the Generated Entity Master (GEM) Semantic Data Core, the TripleMap semantic search interface, and collaboration through knowledge map creation and sharing.

TripleMap has a number of features that distinguish it from more traditional enterprise text-based search systems. In particular, it enables collaborative knowledge sharing and makes it possible for organizations to easily build an instance behind their firewall for their internal data. These features are outlined here.

- Text search equivalent to the current best of breed systems: searching in the TripleMap system is comprehensive including the entire contents of all documents in the system. It also provides relevancy scoring in a manner similar to text search engines such as Google, FAST, and Endeca.

- Next-generation semantic search: as a next-generation semantic search platform, TripleMap goes beyond the capabilities of text search engines by giving users the ability to search not only for text in documents but also for entities, their associations, and their meta-data.

- Display and navigation of entity to entity associations: upon finding a given entity, the user is able to easily identify other entities which are associated with their original input. Furthermore, the user is able to navigate through associations and identify novel associations as they move through information space.

- Automated extraction of entity properties, labels, and associations: TripleMap is able to automatically derive the properties, labels, and associations for entities from the data it is given. Instead of having to laboriously create all of the data connections the system is going to use, administrators are able to provide the system with data sources and some basic information about those data sources and the system automatically derives and interconnects all of the data that it is given. The system is able to conduct this process of automated property, label, and association derivation continuously as novel data become available in all of the originating sources it is monitoring.

- Derive data from numerous sources: TripleMap is able to derive data from numerous sources including static RDF files, dynamic remote SPARQL endpoints, relational databases, XML files, tab-delimited text files, Sharepoint TeamSites, RSS feeds, Documentum repositories and networked file system locations.

- Entity sets as persistent associative models or 'maps': TripleMap users are able to create maps of entities as they conduct searches. These maps are structured data representations of everything known in the system including entity properties, entity–entity associations and relevant documents. These maps can be saved and thus persisted across usage sessions in the system.

- Maps as composite search strings against all information streams: any map a user creates is automatically used by the system as a composite search against all information streams that the system monitors. The system constantly scans these information streams (e.g. documents, RSS feeds, wikis, patent literature) for mentions of any of the entities in a user's saved maps and alerts the user if novel information is detected. Because the system is able to handle synonyms, symbols, and many names for any given entity, all mentions of a given entity, no matter what the name, are tied back to the 'master entity' representation of that entity.

- Automated entity alerting: the user is alerted to the appearance of novel information relating to any of the entities in any of their maps. Alerting is available through email, RSS feeds and by viewing maps in the system.

- One data model, multiple views: TripleMap provides multiple views of the map data and each view allows for an alternative visualization of the interconnection of entities in a map.

- Set sharing and searching: users are able to both share the maps that they create with colleagues and easily search through all available maps in order to identify colleagues who are currently working on similar maps or have created and saved similar maps in the past.

- Collaborative knowledge building through de novo user-defined associations: users are able to create novel associations between entities in the system and set the visibility of those de novo associations to private, group-based, or public. The appropriate path back to the originating creator of an association is visible to all users to allow for validation of user-generated associations.

- Web-based application: the system employs novel software technologies that allow for the provision of rich user experiences through a web-based application. Web-based applications provide a number of key advantages including ease of administration, deployment, and updating.

- Data-driven platform: the system is data-driven and flexible enough so that any given instance of the system can be tailored to the specific interests and needs of the administrator.

- Massively scalable: TripleMap scales both in terms of its ability to support large-scale data sets and its ability to support large numbers of concurrent users. A range of high-performance technologies such as computer clustering and high-performance indexing are employed to enable this.

- Advanced association analytics: the system allows users to perform advanced analytics against the full underlying data network provided by all entities and all of the associations between entities available in the system. For example, users are able to infer high-order connections between any two entities in the system.

19.5 TripleMap Generated Entity Master ('GEM') semantic data core

At the heart of the TripleMap system is a large-scale, high-performance data core referred to as the TripleMap Generated Entity Master ('GEM') (the TripleMap architecture with the central GEM data core is shown in Figure 19.1). The data GEM contains the entire master data network available within the TripleMap system and can be continually updated and enhanced by TripleMap administrators as they identify more data sources for integration. The GEM controller is used to aggregate and integrate data from a variety of sources (e.g. RDF, flat file, relational databases, Sharepoint, RSS feeds, patent literature). TripleMap users interact with this data core by running semantic searches and then storing results as sets, or maps of entities and the associations between entities.

Rich Web-Based
Visualization & Analysis
Interface Deployed via
Standard Web
Browsers

Generated Entity Master
('GEM') Controller:
Semantic Technologies
Data Integration Layer
with Control Interface

High Content
Imagery

RDF Data
Sources
(e.g.
LODD)

Relational
Databases

Sharepoint
& Other
Text
Sources

Figure 19.1 The TripleMap architecture. The GEM is the central
semantic data core of any given TripleMap instance

These sets are sub-networks of data derived from the full master data
network based on the user's interests. Maps can be shared and dynamically
updated thereby enabling collaborative knowledge sharing.

TripleMap proprietary GEM technology is based on a combination of
systems for integrating data across data sources and indexing those data
for high-performance search results, along with the use of a standards-
compliant triple store for storage of the originating data. Any of a number
of triple stores including bigdata®, Sesame, Allegrograph or 4Store can
be used with TripleMap. In current installations, the TripleMap
architecture in association with a standards-compliant triple store is
capable of handling >1 billion triples with millisecond time performance
for the return of search results.

In building a new instance of TripleMap, an administrator starts by
importing multiple data sources for integration into the GEM controller.
Once imported, the administrator determines which entities in the
originating data sources should be combined and represented in TripleMap.
The administrator then creates 'master entities' and through a simple
application interface ties entities from each data source through to the
appropriate 'master entities'. Once this data modeling and linkage process

is complete, the administrator clicks the 'Generate GEM' button and TripleMap initiates a completely automated, comprehensive process referred to as 'data stitching'. During data stitching, entities from the originating sources are integrated based on a set of factors including labels and associative predicates. TripleMap is able to automatically derive and generate all of the integrated representations of entity properties, the associations between entities and the full set of names, synonyms, and symbols for each entity. This process of data import, integration, and stitching can be conducted at any time for a TripleMap instance, and can be run whenever new data become available, thereby allowing administrators to continuously and iteratively update the master data network for users. This full process of GEM creation and administration is performed routinely for the public instance of TripleMap at *www.triplemap.com*.

One of the key advantages of the TripleMap GEM data core is that it provides a mechanism through which administrators can set up and maintain a large-scale, 'master' network of integrated data (Figure 19.2). TripleMap users search and collaborate using this master data network. This concept of a centralized data network differs from applications where users upload data on a use-by-use basis in order to analyze and interact with the data. Instead TripleMap is a comprehensive, continuously updating system that provides a means of searching and interacting with a large-scale, pervasive network of entity-relevant data and associations between entities.

19.6 TripleMap semantic search interface

TripleMap is designed as a semantic search interface with the added capability for users to store and share sets, or maps, of structured search results. The sets of entities that are stored by users have both meta-data properties and entity to entity associations and hence can also be viewed as maps of those entities and the associations between them (Figure 19.3). The left-hand panel of the web application contains a semantic search interface including entity search, faceted filtering and quick link capabilities. Users can drag and drop entities from their search results into the knowledge map view in the right-hand panel. This interplay between searching and collecting sets of entities in knowledge maps is at the heart of the TripleMap usage paradigm. With one or more entities selected, further connections are shown as icons and counts to the right of the search panel. Users can see a list of all of their maps with the Maps List view and also a printout of all properties of each entity as a 'Semantic

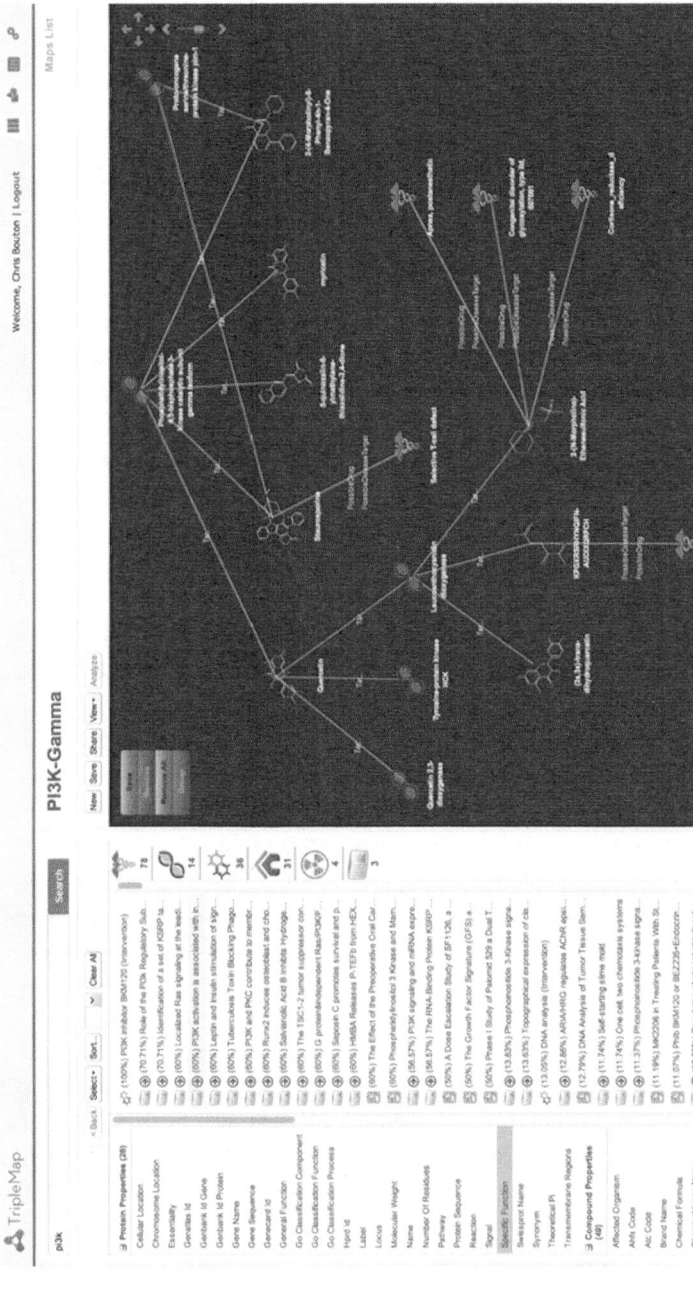

Figure 19.2 Entities and their associations comprise the GEM data network. The TripleMap GEM data network is generated through the representation of entities (e.g. proteins, genes, diseases, assays) and the associations between those entities (e.g. activation, inhibition, associations). Each knowledge map created by a user is a subnetwork of the master entity data network created within the GEM semantic data core

Knowledge Review' in the right-hand panel. Using the Maps List or Map view interfaces, users can securely share maps with each other via email invites and receive alerts (either email or RSS) to new information that is published or added to the GEM data core about interesting entities.

19.7 TripleMap collaborative, dynamic knowledge maps

The rapid growth of the internet has fostered the development of novel 'social media' technologies. These technologies enable large communities of individuals to share information, communicate and interact. The rapid adoption of these systems by user communities speaks to their value for human interaction. As successful as these systems are however, many have very limited semantic capabilities. Twitter [17], for example, allows for nothing more than the sharing and forwarding of small snippets of text (140 characters to be exact) with the 'hashtag' ('#') serving as the only meta-data feature. Facebook [18], another highly successful social system, allows for the simple sharing of personal information and photos. YouTube [19] allows for the sharing of video and LinkedIn [20] allows for the sharing of one's professional information. These are all fairly simple systems and yet they have become massively pervasive, impactful, and successful. As discussed in the chapters by Alquier (Chapter 16), Wild (Chapter 18) and Harland et al. (Chapter 17), we share the vision that as semantic technologies establish themselves, the next generation of collaboration tools will evolve through the more sophisticated capabilities enabled by computable knowledge representation. With that in mind we have designed TripleMap so that it provides an extension of the collaborative capabilities of platforms such as wikis, Twitter, and social networks by utilizing semantic technologies to enable users to share and collaborate around the creation of structured data representations or maps of entities and the associations between entities. The maps in TripleMap are structured data representations of what is known about the entities in a given domain space (e.g. drugs, diseases, targets, clinical trials, documents, people, organizations, the meta-data for those entities, and the associations between them). Maps are created as users search for and identify the entities that are of interest to them. Furthermore, we refer to the maps that are created by users as 'dynamic knowledge maps' (Figure 19.3) because once created by a user any given map is continuously and automatically updated with the latest information being published from structured (e.g. databases) and unstructured (e.g. document) sources.

Knowledge maps promote collaboration because once built they can be shared with colleagues in private groups or can be shared publicly with the entire TripleMap community. Each knowledge map provides a 'bird's eye view' of the things in which a user is most interested (Figure 19.3). Users are able to save and share the knowledge maps that they create. The creation and saving of maps is valuable for several reasons. First, in the act of creating maps, the associations, including unexpected associations, between things are discovered because the system prompts the user with all associations between entities that are stored in the master data network of the GEM data core. Second, maps can be shared with others, thereby allowing users to share structured data representations of the information in which they are interested with their colleagues in groups or more broadly with the wider research community. Third, users can search through all available maps to identify other users working on maps similar to theirs. Finally, maps can be used

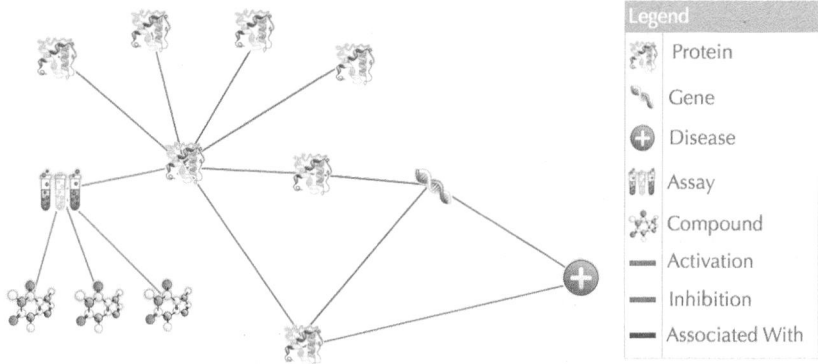

Figure 19.3 TripleMap web application with knowledge maps. The TripleMap web application interface is shown. Knowledge maps are created from search results and then saved and shared by users of the TripleMap system. Maps are used by the system as automatic scans against all unstructured (e.g. documents, journal articles, patent text, RSS feeds) sources available to the system such that novel information about anything in any map is highlighted and the user is alerted (via email or RSS) once it is detected

for auto-searching against all new information being scanned by the system and alerting a user if novel information is available for one or more of the things in which they are interested.

Knowledge building through the sharing of knowledge maps is similar to what is seen in wiki communities where users collaborate around the development of pages of text that describe various topics. The difference between shared map building and shared wiki page building is in the structuring of the data which users are pulling from to create maps in the first place. TripleMap knowledge maps are effectively subnetworks of structured, interlinked data describing things and the associations between them. These subnetworks are portions of the entire master data network integrated and continuously updated in the TripleMap GEM. In the wiki scenario there is less structuring of the content generated and it is more difficult to identify interrelationships between the things mentioned in pages. Despite the differences between TripleMap and wikis, there are valuable ways in which these two types of systems can be linked together in order to provide a more comprehensive collaborative platform for biomedical research. For example, integration of a system like the Targetpedia (see Harland et al., Chapter 17) with TripleMap allows users to interact with information about targets both as a text-based wiki page and then also link out to knowledge maps related to each target within one integrated system.

19.8 Comparison and integration with third-party systems

When used in an enterprise context, TripleMap is not redundant with general-purpose document search systems such as FAST or Google (both are discussed above). Instead TripleMap provides a framework and application for the semantic search that enhances the utility of enterprise document search systems and can be used in conjunction with enterprise search frameworks such as Google or the FAST search engine.

Additionally, TripleMap is not redundant with applications focused on quantitative data analytics such as Spotfire [21] or Tableau [22]. Instead TripleMap provides an environment in which users can conduct semantic searches across a large-scale shared master network of data in order to identify and build 'bird's eye view' representations of what is known in a given information space. Software bridges to specialized systems such as Spotfire or Tableau through available third-party API bridges are being developed so that users can conduct further quantitative data analysis on entities identified and monitored in TripleMap.

19.9 Conclusions

TripleMap is a next-generation semantic search, knowledge discovery and collaboration platform. We have provided an instance of TripleMap based on data from the Linking Open Drug Data sets for free to the biomedical research community at *www.triplemap.com*. Organizations can also bring TripleMap internally for use with proprietary data using the TripleMap Enterprise platform. This has indeed been the case with a number of commercial life science companies who are using the technology to integrate their internal data and documents alongside public content. TripleMap is built on and extends open standards and semantic technologies developed primarily by the W3C. These open standards are critically important for the uptake of solutions based on these standards and the next-generation data-handling capabilities that they enable. We believe that the future holds tremendous promise for the derivation of insights from the vast troves of Big Data available to humanity. Furthermore, we believe that the design, development, and deployment of software systems that enable this derivation of insight from Big Data sources is crucial. Our goal in making the *www.triplemap.com* instance of TripleMap available for free to the biomedical research community is to foster collaborative interaction and discovery in the pursuit of the development of fundamental, ground-breaking treatments for human diseases.

19.10 References

[1]   *http://www.triplemap.com*
[2]   *http://www.w3.org/wiki/HCLSIG/LODD*
[3]   Gantz BJ, Reinsel D. *Extracting Value from Chaos State of the Universe?: An Executive Summary.* 2011;(June):1–12.
[4]   Hal R, Varian PL. How Much Information? 2003. Retrieved from *http://www2.sims.berkeley.edu/research/projects/how-much-info-2003/*
[5]   Hilbert M, López P. The world's technological capacity to store, communicate, and compute information. *Science (New York, N.Y.)* 2011;332(6025):60–5. doi:10.112/science.1200970
[6]   Manyika J, Chui M, Brown B, Bughin J, Dobbs R, Roxburgh C. *Big data: The next frontier for innovation, competition, and productivity.* 2011;(May).

[7]   Lander ES, Linton LM, Birren B, et al. Initial sequencing and analysis of the human genome. *Nature* 2001;409(6822):860–921. doi:10.1038/35057062

[8]   *http://www.google.com*

[9]   *http://www.endeca.com*

[10] *http://www.microsoft.com/enterprisesearch/en/us/fast-customer.aspx*

[11] *http://lucene.apache.org/java/docs/index.html*

[12] Berners-Lee T, Hendler J, Lassila O. The Semantic Web. *Scientific American* 2001.

[13] *http://www.w3.org/RDF/*

[14] http://www.w3.org/TR/rdf-sparql-query/

[15] *http://www.w3.org/wiki/HCLSIG*

[16] Samwald M, Jentzsch A, Bouton C, et al. Linking Open Drug Data for pharmaceutical research and development. *Journal of Cheminformatics* 2011;3(1):19. Chemistry Central Ltd. doi:10.1186/1758-2946-3-19

[17] *http://www.twitter.com*

[18] *http://www.facebook.com*

[19] *http://www.youtube.com*

[20] *http://www.linkedin.com*

[21] *http://spotfire.tibco.com/*

[22] *http://www.tableausoftware.com/*

Extreme scale clinical analytics with open source software

Kirk Elder and Brian Ellenberger

Abstract: Knowledge is at the root of understanding all symptoms, diagnosing every ailment, and curing every disease. This knowledge comes from the deep studies performed by research organizations and diligent healthcare workers who contribute to documenting and responsibly sharing their observations. Through the American Recovery and Reinvestment Act of 2009 (ARRA [1]), the industry was incented to implement electronic medical record systems that capture more information than ever before. When billions of medical records converge within a secure network, the baton will be handed to analytics systems to make use of the data; are they ready? This chapter explores what the next-generation software infrastructure for clinical analytics looks like. We discuss integration frameworks, workflow pipelines, and 'Big Data' storage and processing solutions such as NoSQL and Hadoop, and conclude with a vision of how clinical analytics must evolve if it is to handle the recent explosion in human health data.

Key words: clinical data; ICD10; HL7; electronic health records; Big Data, NoSQL.

20.1 Introduction

One of the largest problems in clinical analytics is that the immense breadth of available services results in a very diverse set of implementation details. Standardizing every scenario quickly becomes impossible. The future of clinical analytics therefore needs an architecture that admits that the sources of data will never be perfectly normalized. Open source communities have been the quickest to address these types of problems. Architectural patterns created and contributed to the community by the thought leaders in internet-scale computing can be leveraged to solve this problem. In fact, solving these problems with these tools will one day become the cheapest, simplest, and most highly scalable solution.

This chapter will introduce a prototypical architecture that can currently satisfy various simple real-world scenarios but has not yet been fully leveraged. There are several good reasons for this. Most of the trends and capabilities introduced here have emerged in just the past few years. Yet experience suggests that the total cost of development and ownership will be orders of magnitude cheaper when these solutions hit their prime. Figure 20.1 demonstrates a generic information technology view of the components of the stack required to fulfill this domain. Most architects have drawn this picture many times. To the left the integration layer brings in detailed transactional information, not suitable for reporting. In the middle we extract, transform, and load (ETL) these data into a data warehouse, where data can be mined using the many reporting tools available.

Many architects have also come to understand that the traditional model described in Figure 20.1 is very expensive to scale to extremely large volumes. In response, this chapter will survey many open source technologies that describe a potential solution addressing both clinical requirements and extreme scales. Throughout the text, a simple use-case of researching potential factors of acute heart disease is presented, to provide a sense of how the technology would work in the real world. The

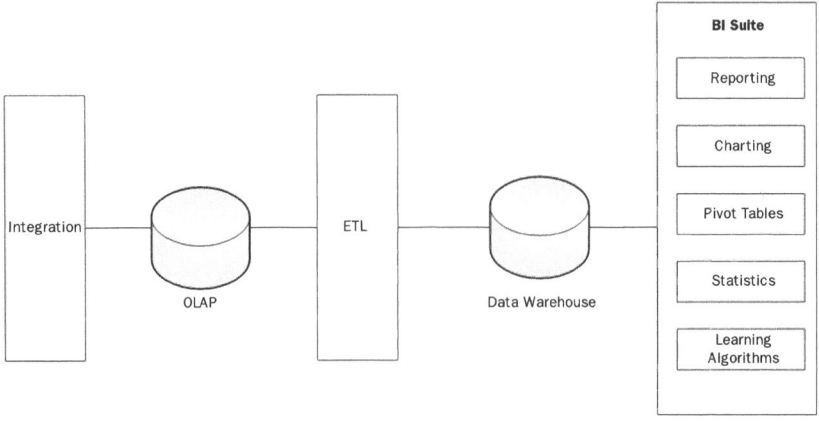

Figure 20.1　Architecture for analytical processing

example will only lightly touch issues of Health Insurance Portability and Accountability Act (HIPPA) security including the agreements, auditing, and proper practices that need to be in place to allow end-users to operate on the data for purposes of healthcare operations (e.g. large insurance company), or public health (e.g. Center For Disease Control). Although non-trivial, such considerations apply to any technological solution and form a discussion out of scope for the main emphasis of this chapter.

20.2 Interoperability

The first step in clinical analytics is to carefully gather as much meaningful data as possible and process it into a form amenable to downstream analysis. The diagnosis and treatment of patients generates a tremendous amount of data from a large number of sources. A typical hospital will have any number of different systems such as health information systems (HIS [2]), radiology information systems (RIS), computerized physician order entry (CPOE), and electronic health records (EHR). Many hospitals may even have multiple systems from different vendors. How do they make sense of all of this complex health information coming from various different sources?

This was the reason that Health Level Seven International (HL7 [3, 4]) was founded. It is a not-for-profit organization developing standards 'for the exchange, integration, sharing, and retrieval of electronic health information that supports clinical practice and the management, delivery and evaluation of health services' [3]. Its standards attempt to navigate the thin line between enabling interoperability and allowing vendors and hospitals flexibility in a complex domain. This is an important point because the data feeds from the individual systems may require transformation in order to be consistent in the analytical data store. HL7 covers a broad range of topics, including application interoperability at the user interface (UI) level (e.g. Clinical Context Object Workgroup or CCOW) and Medical Logic Modules. The two aspects most relevant for analytics are the messaging standard and Clinical Document Architecture (CDA). HL7 V2.x is still the most widely used set of standards. The V2.x message is a character string with delimiters differentiating parts of the message. Messages are made up of segments as shown in Figure 20.2.

PID|||921342||SMITH^JOHN<cr>

Figure 20.2   HL7 V2.x message sample

Specifically:

- PID identifies this as a patient identifier segment.
- Segments are composed of fields, which are delimited by the pipe (|) character.
- Fields one and two are blank, which is why there are three pipe characters in a row.
- Field three is the patient identifier, which in this example is '921342'. This field may have additional information as to how this identifier was generated.
- Field five is the patient name. This field is of type XPN, or extended person name. This type of field can be further broken down into components (separated by the caret symbol '^' for components and the ampersand '&' for subcomponents). With this type the first component is family name and the second given name.

This is an extremely simple example of an HL7 V2.x message. The entire specification for V2.6 messaging spans 27 documents. HL7 Version 3 (V3) has made a number of improvements, including embracing XML for data communication over delimiters. It also went beyond cosmetic changes and developed a semantic ontology for the transmission of medical data between systems. The HL7 reference implementation model (RIM) defines a core set of objects and message life cycles which they intend to use as a base model for any message communication between clinical systems, similar to a set of mathematical axioms.

The RIM is not simply an academic exercise, but is critical to interoperability. This is because in any sort of complex communication, context is vital. In verbal communication, words and phrases taken out of context completely lose their meaning. The same is true in communication between medical systems. For example, consider the medication associated with a patient. Does this mean that the patient is currently taking this medication, or has this medication been ordered for this patient, or does this patient have an allergy to this medication? Simply associating patient and medication together without context is meaningless. The RIM forms the core foundation to give context to these data points.

One of the more interesting standards coming out of HL7 V3 is the CDA, especially for its application in analytics. Previously, most of the clinical information has been kept in narrative form in clinical documents such as history and physicals, operative reports, discharge summaries, and so on. As these documents are in text form, it is difficult to derive

```
<ClinicalDocument ... />
...
...
{1}<recordTarget>
 <patientRole>
   <id extension="921342" ... />
     <patient>
       <name>
         <given>John</given>
         <family>Smith</family>
       </name>
     </patient>
 </patientRole>
</recordTarget>
...
...
<component>
 <section>
   <code code="10155-0" ... codeSystemName="LOINC"/>
   {2}<title>Allergies</title>
       <text>Patient is allergic to penicillin resulting in hives.</text>
   <entry>
     {3}<observation classCode="OBS" ...>
     {4}<code xsi:type="CD" code="247472004" ... codeSystemName="SNOMED CT"

< displayName="Hives"/>
     ....
     {5}<entryRelationship typeCode="MFST">
       <observation classCode="OBS" ....>
       {6}<code code="62014003" ... codeSystemName="SNOMED CT" displayName="Adverse reaction">
         <qualifier>
           {7}<name code="246075003" ... codeSystemName="SNOMED CT" displayName="causative agent"/>
             <value code="6369005" ... codeSystemName="SNOMED CT" displayName="penicillin"/>
         </qualifier>
       </code>
     </observation>
   </entry>
 ...
 ...
<ClinicalDocument>
```

Numbers contained within {} are for reference and are not apart of the document.
Ellipses appearing in the document are portions removed for brevity

Figure 20.3 HL7 V3.x CDA sample

meaningful analytics from them. In order to do analytical analysis, these data need to be in a machine-understandable format. Subsequently, the data can be stored in discrete chunks by a computer in a format that has one and only one semantic meaning. The CDA is meant to solve this problem by providing a way to exchange clinical documentation and retain both the narrative text and the machine-understandable format. Figure 20.3 shows an example of a portion of a CDA document.

The message is made from a number of components highlighted in the figure:

1. the patient record from HL7 V2.x in the V3 format for CDA;

2. the section of this document has a title of Allergies and a text indicating a patient allergy. This is narrative for human use;

3. the machine-processable portion, beginning with an observation;

4. the observation has an associated code that indicates that the patient contracts hives. CDA relies on external vocabularies for describing specific medical concepts. This example is using a code from SNOMED CT;

5. next, the CDA document describes why the patient gets hives. The entryRelationship relates two entities, with the typeCode MFST meaning manifestation;

6. another observation, this time of an adverse reaction;

7. finally, a name value pair indicating penicillin as the cause.

20.3 Mirth

With both HL7 V2 and V3, it is tempting to build individual scripts or programs to handle each of the different interactions. There are open source toolkits in a variety of languages for parsing and manipulating HL7 messages. For example, the Perl HL7 toolkit [4] could be used to write a script that reads an HL7 orders feed, processes it, and outputs the results needed into an analytics database. The problem is that this sort of solution is very difficult to scale out to meet the needs of a modern hospital for the following reasons:

1. Large effort to write and maintain
   - Implementers should consider not only the time it takes to write all of those scripts/programs but also maintaining the scripts. This includes keeping some form of version control for each of the scripts and updating them as applications change.

2. Duplication of effort
   - Implementers often spend a large amount of time on 'housekeeping' items such as file and database processing.
   - Many of the scripts/programs will be very similar with slight differences in the business logic.

3. Fragility
   - Implementers have to account for connectivity and application failures in each of the scripts/programs.
   - There will be no way to monitor all of the scripts/programs for errors or failures unless a monitoring framework is implemented.

Individual point solutions usually collapse under their own weight. They are fragile and require so much time and effort to enhance and maintain that they ultimately do not accomplish the fundamental task. To use an analogy from the physical world, a better solution would be a series of pipes to connect the various sources to our analytical database. However, as previously mentioned, HL7 standards have a large amount of leeway in their implementation. Continuing the analogy, these pipes will not necessarily fit together without adaptors.

Mirth Connect [5] is an open source solution that provides the plumbing and adaptors to connect clinical systems to an analytical store. It provides the pipes in the form of what Mirth refers to as channels. The channels are attached to various inputs and outputs via connectors (Figure 20.4). Filters can be added to the channels to allow for filtering out any data that are not relevant. Transformers allow for the selection and transformation of data.

20.3.1 Connector creation

Mirth supports a wide variety of input connectors, from more general protocols such as TCP or HTTP to more domain-specific ones such as HL7 Lower Level Protocol (LLP) and DICOM. Mirth also supports attaching channels to one another via connectors. The input connections support a wide variety of customization and error handling specific to the protocol that is being used, from timeout and buffer sizes for network protocols to automatic file deletion or movement for the file reader. This is detailed work that is often missed when writing custom scripts. The other important aspect to setting up a connector is setting the data types. This awareness of medical data types is what separates Mirth from a more general message processing system. Mirth can process HL7 v2.x, HL7 v3.0, X12, EDI, XML, NCPDP, DICOM, and general delimited text. There is a separate set of data types for the inbound and outbound source connector as well as the destination connectors.

Depending on the purpose of the analytical data store, an implementer may only be interested in a subset of the data coming from the source systems. Mirth allows for the creation of filters to filter out any data that are not relevant. Filters can be created both on the source connector and the destination connector. Filtering at the destination connector is useful if there are multiple destinations with a subset of the data going

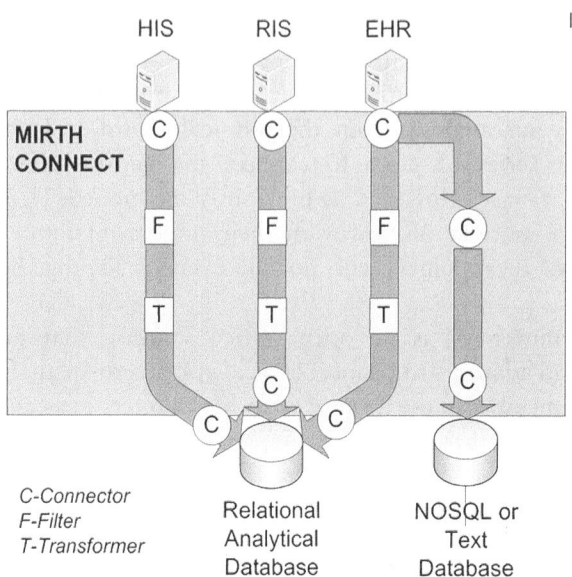

HIS   RIS   EHR

MIRTH
CONNECT

C-Connector
F-Filter
T-Transformer

Relational
Analytical
Database

NOSQL or
Text
Database

Figure 20.4 Mirth Connect showing the channels from the data sources to the databases. Note that channels can connect to other channels. In the example, the EHR connection is split such that both filtered/transformed and raw data can be inserted into independent data stores

to each. Filters are broken up into a set of rules, which are connected to each other by boolean operators. Rules can be written either using a Rule Builder or by executing a small Javascript for more complex rules.

Mirth uses a facility's own sample messages to assist in building filters and transformations. Once a sample message is loaded either from a file or by pasting it into the window, as shown in Figure 20.5, a message tree is built with the message hierarchy from the data type based on the sample data. This message tree is shown in Figure 20.6. Implementers can then drag the data from this tree into either the Rule Builder or Javascript, and Mirth will translate it into the correct variable from the message.

Transformers have a similar interface to filters. Like the filters, the sample input and output messages create a message tree that is used to build a list of transformer steps. There are a couple of different options

Inbound Message Template

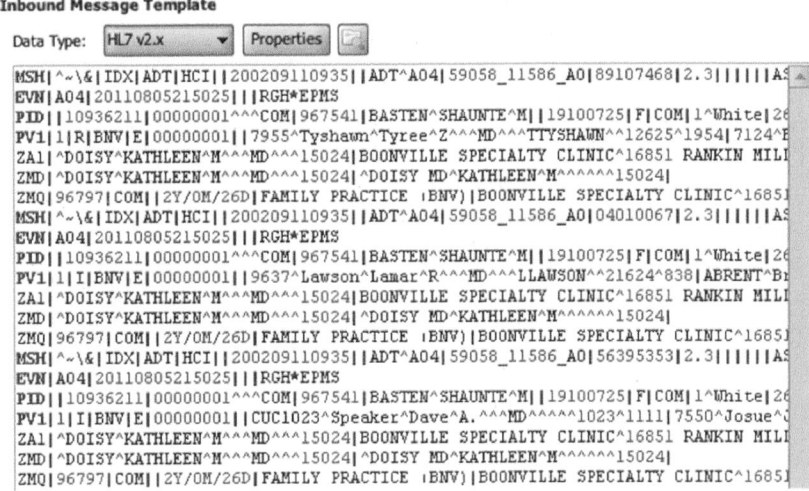

Figure 20.5 Screenshot of Mirth loading template

for the transforms. The Message Builder interface takes an element from the inbound message and puts it in an element in the outbound, with the ability to do string replacement on the element. The Mapper interface is used to map elements to a variable for use later. For analytics, this is useful because fields can be mapped to variables that can then be mapped to database statements. For XML messages, there is the option of executing an XSLT script. Finally, custom Javascript code can be written to perform more complex transformations.

Once the data are transformed, they need to be inserted into the analytical database. If using a relational database, such as MySQL [6], the database writer can be used to insert the data. After inputting the driver and connection information, Mirth can generate sample SQL insert statements. Afterwards, variables mapped during the transform can be used with the SQL insert. If a database is not listed, Mirth does allow the installation of custom database drivers. However, for non-traditional NoSQL databases such as CouchDB [7], this database connection mechanism will not work. Mirth does support the same variety of connection types that the input connector supports and thus, either Javascript or HTTP can be used to communicate with a NoSQL database.

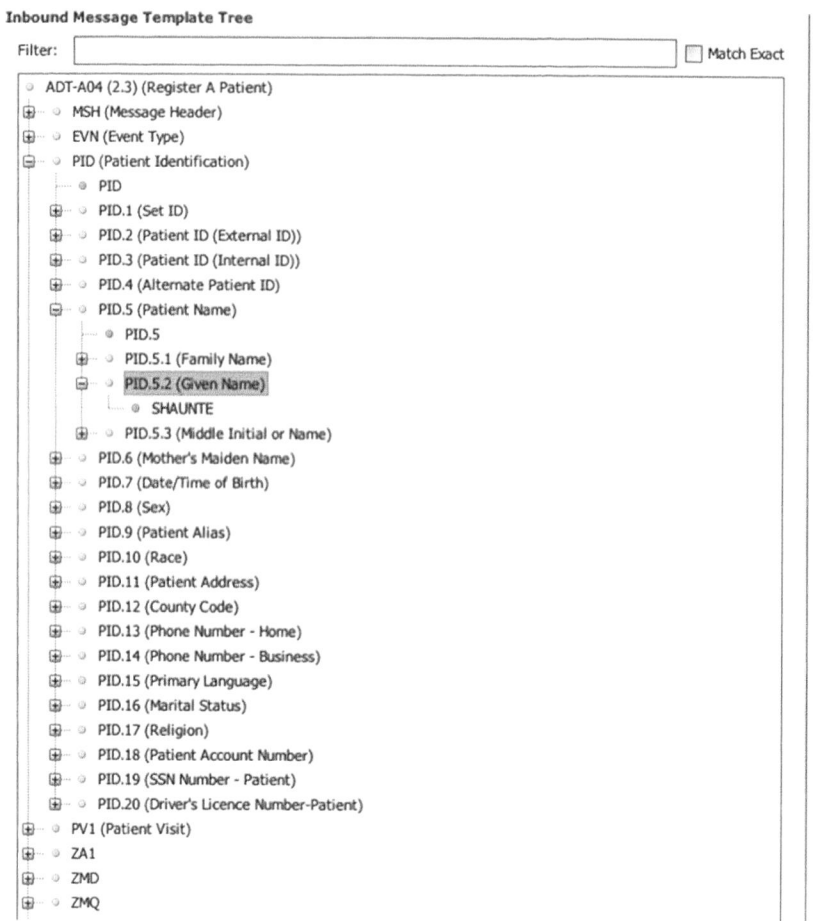

Inbound Message Template Tree

Filter: [   ] ☐ Match Exact

- ADT-A04 (2.3) (Register A Patient)
  - ⊞ MSH (Message Header)
  - ⊞ EVN (Event Type)
  - ⊟ PID (Patient Identification)
    - PID
    - ⊞ PID.1 (Set ID)
    - ⊞ PID.2 (Patient ID (External ID))
    - ⊞ PID.3 (Patient ID (Internal ID))
    - ⊞ PID.4 (Alternate Patient ID)
    - ⊟ PID.5 (Patient Name)
      - PID.5
      - ⊞ PID.5.1 (Family Name)
      - ⊟ PID.5.2 (Given Name)
        - SHAUNTE
      - ⊞ PID.5.3 (Middle Initial or Name)
    - ⊞ PID.6 (Mother's Maiden Name)
    - ⊞ PID.7 (Date/Time of Birth)
    - ⊞ PID.8 (Sex)
    - ⊞ PID.9 (Patient Alias)
    - ⊞ PID.10 (Race)
    - ⊞ PID.11 (Patient Address)
    - ⊞ PID.12 (County Code)
    - ⊞ PID.13 (Phone Number - Home)
    - ⊞ PID.14 (Phone Number - Business)
    - ⊞ PID.15 (Primary Language)
    - ⊞ PID.16 (Marital Status)
    - ⊞ PID.17 (Religion)
    - ⊞ PID.18 (Patient Account Number)
    - ⊞ PID.19 (SSN Number - Patient)
    - ⊞ PID.20 (Driver's Licence Number-Patient)
  - ⊞ PV1 (Patient Visit)
  - ⊞ ZA1
  - ⊞ ZMD
  - ⊞ ZMQ

Figure 20.6   The resulting Mirth message tree

20.3.2 Monitoring

Of course, any integration workflow is unlikely to be permanently problem-free. Mirth does an excellent job in not only providing a solution for implementing the integration, but also monitoring the integration during its lifecycle. The first thing users see when starting Mirth is the Dashboard, which shows all the channels and statistics about the channels. From here, users can drill down into a channel and perform queries on the messages Mirth has processed. Mirth will store messages for a configurable period of time with optional encryption available. It also allows for viewing the raw message, the transformed message, the final message, and any errors that occurred. Finally, alerts may be created to notify individuals via email, specifically customized based on error type and channel where required.

20.4 Mule ESB

As we have seen, Mirth is an excellent solution for integrating medical systems into an analytical data store. However, what about data from sources such as an Enterprise Resource Planning (ERP) system or a spreadsheet containing nursing schedules? These are examples of data that may be relevant to analytical queries but are not necessarily clinical in nature. For the same reasons as before, creating individual point solutions is less favorable than integrated approaches that consistently handle the details around data transport, error handling, monitoring, etc. Enterprise Service Buses (ESB) are commonly used to enable the integration of a heterogeneous set of applications. One of the most popular open source ESBs is Mule ESB [8]. As previously mentioned, Mule ESB is one of the foundational components for Mirth.

ESBs are typically used within a Service Oriented Architecture (SOA), enabling interoperability between a set of loosely coupled services. Each of the services will define a contract, a way of interacting with the service. With HL7 services the contract is predetermined thanks to the inherent standards, but for non-HL7 services this must be created. For example, an administrator may define a contract for querying financial information from an ERP. After defining such services, they are then registered with Mule ESB, which provides the plumbing to connect our services to one another. Mule handles the transport, routing, delivery, and transformation of the message as well handling alerts and monitoring. However, it requires more work in setup and configuration than Mirth because Mule ESB does not have any specialized knowledge of the various systems with which it interfaces.

20.5 Unified Medical Language System (UMLS)

The term heart attack properly refers to a myocardial infarction. However, non-medical personnel often use the term to mean any sudden cardiac arrest, even if it is due to a cause such as long QT syndrome. A medical billing coder may code the event in the ICD-9 [9] vocabulary as 410.02 'Acute myocardial infarction of anterolateral wall subsequent episode of care', whereas a cardiologist may classify the infarction as an 'acute Q wave infarction – anterolateral' which has a SNOMED CT [10] code of 233827001. Obviously, the cardiologist needs a more detailed and descriptive vocabulary than the billing coder. HL7 refers to these different sets of vocabularies as code systems. Where possible, HL7 does not duplicate these code sets but instead relies on the code sets already developed by other organizations. In HL7 V3.x, when using a code it is necessary to refer to the code system from which it originated. For example, take the following XML snippet:

```
<code code='233827001' codeSystem='2.16.840.1.113883.6.96'
codeSystemName='SNOMED CT' displayName='acute Q wave infarction –
anterolateral'>
```

Most of this is self-explanatory, except for the code system. This is an object identifier (OID), which is a code used to refer to the SNOMED CT vocabulary itself. Another dimension relevant to analytics is that systems are not flat but hierarchical in nature. An acute Q wave infarction is a type of acute myocardial infarction, which is a type of acute heart disease, and so forth. To be useful, an analytical query on heart disease will need to contain all records that contain references to any type of heart disease.

The Unified Medical Language System (UMLS) is a project from the US National Library of Medicine (NLM) to bring together various medical vocabularies to enable interoperability [11]. UMLS is organized into a concept hierarchy, starting with a unique medical concept. The unique concept is then related to all known lexical and string variants of the concept. These are then related to individual codes from various clinical vocabularies. Although UMLS is not strictly open source, there is no charge associated with licensing UMLS and the actual data provided can be queried without special tools. However, there may be separate license fees associated with the use of specific terminology sets such as SNOMED CT. The NLM does require that users give a brief report annually of the usefulness of the Metathesaurus.

The UMLS data set is broken up into three areas:

1. Metathesaurus, which contains the vocabularies and relationships;

2. Semantic Network, which contains a set of semantic types (ex. Anatomical Structure, Substance, Finding) and semantic relationships (ex. disrupts, causes, manifestation of), which are used to provide additional meaning to the Metathesaurus;

3. SPECIALIST Lexicon and lexical tools. The SPECIALIST Lexicon adds over 200 000 additional terms from various sources and includes commonly occurring English words. The lexical tools are used to assist in Natural Language Processing.

UMLS bridges the terminology users will use in accessing the analytical data store and the codes contained in the documents. For example, a user wants to find all documents related to 'Acute Myocardial infarction' in a clinical data store with documents coded using SNOMED CT. With UMLS, users can find a mapping from the English term to the SNOMED CT code, and then do a second query to find all SNOMED CT codes whose ancestor is 'Acute Myocardial infarction'. The results of this second query can be used as a filter in the analytical data store.

UMLS does not solve all text matching and text scrubbing problems. Our experience tells us that the last mile of matching is a continuous refinement and build-up of rules and samples that can be matched as time progresses. If made configurable, end-users can populate the queries that help with the mappings and data extraction.

20.6 Open source databases

The next step is data storage. Our use-case poses several challenges on choice of technologies. First, it is becoming increasingly difficult to build a single system that supports the myriad of implementation details of even small regional sets of healthcare providers. CDA is flexible and extensible, so similarly flexible mechanisms to store a complete set of disparate, raw data are required. Second, the volume of data is expected to be extremely large. Medical records, radiology images, and lab or research data are notorious for large file components of high-fidelity information that contain more information than is immediately usable given any immediate questions. These requirements generally wreak havoc on traditional application development. Third, as our understanding of healthcare and the human body evolves, we need to support new questions being asked of old data. The goal, therefore, is to evaluate open source technology's ability to meet the following requirements:

1. ability to store extreme amounts of data in a flexible schema;
2. ability to re-process this data on-demand with new business rules;
3. ability to re-process data to create marts or data-cubes that allow ad hoc analysis on new questions.

The tendency to utilize expensive closed source databases is waning, and more and more people are learning that robust, highly scalable, distributed data storage solutions are available in the open source realm. There has been an even more dramatic shift, though. The ecosystem of application development patterns, using SQL and relying on overly complicated joins of normalized tables, is giving way to the convention and ease of NoSQL databases. The industry has learnt that if the application development paradigms of the early 2000s are re-designed, without relying on SQL, then linearly scalable data storage will be within our reach. In our interoperable analytics architecture we are going to study implementation tradeoffs between four distinct flavors of schema design.

The traditional manner of approaching application development and data storage is through Relational Database Management Systems (RDBMS). These technologies offer widely understood Structured Query Language (SQL) interfaces through application level protocols such as Open DataBase Connectivity (ODBC) and Java DataBase Connectivity (JDBC). We will briefly discuss applying the most popular open source RDBMS, MySQL, to our acute heart disease analytics scenario. MySQL has a long history. Michael Widenius and David Axmark originally developed it in the mid-1990s [12], and it is now owned by the Oracle corporation. MySQL has claimed many millions of installations, it supports over 20 platforms, and is used by some of the largest internet sites in the world. MySQL is a safe choice for open source database storage. PostGresSQL is another mature, stable, and safe choice. Whereas MySQL supports a number of storage back-ends, PostGresSQL is famous for offering pluggable procedural language utilities, which offer some alternative advantages that we will discuss.

Using a RDBMS to handle our analytical requirements would require traditional data schema design and data transformation. A relational schema would be created that supports storing each and every data field from the clinical document. CDA is quite extensive and flexible, so to cover longitudinal scenarios, across a spectrum of providers and EMR systems, a generalized approach is needed; indeed, a generic table structure that allows new data items to be stored as rows could be created. A meta-data model would then need to be designed and administered to define what each row meant, such as that in Figure 20.7.

Such a data model is extremely flexible in the face of many requirements and an ever-changing landscape of healthcare data. The problem with it is that at large scales it will not perform well, and it will be cumbersome to work with. The variation of the data that would be held in the 'CDA

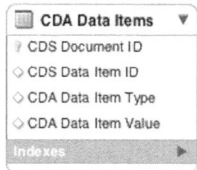

Figure 20.7 CDA data model

Data Item Value' column and the number of rows produced by un-marshaling the CDA XML will prove to be a challenge for scalability and a productivity hit for developers. It seems that to get the level of flexibility desired for longitudinal medical records sacrifices the very reason for having a relational database!

An alternative design would be to store the entire clinical document directly in the database. Many commercial relational databases provide extensive XML support as a native data type. This allows easy storage, intelligent search, and manipulation of XML using powerful and familiar standards such as XPath and XQuery. MySQL does not support XML data types natively, but does support some basic procedural functions to help translate XML to and from the data model. When storing raw XML in MySQL, the best option is storing it in the TEXT data type and manipulating it with a programming language. Using the TEXT data type allows use of MySQL's full-text search capabilities to find specific content en masse. However, this approach may cause issues with the accuracy of results compared to a search mechanism that understands the differences between the tags, the attributes, and the data. A better alternative for storing raw XML would be a specialized database such as BaseX, or eXist-db. These databases can be paired with MySQL such that the raw XML is stored in an alternative location. However, an easier alternative is to skip the use of a relational database altogether and use a NoSQL database instead.

20.6.1 A NoSQL approach

NoSQL databases offer significant application development productivity benefits due to the ease of access from an object-oriented development environment and due to the lack of rigidity required in the definition of the schema. Specifically, in clinical analytics, they offer the ability to store longitudinal clinical documentation more efficiently as the intersection of many different EMR data models does not need to be pre-established.

Additionally, these databases are ready for the massive scalability and redundancy required to handle an entire region's clinical documentation en masse. Certainly relational databases can scale, but the total cost of ownership with these next-generation databases is demonstrably less. In fact, we would argue that NoSQL databases are the best for both simple agile projects and applications, and for extreme scale or highly distributed data stores. Medium-sized platforms with plenty of rich transactional use-cases and rich reporting will probably remain best suited for a relational database.

The first major success of the NoSQL storage paradigm was by Google. Google built a scalable distributed storage and processing framework called Google File System, BigTable and MapReduce [13–16] for storing and indexing web pages accessible through their search interface. These were all donated to Apache for open source development as a set of Hadoop [17] frameworks. BigTable is simply an extremely de-normalized, flat, and wide table, that allows any type of data to be stored in any column. This schema provides the ability to retrieve all columns by a single key, and each row retrieved could be of a different columnar form. This is similar to pivoting the relational model discussed earlier.

MapReduce is a powerful framework that combines a master process, a map function, and a reduce function to process huge amounts of data in parallel across distributed server nodes. Because the map functions can contain arbitrary code, they can be used to perform extremely expensive and complicated functions. However, due to the framework they must return results via an emit function to a data (or value) reduction phase. The reduction phase can also process the intermediary data with any logic it wishes so long as it produces a singular (but possibly large) answer. Figure 20.8 shows a simple data flow where a map function reviews CDA for large segments of population. Using UMLS one could look for all medical codes that imply the patient has an acute heart disease. Each map function would then put each resulting data set to an intermediary location. The master process would then coordinate the hand-off to reduce functions that then combine the intermediary data set into one final data set.

The success of this architectural breakthrough led to many other uses within their product suite. Seemingly in parallel, all the largest major internet sites that handled 'Big Data' had approached this problem by building or utilizing (and subsequently making famous) various products that exist in this realm. Public recognition accelerated when Google and Facebook [18] donated their inventions to the open source community via Apache. Finally, Amazon [19] paid the methodology a final dose of

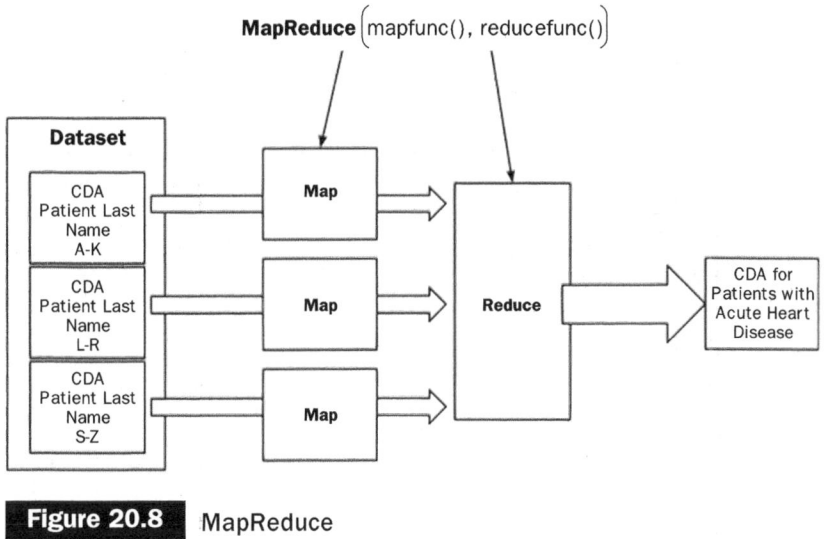

Figure 20.8 MapReduce

respect by allowing the frameworks to be accessed publicly, for a fee, via Amazon Web Services. Now the landscape is ripe with tools and technologies targeted at the NoSQL and Big Data paradigm. Table 20.1 summarizes the database technologies discussed here, and some of their common characteristics.

Each of the four NoSQL databases store data using a global primary key/value pair. Each key is scoped to a namespace (also called a bucket or a partition), which allows the system to identify each piece of data uniquely. This is logically analogous to the 'CDA Document ID' or the 'CDA Data Item ID' in the 'CDA Data Items' table, in Figure 20.7. However, the database's intrinsic understanding of this allows each of these engines to distribute the data for native sharding across distributed disks and distributed server nodes. This is similar to implementing table partitioning across the 'CDA Data Item', although in the case of the NoSQL clustering capabilities, it is built in, easy to administer, and native to the basic clustering methodology. This also mandates extremely simple query interfaces for storing and retrieving groups of data. These interfaces do not allow the messy unpredictable joins, and thus provides a very uniform buffering and disk IO usage pattern that allows the databases to be tuned and engineered for high reads. Finally, many of them provide lightning fast writes due the simplification of the locking mechanisms by requiring writes to simply be an append operation while versioning the entire set of data.

Table 20.1 Database comparisons

	PostgreSQL	MySQL	Cassandra	Riak	CouchDB	HBase	Hive
Type	SQL	SQL	NoSQL	NoSQL	NoSQL	NoSQL	NoSQL
Access Protocol	ODBC, JDBC, Native API	ODBC, JDBC, Native API	Thrift-based Client APIs Native APIs	REST JSON	REST JSON	Thrift and REST	HiveQL JDBC
Processing	Pluggable (e.g. plpgsql, pltcl, plperl)	Procedural SQL	Can be paired with Hadoop	Internal JSON Map/ Reduce	Internal JavaScript Map/Reduce	Native API Map Reduce	HiveQL compiled to Map Reduce
Data Schema	SQL DDL	SQL DDL	'Big Table' Column Store	Key Value	Document Style	'Big Table' Column Store	Semi-Structured Data
Storage Engine	PostGresSQL	Pluggable (ISAM, InnoDB)	SSTable	BitCask	Native FileSystem BLOB	Hadoop HDFS	Hadoop HDFS
Cluster	Many options	NDB Storage Engine	Built-in Dynamo Based	Built-in Dynamo Based	Lounge and Sharding	Proxy	Hadoop
OLTP	Yes	Yes	Yes	Yes	Yes	Yes	No
OLAP	Yes	Yes	No	No	No	No	No
Cubing	Yes	No	No	No	No	No	No

With all of these advantages, the tradeoff is a modified, but surprisingly simple, application development architecture. Use-cases become more modular and web interfaces become mash-ups of many different services hitting many different database clusters. Fat use-cases begin to disappear because the complicated joins that produce them cannot be accomplished. (Note: this is also partially true for the 'CDA Data Items' table in the RDBMS example.) The methodology also has a major drawback of not allowing traditional widespread access to the data by non-programmers. One solution to provide traditional SQL access for non-programmers is to use Hive on top of the HBase system. It does not fully replace the capabilities of RDBMS, but at least it gives a familiar entry point.

To further understand the value of these approaches in this domain, we will explore two systems that provide good insight into the power of NoSQL, namely Cassandra [20] and Riak [21]. Cassandra was contributed to Apache from Facebook, and is an example of a NoSQL column store database. Like the others it is still driven primarily by key/value access, but the value is built of a structured but extremely 'de-normalized' schema to be stored under each key, called a ColumnFamily. The most powerful aspect of this being that the number of practically usable columns is not fixed; in fact the maximum number of columns supported is over two billion! Each of these columns can be created ad hoc, on the fly, per transaction. This far exceeds the level of flexibility of new data items that might be expected in clinical documentation. Cassandra takes this one step further and allows for SuperColumns. A SuperColumn is essentially a column that supports additional columns within it. For example, this allows the ability to specify the patient as a SuperColumn, with the first name and the last name being subcolumns. The key/value, ColumnFamily, SuperColumn model provides a nice mix of highly scalable, highly flexible storage and indexable, multidimensional, organized data.

Like most if not all NoSQL databases, Cassandra scales very easily. What sets Cassandra apart is its ability to give the developer control over the tradeoffs of consistency, availability, and partitioning (CAP). The CAP theorem [22], first proposed by Eric Brewer [23] at Inktomi, submits that in any single large-scale distributed system, one can pick any two of the three fundamental goals of highly available, scalable, distributed data storage. The designers of Cassandra prioritized partitioning and availability, and allowed consistency to be selected by the application developer at a cost of latency. The design decision to allow these tradeoffs to be tuned by the developer was ingenious, and more and more architectures are moving this way. Medical use-cases typically experience low concurrency requirements (medical records are updated every few hours rather than every second). Therefore a tolerance to consistency issues should be reasonable.

Cassandra is based on a distributed storage scaling architecture made famous by Amazon, called Dynamo [24]. Cassandra provides clustering and data replication with a peer-to-peer architecture and easily configured parameters (even per query) for the numbers of nodes to successfully respond with a successful read or write. Each node does not need to have the correct replica of each data item. Versions are automatically tracked via timestamps, and, on reads, inconsistent nodes are repaired automatically. These settings can be applied per query and can be tailored based on how critical it is that the data are guaranteed to be the latest or guaranteed to be written.

Riak is another NoSQL database based on the dynamo clustering architecture, and thus the clustering and consistency is similar to Cassandra. Riak is a polished, but young, open source technology produced by Basho that improves on the Cassandra capabilities by offering simpler and easier to understand programmatic access and clustering capabilities. In fact, Riak offers an extremely easy to use RESTful/JSON access mechanism that greatly simplifies web development. Although Riak can support all content types, the native indexing and awareness of JSON for structured data will lead to a natural tendency to standardize on JSON throughout an entire application. For CDA though, this means that an XML to JSON converter must be used to take full advantage of this.

Although Riak's JSON support allows it to support constructs like the Column and SuperColumn features, it furthers this ability by supporting links. Links provide innate relationships between the flexibly structured key/value pairs. The ability to add links to the data supports an ability to model named (or typed) relationships between entities. On top of this modeling, Riak supports link walking within MapReduce functions, which allows semi-relational capabilities. Figure 20.9 shows an example of a simple Riak http query that an application tier might submit to the database. The first portion of the URL points to the Riak Cluster domain name, specifies to use the Riak bucket called 'patients', and specifies to retrieve the patient identified by 'uniquepatientid'. The second part of the http query is a Riak convention that supports walking links. The template for this link walking section is '_,_,_', with an underscore being convention to not filter on this field. The first underscore filters on the bucket, the second portion filters on the link type, and the third portion filters on the actual id. In this example, patient demographics and three

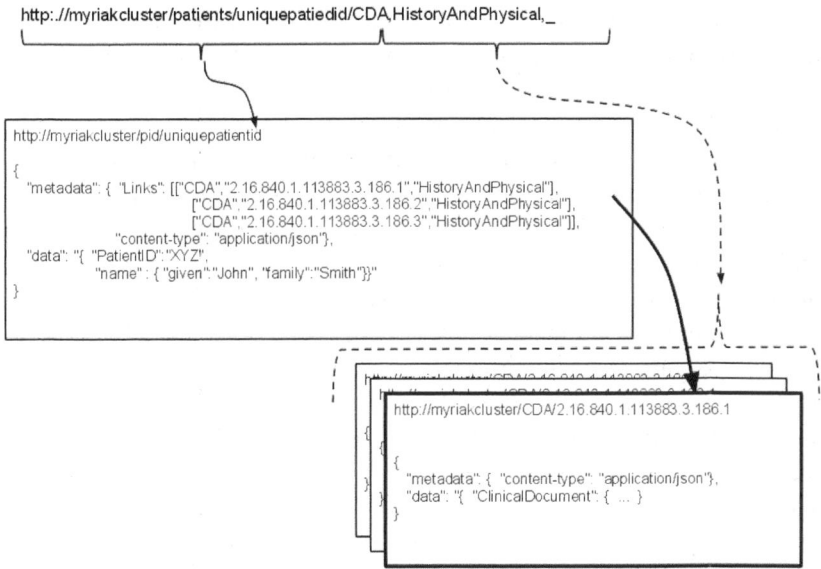

Figure 20.9 Riak RESTful API query

CDA documents corresponding to history and physicals are retrieved, using an OID to identify each unique document held by the system. This also highlights the simplicity and power of querying via a RESTful API on flexible data structures.

In this case, a MapReduce function would walk these links and inspect each clinical document data segment. Algorithms would then inspect the data for all medical codes, and map each medical code through the UMLS vocabulary set. All documents with indications of acute heart disease and the related data required would then be delivered to the reduction phase, ready for processing via analytics engines.

20.7 Analytics

As discussed above, there are many architectural options for storing CDA in flexible schemas in large scalable distributed databases. A range of transactional methodologies for working with these data are also possible. The next challenge is to make sense of the wealth of content, how do we visualize it, and how do we become confident of the contents of potentially petabyte storage?

Clinical analytics is fraught with complexities in this area, and indeed the market is in its infancy, especially in open source. The mathematical and programmatic foundations are strong and companies like Google and Twitter [25] are bringing meaning to massive amounts of dirty data. A smorgasbord of topics like population health, pay for performance, accountable care organizations, clinical trials, risk adjustment, disease management, and utilization statistics are about to undergo a golden age of understanding, intelligence, and adoption. The rollout and standardization of medical records through initiatives like HL7, IHE, and Health Story are producing immense amounts of structured data exactly for this purpose. It is time to start analyzing these data in aggregate!

Data in aggregate come with a large set of challenges. Mathematicians and statisticians have been hard at work for centuries developing techniques to understand how to interpret small and large sets of data. The following is a short list of common capabilities and methodologies needed when studying clinical analytics.

- Linear/non-linear/curvilinear regressions – different levels of standard statistical methods to study the interdependency between variables. Examples: what is the relationship between smoking and lung cancer? What is the relationship between birth weight and health outcomes?

- Statistical classification – algorithms for specification of new sets based on patterns learned from previous sets of data. Examples: used in natural language processing of medical text. A new disease outbreak has been discovered, what known patterns does it exhibit?

- Clustering – algorithms to identify subpopulations and to study the identities and characteristics of those subpopulations. Example: what are the different lifestyles, cultures, or age groups associated with different types of diabetes?

- Pattern-mining – a method of data-mining to find repeatable situations. Example: what behavior patterns exhibit a likelihood of alcoholism?

- Dimension reduction – process of removing dimensions to simplify a problem for study. Examples: when determining the relative cost of a disease, remove variables associated with geography or wealth. Uncovering the molecular variation of certain cancers.

- Evolutionary algorithms – understanding impacts of repeated conditions. Examples: what impact on health do certain policy changes cause? Could a certain method of continual treatment have been a contributing factor over time?

- Recommenders/collaborative filtering – methods to identify likelihood of one group to share common traits with other groups who exhibit certain traits. Example: potential for understanding reactions to drugs based on the reactions of other similar users.

But how should analysts bring these algorithms and the massive amounts of flexibly structured data together? This is an exciting field and quite immature, but below we review how the industry has begun to piece together the constituent parts to make this happen. There are several options and the Business Intelligence (BI) stack companies, Pentaho [26] and JasperSoft [27], have begun to assemble SQL and NoSQL connectors, statistics packages, analysis packages, ETL packages, and Cubing and OLAP packages. Table 20.2 shows several options for different capabilities in different stacks and platforms.

Pentaho Open BI Suite is a complete packaged suite for business-enabled reporting solutions. The community edition is open source and the breadth of functionality is very thorough. The solution is very business-oriented, and its target use-cases are guided by a philosophy of actionable reporting through complete round-trip business processes that supports alerts and scheduling. From a visualization perspective, Pentaho delivers a portal-based product that allows administrators, reports designers, and reports users to define, design, and deliver reports and charts. It has a robust plug-in capability that allows others to replace or add to many of the components currently embedded. A central meta-data repository maps the physical database from the logical visualization tier. The portal and the meta-data model work together to allow users to define reports and charts via JFreeReports and JFreeCharts.

To handle extreme scales of future clinical data, many different slices of the data will be produced, based on the clinical problem being solved. For example, a database whose data model is only focused on the factor of heart disease may be created and loaded with information only from medical records from patients with this condition. Data would be extracted from the flexible data store, subjected to semantic and clinical normalization, and transformed into a view that can more easily be navigated. As discussed above, the desire is to create an architecture that allows the raw CDA to be re-processed based on the current rule set of interest. In the study on acute heart disease, analysts might later find that a new factor exists and might want to re-build the data mart to include this. Thus, the best path is a tradeoff between processing power and finding the perfect data model, and in our experience, the quest for the latter only delays projects.

Table 20.2 Comparison of open source BI frameworks

	Pentaho	Jasper	Other components
Reports and charts	JFreeReports, JFreeCharts	JasperReports, JFreeCharts	
Ad-hoc analysis (drilldowns, etc.)	Pentaho Analyzer, CDF, Weka, Saiku	JPivot	
Workflow	Shark	'NA (Spring Web Flow for UI WF) Scheduling: Quartz'	
Dashboards	CDF	Spring Web Flow, SiteMesh, Spring Security	
APIs	JBoss/WSDL/ SOAP	AXIS	RServe, JRI
Data integration/ ETL	Kettle	Talend	
Data quality/ MDM	None	Talend	
Big Data integration	Pentaho data integrator for Hadoop	JasperSoft Connectors for Big Data	
Cubing	Mondrian	Mondrian	
Statistics	'Univariate Statistics plug-in Weka OpenBI R Analytics plug-in'	RevoConnectR for JasperReports Server	R
Machine-learning/AI	'Weka OpenBI R Analytics plug-in'	RevoConnectR for JasperReports Server	R, Mahout

There are several open source options for ETL. The two predominant ones would be the Pentaho Data Integrator, which is built on Kettle (sponsored by Pentaho Corporation), or Talend, which is used by JasperSoft. The Pentaho Data integrator seems like a very good option for this problem set. Not only has it been proven to be a solid choice for relational mapping, it seems to lead the pack in support for Hadoop and MapReduce integration. This framework allows for producing new mappings or rule sets to evaluate against raw data sources, and economically scaling this across a cloud-based compute cluster. We would recommend the following staged process:

1. UMLS selection of all code sets indicative of acute heart disease;
2. map selection of raw data set from NoSQL;
3. MPI normalization (or other if not focused on patient centricity);
4. reduce selections to a single data set;
5. import into Data Mart.

Talend would be another good choice for this, although its native support for NoSQL databases and MapReduce processing is not very strong. JasperSoft integrates Talend into their suite for ETL, Master Data Management, and Data Quality. This partnership forms a strong data management solution. JasperSoft has recently released a set of NoSQL connectors for native reporting directly from NoSQL databases; however, it is unclear how well this works or scales.

Once the data are in the required form, they can be analyzed. There are a few really good open source options for this. The R Programming Language is an open source statistics, mathematics, and visualization toolset, on par with SAS and Stata capabilities. Its for-profit sponsorship comes from a company called Revolution Analytics [28]. The basics of R support datafile manipulation, text manipulation, probability, math, statistics, set manipulation, indexing, and plotting functions. The R community have created numerous frameworks that form a large suite of capabilities. Two important frameworks used by Open BI to produce a Pentaho plug-in are RServe, a TCP/IP server to R the environment, and JRI, a Java to R language interface. Some plug-ins and utilities for MapReduce and Hadoop have also been created; however, many are not very active. With the Java R Interface it is easy enough to include these into NoSQL MapReduce programs. Both Pentaho and Jasper provide R plug-ins for advanced statistical analysis within their portals. Although R is arguably the most mature technology in this category, Pentaho sponsors Weka [29]. Weka is a set of Java libraries that perform various statistics, machine-learning, and predictive algorithms. Weka is a strong, well documented, but separately built utility. We expect to see powerful integrations with the Pentaho suite in future.

With the capabilities of the Hadoop suite under its wing, one would fully expect Apache to embrace the statistics and machine-learning algorithms, which are indeed provided by Mahout [30]. Written to integrate with MapReduce algorithms, Mahout is a very promising statistics and machine-learning library. However, currently it does not yet have broad coverage of standard algorithms. The trend is clear, as we get further away from the standard stack, available solutions begin to thin out. To fill out the stack it is critical to consider online analytical processing (OLAP) technologies that support the ability to build cubes. Cubes are multidimensional data structures that support pre-computed aggregations and algorithmic slices of data. The only viable option for open source OLAP databases is Mondrian [31], sponsored by Pentaho, and it has a broad user base. Mondrian is also used by JasperSoft. For bringing content from cubes into web-based pivot tables, one should consider Saiku [32], which has beautiful usability, supports RESTful OLAP Queries, and a Pentaho plug-in.

20.8 Final architectural overview

This chapter gives an overview of the state of the technology industry when it comes to scaling clinical analytics using open source software. Reviewing the architecture in Figure 20.10, we observe that there are several standard toolsets on the back-end that have risen to become standards. Mirth, UMLS, and MySQL are mature and capable. Moving to the right through the figure shows products and/or integrations that are younger but promising. NoSQL and MapReduce are proven by the giants, but they do require capital, engineering sophistication, and effort to make use of them. Some of the open source BI tools are mature,

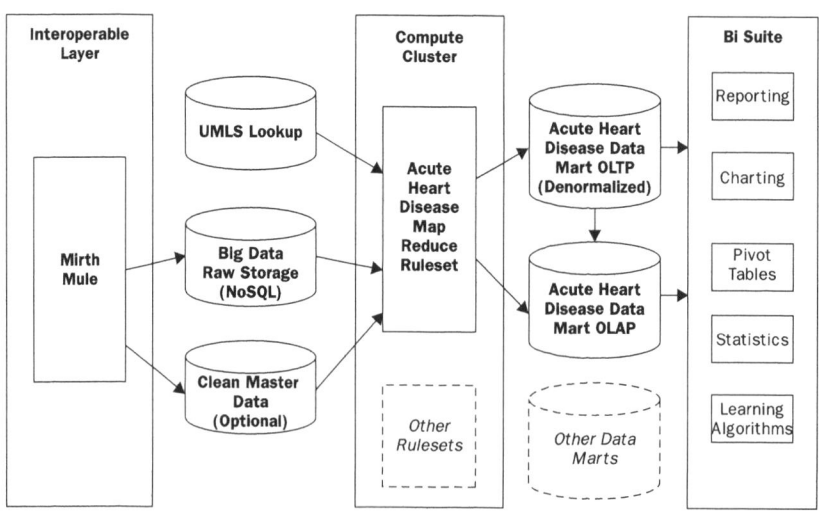

Figure 20.10 Complete analytic architecture

whereas others are not, and it is not clear which integrations will be successful and active in the community.

Seamless generalized use-cases across this architecture still require a large amount of engineering and are best left to the product companies. Most information technology organizations should continue to select narrow deployments for their BI tools, but should start to utilize some of these toolsets to begin to understand how they best fit into their organizations. If an application requires adherence to 21 CFR Part 11 validation, one must tread carefully as the abilities of these tools may cause an insurmountable testing challenge, something discussed in more detail in Chapter 21 by Stokes. Focusing on narrow use-cases that may turn out to exhibit an immediate return and an amazingly low total cost of ownership is always a good approach.

Applications in a position for high-growth volume and a potential for constant performance ceilings should deploy in the cloud and be prepared with tools that scale up with ease. Prioritize the ability to be agile at large scales. Accept that failures will occur and that mistakes will happen and use tools that will allow for recovery from issues. We believe, as many of the largest scaling internet companies do, that the open source toolsets discussed here meet that mark.

A final note on the Semantic Web, discussed in detail in the chapters by Wild (Chapter 18) and Bildtsen et al. (Chapter 19). Although perhaps not immediately apparent, in many respects these two technologies are compatible. Analytics can help gather insight into trends or population conditions that can help steer clinical research. The Semantic Web is seen as the technology that will allow us to codify artificial intelligence that is gleaned from the raw data. Eventually, semantic-based rules could be embedded into our MapReduce or machine-learning statistical utilities to speed up the data selection and interpretation.

20.9 References

[1]   http://www.irs.gov/newsroom/article/0,,id=204335,00.html
[2]   http://www.emrconsultant.com/education/hospital-information-systems
[3]   http://www.hl7.org/newsroom/HL7backgrounderbrief.cfm
[4]   http://hl7toolkit.sourceforge.net
[5]   http://www.mirthcorp.com/community/mirth-connect
[6]   http://dev.mysql.com
[7]   http://couchdb.apache.org
[8]   http://www.mulesoft.org
[9]   https://www.cms.gov/icd9providerdiagnosticcodes

[10] http://en.wikipedia.org/wiki/SNOMED_CT
[11] http://www.nlm.nih.gov/research/umls/Snomed/snomed_main.html
[12] http://www.mysql.com/
[13] http://code.google.com/opensource
[14] http://labs.google.com/papers/gfs.html
[15] http://labs.google.com/papers/bigtable.html
[16] http://labs.google.com/papers/mapreduce.html
[17] http://hadoop.apache.org
[18] https://developers.facebook.com/opensource
[19] http://aws.amazon.com/what-is-aws
[20] http://wiki.apache.org/cassandra
[21] http://wiki.basho.com
[22] http://www.cs.berkeley.edu/~brewer/cs262b-2004
[23] http://en.wikipedia.org/wiki/Eric_Brewer_(scientist)
[24] http://www.allthingsdistributed.com/2007/10/amazons_dynamo.html
[25] http://twitter.com/about/opensource
[26] http://community.pentaho.com
[27] http://jasperforge.org
[28] http://www.revolutionanalytics.com
[29] http://www.cs.waikato.ac.nz/ml/weka
[30] http://mahout.apache.org
[31] http://mondrian.pentaho.com/
[32] http://analytical-labs.com

20.10 Bibliography

Cloudera Hadoop. Available at: *http://www.cloudera.com/company/open-source/*. Accessed August 23, 2011.

Couchdb Wiki. Available at: *http://wiki.apache.org/couchdb/*. Accessed August 23, 2011.

IHTSDO: International Health Terminology Standards Development Organisation. Available at: *http://www.ihtsdo.org/*. Accessed August 23, 2011.

No-SQL Reference. Available at: *http://nosql-database.org/*. Accessed 12/11/11.

## 21

Validation and regulatory compliance of free/open source software

David Stokes

Abstract: Open source systems offer a number of advantages, but the need to formally validate some open source applications can be a challenge where there is no clearly defined 'software vendor'. In these cases the regulated company must assume responsibility for controlling a validated open source application that is subject to ongoing change in the wider software development community. Key to this is knowing which open source applications require validation, identifying the additional risks posed by the use of open source software and understanding how standard risk-based validation models need to be adapted for use with software that is subject to ongoing refinement.

Key words: validation; verification; GAMP®; risk; compliance.

21.1 Introduction

There is no doubt that the use of free/libre open source software (FLOSS) can offer some significant advantages. Within the life sciences industry these include all of the advantages available to other industries such as the ability to use low or no cost software, but the use of open source software can also provide:

- the ability to access and use new and innovative software applications in timescales that can be significantly advanced when compared to software developed by commercial vendors;

Published by Woodhead Publishing Limited, 2012

- the ability to collaborate with other industry specialists on the development of software to address the general non-competitive needs of interest to the industry.

As significant changes are taking place within the life sciences industry (such as the move towards greater outsourcing and collaboration, the greater need to innovate and reduce the time to bring products to market), these advantages can be particularly beneficial in areas such as:

- the collaborative development of drug candidates by 'Big Pharma' working with smaller biotechnology start-ups and or academia (using open source content management and collaborative working) solutions;
- the development of innovative medical devices such as pharmacotherapeutic devices or devices using multiple technologies using open source design, modelling and simulation tools (e.g. Simulations Plus [1]);
- the collaborative conduct of clinical trials including the recruitment of subjects using open source relationship management software (e.g. OpenCRX [2]) or cloud-based social networking tools such as LinkedIn [3];
- the analysis of clinical trial data, the detection of safety signals from adverse events data and the monitoring of aggregated physician (marketing) spend data using open source statistical programming and analysis tools;
- the development of process analytical technology (PAT) control schemes, facilitated by open source data analysis and graphical display tools;
- the analysis of drug discovery candidate data or manufacturing product quality data, using open source data analysis tools (e.g. for mass spectrometry, see Chapter 4).

21.2 The need to validate open source applications

When such applications have a potential impact on product quality, patient safety or the integrity of regulatory critical data, some of these advantages can be offset by the need to appropriately validate the software.

There is a clear regulatory expectation that such applications are validated – stated in national regulations (e.g. US 21CFR Part 211 [4] and Part 820 [5]), regulatory guidance (e.g. EU Eudralex Volume 4 Annex 11 [6], PIC/S PI-011 [7]) and applicable industry standards (e.g. ISO 13485 [8]). These documents require that such systems are validated, but do not provide detailed requirements on how this is achieved.

Computer system (or software) validation is a process whereby clear and unambiguous user and functional requirements are defined and are formally verified using techniques such as design review, code review and software testing. This should all be governed as part of a controlled process (software development life cycle) defined in a validation plan, supported by appropriate policies and procedures and summarised in a validation report.

Although some regulatory guidance documents (i.e. FDA General Principles of Software Validation [9], and PIC/S PI-011) do provide a high-level overview of the process of computer system validation, industry has generally provided detailed guidance and good practices through bodies such as the International Society for Pharmaceutical Engineering (ISPE) and the GAMP® Community of Practice.

The need to validate such software or applications is not dependent on the nature of the software or how it is sourced – it is purely based on what the software or application does. If an application supports or controls a process or manipulates regulatory significant data that is within the scope of regulations (e.g. clinical trial data, product specifications, manufacturing and quality records, adverse events reporting data) or provides functionality with the potential to impact patient safety or product quality, there is usually a requirement to validate the software.

This also includes an expectation to ensure that the application or software fulfils the users stated requirements, which should, of course, include requirements to ensure compliance with regulations. Validating the software or application therefore provides a reasonably high degree of assurance that:

- the software will operate in accordance with regulatory requirements;
- the software fulfils such requirements in a reliable, robust and repeatable manner.

Not all open source software has regulatory significance, but where this is the case failure to validate such software can have serious consequences, which includes regulatory enforcement action. There have been cases where a serious failure to appropriately validate a computerised system has directly led to or has partly contributed to enforcement actions including failure to issue product licences, forced recall of products, import restrictions being imposed or, in the most serious cases, US FDA Consent Decrees [10], which have required life sciences companies to implement expensive corrective and preventative actions and pay penalties in the range of hundreds of millions of dollars for non-compliance.

More importantly, the failure to appropriately validate a computerised system may place product quality and patient safety at risk and there are several known instances where faulty software has led to indirect or direct harm to patients.

21.3 Who should validate open source software?

Although the regulations do not provide details on how software should be validated, those same regulatory expectations (and numerous enforcement actions by various regulatory agencies) make it clear that regulated companies (i.e. pharmaceutical, medical device, biotechnology, biological and biomedical manufacturers, CROs, etc.) are accountable for the appropriate validation of such software applications.

Industry guidance such as the widely referenced GAMP® Guide [11] is increasingly looking for ways in which regulated companies can exercise such accountabilities while at the same time leveraging the activities and documentation of their software vendors or professional services providers (systems integrators, engineering companies, etc.). The GAMP® Guide provides pragmatic good practice for the risk-based validation of computerised systems and is accompanied by a number of companion Good Practice Guides. Although all of these recommended approaches can be applied to the validation of open source software, there is very little specific guidance on the specific validation of such software.

Following one of the key concepts of the GAMP® Guide – leveraging supplier involvement, activities and documentation – presents a significant challenge for the validation of open source applications where there is often no software vendor with which to establish a contractual relationship and often no professional services provider who will assume responsibility for implementing and supporting open source software.

There are examples of third-party organisations providing implementation and support services for open source infrastructure (e.g. Linux Red Hat) or applications. This may not, however, be the case for many new or innovative applications which are at the leading edge of functionality and which may offer significant advantages to a life sciences organisation.

In most cases it is the regulated company which is not only accountable for the validation of the open source application but must also assume day-to-day responsibility for ensuring that the application is regulatory-compliant, for validating the application and for maintaining the application in a validated state with respect to their own operational use of the software. In the case of the same instance of open source software used by multiple regulated companies (e.g. Software as a Service running in a community cloud), it is theoretically possible for a number of regulated companies to form their own open source community and share responsibility for the validated state of the shared software, but this is a relatively new and untested model.

21.4 Validation planning

As with any software or application, the validation process starts with validation planning. Many regulated companies have resources to address this in-house, whereas small to medium organisations may prefer to take the acceptable step of using qualified third parties such as consultants.

Thorough and thoughtful validation planning is the key to successfully validating open source software. The use of open source software is different from the validation of commercially developed and supported applications, and the validation planning should be undertaken by experienced resources that understand and appreciate the differences and can plan the validation accordingly.

21.4.1 Package assessment

Validation planning should start with an assessment of the open source software package (Figure 21.1) and will ask two important questions.

- Does (or can) the application deliver functionality that complies with our regulatory requirements?
- Is it possible to cost-effectively validate the open source software in our organisation?

Like commercial software, open source software can be categorised as GAMP® software category 1, 3, 4 or 5 or a combination thereof (see Table 21.1).

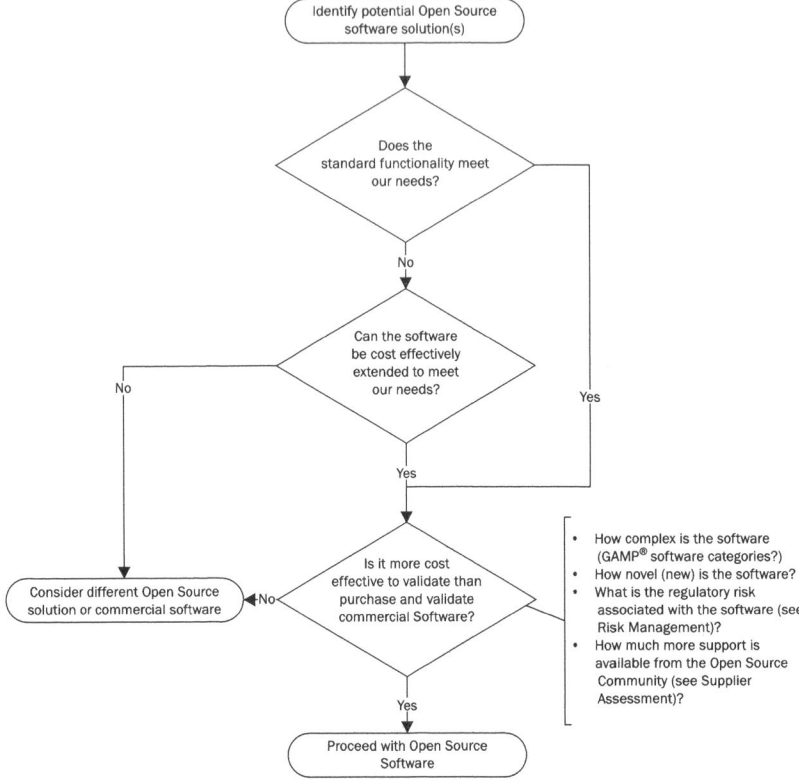

Figure 21.1 Assess the open source software package

Table 21.1 GAMP® 5 software categories

GAMP® 5 software category	Description
1	Infrastructure software (operating system, middleware, database managers, development tools, utilities, etc.)
2	No longer used in GAMP® 5
3	Non-configured software (little/no configurability)
4	Configured software (configured by the parameterisation of the application for use by the regulated company)
5	Custom software (custom developed for the regulated company)

Open source software may come in many forms. At the simplest it may be GAMP® software category 3, with fixed functionality with the possibility to enter only simple parameters. It may be GAMP® software category 4, meaning that it is a configurable off-the-shelf (COTS) application or, because of the nature of open source software it may be customisable – GAMP® software category 5. It may, of course, also combine elements of GAMP® software categories 4 and 5, meaning that a standard configurable application has been functionally extended for specific use by the regulated company or possibly by the wider open source community – the ease with which functionality can be extended is, of course, one of the benefits of open source software.

From a validation perspective, such project-specific development should be considered as customisation (GAMP® software category 5) or at least as novel configurable software (GAMP® software category 4). Categorisation of open source software as GAMP® category 4 or 5 will significantly extend the rigour of the required validation process, which may in turn further erode the cost/benefit argument for using open source software. A question well worth asking is at what point does the further development and validation of an open source software cost more in terms of internal resource time and effort than acquiring an equivalent commercial product?

This is specifically a problem with generic open source software which is not specifically aimed for use in the life sciences industry. An example might be an open source contact management application that is extended to allow sales representatives to track samples of pharmaceutical product left with physicians. In this case the extension to the open source software to provide this functionality may require a significant validation effort, which could cost more than purchasing a more easily validated commercial CRM system designed for the pharmaceutical industry and provisioned as Software as a Service. For software that is specific to the life sciences industry (e.g. an open source clinical trials data entry package), there is always the possibility of further development being undertaken by the wider open source community, which can be a more cost-effective option.

There is then the question of whether the open source software can successfully be validated. Although this usually means 'Can the software be validated cost-effectively?', there are examples of open source software that cannot be placed under effective control in the operational environment (see below). This should be assessed early in the validation planning process, so that time and money is not wasted trying to validate an open source application that cannot be maintained in a validated state. This step should also include an initial assessment of the software risk severity (see below) to determine what will be the appropriate scope and rigour of the risk-based validation.

The answer to these questions will determine whether validation is possible and whether it is still cost-effective to leverage an open source solution. Assuming that this is the case, it is then necessary to determine whether it is possible to leverage any validation support from the software 'supplier'.

21.4.2 Supplier assessment

There is a clear regulatory expectation to assess suppliers. Such assessments (including audits) are potentially subject to regulatory inspection (Eudralex Volume 4, Annex 11). Given the additional risks associated with open source software, it is essential that such an assessment is carefully planned, executed and documented. An overview of the approach is presented in Figure 21.2.

In most cases with open source software there is no supplier to assess, so a traditional supplier audit of a commercial supplier is out of the question. It is, however, possible to assess the support available from the open source community, even if this is not available on a contractual basis.

Key questions to consider and document the answers to include:

- Is there a formal community supporting the software or is it just a loose collection of individuals?

- Does the community have any formal rules or charter that provide a degree of assurance with respect to support for the software?

- How mature is the software? How likely is it that the open source community will remain interested in the development of the software once the immediate development activities are complete?

- What level of documentation is available within the community? How up-to-date is the documentation compared to the software?

- How does the community respond to identified software bugs? Are these fixed in a timely manner and are the fixes reliable?

- What level of testing is undertaken by the community? Is this documented and can it be relied upon in lieu of testing by the regulated company?

- What level of involvement are we willing to play in the community? Will we only leverage the software outputs, or actively support the development?

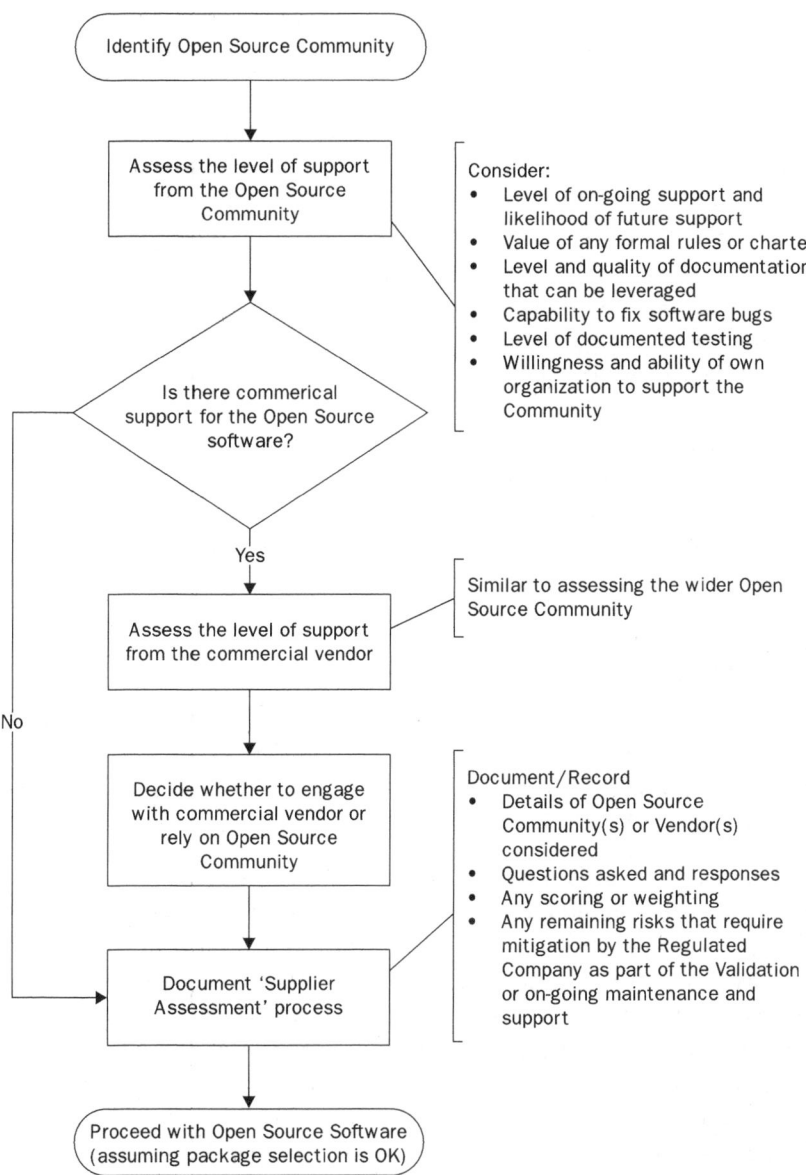

Figure 21.2 Assess the open source community

Although it is not possible to formally audit any supplier, it is possible to consider these questions and form a reasonably accurate set of answers from talking to other regulated companies who are already using the software and by visiting online forums.

In many cases, open source software is used for drug discovery and product development. Examples of such software include general data analytic tools as well as industry-specific bioinformatic and cheminformatic and pharmacogenomic tools (examples include KNIME [12], see Chapter 6, JAS3 [13] and RDKit [14], Bioconductor [15], BioPerl [16], BioJava [17], Bioclipse [18], EMBOSS [19], Taverna workbench [20], and UGENE [21], to name a few among many). Software applications used in drug discovery require no such formal validation, which is just as well as the support from different development communities can vary significantly.

However, in some other cases the level of support is very good and allows such open source software to be used for applications where validation is required. In one example of good practice, the R Foundation [22] not only provides good documentation for their open source statistical programming application, but has also worked in conjunction with compliance and validation experts and the US FDA to:

- produce white papers providing guidance on how to leverage community documentation and software tools to help validate their software;
- provide guidance on how to control their software in an operational environment and maintain the validated state;
- address questions on the applicable scope of Electronic Records and Signatures (US 21CFR Part 11 [23]).

Although not solely intended for use in the life sciences industry, these issues were of common concern to a significant number of users within the open source community, allowing people to work together to provide the necessary processes and guidance to support the validation of the software. This level of support has allowed the use of this open source software to move from non-validated use in drug discovery, into validated areas such as clinical trials and manufacturing. In other cases the open source community may only be loosely organised and may have no significant interest in supporting users in the life sciences industry. Although this does not rule out the validation and use of such software, regulated companies should realise that they will incur a significantly higher cost of validation when compared to more organised and supportive communities.

The cost of such validation support needs to be evaluated and this is best achieved by identifying the 'gaps' left by the community and estimating the activities, documentation and costs required to initially validate the software and to maintain the validated state. Such gaps should be measured against the requirements identified below.

Following the package assessment, initial risk assessment, supplier assessment and gap analysis, it will be possible to develop a suitable validation plan. This will often be quite unlike the validation plan prepared for a traditional commercial software package and regulated companies need to be wary of trying to use a traditional validation plan template, which may be unsuited to the purpose. It is important that the validation plan reflects the specific nature of and risks associated with open source software, which is also why it is so important to involve experienced validation resources in the planning.

21.5 Risk management and open source software

As with any software application, in order to be cost-effective the validation of open source software should take a risk-based approach. This should follow established industry guidance such as ICH Q9, ISO 14971 or specifically the risk management approach described in appendix M3 of the GAMP® Guide.

An initial risk assessment should be conducted, as outlined in Figure 21.3, to facilitate the validation planning as described above. This should determine the overall risk severity of the open source software (with a focus on risks to patient safety, product quality and data integrity) and should consider the risks that result from both the package and supplier assessment.

As part of the implementation or adoption of open source software, a detailed functional risk assessment should be conducted as described in the GAMP® Guide. This is the same process as for any other software package and should focus on the risk severity of specific software functions. This is in order to focus verification activities on those software functions that pose the highest risk.

However, with open source software, additional package and supplier risks may need to be considered as outlined above. Although the nature of the open source software will not change the risk severity, there may be specific issues that affect risk probability or risk detectability. These include:

- an increased risk probability
  - where novel open source software is being used,
  - due to poor (or unknown) software quality,

- o due to poor documentation,
- o resulting from difficulties in controlling the production environment (see below),
- o due to compatibility issues with unverified IT infrastructure;

- ■ a decreased risk detectability

  - o where open source software lacks error trapping, reporting or alerting functions.

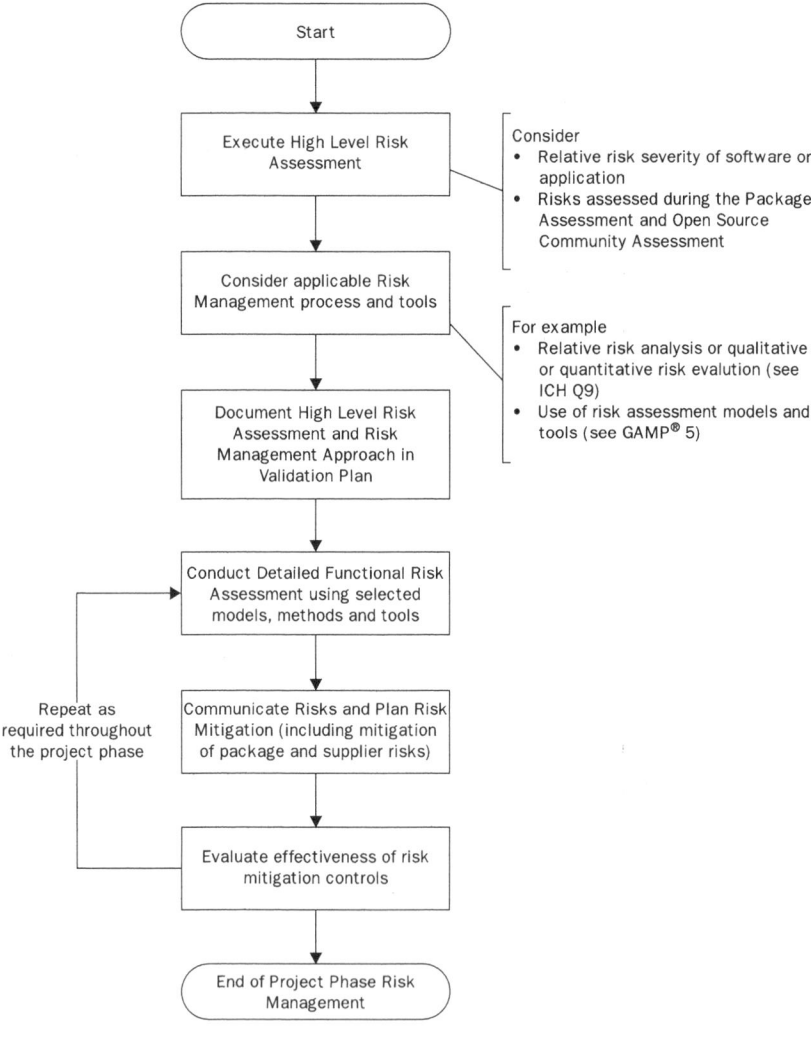

Figure 21.3 Risk management process

It is likely that any increased risk probability and decreased risk detectability will increase the resulting risk priority, requiring additional scope and rigour with respect to verification activities (design reviews, testing, etc.). For each of these risks appropriate controls need to be identified, understanding that it may not be possible to leverage any supplier activities and that the regulated company will be responsible for any risk control activities.

Additional risk controls for open source software may include:

- more rigorous testing of standard software functions, where no formal supplier testing can be leveraged;
- development of additional documentation, including user manuals, functional and technical specifications;
- development of rigorous change control and configuration management processes including detailed software release processes;
- more frequent periodic reviews and revalidation.

The effectiveness of these risk controls should be monitored and reviewed to ensure that any risks are mitigated.

21.6 Key validation activities

Whatever the nature (categorisation) of the software, it is important that the regulated user defines and most probably documents clear requirements. Where there is little or no ability to change the functionality, it is acceptable for regulated company users to assess the package and confirm that it meets their requirements (Eudralex Volume 4 Annex 11). Even this will require a clear understanding of the business requirements, even if these are not documented in extensive user requirement specifications. It is recommended that the business process and data flows are documented as a minimum, against which the package can be assessed and this can be done at the package assessment stage.

For open source systems that are highly configurable, it is recommended that formal user requirements are documented and key areas are highlighted in Figure 21.4. It will be important to maintain these requirements in an up-to-date state as the use of the software changes over time. These may record user requirements in a variety of applicable formats including business process flows, use-cases, business rules and traditional singly stated user requirements ('The system shall . . .'). It is

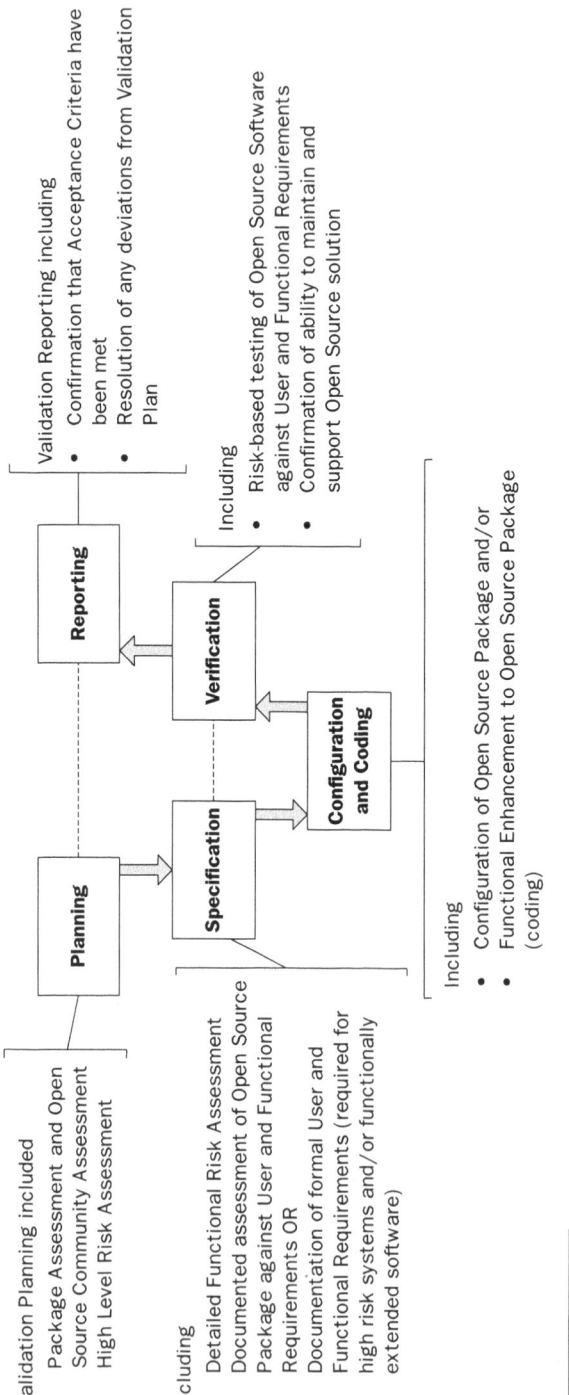

Figure 21.4 Typical validation activities

Validation Planning included
- Package Assessment and Open Source Community Assessment
- High Level Risk Assessment

Validation Reporting including
- Confirmation that Acceptance Criteria have been met
- Resolution of any deviations from Validation Plan

Including
- Risk-based testing of Open Source Software against User and Functional Requirements
- Confirmation of ability to maintain and support Open Source solution

Including
- Detailed Functional Risk Assessment
- Documented assessment of Open Source Package against User and Functional Requirements OR
- Documentation of formal User and Functional Requirements (required for high risk systems and/or functionally extended software)

Including
- Configuration of Open Source Package and/or
- Functional Enhancement to Open Source Package (coding)

Planning

Specification

Configuration and Coding

Verification

Reporting

usual to conduct a detailed functional risk assessment at this stage, examining each function (business process, use-case, etc.) to determine:

- whether or not the function requires formal validation, that is does it support regulatory significant requirements?
- the risk severity of the specific function – usually assessed of a relative basis, but potentially using qualitative or quantitative risk assessment for software with a higher overall risk severity (see ICH Q9).

For functions that are determined to be regulatory significant and requiring validation, specifications will also need to be developed in order to document the setup (configuration) of the open source software by the specific regulated company. For COTS software this will include a record of the actual configuration settings (package configuration specification or application setup document). Where custom development is required this should also be documented in appropriate functional and technical design specifications.

It is worth noting here that where specific functions are developed by the open source community to meet the needs of a specific project, and where the community does not provide functional and technical design specifications, it is recommended that the regulated company's involvement in the development includes the production of these documents. Not only will these provide the basis for better support and fault finding in the future, but they will also be of use to other life sciences companies wishing to use the same functional enhancements.

The nature of the open source movement implies a moral responsibility to share such developments with the wider community and it is common practice for developers in the open source community to 'post' copies of their software enhancements online for use by other users. While it is possible for life sciences companies to develop enhancements to open source software for their own sole benefit, this goes against the ethos of the movement and in some cases the open source licence agreement may require that such developments are made more widely available.

Where such functional enhancements require validation, the associated documentation (validation plans, requirements, specifications, test cases, etc.) is almost as useful as the software. Sharing such validation documentation (devoid of any company confidential or commercial details) fits well with the ethos of the open source movement and such documentation may be shared via open source forums or online repositories. Most open source software sites have a location where documentation can be shared and this practice is to be encouraged in line with the GAMP® key concept of leveraging 'supplier' documentation.

The open source community includes developers with widely varying experience. At one end there are professional software developers choosing to apply formal methods to the open source platform, and at the other are 'amateurs', developing software on a trial and error basis in their spare time, with little or no formal training and with little or no documentation.

Although all parties can make an active and valuable contribution to the development of open source software, there are inherent risks when regulatory significant applications are developed by those not formally trained in good software development practices. These risks may include:

- requirements are not fully understood, defined or documented, making it impossible to validate the software;
- the developed functionality may not fulfil the requirements due to poor specification and design;
- code may be excessively error-prone and inefficient;
- important non-functional requirements may be missed, such as user authentication and secure password management, data security and data integrity;
- errors may not be identified due to insufficient or inappropriate testing.

Actual examples of 'bad practices' encountered in open source software developed in this way include:

- failure to provide even basic user authentication in a web application accessing clinical trial data, allowing anyone with the URL to view personally identifiable health information;
- the storage of user IDs and passwords in an unprotected, unencrypted file;
- the ability of a laboratory data capture routine to enter an infinite wait state with no timeout mechanism and no way of halting the program by normal means;
- the ability for any user to access and change analytical data processed by an open source bioinformatics toolkit;
- complete lack of error handling at the applications level in a statistical program, allowing a 'divide by zero' error to halt the program with no indication of the problem.

Although it is possible that part-time, untrained developers can develop software that can be validated, this requires an understanding of both good software development practices and an understanding of software validation in the life sciences industry. Experience is that the development and validation of regulatory significant applications is best accomplished by real subject matter experts.

Where business process owners wish to contribute to the development of open source solutions, this is perhaps best achieved by defining clear requirements and executing user acceptance testing. In many cases the setup and use of the software may follow a 'trial and error' process – this may often be the case where documentation is poor or out of date. It is acceptable to follow such a prototyping process, but all requirements and specifications should be finalised and approved before moving into formal testing of the software.

Prior to formal testing, the software and associated documentation should be placed under change control, document and configuration management. It is unlikely that open source community controls will be adequate to provide the level of control required in the regulated life sciences industry (see below) and the regulated company will usually need to take responsibility for this.

It is worth noting that there are a small number of open source software tools on the market providing basic change control, configuration management, document management and test management functionality (e.g. Open Technology Real Services [24], and Project Open OSS [25] IT Service Management suites). Although these do not require formal validation, they should be assessed for suitability to ensure that they function as required, are secure and will be supported by the development community for the foreseeable future. Important areas to consider here are described in Figure 21.5.

Depending on the outcome of the risk assessment and the nature of the software, testing may consist of:

- unit and integration testing of any custom developed functions. This may take several forms including:
  - white box testing (suitable where the custom function has been developed by the regulated company and the structure of the software is known),
  - black box testing (where the custom function has been developed in the open source community and the structure of the software is not known – only input conditions and outputs can be tested),
  - positive case testing, and where indicated as applicable by risk-assessment, negative case, challenge or stress testing;

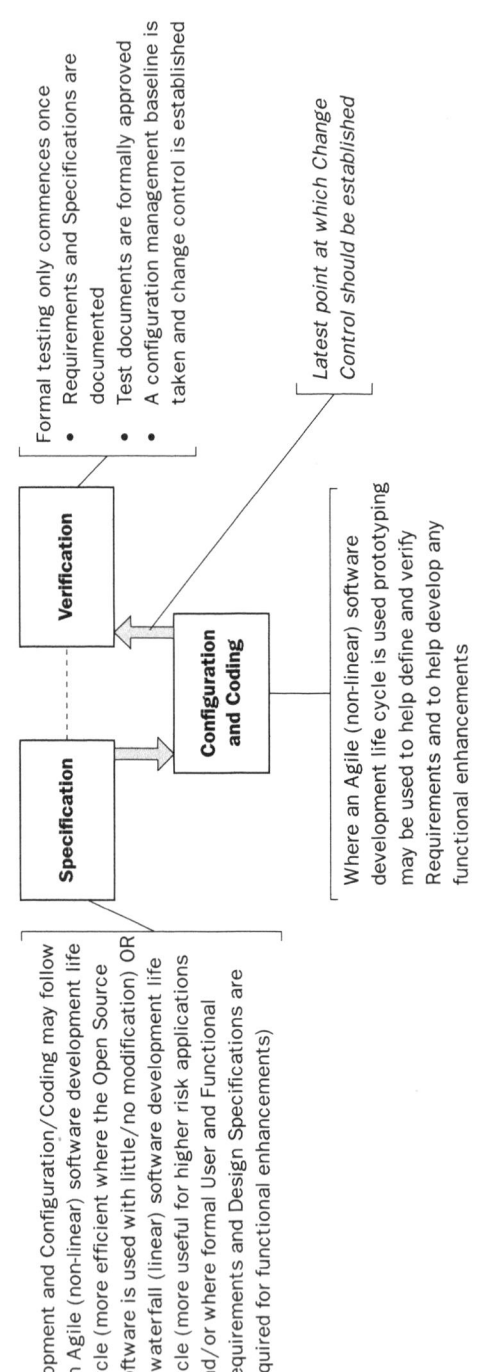

Development and Configuration/Coding may follow

- An Agile (non-linear) software development life cycle (more efficient where the Open Source software is used with little/no modification) OR
- A waterfall (linear) software development life cycle (more useful for higher risk applications and/or where formal User and Functional Requirements and Design Specifications are required for functional enhancements)

Where an Agile (non-linear) software development life cycle is used prototyping may be used to help define and verify Requirements and to help develop any functional enhancements

Formal testing only commences once

- Requirements and Specifications are documented
- Test documents are formally approved
- A configuration management baseline is taken and change control is established

Latest point at which Change Control should be established

Specification

Verification

Configuration and Coding

Figure 21.5 Software development, change control and testing

- performance testing, where there are concerns about the performance or scalability of the software;
- functional testing, including:
  - appropriate testing of all requirements, usually by the development team (which may be an internal IT group or a third party),
  - user acceptance testing by the process owners (users), against their documented requirements (may not be required where a previous assessment has been made of a standard package that has not been altered in any way).

The use of open source software does not infer that there is a need to test to a lower standard. If anything, some of the additional risks arising from using open source software means that testing by the regulated company will be more extensive than in cases where testing from a commercial vendor (confirmed by a traditional supplier audit) can be relied on. All testing should follow accepted good practices in the life sciences industry, best covered in the GAMP® 'Testing of GxP Systems' Good Practice Guide. In some cases requirements cannot be verified by formal testing and other forms of verification should be leveraged as appropriate (i.e. design review, source code review, visual inspection, etc.). Supporting IT infrastructure (servers, operating systems, patches, database, storage, network components, etc.) should be qualified in accordance with the regulated company's usual infrastructure qualification processes, and as required by Eudralex Volume 4, Annex 11.

All of the above should be defined in a suitable validation plan and reported in a corresponding validation report. Based on accepted principles, any of the above documentation may be combined into fewer documents for small or simple software applications, or alternative, equivalent documentation developed according to the applicable open source documentation conventions. Where the software is used within the European Union or used in a facility that complies with EU regulations – and soon to be any country that is a member of the Pharmaceutical Inspection Cooperation Scheme (PIC/S) – the documentation should also include a clear System Description (Eudralex Volume 4 Annex 11). The use of the open source software as a validated application also needs to be recorded in the regulated company's inventory of software applications (Eudralex Volume 4, Annex 11).

In summary, the validation of open source software should follow a scalable, risk-based approach, just as any commercial software package. The only difference is that the inability to rely on a commercial vendor means that the regulated company may need to validate with broader scope and greater rigour than would otherwise be the case. This is likely to increase the cost of validation, which will offset some of the cost savings inherent in the use of open source software. This is why this needs to be assessed after the package and 'supplier' assessment – to confirm that the financial advantages for using open source software are still valid, despite the cost of validation.

21.7 Ongoing validation and compliance

Once the software has been initially validated, it is essential that it is maintained in a validated state and subject to periodic review and, in the case of any changes, appropriate re-validation (which may include regression testing of unchanged functions). This follows well-defined principles and processes best defined in the GAMP® Good Practice Guide 'A Risk-Based Approach to Operation of GxP Computerized Systems'. Once again, the use of open source software does not provide a rationale for not following the processes defined in this guidance document, which are:

- handover;
- support services;
- performance monitoring;
- incident management;
- corrective and preventative action;
- operational change and configuration management;
- repair;
- periodic review;
- backup and restore;
- business continuity;
- security;
- system administration;
- data migration;
- retirement, decommissioning and disposal.

It is, of course, acceptable to scale these processes based on the risk severity, size and complexity of the open source software.

Key to maintaining the validated state is the use of an effective change control and configuration management process. This is often seen as contrary to the ethos of some parts of the open source community, where it is possible (and sometimes encouraged) for users to download and install the latest functional enhancements, patches and bug fixes – this speed of enhancement and the ability to leverage the efforts of other developers is undoubtedly one of the attractions of using open source software, especially in innovative applications. However, if this is allowed to happen the validated state of the software is immediately placed at risk and appropriate controls must be established. It is possible to allow this flexibility, but there are additional costs involved.

This requires a controlled set of environments to be established and these are outlined in Figure 21.6.

- A development environment for use by users and developers who wish to try out new functions, or even play a more participatory role in the open source community. This can be used to try out new functions available in the open source community or to develop new functionality, which may in turn be shared with the wider community.

- A test environment where new functions are tested – and old functions regression tested – prior to release into the production environment. This environment should be under formal change control, should be qualified in accordance with industry good practice and may be maintained and supported using formal processes and procedures.

- An operational environment – which is under formal change control and subject to clearly defined release management processes. This environment also should be under formal change control, should be qualified in accordance with industry good practice and should be maintained and supported using formal processes and procedures.

Although it is possible for very frequent updates to be made to the development environment (including the download of new functions, bug fixes and patches) – sometimes weekly or daily – the release of such software to the test environment should have a clear scope, defined in terms of which updated requirements will be fulfilled. Release from the development environment to the test environment should be subject to defined version management and release management processes and will occur with a much lower frequency (every two weeks, every month or every three months depending on how much the open source software is changing).

The test environment should be formally qualified and subject to change control. This is to ensure that the test environment is functionally

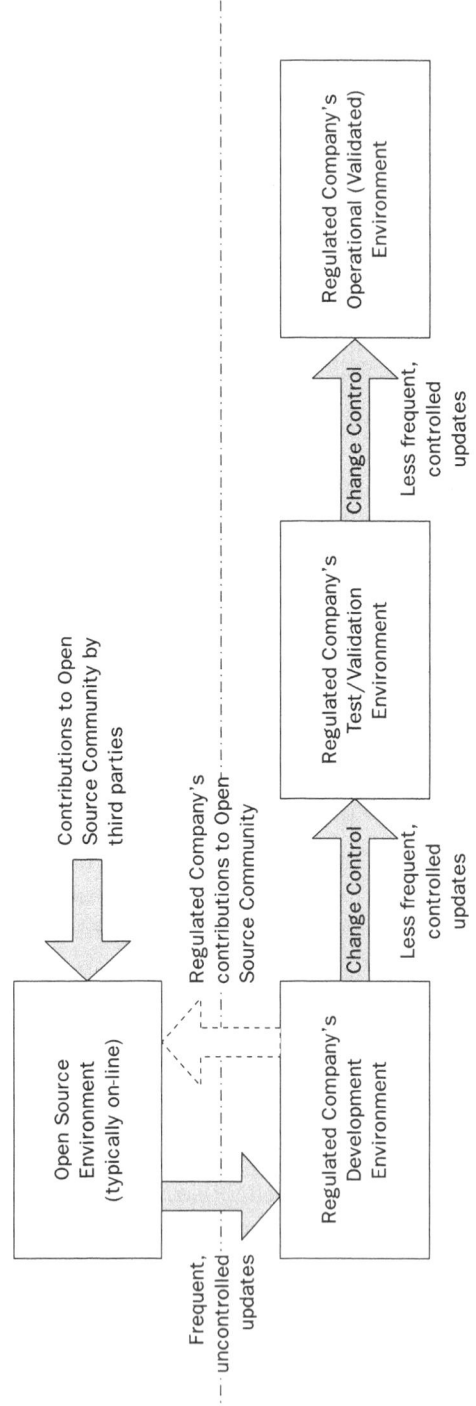

Figure 21.6 Development environments and release cycles

equivalent to the production environment and that testing in the test environment is representative of the production environment. Release of tested software should also be subject to defined version, release and change management processes and will again be on a less frequent basis. Although it is possible to 'fast track' new functions, the need to regression test existing, unchanged functionality places limits on exactly how fast this can be achieved. The production instance should, of course, also be qualified.

21.8 Conclusions

The validation process and controls described above apply only to validated open source software – those supporting regulatory significant processes or managing regulatory significant data or records. Non-validated open source software does not always need such control, although the prudent mitigation of business risks suggests that appropriate processes and controls should nevertheless be established.

The use of open source software does provide a number of advantages to life sciences companies and these include speed to implement new solutions, the ability to deploy ground-breaking software and the opportunity to participate in and benefit from a broad collaborative community interested in solving similar problems. However, the need to assure regulatory compliance and the requirement to validate and control such software does bring additional costs and add time to the process. Although this should not completely negate the advantages of open source software, the cost- and time-saving benefits may not be as great as in other non-regulated industries. Life sciences companies should therefore ensure that initiatives to leverage open source software are properly defined and managed, and that the relevant IT and QA groups are involved at an early stage. This will allow relevant package and supplier assessments to be conducted and the true costs and return on investment to be estimated.

Although much of the validation process is similar to the process used to validate commercial software, the nature of open source software means that there are some important differences, and the use of experienced subject matter experts to support the validation planning is highly recommended. Effective controls can be established to maintain the validated state and this will again require a common understanding of processes and close cooperation between the process owners (users), system owners (IT) and QA stakeholders.

In conclusion, there is no reason why free/open source software cannot be validated and used in support of regulatory significant business operations – it just may not be as simple, quick and inexpensive as it first appears.

21.9 References

[1] See *http://www.simulations-plus.com/*
[2] See *http://www.opencrx.org/*
[3] See *http://www.linkedin.com/*
[4] US Code of Federal Regulations, Title 21 Part 211 '*Current Good Manufacturing Practice for Finished Pharmaceuticals*'.
[5] US Code of Federal Regulations, Title 21 Part 820 '*Quality System Regulation*' *(Medical Devices)*.
[6] European Union Eudralex Vol 4 *(Good manufacturing practice (GMP) Guidelines), Annex 11 (Computerised Systems)*.
[7] Pharmaceutical Inspection Co-operation Scheme PI-011 '*Good Practices For Computerised Systems In Regulated GxP Environments*' (See *http://www.picscheme.org/*).
[8] Ref ISO 13485:2003 '*Medical devices — Quality management systems — Requirements for regulatory purposes*' (see *http://www.iso.org/*).
[9] US FDA '*General Principles of Software Validation*', 11 January 2002 (see *http://www.fda.gov*).
[10] See US FDA website *at http://www.fda.gov/* for examples of Consent Decrees.
[11] *GAMP® 5: A Risk-Based Approach to Compliant GxP Computerized Systems*, February 2008 (published by ISPE – see *http://www.ispe.org/*)
[12] See *http://www.knime.org/*
[13] See *http://jas.freehep.org/jas3/*
[14] See *http://rdkit.org/*
[15] See *http://www.bioconductor.org/*
[16] See *http://www.bioperl.org/*
[17] See *http://biojava.org/*
[18] See *http://bioclipse.net/*
[19] See *http://emboss.sourceforge.net/*
[20] See *http://www.taverna.org.uk/*
[21] See *http://ugene.unipro.ru/*
[22] The R Foundation for Statistical Computing – see *http://www.r-project.org/foundation/*
[23] US Code of Federal Regulations, Title 21 Part 11 '*Electronic Records and Signatures*'.
[24] See *http://otrs.org/*
[25] See *http://www.project-open.com/en/solutions/itsm/index.html*

22

The economics of free/open source software in industry

Simon Thornber

Abstract: Free and open source software has many attractive qualities, perhaps none more so than the price tag. However, does 'free' really mean free? In this chapter, I consider the process of implementing FLOSS systems within an enterprise environment. I highlight the hidden costs of such deployments that must be considered and contrasted with commercial alternatives. I also describe potential business models that would support the adoption of FLOSS within industry by providing support, training and bespoke customisation. Finally, the role of pre-competitive initiatives and their relevance to supporting open source initiatives is presented.

Key words: pharmaceuticals; economics; enterprise deployment; pre-competitive; cloud computing; Pistoia.

22.1 Introduction

In the last few years use of free/libre open source (FLOSS) software has exploded. Versions of Linux are found in millions of television set-top boxes, Google's Android is arguably the leading smart phone operating system, countless websites are powered by Apache, and millions of Blogs by WordPress. However, for total number of users Mozilla's Firefox browser is hard to beat, with an estimated 270 million active users (as measured by active security pings received by Mozilla [1]), accounting for around 20% of web traffic in general surveys of such things. Firefox is easy to use, highly standards-compliant, secure, well supported, easy to extend with an ecosystem of thousands of plug-ins, and of course it is free to download, use or distribute.

It therefore seemed a perfectly reasonable request when, at a department town-hall meeting in 2009, Jim Finkle asked the US Secretary of State, Hilary Clinton [2] 'Can you please let the staff use an alternative web browser called Firefox?'. The reply, which came from under-secretary Kennedy, that '. . It is an expense question . .' was interrupted from the audience with a shout of 'it's free'. Kennedy continued 'Nothing is free. It's a question of the resources to manage multiple systems. . . .'. He is, of course, absolutely right, that in a large organisation, be that governmental, academic or corporate there are many hidden costs in rolling out, securing, insuring, maintaining and supporting any software. These costs are incurred, regardless of whether the software is open or closed source, free or expensive.

In the following chapter, I will explore some of these costs in more detail, and analyse a real-world example of how parts of the life sciences industry have come together to try and find new ways they can use open source software in the fight against disease. Before delving into the practical use within industry, I will provide an introduction to the technologies and massive data volumes under consideration by exploring recent advances in human genetics.

22.2 Background

22.2.1 *The Human Genome Project*

The Human Genome Project is in many ways one of the wonders of the modern world. Thousands of scientists from laboratories around the globe spent a decade and $3 billion to produce what then US President Bill Clinton described as '. . . without a doubt this is the most important, most wondrous map, ever produced by humankind' [3]. There was great excitement and anticipation that its publication would herald a new era of medical breakthroughs, and it would only be a matter of time until diseases like cancer would only be heard about in history books. I do not intend to go into detail about how the Human Genome Project was delivered, if you are curious its Wikipedia page [4] is an excellent starting point. It is, however, worth mentioning that the human genome is essentially an encyclopaedia written in 46 volumes (known as chromosomes), and written in the language of DNA. The language of

Published by Woodhead Publishing Limited, 2012

DNA is made up of an alphabet of four letters, rather than the 26 we use in English. These DNA letters are called 'base pairs'. The human genome is about 3 billion DNA base pairs in length. This makes the maths of working out the cost of sequencing each base pair a remarkably simple affair of dividing the total $3 billion project cost by 3 billion, giving us a dollar per base pair. Given that each base pair can be coded by 2 bits of information then the whole genome could be stored on about 750 MB of disk space. The cost of storing this much information, even 10 years ago would only be a few dollars, making the data storage costs a microscopic percentage of the overall project costs.

22.2.2 Things get a little more complex

There were huge celebrations in the scientific community when the human genome was published, but this was really only the beginning of the work that needed to be done to get value from the genome. The next step was to read and understand the information in the 46 volumes. Fortunately, just like in an encyclopaedia the genome is broken down into a number of entries or articles. In the world of genetics these entries are called genes. As humans are quite complicated animals, it was expected that we would have a large number of genes, many more than simpler organisms such as worms, flies, plants, etc. All we had to do was work out which gene corresponded to a given human trait and we would have unlocked the secrets to human biology, with hopefully fantastic benefits to healthcare. Unfortunately, there was a problem, it turned out that there were far fewer genes in humans than originally expected. The total number of genes in the human genome was found to be around 23 000. To put this in perspective that is about 9000 genes less than the corn that grows in our fields has, and only just 7000 more than a very simple worm.

As scientists investigated further, new levels of complexity were discovered. If we keep with the analogy of the genome as an encyclopaedia, genes can be thought of as the topics within it. Researchers discovered that paragraphs within each topic could be read or ignored, depending on circumstances, which would change the meaning of the article (this is known as splice variants in genetics). They also found many spelling changes, small additions or omissions from one person's genome to the next. It soon became apparent that these differences were the key to unlocking many more secrets of the human genome. Initially it was only possible to investigate these differences on a very small scale and on very small parts of the genome. However, in the last couple of years huge advances have been made in the science of DNA sequencing; collectively called NGS (next-generation sequencing [5]).

There are several types of NGS technology, and heavy innovation and competition in the area is driving down sequencing costs rapidly. Most NGS technology relies on a process known as 'shotgun' sequencing. In simple terms, the DNA to be sequenced is broken up into millions of small pieces. Through a number of innovative processes, the sequences of these short pieces of DNA are derived. However, these overlapping short 'reads' must be re-assembled into the correct order, to produce a final full-length sequence of DNA. If each of the short sequences is 100 bp long, that would mean an entire human genome could be constructed from 30 million of these. However, to successfully perform the assembly, overlapping sequences are required, allowing gradual extension of the master DNA sequence. Thus, at least one more genomes' worth of short sequences is required, and in practice this is usually somewhere between five and ten genome's worth – known as the coverage level. At ten-fold coverage, there would be a total of 300 000 000 short sequences or 30 000 000 000 base pairs to align. Even that is not the end of the story, each of those 30 billion base pairs has confidence data attached to it; a statistical assessment of whether the sequencing machine was able to identify the base pair correctly. This information is critical in shielding the assembly process from false variations that are due to the sequencing process rather than real human genetic variation. For an informatician, the result of all of this is that each sample run through the NGS procedure generates a vast amount of data (~100 GB), ultimately creating experimental data sets in excess of TB of data. Efficient storage and processing of these data is of great importance to academic and industrial researchers alike (see also Chapter 10 by Holdstock and Chapter 11 by Burrell and MacLean for other perspectives).

22.3 Open source innovation

Many of the large projects that are starting to tackle the challenge of unravelling human genetic variation are publicly funded. A great example of such an effort is the '1000 genomes project' [6], which is looking to sequence the genomes of well over a thousand people from many ethnic backgrounds in order to give humanity its best insight yet into the genetic variation of our species. Collecting the samples, preparing them and running them through the NGS platforms, has been a huge logistics and project management operation. However, the part of the project of most relevance here is the informatics task of managing this data avalanche. In particular, analysing and assembling the results into useful information that biologists are able mine and use to generate new insights into human variation. A project of this scale had never been attempted previously in the world of biology, new technology has been developed and iteratively improved by some of the world's most prestigious research institutes. As many of these institutes receive public funding, there has been a long tradition of publishing their software under open source licences. This has many benefits.

- By definition open source code has nothing hidden, so its function is completely transparent and easy to examine and analyse. This allows for scientific peer review of a method, easy repetition of experiments and for scientific debate about the best way to achieve a given goal.

- This transparency allows other scientific groups not only to replicate experiments, but also to add new features, or indeed suggest code improvements without fear of a team of lawyers knocking on their door.

- A project like the 1000 genomes is for the benefit of all mankind and as soon as the data are available in a suitable format, they are put up on an FTP site for anyone in the world to download. There are no commercial considerations, no intellectual property to collect licence fees on, no shareholders looking for a return on investment. Everything that is done is available to everyone.

Often open/community approaches can be criticised, rightly or wrongly, for not being as 'good', well supported or feature-rich as their commercial rivals. However, the sheer scale of investment and intellect that is focused on challenges such as the 1000 genomes produces science of the very highest quality. Software that has been built and tested against the biggest and most demanding data sets in history, and consequently really sets a gold standard. The technology developed for projects such as the 1000 genomes is being re-used in many additional projects; from helping sub-Saharan farmers by understanding the relationship of genetics to viability of their cattle, to helping answer questions about the nature of human migration out of Africa some 10000 years ago. However, by far and away the biggest value of the utilisation of this technology will be in the battle against disease.

22.4 Open source software in the pharmaceutical industry

It takes teams of thousands of scientists, billions of dollars and many years of hard work to produce a new medicine. The risks are huge, and competition between pharmaceutical companies is intense, as there are no prizes for being second to patenting the same molecule. As a consequence, pharmaceutical companies are as security-conscious as those in the defence industry. Security is not the only concern for IT departments in the pharmaceutical industry. As soon as a potential new pharmaceutical product reaches clinical testing, there are a huge set of rules that must be followed to ensure that everything is done reproducibly and correctly. A huge amount of effort is put into validating these systems (see Chapter 21 by Stokes), which would need to be re-done at great expense if things change. This means that there must be an extensive support network behind these systems, and with that appropriate contacts and guarantees.

An IT manager is much more confident of getting support from a large software company, which offers teams of experts and global support for its software for a decade or more, than the hope that someone on a support Wiki might be able to offer a suggestion when something goes wrong. These large software companies have strong patent protection on their software, which enables them to make money from their work. This enables them to grow into global corporations, capable of offering 24/7 support teams of experts in many languages, and legal protection of millions of dollars to their customers if data were to leak or service be interrupted. As such, the pharmaceutical industry has generally been slow to adopt open source software, simply because it does not offer the protections and support guarantees of a commercial counterpart. However, where those protections do exist (for example, from RedHat for its version of Linux), open source software can quickly appear in the data centres and scientific workstations of pharmaceutical companies.

22.5 Open source as a catalyst for pre-competitive collaboration in the pharmaceutical industry

The past few pages have outlined how the pharmaceutical industry has recently found itself pulled in two contradictory directions. On the one hand there is the need for a reliable, supported, secure suite of software tools, and on the other is a wealth of high-quality, powerful open source software written to analyse and manage the torrent of data produced by the latest DNA sequencing technology. Traditionally when faced with this type of challenge, pharmaceutical companies would turn to their vast internal IT departments, who would internally develop the skills needed to manage and support the software. However, this is an expensive and wasteful approach, with each pharmaceutical company having to build an internal platform, integrate it with existing systems and provide the high level of expert support demanded by the scientists in their company. The skills needed to build such a platform are a blend of world-class IT with cutting-edge scientific knowledge, which is rare, valuable and in high demand, which in turn makes it expensive.

However, in many instances, even when pharmaceutical companies do make significant internal investments in building and supporting IT, it is arguable that they provide any competitive advantage over their peers [7]. This is particularly true in the DNA sequence domain, where many of the core algorithms and software are free and open for anyone to use. It has quickly become apparent that the competitive value between companies came from the results generated by these systems and not from building the systems themselves. It is from this type of realisation that organisations such as the Pistoia Alliance [8] were conceived.

The Pistoia Alliance is a non-profit organisation that was set up in recognition of the great deal of important, yet pre-competitive work being done in the life sciences industry. The Pistoia Alliance has members ranging from giant Fortune 500 companies, through to new start-ups, and it spans both life sciences companies, and IT vendors who supply all sorts of the infrastructure used in modern pharmaceutical research. The stated goal of the Pistoia Alliance is to 'lower the barriers to innovation in Life-sciences research'. One excellent way of achieving this is to develop, champion and adopt open standards or technologies that make it easy to transfer data and collaborate with other companies, institutions and researchers. The Pistoia Alliance has been running several projects ranging from open standards for transferring scientific observations from electronic lab notebooks, to developing new open technology to allow better queries across published scientific literature. However, the project that illustrates how industry can work in a more open environment in the area of DNA sequence technology is known as 'Sequence Services'.

Published by Woodhead Publishing Limited, 2012

22.6 The Pistoia Alliance Sequence Services Project

NGS has produced a boom in commercial opportunities; from companies that build the robots to do the work, and reagents needed to fuel the process, to high-tech sequencing centres with the capacity and expertise to service the sequencing needs of many life science companies. Consequently, a great deal of sequencing is outsourced to these centres. Once the raw sequencing has been completed, the complex process of assembling the DNA jigsaw starts, prior to the next level of analysis where patterns in variation are identified. These are correlated with other observable traits (know as phenotypes to geneticists) such as how well someone in a clinical trial responds to the treatment. Increasingly, it is likely that some or all of this analysis may be done by specialist companies, external experts or via research agreements with universities. All of this work is done under complex legal contracts and conditions, which ensure appropriately high ethical and security standards are always adhered to.

Managing these complex collaborations can produce a huge logistical and security challenge for IT departments. The first of these problems is in shipping the data to where they are needed. As the data sets tend to be very large, sending the data over the internet (usually a process known as Secure File Transfer Protocol, or SFTP) can take a very long time. So long, that it is often far faster to use what is jokingly called the 'sneaker net', which in reality tends to be an overnight courier service transporting encrypted hard drives. This then involves the complexity of getting the decryption keys to the recipients of the hard drives, and therefore the setting up of a key server infrastructure. Once the data have been decrypted, they inevitably get moved onto a set of servers where they can be assembled and analysed. Controlling who has access to the data, and managing copies becomes hugely complex, and expensive to manage.

The past few years has seen a new phrase enter the IT professional's vocabulary, 'cloud computing'. By 2011 almost every major IT company offers its own cloud platform, and hundreds of analysts' reports recommend that corporations leverage the cloud as part of their IT infrastructure strategies. In theory, the cloud was the perfect platform to answer the challenge of NGS data. The cloud is fantastically scalable, so IT managers would not have to estimate data volumes at the beginning of a budget cycle, then have spare capacity for most of the year, only to have to scramble for more at the end of the cycle if use had been underestimated. The cloud also has huge on-demand computational

Published by Woodhead Publishing Limited, 2012

resources, again meaning you only pay for the cycles you use, not for a supercomputer that sits idle for 80% of the year. The cloud also sits outside a corporation's data centres, which is excellent for supporting collaboration with third parties, and prevents corporate fire walls having holes punched through them to allow third-party access. It is also expected that the sheer scale of cloud operations should offer considerable cost-savings over replicating the same functionality in a private data centre. In fact, open source software lends itself very nicely to cloud deployment. First, there are no licencing issues about running the software on hardware that is outside of an organisation, and, second, there are no restrictions about the number of CPUs that can accommodate the workload (other than justifying the actual running costs of course). In contrast, there are often restrictions placed on such things with commercial software depending on individual business models.

There are, however, also risks in using open source software especially as part of a commercial process. For open source software to be a key part of such a process the following questions would need to be addressed.

1. Is the software written in a way that prevents malicious use, in a nutshell is it secure and hack-resistant?

2. How would a scientist get support? As part of a key platform support has to be more than posting a question on a wiki and hoping someone responds in a reasonable time with the right answer. Also in a modern global pharmaceutical company these scientists could be in any time zone.

3. How would users get training? Again this is expected to be more than a FAQ sheet, and depending on the software could range from instructor-led classes, through to interactive online training.

4. How would the system receive updates and upgrades? How quickly would this follow release? How would it be tested? What are the change control mechanisms?

5. Are there any restrictions or incompatibilities with the various open source licences, and the use of other software and services that may restrict how things like extra security (e.g. user authentication) can be wrapped round an open source package?

It was with these questions in mind that the Pistoia Sequence Services Project was launched. The ultimate aim was to host DNA sequence software in the cloud, learning what works and addressing some of the questions above. Indeed, two of the above questions were quickly resolved by the 'vendor supported' model that Pistoia is exploring. First, for question 2, commercial vendors would agree to host open source software and use their current support solutions to address the issues of their industry users. In particular, ensuring that an expert inside their organisations would always be able to respond to support requests. In terms of question 3 concerning training, the Sequence Services Project selected software to host that already had excellent online training modules that had been written by the public institute that wrote the software. Finally, question 4, the challenge of updating the software and its underlying data when a new production version was released, is also not insurmountable. Indeed, a one month lag from public release to hosted service is perfectly acceptable while whatever testing was undertaken by the hosting provider. Thus, the vendor-hosted model of free/open source software addresses some of the key initial challenges to industry adoption. However, the project's aim was to go further and investigate the remaining issues.

Perhaps surprisingly, there can be some challenges to the adoption of open source software in a commercial environment due to the licence terms that are used when releasing software. Take the following statement which has been taken from an open source licence (but anonymised).

> You agree that you will deliver, and you will cause all sublicensees to deliver, to XXX copies of all Modifications that are made to the Software, on or before the date on which the Modifications are distributed to any third party, and you hereby grant to XXX, and will cause all of your sublicensees to grant to XXX, a non-exclusive, royalty-free, irrevocable, non-terminable license to use, reproduce, make derivative works of, display, perform, and distribute any and all Modifications in source code and binary code form for any purpose and to sublicense such rights to others.

From the point of view of the author of the software, it is not unreasonable to expect people who are using and benefiting from your hard work to contribute back any improvements they add. However, this may mean that the software could not be easily used with any commercial tools. It may be completely reasonable for a company to want to put some user authentication process into the software, or perhaps a plug-in that allows the use of commercial visualisation software. At the very minimum, lawyers now need to be involved (probably on both sides) to agree what constitutes a 'modification', and even if a third-party developer could be commissioned by a company to do any work on the code without forcing the original author into the development process (and into a confidentiality agreement).

There are a great number of open source licences available, but they are certainly not all the same, and indeed not always compatible with each other. For example, when the open source 'SpamAssassin' software was moved into Apache, project organisers spent months getting permission for the move from all the licence holders (around 100 in total) [9]. Not all contributors could be tracked down, and some software had to be rewritten to allow it all to move to the Apache licence. This demonstrates how important it is to understand any reach through claims and restrictions that are associated with a given open source licence. In a large company this would almost always involve taking legal advice, and producing rules on how to use the software safely within the licence; another huge potential hidden cost of using open source software. For anyone who wants to know more about the many and various open source licences an excellent resource from the 'Open Source Software Advisory Service' is available [10] and the subject was covered in a recent book chapter [11].

One of the biggest challenges in deploying open source software in a commercial environment is ensuring that it is secure. Each year there are a huge number of malicious hackings of corporations, with many well-known companies suffering significant damage to their businesses and reputations. The IT vendors who joined the sequence services' team all knew how to build secure platforms. Many work closely with customers in the financial and defence industries, and are consequently experts in building secure IT platforms. When building these systems, security is thought about and tested at every step. As open source software is often not built with commercial considerations, the developers quite rightly primarily focus on the functional aspects of the software. Although there is some highly secure open source software (such as EnGarde Linux), often the developer's focus is not necessarily securing the software from skilled 'blackhat' hackers and crackers intent on breaking in. Traditionally this has not been a huge problem. Where the software is hosted and available on the internet, it either contains no data or only already publicly available data. Given that the code is already open source, there is little incentive for anyone to hack into the system other than pure vandalism, and this means most sites come under nothing more than basic 'script kiddie' attack. Where a large commercial company uses the software it would be normal for it to be hosted inside their own data centres on the internal intranet, which is safely protected from the external world by large, sophisticated corporate firewalls.

This intranet solution (shown in Figure 22.1) does, of course, have a number of drawbacks. First, there is the cost involved of hosting the software, testing the software and managing updates, as well as providing training to the IT helpdesk, and the users. These costs are not insignificant, and in many cases make open source software as expensive to run and support as commercial alternatives, especially when training material of a high standard is not already available. This cost is then duplicated many times at other corporations that are doing similar research. On top of that, as many companies begin to work in more collaborative ways, they open B2B (business to business) connections through their firewalls. This has led to more concern about the security vulnerabilities of the software hosted internally.

The vision of the Sequence Services Project is to try to reduce the total costs of running and maintaining such a system, while ensuring that

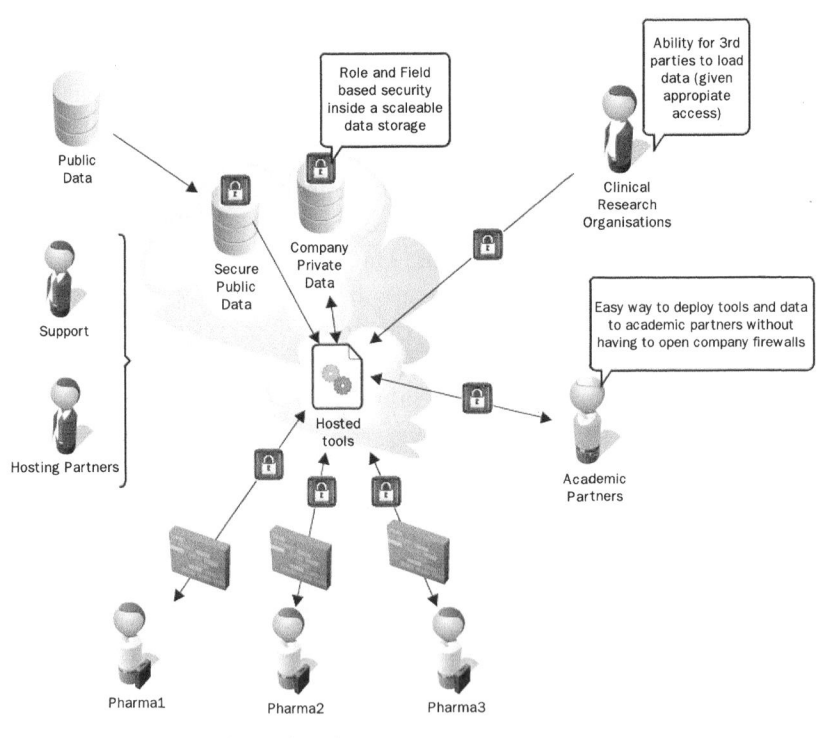

Figure 22.1 Deploying open source software and data inside the data centres of corporations

Published by Woodhead Publishing Limited, 2012

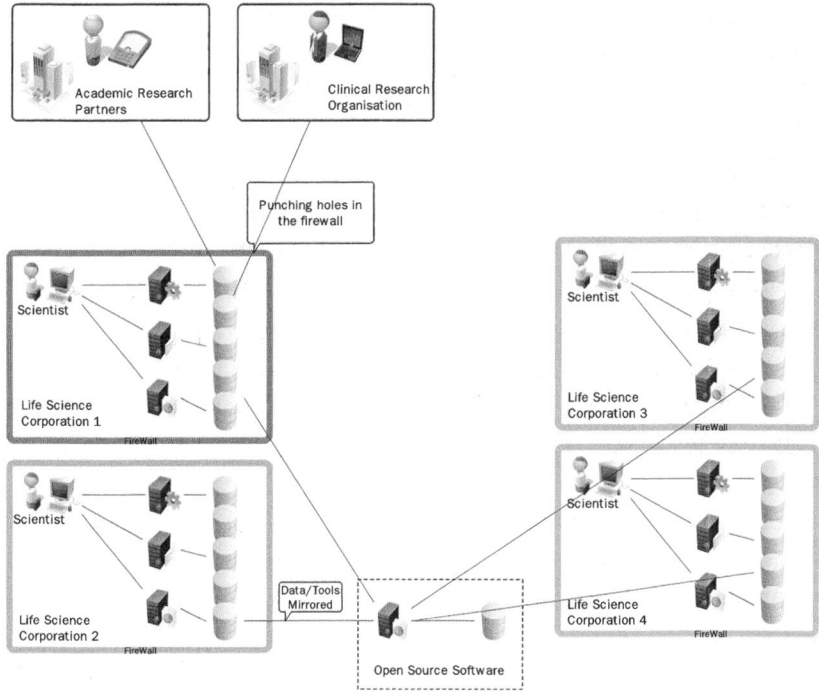

Figure 22.2 Vision for a new cloud-based shared architecture

functionality, performance and, above all, security are not compromised. This new vision is shown in Figure 22.2, where the services are hosted and maintained by a third-party vendor.

Pistoia chose an open source suite of software that is well known in the bioinformatics world as its test software for the proof of concept work. Before starting there were discussions with the institute that wrote the software, building an agreement to communicate the results of the project and, in particular, share any recommendations or security vulnerabilities that were discovered.

In order to test and maintain security a number of approaches were taken. The first was to run an extensive ethical hack of the chosen open source software. Pistoia did this by employing the professional services of a global IT company with a specialist security division, producing an extensive report containing a breakdown of vulnerabilities. Although I cannot list specific security issues found in the Pistoia project, I would like to highlight the most common vulnerabilities found in such software. For those that are particularly interested in this topic, a valuable resource in this area plus a list of the current top 10 vulnerabilities can be found at the Open Web Application Security Project (OWASP) website [12].

- Application allows uploading of malware

  Whenever a file is being uploaded, at minimum the software should check if this is the type of file it is expecting, and stop all others. A further step could be taken by running a virus check on the file.

- Insufficient account lockout

  When a user has authenticated and gained access to the system there should be a method to log out if the user has been inactive for a set period of time.

- Unauthorised read access

  A common example of this is the ability once you have been given access to a system to be able to find access areas that you shouldn't by manually entering a URL.

- Cross site framing/scripting

  This is when an attacker can use a bug to re-direct a user to a website that the attacker controls. This will usually be made to look like the original site, and can be used to harvest information like login IDs and passwords.

- Autocomplete not disabled

  The system should not offer to remember passwords or similar security tokens.

- Web server directory indexing enabled

  This is often left on by default, and can give an attacker some indication of where weaknesses exist or where sensitive data may reside.

- Sensitive information disclosed in URL

  Some web-hosted software will place certain pieces of information, such as search terms, in a URL (. . . . search=web+site+security+ . . .). This could be viewed by an attacker if the connection is not secured.

- Verbose error messages

  An error message should ideally be written so it is clear to the user an error has occurred; however, it should not contain any information that an attacker may find useful (such as the user's login ID, software versions, etc.) This can all be put in an error log and viewed securely by a support team.

- SQL injection

  This is a technique whereby an attacker can run a database search that the software would not normally allow, by manipulating the input to the application. This is not difficult to prevent if developers are thinking about security, so that the system only ever runs validated input that it is expecting.

- Concurrent sessions

  Users should not be able to log into the system more than once at any given time.

- Web server advertises version information in headers

  Revealing the version of the server software currently running makes it easier for hackers to search for vulnerabilities for that version.

Even if a system has been found to have no known vulnerabilities during testing, there are further steps that can to be taken to minimise any risks.

- Regular security re-testing, at least at each version release – this, of course, costs money to maintain.

- Use of virtual private clouds, where private data are clearly segregated.

- Use of IP filtering at the firewall to reject any IP address other than the approved ones. Although IP addresses can be spoofed, a hacker would need to know the correct ones to spoof.

- Use of appropriate validated authentication standards, obviating the need for services to provide their own ad hoc systems.

- Use of remote encryption key servers, which would prevent even the vendor hosting the software from being able to view the private data they are hosting.

- Use of two factor authentication. Many users will use the same short password for every service, meaning that once an attacker has found a user name and password at one site, they can quickly try it at many others. Two factor authentication usually takes the form of a device that can give a seemingly random number that changes on a regular basis which would need to be entered along with the traditional user name and password.

- The use of appropriate insurance to provide some financial compensation if a service is down or breached.

22.7 Conclusion

As I hope I have highlighted, just as under-secretary Kennedy explained in his answer at the beginning of this chapter, there are many costs associated with the use of open source software in government departments or in industry. As soon as there is a need to store or process sensitive data such as medical records, then the costs and complexity can increase considerably. The cost-benefits of open source software being free to download and use can quickly be lost once training, support, security, legal reviews, update management, insurance and so on are taken into account.

However, in many areas of scientific research, open source software projects are truly world leading. For industry to get maximum value from open source software it needs to actively participate in the development process. Organisations like Pistoia show one way that this participation can be done, where many companies come together in areas that are important (yet pre-competitive). This approach has many benefits.

- Shared requirements – allowing industry to speak with one voice, rather than many different ones. This can obviously help with understanding how software might be used, in prioritising future work and in improving and adopting open standards.

- Shared costs and risks – with many companies all sharing the cost of things like expert security reviews of software, the share of the cost to each company can quickly become very reasonable.

- Feedback and collaboration – even if industry is not participating directly with the writing of a given open source software package, it can help to build relationships with the authors. This way the authors can better understand industry requirements. Importantly, authors will not see a list of security vulnerabilities merely as a criticism, but rather as a contribution to making the software better for everyone.

Open source software is very likely to play an important role in the life sciences industry in the future, with companies not only using open source software, but also actively contributing back to the community. Indeed, earlier in this book (Chapter 1) we saw a great example of exactly this from Claus Stie Kallesøe's description of the LSP4All software. Another exciting example of industry using open source software to help drive innovation is demonstrated by the Pistoia Alliance's recent 'Sequence Squeeze' competition. The Pistoia Alliance advertised a challenge to identify improved algorithms to compress the huge volumes of data produced by NGS. In order to find an answer they reached out to the world by offering a $15000 prize to the best new algorithm. All algorithms had to be submitted via the sourceforge website, and under the BSD2 licence. The BSD2 licence was deliberately chosen by Pistoia as it has minimal requirements about how the software can be used and re-distributed – there are no requirements on users to share back any modifications (although, obviously, it would be nice if they did). This should lower the legal barriers within companies to adopting the software, and is a great example of industry embracing open source standards to help drive innovation. The challenge was open to anyone and entries were ranked on various statistics concerning data compression and performance. A detailed breakdown of the results of the challenge is available from [13].

In conclusion, it is clear that free and open source software is not really 'cost-free' in an industrial setting. It is also true that there can be issues that make deployment difficult, particularly where sensitive data are concerned. However, by developing new models for hosting and supporting such software we could be witnessing a new direction for the use of free and open source software in commercial environments.

22.8 References

[1] http://weblogs.mozillazine.org/asa/archives/2009/05/firefox_at_270.html
[2] http://www.state.gov/secretary/rm/2009a/july/125949.htm
[3] http://www.dnalc.org/view/15073-Completion-of-a-draft-of-the-human-genome-Bill-Clinton.html
[4] http://en.wikipedia.org/wiki/Human_Genome_Project
[5] http://en.wikipedia.org/wiki/DNA_sequencing
[6] The 1000 Genomes Project Consortium, A map of human genome variation from population-scale sequencing. Nature 2010;467:1061–73.
[7] Barnes, MR et al. Lowering industry firewalls: pre-competitive informatics initiatives in drug discovery. Nature Reviews Drug Discovery 2009;8:701–8.
[8] http://www.pistoiaalliance.org
[9] http://en.wikipedia.org/wiki/SpamAssassin
[10] http://www.OSS-watch.ac.uk
[11] Wilbanks, J. Intellectual Property Aspects Of Collaboration in Collaborative Computational Technologies for Biomedical Research (Ekins S, Hupcey MAZ, Williams AJ eds.) Wiley (London) ISBN0470638036
[12] https://www.owasp.org/index.php/Main_Page
[13] http://www.sequencesqueeze.org/

Index

Published by Woodhead Publishing Limited, 2012

Published by Woodhead Publishing Limited, 2012